Lecture Notes in Artificial Intelligence   **12299**

Subseries of Lecture Notes in Computer Science

Martin Michalowski · Robert Moskovitch (Eds.)

Artificial Intelligence in Medicine

18th International Conference on Artificial Intelligence in Medicine, AIME 2020
Minneapolis, MN, USA, August 25–28, 2020
Proceedings

 Springer

Editors
Martin Michalowski
School of Nursing
University of Minnesota
Minneapolis, MN, USA

Robert Moskovitch
Ben-Gurion University of the Negev
Tonawanda, NY, USA

ISSN 0302-9743   ISSN 1611-3349   (electronic)
Lecture Notes in Artificial Intelligence
ISBN 978-3-030-59136-6   ISBN 978-3-030-59137-3   (eBook)
https://doi.org/10.1007/978-3-030-59137-3

LNCS Sublibrary: SL7 – Artificial Intelligence

This Springer imprint is published by the registered company Springer Nature Switzerland AG
The registered company address is: Gewerbestrasse 11, 6330 Cham, Switzerland

Preface

The European Society for Artificial Intelligence in Medicine (AIME) was established in 1986 following a very successful workshop held in Pavia, Italy, the year before. The principal aims of AIME are to foster fundamental and applied research in the application of artificial intelligence (AI) techniques to medical care and medical research, and to provide a forum at biennial conferences for discussing any progress made. Hence, the main activity of the society thus far is the organization of a series of biennial conferences, held in Marseilles, France (1987), London, UK (1989), Maastricht, The Netherlands (1991), Munich, Germany (1993), Pavia, Italy (1995), Grenoble, France (1997), Aalborg, Denmark (1999), Cascais, Portugal (2001), Protaras, Cyprus (2003), Aberdeen, UK (2005), Amsterdam, The Netherlands (2007), Verona, Italy (2009), Bled, Slovenia (2011), Murcia, Spain (2013), Pavia, Italy (2015), Vienna, Austria (2017), and Poznan, Poland (2019).

In the last decade, there were discussions within the AIME society to make the AIME conference truly international (initially it was a European conference) and host it in North America, but those plans could not be realized. For the first time in the conference's history, it was supposed to be hosted in North America in 2020. This event came with real excitement and was a huge step for the conference and the community. However, due to the global COVID-19 pandemic, the decision was made to forgo an in-person meeting for a virtual one. This volume contains the proceedings of AIME 2020, the International Conference on Artificial Intelligence in Medicine, hosted virtually by the University of Minnesota in Minneapolis, USA, August 25–28, 2020.

The AIME 2020 goals were to present and consolidate the international state of the art of AI in biomedical research from the perspectives of theory, methodology, systems, and applications. The conference included two invited keynotes, full and short papers, tutorials, a COVID-19 themed workshop, a plenary session panel, and a doctoral consortium.

In the conference announcement, authors were invited to submit original contributions regarding the development of theory, methods, systems, and applications for solving problems in the biomedical field, including AI approaches in biomedical informatics, molecular medicine, and health-care organizational aspects. Authors of papers addressing theory were requested to describe the properties of novel AI models potentially useful for solving biomedical problems. Authors of papers addressing theory and methods were asked to describe the development or the extension of AI methods, to address the assumptions and limitations of the proposed techniques, and to discuss their novelty with respect to the state of the art. Authors of papers addressing systems and applications were asked to describe the development, implementation, or evaluation of new AI-inspired tools and systems in the biomedical field. They were asked to link their work to underlying theory, and either analyze the potential benefits to solve biomedical problems or present empirical evidence of benefits in clinical practice. All authors were asked to highlight the value their work created for the patient, provider, and institution through its clinical relevance.

AIME 2020 received 103 submissions across all types of paper categories. Submissions came from 28 countries, including submissions from Europe, North and South America, Asia, Australia, and Africa. All papers were carefully peer reviewed by experts from the Program Committee, with the support of additional reviewers, and by members of the Senior Program Committee Committee – a new review layer introduced in AIME 2020. Each submission was reviewed in most cases by three reviewers, and all papers by at least two reviewers. The reviewers judged the overall quality of the submitted papers, together with their relevance to the AIME conference, originality, impact, technical correctness, methodology, scholarship, and quality of presentation. In addition, the reviewers provided detailed written comments on each paper, and stated their confidence in the subject area. One Senior Program Committee member was assigned to each paper and they wrote a meta-review and provided a recommendation to the scientific chair.

A small committee consisting of the AIME 2020 scientific chair, Dr. Nitesh Chwala, and the conference co-chairs, Dr. Martin Michalowski and Dr. Robert Moskovitch, made the final decisions regarding the AIME 2020 scientific program. This process began with virtual meetings starting in June 2020. As a result, 34 long papers (an acceptance rate of 33%) and 9 short papers were accepted. Each long paper was presented in a 20-minute oral presentation during the conference. Each regular short paper was presented in a 5-minute presentation and by a poster. The papers were organized according to their topics in the following main themes: (1) Deep Learning; (2) Natural Language Processing; (3) Predictive Modeling; (4) Image Processing; (5) Unsupervised Learning; (6) Temporal Data Analysis; (7) Clinical Practice Guidelines; (8) Information Retrieval; and (9) Bioinformatics.

AIME 2020 had the privilege of hosting two invited keynote speakers: Dr. Edward H. Shortliffe, Chair Emeritus and Adjunct Professor in the Department of Biomedical Informatics at Columbia University, USA, giving the keynote entitled "AI Today: Are We Forgetting Our Roots?," and Dr. Vimla L. Patel, Senior Research Scientist and Director for the Center for Cognitive Studies in Medicine and Public Health at The New York Academy of Science, USA, describing "Human Cognition: A Guide to the Evolution of AI in Medicine." AIME 2020 also hosted an invited panel during the main conference that focused on the effects of the global pandemic on AI in medicine research and implementation. Panelists were academic and industry experts from the USA, Canada, China, the UK, and Israel.

The doctoral consortium received six PhD proposals that were peer reviewed. AIME 2020 provided an opportunity for these PhD students to present their research goals, proposed methods, and preliminary results. A scientific panel consisting of experienced researchers in the field provided constructive feedback to the students in an informal atmosphere. The doctoral consortium was chaired by Dr. Mary Regina Boland.

One workshop was organized before the AIME 2020 main conference. This workshop focused on the challenges and problems data science and AI can address related to the global pandemic, and relevant deployments and experiences in gearing AI to cope with COVID-19. The workshop was chaired by Dr. Martin Michalowski and Dr. Robert Moskovitch, with invited extended abstracts presented by experts in the public and private sectors from around the globe. The work from this workshop will be extended and presented in a special journal issue devoted to the topic.

In addition to the workshops, three interactive half-day tutorials were presented prior to the AIME 2020 main conference: (1) Methods and Applications of Natural Language Processing in Medicine (Rui Zhang, University of Minnesota, USA); (2) Large Scale Ensembled NLP Systems with Docker and Kubernetes (Raymond Finzel, University of Minnesota, USA), (3) The Overview Effect: Clinical Medicine and Healthcare Concepts for the Data Scientist (Anthony Chang, Children's Hospital of Orange County, USA).

We would like to thank everyone who contributed to AIME 2020. First of all, we would like to thank the authors of the papers submitted and the members of the Program Committee together with the additional reviewers. Thank you to the Senior Program Committee for writing meta-reviews and to members of the Senior Advisory Committee for providing guidance during conference organization. Thanks are also due to the invited speakers and panelists, as well as to the organizers of the tutorials and doctoral consortium panel. Many thanks go to the Local Organizing Committee, who helped plan this conference and transition it to a virtual one. The free EasyChair conference system (http://www.easychair.org/) was an important tool supporting us in the management of submissions, reviews, selection of accepted papers, and preparation of the overall material for the final proceedings. We would like to thank Springer and the *Artificial Intelligence Journal* (AIJ) for sponsoring the conference and the Association for the Advancement of Artificial Intelligence (AAAI) for establishing a cooperative agreement with AIME 2020. Finally, we thank the Springer team for helping us in the final preparation of this LNAI book.

July 2020

Martin Michalowski
Robert Moskovitch

Organization

General Chairs

Martin Michalowski	University of Minnesota, USA
Robert Moskovitch	Ben-Gurion University, Israel

Program Committee Chairs

Nitesh Chawla (Scientific Program)	University of Notre Dame, USA
Mary Regina Boland (Doctoral Consortium)	University of Pennsylvania, USA
Fei Wang (Tutorial Program)	Weill Cornell Medicine, USA
Anthony C. Chang (Demonstration)	CHOC Children's, USA

Local Organizing Committee

Martin Michalowski (Local Organization Chair)	University of Minnesota, USA
Karen Monsen	University of Minnesota, USA
Tom Clancy	University of Minnesota, USA
Lisiane Pruinelli	University of Minnesota, USA
Maria Gini	University of Minnesota, USA
Serguei Pakhomov	University of Minnesota, USA
Steven Rudolph (Director of Communications & Marketing)	Events & Social Media Coordinator
Nikayla Speltz (Events & Social Media Coordinator)	Events & Social Media Coordinator
Breanne Krzyzanowski (Cooperative Associate)	Events & Social Media Coordinator

Senior Advisory Committee

Riccardo Bellazzi	Università degli Studi di Pavia, Italy
Carlo Combi	Università degli Studi di Verona, Italy
Noemie Elhadad	Columbia University, USA
Deborah Estrin	Cornell Tech, USA
Milos Hauskrecht	University of Pittsburgh, USA

John H. Holmes	University of Pennsylvania, USA
George Hripcsak	Columbia University, USA
Mark Musen	Stanford University, USA
Mor Peleg	University of Haifa, Israel
Yuval Shahar	Ben-Gurion University, Israel
Peter Szolovits	Massachusetts Institute of Technology, USA
Lei Xing	Stanford University, USA
Blaž Zupan	University of Ljubljana, Slovenia

Senior Program Committee

Riccardo Bellazzi	Università degli Studi di Pavia, Italy
Prithwish Chakraborty	IBM, USA
Jake Chen	University of Alabama at Birmingham, USA
Carlo Combi	Università degli Studi di Verona, Italy
Michel Dojat	INSERM, France
Joseph Finkelstein	Icahn School of Medicine at Mount Sinai, USA
Adela Grando	Arizona State University, USA
Milos Hauskrecht	University of Pittsburgh, USA
John Holmes	University of Pennsylvania, USA
Meng Jiang	University of Notre Dame, USA
Elpida Keravnou-Papailiou	University of Cyprus, Cyprus
Nada Lavrač	Jozef Stefan Institute, Slovenia
Peter Lucas	Leiden University, The Netherlands
Silvia Miksch	Vienna University of Technology, Austria
Niels Peek	The University of Manchester, UK
B. Aditya Prakash	Georgia Institute of Technology USA
Silvana Quaglini	University of Pavia, Italy
Chandan K. Reddy	Virginia Tech, USA
David Riaño	Universitat Rovira i Virgili, Spain
Lucia Sacchi	University of Pavia, Italy
Stefan Schulz	Medical University of Graz, Austria
Arash Shaban-Nejad	University of Tennessee Health Science Center, USA
Yuval Shahar	Ben-Gurion University, Israel
Annette Ten Teije	Vrije Universiteit Amsterdam, The Netherlands
Paolo Terenziani	Università del Piemonte Orientale, Italy
Allan Tucker	Brunel University, UK
Szymon Wilk	Poznan University of Technology, Poland
Ping Zhang	The Ohio State University USA

Program Committee

Samina Abidi	Dalhousie University, Canada
Syed Sibte Raza Abidi	Dalhousie University, Canada
Pedro Henriques Abreu	FCTUC-DEI, CISUC, Portugal
Oguz Akbilgic	Loyola University Chicago, USA

Helena Lindgren	Umeå University, Sweden
Fang Liu	University of Notre Dame, USA
Gang Luo	University of Washington, USA
Beatriz López	University of Girona, Spain
Hiroshi Mamiya	McGill Clinical and Health Informatics, Canada
Mar Marcos	Universitat Jaume I, Spain
Stan Matwin	Dalhousie University, Canada
Paola Mello	University of Bologna, Italy
Wojtek Michalowski	University of Ottawa, Canada
Diego Molla	Macquarie University, Australia
Stefania Montani	Universita del Piemonte Orientale, Italy
Laura Moss	University of Aberdeen, UK
Fleur Mougin	Université de Bordeaux, France
Goran Nenadic	The University of Manchester, UK
Øystein Nytrø	Norwegian University of Science and Technology, Norway
Dympna O'Sullivan	Technological University Dublin, Ireland
Barbara Oliboni	University of Verona, Italy
Serguei Pakhomov	University of Minnesota, USA
Enea Parimbelli	University of Pavia, Italy
Mor Peleg	University of Haifa, Israel
Pedro Pereira Rodrigues	University of Porto, Portugal
Christian Popow	Medical University of Vienna, Austria
Jędrzej Potoniec	Poznan University of Technology, Poland
Lisiane Pruinelli	University of Minnesota, USA
Cédric Pruski	Luxembourg Institute of Science and Technology, Luxembourg
Kirk Roberts	The University of Texas at Houston, USA
Aleksander Sadikov	University of Ljubljana, Slovenia
Abeed Sarker	Emory University, USA
Erez Shalom	Ben-Gurion University, Israel
S. M. Reza Soroushmehr	University of Michigan, USA
Karthik Srinivasan	University of Kansas, USA
Gregor Stiglic	University of Maribor, Slovenia
Samson Tu	Stanford University, USA
Frank Van Harmelen	Vrije Universiteit Amsterdam, The Netherlands
William Van Woensel	Dalhousie University, Canada
Alfredo Vellido	Universitat Politècnica de Catalunya, Spain
Herna Viktor	University of Ottawa, Canada
Dimitrios Vogiatzis	ACG Deree, NCSR Demokritos, Greece
Trang Vopham	Fred Hutchinson Cancer Research Center, USA
Jens Weber	University of Victoria, Canada
Nicole Weiskopf	Oregon Health & Science University, USA
Laura Wiley	University of Colorado, USA
Antje Wulff	TU Braunschweig and Hannover Medical School, Germany

Yuanlin Zhang   Texas Tech University, USA
Yu-Dong Zhang   Université Paris-Saclay, France

Additional Reviewers

Buomsoo Kim   José P. Amorim
Yuanxia Li   Alexandra Kogan
Kyuhan Lee   Pietro Bosoni
Zhengchao Yang   Primoz Kocbek
Aida Kamisalic   Yijun Lin
George Drosatos   Ricardo Cardoso Pereira
Jie Xu   Maurice Webe
Weiwei Duan   Mohsin Ahmed
Zekun Li

Contents

Image Processing

Unsupervised Learning

Clinical Practice Guidelines

Information Retrieval

Bioinformatics

Deep Learning

All-Cause Mortality Prediction in T2D Patients

Pavel Novitski[1(✉)], Cheli Melzer Cohen[2], Avraham Karasik[2], Varda Shalev[2], Gabriel Hodik[2], and Robert Moskovitch[1]

[1] Software and Information Systems Engineering,
Ben Gurion University of the Negev, Beer-Sheva, Israel
`pavelnov@post.bgu.ac.il, robertmo@bgu.ac.il`
[2] Maccabi Data Science Institute, Maccabi Healthcare Services, Tel-Aviv, Israel
`{melzerco_c,shalev_v,hodik_g}@mac.org.il, karasik@post.tau.ac.il`

Abstract. Mortality in elderly population having type II diabetes (T2D) can be prevented sometimes through intervention. For that risk assessment can be performed through predictive modeling. This study is part of a collaboration with Maccabi Healthcare Services' Electronic Health Records (EHR) data, that consists on up to 10 years of 18,000 elderly T2D patients. EHR data is typically heterogeneous and sparse, and for that the use of temporal abstraction and time intervals mining to discover frequent time-interval related patterns (TIRPs) are employed, which then are used as features for a predictive model. However, while the temporal relations between symbolic time intervals in a TIRP are discovered, the temporal relations between TIRPs are not represented. In this paper we introduce a novel TIRPs based patient data representation called Integer-TIRP (iTirp), in which the TIRPs become channels represented by values representing the number of TIRP's instances that were detected. Then, the iTirps representation is fed into a Deep Learning Architecture, which can learn this kind of sequential relations, using a Recurrent Neural Network (RNN) or a Convolutional Neural Network (CNN). Finally, we introduce a predictive model that consists of a committee, in which two inputs were concatenated, a raw data and iTirps data. Our results indicate that iTirps based models, showed superior performance compared to raw data representation and the committee showed even better results, this by taking advantage of each representations.

Keywords: Pattern mining · Deep Learning · Temporal data prediction

1 Introduction

Diabetes is a major chronic disease in the western society and its prevalence is on the rise worldwide. Type 2 diabetes (T2D) patients often suffer from heart disease and the prevalence of coronary artery disease and heart failure is also

© Springer Nature Switzerland AG 2020
M. Michalowski and R. Moskovitch (Eds.): AIME 2020, LNAI 12299, pp. 3–13, 2020.
https://doi.org/10.1007/978-3-030-59137-3_1

much higher among diabetic patients [4]. Moreover, cardiac related in-hospital mortality is also much higher among patients with diabetes [12].

Israel's HMOs had implemented disease management programs to improve quality of care for diabetes and prevent those complications through risk reduction [10]. These programs aim at achieving centrally controlled documented multifactorial risk reduction that are implemented mainly by primary care givers. To date the effect of these programs on cardiac morbidity and mortality were not assessed. A potential deficiency in these plans may be the lack of targeted case management for high-risk patients. Such identification may lead to on time intensive intervention that may reduce morbidity and mortality. Namely, prevent hospitalization for cardiac disease and lower cardiac mortality. To this end it is desirable to develop a predictive model that will help to identify the patients that are more prone to cardiac deterioration. This will form the basis for intervention aimed at prevention of costly and lethal consequences. For that purpose, in this paper the focus is on prediction of All-Cause Mortality in T2D patients.

In this paper we introduce for the first time iTirps, which are temporal patterns based representation that can be later fed into temporal architectures of Artificial Neural Networks (ANNs), which we use to learn predictive models for outcomes, which in our study is all cause mortality in T2D patients. To have the iTirps representation, first temporal abstraction is used [19] and time-intervals mining to discover TIRPs [19]. Then these are transformed into a new representation, called integer-TIRPs, which are described in greater details later. The contributions of the paper are the following: 1. iTirps, a novel representation for temporal data consisting on frequent TIRPs instances, which enable to represent a time period according to the relations among the temporal variables along time, and their appearances, which are hard to represent by TIRPs, nor by temporal ANNs. 2. A rigorous evaluation on a large real-life data of T2D patients, using iTirps for the prediction of all-cause mortality.

2   Related Work

We start with a review of the use of data science in diabetic patients' data. Then we proceed with discussing time intervals related patterns mining in heterogeneous multivariate temporal data and their use for classification, and then we go over approaches in the field of ANN for time series classification.

2.1   Outcomes Prediction in Diabetes

The use of data mining and machine learning methods in diabetes related research is constantly increasing [11]. There is a relatively small number of studies that intend to predict mortality in T2D patient. For example, prediction of ICU mortality of diabetic patient by applying several classifiers on aggregated data and showed good results on predict risk of mortality [1]. Most of current research that assesses mortality risk in diabetic patients, are using Cox proportional hazards model, in [16] used Cox model to create risk equations for all cause, cardiovascular, and non-cardiovascular mortality diagnosed of type 2 diabetes patients. In [5] a Cox model was used specifically for prediction of mortality in adults population. The use of ANN in diabetes related research was not very extensive, and most of the work use feed forward (FF) network on a temporal data [9].

2.2 Temporal Abstraction, TIRPs Discovery and TIRPs Based Classification

A major challenge in analyzing EHR data is the heterogeneity of the sampling forms of the data. Additionally, challenges may include sparsity, and exploiting the temporal information. Therefore, increase usage of temporal abstraction (TA) and time intervals mining is being reported [19]. In order to transform the heterogeneous temporal variables into a uniform representation, state TA is used, in which the time point series are transformed into symbolic time intervals (STIs), given a set of cutoffs. The cutoffs can be knowledge based [20], or data driven, based on discretization methods, such as Symbolic Aggregate approXimation [14] or the Temporal Discretization for Classification (TD4C) [19,21]. Another type of temporal abstraction is gradient abstraction, which segments the data based on the first derivative into periods of time, in which the variable is increasing or decreasing [20]. Once symbolic time intervals series are created, frequent Time Intervals Related Patterns (TIRPs) can be discovered. Several methods for TIRPs discovery were proposed in the past [19,21], mostly consisting on Allen's temporal relations [17] which include seven relations such as before, meet, overlap, and more, and their inverse. Beyond temporal knowledge discovery, frequent TIRPs were shown to be effective for classification and prediction in electronic health records [3,17,18,22]. However, incorporating the use of TIRPs to represent the temporal relations between heterogeneous temporal variables in ANNs based architectures is still a challenge, which we explore in this paper.

2.3 Artificial Neural Networks for Temporal Data

ANNs designed for temporal data were successfully used in several domains and tasks. For example, RNN can store information about previous inputs in internal memory (hidden states), that abstract and carry information from earlier time stamps and CNN are achieving state-of-the-art results in a high variety of tasks including computer vision tasks and more [13]. These methods are increasingly employed also in clinical data. RNN based methods showed superior results than the use of classical algorithm like logistic regression (LR) and multilayer perceptron with hand-engineered features in predicting diagnosis codes [8,15]. RNN based method for missing values imputation in temporal data, called GRU-D [6], showed better performance than traditional methods, such as mean-imputation, imputation with k-nearest neighbor and other. Modified CNN, for capturing temporal relations, was trained CNN on temporal matrix representation of medical codes for outcome prediction and showed better results than LR using aggregated clinical features on real world EHR data [7].

3   Methods

We present here the framework for the development of the iTirps and the iTirpsMap representation which is used later with temporal ANNs for classification and specifically prediction in this study. With iTirps representation we preserve more temporal information then the regular TIRP representation. iTirps hold the information about the starting and ending time of each TIRP, and through that the duration of each TIRP instance is captured, including its relative location in the time series. This information can be learned by ANN and improve the classification performance.

3.1   iTirps and iTirpsMap

Figure 1 presents the steps in the creation of iTirps and a corresponding iTirpsMap. First the multivariate temporal data abstracted and transformed into a uniform representation of symbolic time intervals [19]. Then, frequent TIRPs are discovered by mining the symbolic time intervals. For the mining process, we use the KarmaLego algorithm [21]. The result is a bag of frequent TIRPs. Next, the TIRPs are detected and transformed to iTirp, that are passed in the form of iTirpsMap as input to a CNN/RNN.

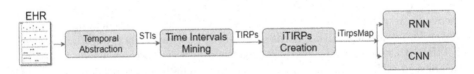

Fig. 1. iTirps and iTirpsMap based classification

Temporal Abstraction. In this study we perform state abstraction using Symbolic Aggregate approXimation (SAX) [14], in which the states are derived from the Gaussian distribution of the values, and Temporal Discretization for Classification (TD4C) [19] that determines the cutoffs in a supervised manner, so that the states distribution are most different among the classes. The result of the Temporal Abstraction process is a uniform representation of the temporal variables as symbolic time intervals. A symbolic time interval, $I = <s, e, sym>$, is an ordered pair of time points, start-time (s) and end-time (e), and a symbol (sym) that represents one of the domain's symbolic concepts, which in our study can be laboratory results that went through abstraction, conditions or procedures. As mentioned in the background, once the data is transformed into a uniform representation of symbolic time intervals, TIRPs can be discovered.

TIRPs Discovery. To discover TIRPs, the KarmaLego algorithm [19] is used, which uses Allen's temporal relations, such as starts, meets, overlap, contains, and more, and their inverse [20] to represent the temporal relation among a pair of symbolic time intervals. In this study a set of generalized temporal relations, which are the disjunction of part of Allen's seven relations were used.

These include: BEFORE based on $before\|meets$; OVERLAP based on $overlaps$; and CONTAIN based on $\{starts \parallel contains \parallel finish - by \parallel equal\}$. In addition, a maximum allowed gap duration is set for the before relation [20]. A non-ambiguous TIRP P is defined as $P = I, R$, where $I = I1, I2, .., Ik$ is a set of k ordered symbolic time intervals and the conjunction of all their pairwise temporal relations among each of the $(k2 - k)/2$ pairs of the symbolic time intervals in I, $R = U_{i=1}^{k-1}U_{j=i+1}^{k}r(I^i, I^j)r_{1,2}(I^1, I^2), .., r_{1,k}(I^1, I^k), ..., r_{k-1,k}(I^{k-1}I^k)$. Thus, given a database of entities (i.e., patients), the vertical support of a TIRP P (frequency in the database) is denoted by the cardinality of the distinct entities having P, relative to the size of the database. However, in this study we propose a novel use of the TIRPs, which become channels, and called iTirps.

iTirp and iTirpsMap Creation. We introduce here iTirps, a new temporal representation of multivariate temporal data through TIRPs' instances that results in a numeric matrix representation of the appearance of the TIRPs along time, which can be later fed to various methods, such as RNN/CNN as happens in this study. In previous studies TIRPs were used as features for classifiers [3,19], however, in order to represent them explicitly along time, we present iTirpsMap. Figure 2 illustrates the process of the iTirps and iTirpsMap creation. The description starts at the bottom and goes up. The x-axis is the time by months along 12 months. Starting with the symbolic time intervals at the bottom, which can be raw concepts, such as drug exposers, conditions, or a state abstraction of time point series, such as lab tests.

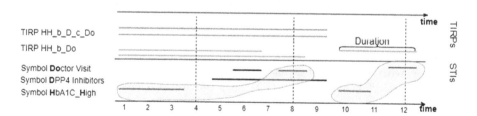

Fig. 2. iTirps and iTirpsMap creation. Starting at the bottom with the STIs series data. Above two TIRPs are shown, and their corresponding instances' durations, which later will be counted and become vectors of the number of TIRPs occurrences in each time stamp.

In Fig. 2 there are three STIs at the bottom, HbA1c_High, which represents their measurements abstracted along three months (which a HbA1c test is valid for), Dipeptidyl peptidase-4 (DPP4) Inhibitors, which are a class of medications that decrease high blood glucose, and Doctor Visit events. Two TIRPs examples are presented above: TIRP HH b Do (HH before Do) appears three times in the periods of 16, 1–8, and 10–12; and the TIRP HH b D c Do (HbA1c before DPP4 Inhibitors, HbA1c High before Doctor Visit, and DPP4 Inhibitors contains Doctor Visit) which appears twice (since there two Doctor Visits) during 1–9. In fact, the HH_b_Do TIRP is shown in the bottom illustration of the STIs surrounding the relevant STIs which include each HH_b_Do instance – there are two. To create iTirps that construct the iTirpsMap, we have two steps, in the first step each TIRP instance becomes a time series of one and zero values (one values are placed from TIRP starting point to ending point). In the second step, we aggregate the TIRPs to create iTIRPs. Thus, for example, iTirp HH b Do value is 2 in time stamps 1–6, since there are two instances of the TIRP during this time stamp. Thus, an iTirp represents the number of the TIRP appearances in each time stamp. Eventually, the entire set of iTirps are combined into an iTirpsMap 3-dimensional matrix (of the Entities, the time axis, and the TIRPs' channels) representation.

Artificial Neural Network. In this paper the purpose of iTirps and iTirpsMap is to enable to combine the advantages of TIRPs in capturing temporal relations between heterogeneous temporal variables and the advantages of neural learning, specifically when using temporal versions of ANNs, such as the RNN, CNN and their ensemble. **RNN-ALSTM.** Our RNN architecture is an Attention block followed by a LSTM (ALSTM). The attention mechanism enables the network to better learn long-term dependencies for the prediction task proposed by [2]. Long Short-Term Memory (LSTM) [13] is variation of RNN, that can overcome RNN's limitations like vanishing and exploding gradient by a gating mechanism that regulates the information flow. **Encoder-CNN.** For the CNN architecture the Encoder is used. In Encoder the first three layers are CNN that followed by attention mechanism, that summarize the temporal dimension, proposed in [23]. To map the network output to a probability distribution the last layer is SoftMax, for both networks. **Committee.** We experiment also with a committee of two classifiers, in which the first classifier is based on the raw data, and the second classifier is based on the iTirpsMap input. First, we train each model separately, with the different inputs, one with raw data and the second with the iTirpsMap, based on some type of TA. Then the SoftMax layer is removed from both models, while the last layers of the network are concatenated, and a new SoftMax added as the last layer.

4   Evaluation and Results

We first state our research questions, and then we describe the data, and the experiments that were designed to answer the questions, and the results.

4.1 Research Questions

1. What type of temporal abstraction is best for classification? 2. What are the best prediction time periods? 3. Which ANN with iTirps performs best, in comparison to the use of raw data? 4. What 'Committee' of ANNs is the best for outcome prediction?

4.2 Dataset

The diabetes dataset of Maccabi Healthcare Services contains data of up to 10 years on 18,000 elderly patients with T2D. The dataset includes 9,000 cases, which are T2D patients who experienced an outcome, defined as all-cause mortality. The data collected from the years 2008–2018. Cohort Inclusion criteria: (1) Patients with diabetes according to the diabetes registry (and not defined as type 1). (2) Experiences an outcome from 2011–2018. 9,000 controls, which are patients without the outcome that were matched according 2 parameters Age and Gender to control patients. Control (Matched patients) will be defined as followed: (1) Patients with diabetes according to the diabetes registry (and not defined as type 1). (2) Being Maccabi Health Services members and without recorded outcomes during the outcome period. Patients included in cancer registry prior to outcome are excluded from the dataset. Control patients, outcome date will be defined as January 1st. The variables include Demographic data, Therapies (medication), Co-morbidities indicators, Lab results, hospitalizations and inpatients and outpatient visits.

4.3 Experimental Setup

To answer the research questions, while reflecting a real application conditions for continuous prediction using a sliding window, the most suitable study design is case-crossover-control. Thus, observation time windows are extracted from the cases, and the matched controls. In the cases, the latest observation time window, which is located a prediction time period prior to the outcome is labeled as positive, while the earlier observation time windows in the cases are labeled as negative, as well as the observation time windows from the controls' data (taken randomly, since there are no outcomes), which enables to evaluate the method both on cases', or controls' time windows. We report quantitative results using the Receiver Operating Characteristic AUC (ROC-AUC), based on 10-fold grouped cross-validation (CV). Thus, time windows of a specific patient were either at the training or in the testing set. To answer the research questions, two experiments were designed. We used an observation time period of 12 months, and to discover TIRPs, KarmaLego was applied with 55% minimal vertical support, and unlimited maximal gap.

Experiment 1. The goal was to evaluate the iTirps based prediction, using an observation time window of 12 months. For this experiment we have two possible inputs for the ANN architectures: raw data, or the iTirpsMap based on the two types of abstraction (research question 1) using the SAX and TD4C-Cos with 2 states. All inputs with Encoder-CNN and RNN-ALSTM were evaluated on two prediction time periods of 90 and 180 days (research question 2, 3). Figure 3 presents the mean results of the iTirps based on the SAX or TD4C-Cos, in comparison to the use of raw data, with two prediction time periods of 90 and 180 days. Generally predicting 90 days performed better than within 180 days, which makes sense. Using the Encoder-CNN the use of iTirps with SAX performed significantly better than the other. Using the RNN-ALSTM the results were quite similar, and the iTirp with SAX performed best.

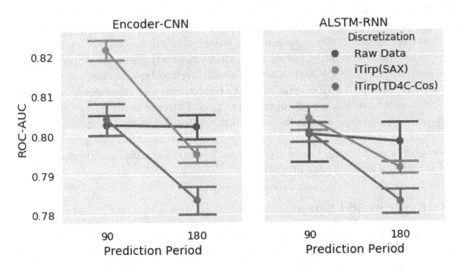

Fig. 3. iTirp(SAX) outperforms significantly, especially with the Encoder-CNN and 90 days prediction period.

Experiment 2. To evaluate what is the best committee (research question 4) the performance of the 'Committee' merged network was evaluated. The committees included a combination of the raw data as input to one classifier, and the second with the iTirpsMap based on SAX or TD4C-Cos TA. Figure 4 presents the performance of the two committees, including for comparison the use of only raw data (as performed in the first experiment, as a baseline). The committee using iTirp with SAX and raw data, was significantly better with Encoder-CNN when predicting 90 days ahead, and also with the other options. Overall, the best performance was 84.5% AUC.

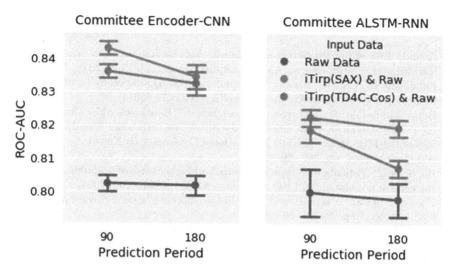

Fig. 4. The iTirp(SAX) committee outperforms significantly, especially with Encoder-CNN and 90 days prediction period.

5   Discussion and Conclusions

Clinically, we focused on the prediction of all-cause mortality in T2D patients, to enable ideally prevention through intervention. In this paper, we introduced a new method for multivariate temporal data representation, called iTirps, which is consists on the discovery of frequent TIRPs, which enables to be fed to RNN or CNN. The use of temporal abstraction and TIRPs as features for classification, is very effective for the analysis of heterogeneous multivariate temporal data, such as often happens in Electronic Health Records. However, in order to employ the advantages of the temporal versions of ANNs, such as CNN or RNN, we propose the iTirps representation. Additionally, we proposed and experimented with the Committee network that combines both raw data and the iTirpsMap. The results of Experiment 1 show that using the iTirp with SAX with the Encoder-CNN had the best performance, and generally the use of the Encoder-CNN and RNN-ALSTM were comparable. In experiment 2, the performance of the committee network was evaluated, in which the committee of the Encoder-CNN with iTirp SAX and the raw data as input, performed best and was significantly higher compared to the raw data. The directions for future research include evaluation of additional discretization methods and number of bins, on a bigger number of SOTA ANN architectures.

Acknowledgements. The authors wish to thank the Israeli Ministry of Science and Technology, who assisted in funding this project with grant 8760521.

References

1. Anand, R.S., et al.: Predicting mortality in diabetic ICU patients using machine learning and severity indices. AMIA Summits Transl. Sci. Proc. **2018**, 310 (2018)
2. Bahdanau, D., Cho, K., Bengio, Y.: Neural machine translation by jointly learning to align and translate. arXiv preprint arXiv:1409.0473 (2014)
3. Batal, I., Fradkin, D., Harrison, J., Moerchen, F., Hauskrecht, M.: Mining recent temporal patterns for event detection in multivariate time series data. In: Proceedings of the 18th ACM SIGKDD International Conference on Knowledge Discovery and Data Mining, pp. 280–288 (2012)
4. Bo, S., et al.: Patients with type 2 diabetes had higher rates of hospitalization than the general population. J. Clin. Epidemiol. **57**(11), 1196–1201 (2004)
5. Chang, Y., et al.: A point-based mortality prediction system for older adults with diabetes. Sci. Rep. **7**(1), 1–10 (2017)
6. Che, Z., Purushotham, S., Cho, K., Sontag, D., Liu, Y.: Recurrent neural networks for multivariate time series with missing values. Sci. Rep. **8**(1), 1–12 (2018)
7. Cheng, Y., Wang, F., Zhang, P., Hu, J.: Risk prediction with electronic health records: a deep learning approach. In: Proceedings of the 2016 SIAM International Conference on Data Mining, pp. 432–440. SIAM (2016)
8. Choi, E., Bahadori, M.T., Schuetz, A., Stewart, W.F., Sun, J.: Doctor AI: predicting clinical events via recurrent neural networks. In: Machine Learning for Healthcare Conference, pp. 301–318 (2016)
9. El_Jerjawi, N.S., Abu-Naser, S.S.: Diabetes prediction using artificial neural network (2018)
10. Heymann, A.D., et al.: The implementation of managed care for diabetes using medical informatics in a large Preferred Provider Organization. Diab. Res. Clin. Pract. **71**(3), 290–298 (2006)
11. Kavakiotis, I., Tsave, O., Salifoglou, A., Maglaveras, N., Vlahavas, I., Chouvarda, I.: Machine learning and data mining methods in diabetes research. Comput. Struct. Biotechnol. J. **15**, 104–116 (2017)
12. Khalid, J., Raluy-Callado, M., Curtis, B., Boye, K., Maguire, A., Reaney, M.: Rates and risk of hospitalisation among patients with type 2 diabetes: retrospective cohort study using the UK General Practice Research Database linked to English hospital episode statistics. Int. J. Clin. Pract. **68**(1), 40–48 (2014)
13. LeCun, Y., Bengio, Y., Hinton, G.: Deep learning. Nature **521**(7553), 436–444 (2015)
14. Lin, J., Keogh, E., Wei, L., Lonardi, S.: Experiencing SAX: a novel symbolic representation of time series. Data Min. Knowl. Discov. **15**(2), 107–144 (2007)
15. Lipton, Z.C., Kale, D.C., Elkan, C., Wetzel, R.: Learning to diagnose with LSTM recurrent neural networks. arXiv preprint arXiv:1511.03677 (2015)
16. McEwen, L.N., et al.: Predictors of mortality over 8 years in type 2 diabetic patients: Translating Research Into Action for Diabetes (triad). Diabetes Care **35**(6), 1301–1309 (2012)
17. Moskovitch, R., Choi, H., Hripcsak, G., Tatonetti, N.P.: Prognosis of clinical outcomes with temporal patterns and experiences with one class feature selection. IEEE/ACM Trans. Comput. Biol. Bioinform. **14**(3), 555–563 (2016)
18. Moskovitch, R., Polubriaginof, F., Weiss, A., Ryan, P., Tatonetti, N.: Procedure prediction from symbolic electronic health records via time intervals analytics. J. Biomed. Inform. **75**(C), 70–82 (2017). https://doi.org/10.1016/j.jbi.2017.07.018

19. Moskovitch, R., Shahar, Y.: Classification-driven temporal discretization of multi-variate time series. Data Min. Knowl. Discov. **29**(4), 871–913 (2014). https://doi.org/10.1007/s10618-014-0380-z
20. Moskovitch, R., Shahar, Y.: Classification of multivariate time series via temporal abstraction and time intervals mining. Knowl. Inf. Syst. **45**(1), 35–74 (2015)
21. Moskovitch, R., Walsh, C., Wang, F., Hripcsak, G., Tatonetti, N.: Outcomes prediction via time intervals related patterns. In: 2015 IEEE International Conference on Data Mining, pp. 919–924. IEEE (2015)
22. Sacchi, L., Larizza, C., Combi, C., Bellazzi, R.: Data mining with temporal abstractions: learning rules from time series. Data Min. Knowl. Discov. **15**(2), 217–247 (2007)
23. Serrà, J., Pascual, S., Karatzoglou, A.: Towards a universal neural network encoder for time series. In: CCIA, pp. 120–129 (2018)

Heterogeneous Graph Embeddings of Electronic Health Records Improve Critical Care Disease Predictions

Tingyi Wanyan[1,2,3,4], Martin Kang[5,6], Marcus A. Badgeley[5,7,8],
Kipp W. Johnson[2], Jessica K. De Freitas[1,2], Fayzan F. Chaudhry[1,2],
Akhil Vaid[1,2], Shan Zhao[1], Riccardo Miotto[1,2], Girish N. Nadkarni[1,2,9,10],
Fei Wang[11], Justin Rousseau[12], Ariful Azad[4], Ying Ding[3,4,12],
and Benjamin S. Glicksberg[1,2(✉)]

[1] Hasso Plattner Institute for Digital Health at Mount Sinai,
Icahn School of Medicine at Mount Sinai, New York, NY, USA
`benjamin.glicksberg@mssm.edu`
[2] Department of Genetics and Genomic Sciences,
Icahn School of Medicine at Mount Sinai, New York, NY, USA
[3] Intelligent System Engineering, Indiana University, Bloomington, IN, USA
[4] School of Information, University of Texas Austin, Austin, TX, USA
[5] nference, Cambridge, MA, USA
[6] Department of Dermatology, Mayo Clinic, Rochester, USA
[7] Department of Anesthesiology, Critical Care and Pain Medicine,
Massachusetts General Hospital, Boston, MA, USA
[8] Department of Brain and Cognitive Sciences,
Massachusetts Institute of Technology, Cambridge, MA, USA
[9] Department of Medicine, Division of Nephrology,
Icahn School of Medicine at Mount Sinai, New York, NY, USA
[10] Charles Bronfman Institute for Personalized Medicine,
Icahn School of Medicine at Mount Sinai, New York, NY, USA
[11] Department of Health Policy and Research, Weill Cornell Medical School,
Cornell University, New York, NY, USA
[12] Dell Medical School, University of Texas, Austin, TX, USA

Abstract. Electronic Health Record (EHR) data is a rich source for powerful biomedical discovery but it consists of a wide variety of data types that are traditionally difficult to model. Furthermore, many machine learning frameworks that utilize these data for predictive tasks do not fully leverage the inter-connectivity structure and therefore may not be fully optimized. In this work, we propose a relational, deep heterogeneous network learning method that operates on EHR data and addresses these limitations. In this model, we used three different node types: patient, lab, and diagnosis. We show that relational graph learning naturally encodes structured relationships in the EHR and outperforms traditional multilayer perceptron models in the prediction of thousands of diseases. We evaluated our model on EHR data derived from MIMIC-III, a public critical care data set, and show that our model has improved prediction of numerous disease diagnoses.

© Springer Nature Switzerland AG 2020
M. Michalowski and R. Moskovitch (Eds.): AIME 2020, LNAI 12299, pp. 14–25, 2020.
https://doi.org/10.1007/978-3-030-59137-3_2

Keywords: Electronic health records · Heterogeneous graph learning · Skip-gram model · Embeddings

1   Introduction

Electrical Health Records (EHRs) have rapidly emerged over the past 10 years as a powerful source for biomedical research [6,7,13]. EHRs consist of clinical data from patient encounters with healthcare systems, which include demographic information, diagnoses, laboratory tests, medications, and clinical notes. EHR data have been used to develop machine learning (ML), and deep learning (DL), models for predicting diagnoses, mortality, length of hospital stay, and future illnesses. However, many of these ML-related solutions to clinical tasks consist of simple rule based models that, while possibly are easier to implement, often do not capture the complex patterns of the data. Some of these solutions are sufficient for certain clinical tasks, but for others they are lacking. For example, one factor for determining priority for transplantation is a model for end-stage liver disease which includes only four variables and was trained on only 231 patients [10]. While there are a number of barriers that need to be overcome for DL to pervade healthcare operations, one particular hurdle is developing more suitable EHR representations for modeling.

Current EHR systems are constructed with numerous medical codes of different types to represent diverse data elements captured in clinical encounters. The performance of DL models on EHR could benefit from accurately capturing and jointly modeling these heterogeneous data [2,12]. The most common approach to handle disparate data types is to treat each patient encounter as an unordered set of features, and concatenate these features together as the input to a DL system [12]. Such an approach is straightforward, intuitive, and easy to manipulate. However, this feature integration approach disregards the graphical structure and inter-connectivity between medical concepts (i.e., how a lab is indicative of disease status) [4]. Furthermore, utilizing graph model encodings can better leverage underlying patient similarity structure that can provide valuable information to enhance cohort analyses, disease subtyping, comparing diagnosis outcomes and treatment effectiveness [14]. Some recent graphical modeling techniques have demonstrated the value of the taking into consideration the inter-connectivity of clinical data, such as being able to agnostically derive a physician's treatment decision procedure [3,4], and a temporal graph model [9] which captures the medical concepts inter-connectivity patterns over time. These techniques, however, undergoes per-patient training which lacks the ability to learn from the information provided within similar patients.

Furthermore, DL modeling is difficult because of issues of data quality [8] due to insufficient patient information, messy data, and missing values among others. Besides that, data diversity and non-uniform length of time series within each patient also create issues for modeling. For example, patient encounter frequency varies in length, ranging from only one encounter to multiple readmissions. Also, length of stay could vary from a few hours to several months. The data sparsity along with data diversity create difficulties for deep learning models such as LSTM system [2,8], which requires abundant training data in order to reach good performance.

In this paper, we propose a Heterogeneous Graph Learning Model (HGM) and apply it to EHR data. It contains various techniques and properties that attempts to overcome the aforementioned problems. Since it has a graph structure, this model could more naturally capture the inter-connectivity between medical concepts. It also connects similar patients by their disease profiles, so that information from a similar patient could be leveraged for encoding in the target patient representation. The graph model learns representations by propagating information through the whole network, so when the data set is sparse, the embedding representation for each patient could be learned from the information traversing the whole network. This model, which utilizes the Skip Gram With Negative Sampling strategy [11], is an efficient way of incorporating all complex data available. We show that with the relational heterogeneous graph learning, we can reach marginal improvement on diagnoses classification accuracy given patients' lab tests against traditional per-patient training strategy using shallow feed forward neural network.

2   Methodology

In this section, we introduce the theoretical construction of our heterogeneous EHR graph model. Please refer to the Supplementary Materials Appendix A for a description of the preliminaries for these models.

2.1   Data Set

For this work, we utilized EHR data from the critical care MIMIC-III de-identified data set. This data set is comprised of various elements relating to patients during their hospital care in an intensive care unit, such as demographics, lab test results, disease diagnoses, among others. We sampled the first 3801 patients in the data set and collected all of their associated lab tests and diagnoses. These patients had received 447 unique lab tests and 2922 unique diagnoses. The limitation on the cohort sample size was due to the RAM required to load all of these data as a graph into memory.

2.2   Data Representation for Graphical Model

We created a graphical model of the EHR data by representing patients, labs, and diagnoses as nodes in a directed graph. Nodes are connected by edges, which come in two flavors and can be represented with the following triples:

$$Lab \xrightarrow{testing} Patient : \{Lab, testing, Patient\}$$

$$Patient \xrightarrow{diagnosed} Diagnosis : \{Patient, diagnosed, Diagnosis\}$$

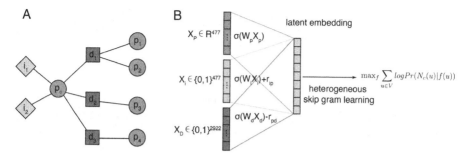

Fig. 1. Model schematics for representing EHR data in a heterogeneous graphical model (A) and dense vectors (B). All graph nodes in (A) have a corresponding vector like those shown in (B). The vector representations can be projected into a shared space with the TransE method, and this projection optimized for retaining relations in the original data in the embedding via skip-gram optimization.

The initial *Patient* node representation is a vector $X_p \in \mathbb{R}^{477}$ containing the measured values from lab tests. We initially represented labs and diagnoses as one-hot encodings: $X_l \in \{0,1\}^{477}$ and $X_d \in \{0,1\}^{2992}$.

With these two types of relationships T_E, we can construct the heterogeneous graph integrating the specified elements of the entire EHR data (Fig. 1A). One *Patient* node could have connection with multiple *Diagnoses* node, and these *Diagnoses* nodes could link to other patients who have the same ICD code.

2.3   Embedding the HGM into a Latent Space

Nodes from a HGM can be embedded into a shared latent space using the TransE method (Fig. 1B) [1]. This method uses a set of 1) projection matrices and 2) relation vectors. After initialization, projections and translations can be optimized end-to-end (see Sect. 2.4).

HGM nodes X_p, X_l, X_d are projected into a shared latent space with trainable projection matrices W_p, W_i, W_d using these nonlinear mappings:

$$c_p = \sigma(W_p \cdot X_p)$$
$$c_i = \sigma(W_i \cdot X_i)$$
$$c_d = \sigma(W_d \cdot X_d)$$

Where σ is a non-linear activation function and c_p, c_i, c_d are the latent representations of each type of node. Despite the EHR-space using different dimensions for different node types X_p, X_i, X_d, all nodes types are projected into the same latent space.

Then we apply translation operations to link these different types of nodes:

$$c_p = c_i + r_{ip}$$
$$c_d = c_p + r_{pd}$$

Where r_{ip} and r_{pd} are the relation vectors connecting patients to labs and diseases, respectively.

2.4 Optimizing the HGM Embedding

With the projection and translation operations we can convert different types of nodes into the same latent space. We then tune these parameterized transforms to increase the proximity between those embedding points whose corresponding graph nodes are often connected. Specifically, we apply Heterogeneous Skip-gram optimization using the optimization model [5]:

$$\max \sum_{u \in V} \sum_{t \in T_V} logPr(N_t(u)|f(u)) \tag{1}$$

Where $N_t(u)$ is the heterogeneous neighborhood vertices of center node u, and $t \in T_V$ is the node type. Here, we learn effective node embeddings by maximizing the probability of correctly predicting the a patient node's associated labs and diagnoses. The prediction probability is modeled as a softmax function:

$$Pr(c_t|f(u)) = \frac{e^{c_t \cdot u}}{Z_u} \tag{2}$$

Where u is the latent representation of patient u, c_t is the latent representation of lab and diagnosis neighbors of node of u, and $c_t \cdot u$ is the inner product of the two embedding vectors representing their similarity. Z_u is the normalization term $Z_u = \sum_{v \in V} e^{v_t \cdot u}$. Where Z_u integrate over all vertices. Therefore, Eq. 1 could be simplified to:

$$\mathcal{L}_s = -\sum_{t \in T} \sum_{u \in V} \left[\sum_{c_t \in N_t(u)} c_t \cdot u - logZ_u \right] \tag{3}$$

Numerical computation of Z_u is intractable for very huge graph with millions of nodes. So we adopt negative sampling strategy [11] to approximate the normalization factor, making the optimization function:

$$\mathcal{L}_s = -\sum_{t \in T} \sum_{u \in V} \left[\sum_{c_t \in N_t(u)} log\sigma(c_t \cdot u) + \sum_{j=1}^{K} E_{c_j \sim P_v(c_j)} log\sigma(-c_j \cdot u) \right] \tag{4}$$

where $\sigma(x) = \frac{1}{1+exp(-x)}$, K is the number of negative samples. $P_v(c_j)$ is the negative sampling distribution. Equation 4 is the final objective function we are using for heterogeneous graph learning.

For training our Heterogeneous Graph Model (HGM), we perform heterogeneous neighborhood sampling by its one-hop connectivity, and pick *Patient* node as the center node, since it has one-hop connections to both *Diagnoses* and *Item_test* nodes. Specifically, for one training center *Patient* node, we uniformly sample 10 *Diagnoses* one-hop direct connected nodes, and 10 *Item_test* one-hop direct connected nodes. From these sampled 10 *Diagnoses* nodes, we sample another 10 *Patient* nodes, each has connection with each of the 10 *Diagnoses* nodes. In this way, we connect the center patient node with its similar other *Patient* nodes by their common diagnoses. For negative sampling [11], we perform uniform sampling through all *Diagnoses* node and *Item_test* nodes that don't have one-hop connections with the center training patient node. Then we project these different nodes into same latent space through TransE model, after unifying the embeddings for different node types, each concept is represented as a point in a Euclidean space. In this space we can measure the similarity between any two points by the angle between vectors between them and the origin.

2.5 Disease Prediction

For diagnosis prediction, we used the HGM embedding vectors to identify similar patients and diagnoses, and evaluate how this approach compares to a classical feed forward neural network approach, specifically a multilayer perceptron (MLP). We record F1 score and AUC score as the evaluation metric for comparison.

We performed 10-fold cross-validation by randomly splitting patients into a group of 2,660 used to fit the MLP and HGM embedding and 1,141 used to evaluate disease prediction. For each patient, we computed the distance between a patient plus the diagnosis translation vector r_{pd} to all diseases.

The baseline MLP model is a shallow feed-forward encoding-decoding neural network structure with a single hidden embedding layer whose dimensionality matches the embeddings produced by the HGM. The decoding part is a softmax layer for classifying correct diagnoses.

3 Results

3.1 Embedded Representation

By learning a heterogeneous graph embedding for each node type and then using transE to translate between type specific embeddings, we generated dense vector representation in a space shared between all node types (Fig. 3). There was a mixed cluster of all node types and several type-selective clusters.

Upon inspection, salient clusters of labs tests can be identified when the embeddings are projected into 2D t-SNE space for visualization. The members of one cluster corresponded to routine comprehensive metabolic panels, while members of another cluster largely consisted of ventilator measurements.

3.2 Diagnosis Prediction Performance Comparison

The HGM outperformed the MLP on many diagnoses. When evaluating both models' diagnosis predictions across all common diagnoses, the HGM had a higher performance than an MLP across all tested latent embedding dimensions (Table 1, Fig. 2). Notably, the performance of HGM remained consistent with

Table 1. Diagnosis classification performance

Model	F1 score	AUC score
100 hidden latent embedding dimension		
MLP-Sigmoid	0.671	0.788
MLP-Tanh	0.517	0.778
MLP-Relu	0.483	0.765
HGM-Sigmoid	**0.739**	**0.834**
HGM-Tanh	0.727	0.829
HGM-Relu	0.713	0.839
200 hidden latent embedding dimension		
MLP-Sigmoid	0.625	0.766
MLP-Tanh	0.447	0.755
MLP-Relu	0.446	0.746
HGM-Sigmoid	**0.741**	**0.835**
HGM-Tanh	0.733	0.828
HGM-Relu	0.739	0.840
500 hidden latent embedding dimension		
MLP-Sigmoid	0.537	0.753
MLP-Tanh	0.377	0.724
MLP-Relu	0.419	0.734
HGM-Sigmoid	**0.751**	**0.834**
HGM-Tanh	0.735	0.829
HGM-Relu	0.743	0.842

larger embeddings, while the performance of MLP degraded with larger embeddings.

The predictive performance of these models varied widely by disease, as shown in (Fig. 3B). The performance of HGM was particularly strong with diseases that were more prevalent in the test set (see Table 2). We observed only one diagnosis, end stage renal diseases (ESRD), where MLP outperformed HGM (MLP F1: 0.606, HGM F1: 0.245) (Fig. 3A).

For the diagnoses with at least one percent prevalence, the median, 25th percentile, and 75th percentile of MLP predictive F1 scores are 0. The range of MLP F1 distribution is 0 to 0.606. For the same set of diagnoses, the median, 25th percentile, and 75th percentile of HGM F1 scores are 0.041, 0.024, 0.081, respectively, and the range of HGM F1 distribution is 0 to 0.562.

4   Discussion

In this work, we present HGM embeddings as a way to naturally represent EHR data relations with dense vectors and an embedding space containing all node

Table 2. Prediction performance on most observed diagnoses

Diagnoses	HGM-sigmoid F1 score	MLP-sigmoid F1 score
Congestive heart failure	0.562	0.406
Unspecified essential hypertension	0.512	0.435
Atrial fibrillation	0.447	0.423
Acute kidney failure	0.455	0.415
Coronary atherosclerosis	0.365	0.163
Other and unspecified hyperlipidemia	0.367	0.297
Acute respiratory failure	0.316	0.067
Esophageal reflux	0.311	0.041
Diabetes mellitus	0.297	0.192
Urinary tract infection	0.276	0.060

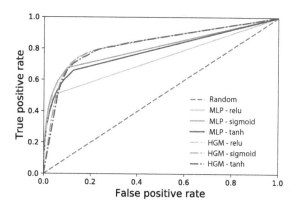

Fig. 2. Binary classification performance of HGM and MLP across common diseases. Each line represents the trade-off of sensitivity and specificity for a classifier. HGM frameworks with larger embedding spaces perform better than the MLP models.

types in the original graph. By measuring distances between patient and disease concepts in this embedding space, we were able to predict which diagnoses a group of hold-out patients had with better performance than a supervised model trained specifically to predict patients' diagnoses from their labs.

Averaged across all diseases, the HGM consistently outperformed the MLP across a range of activation functions and embedding dimension sizes. HGM and MLP had different trends as the dimensionality of the embedding increased. A larger embedding provides more complex representations, but is more likely to be overfit to training data. As the dimensionality of the embedding was increased, the MLP AUC decreased with a dosage effect observed across all activation functions. However, the HGM maintained a stable performance across all embedding

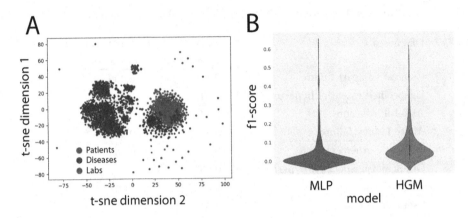

Fig. 3. A) R^2 view of the R^{500} embedding space shared by all data types. Each point represents a graph node colored by its type. The dimensionality of the space is reduced for visual interpretation with t-SNE. B) Distribution of F1 scores for common diagnoses using a HGM or MLP model. Diagnoses with at least 1% prevalence in the test set.

capacities. This suggests that the embeddings learned by HGM are less susceptible to overfitting training data.

MLP outperformed HGM on only one diagnosis, End Stage Renal Diseases (ESRD). This may be because the diagnosis of ESRD can be determined solely by a single lab test, estimated Glomerular Filtration Rate (eGFR). Thus, it is less likely that the prediction of ESRD will benefit from the graphical property of HGM.

One key feature of HGM is that the graphical structure of HGM explicitly declares and takes the sum of information from all patient nodes connecting a given pair of diagnosis and lab test. On the contrary, MLP flattens all features at the patient-level, and performs training on the per-patient basis, allowing only indirect connections between pairs of diagnoses and lab tests and relying only on network parameters to learn the underlying biomedical relationships.

The data set we used to fit HGMs allowed us to develop an EHR-knowledge graph across the compendium of care provided in an intensive care unit. We found that our model was robust to overfitting but there may be bias in lab-disease relationships between patient populations or intensive care practices. Only the sickest patients are admitted to the ICU, so this model should be fine-tuned for other inpatient applications. Another consequence of only having ICU visits is that most people have only one or fewer ICU admissions, which is not suitable for time series models. Other studies applying graph theory to EHRs have been able to perform robust sequential diseases prediction that consistently outperforms non-graph models [3].

5   Conclusion

In this study, we apply a deep heterogeneous graph model (HGM) to learn the representation of EHR data. In the task of diagnosis prediction, HGM embeddings consistently outperformed non-graphical baseline models across diagnoses and appears less susceptible to overfitting of training data. Our findings suggest that HGM is a promising strategy to develop generalizable EHR-knowledge graph. In the future, we expect to apply HGM to other clinically relevant tasks and assess performance across multi-institutional data sets.

A   Appendix

Definition 1 (Heterogeneous Network). *A heterogeneous network is defined as a graph $G = (V, E, T)$, where each node v and each link e are represented by their mapping functions to a specific node and relation type $\phi(v) : V \rightarrow T_V$ and $\phi(e) : E \rightarrow T_E$. Where T_V and T_E denote the sets of node and relation types, and $|T_V| + |T_E| > 2$.*

Definition 2 (Heterogeneous Graph Learning). *Given a heterogeneous network G, the task of heterogeneous graph learning is to learn a function mapping $f : V \rightarrow R^d$, that connects disparate type of nodes into a $d - dimensional$ uniform latent representation $X \in R^{|V| \times d}$, and $d \ll |V|$, that are able to capture the structural and semantic relations between them.*

Definition 3 (One-hop Connectivity). *One-hop connectivity in a heterogeneous network is the local pairwise connection between two consecutive vertices, which directly linked by an edge belongs to a relational type.*

A.1   Skip-Gram Model

The skip-gram model [11] seeks to maximize the probability of observing the context neighborhood nodes given the center node:

$$\max_f \sum_{u \in V} log Pr(N_c(u) | f(u)) \tag{5}$$

Where $N_c(u)$ is the neighborhood context nodes of the center node u, and $f(u)$ is the latent representation of u.

A.2   Heterogeneous Skip-Gram Model

EHR data is heterogeneous, including varies type of vertices, such as lab tests, diagnoses, prescriptions, and patient demographics. Each of these vertices encodes different information. Heterogeneous Skip-gram model [5] learns the latent expression of these different type of nodes by maximizing the probability of observing heterogeneous neighborhood given a center node:

$$\max \sum_{u \in V} \sum_{t \in T_V} log Pr(N_t(u)|f(u)) \qquad (6)$$

Where $N_t(u)$ is the heterogeneous neighborhood vertices of center node u, and $t \in T_V$ is the node type.

A.3   TransE

TransE model [1] aims to relate different type of nodes by their relationship type. Specifically, two different types of nodes are connected by a relation type would be represented as a triple (head, relation, tail), denoted as (h, l, t). For example, one triple from EHR data could be $(patient, diagnosed, ICD)$, where $patient$ is the head node, ICD is the specific diagnosis code attributed to the patient, and the relation between these two vertices is $diagnosed$.

This TransE model leverages the procedure by first projecting different type of node with different initial representation dimension into a same latent dimension space (where the dimension of this latent space can be customized), and these two different type projected nodes are linked by a relation type which is represented as a translation vector in that latent space. Both the projection matrix and the relational translation vector are learnable parameters in the deep learning system.

References

1. Bordes, A., Usunier, N., Garcia-Duran, A., Weston, J., Yakhnenko, O.: Translating embeddings for modeling multi-relational data. In: Advances in Neural Information Processing Systems, pp. 2787–2795 (2013)
2. Choi, E., Bahadori, M.T., Schuetz, A., Stewart, W.F., Sun, J.: Doctor AI: predicting clinical events via recurrent neural networks. In: Machine Learning for Healthcare Conference, pp. 301–318 (2016)
3. Choi, E., Xiao, C., Stewart, W., Sun, J.: Mime: multilevel medical embedding of electronic health records for predictive healthcare. In: Advances in Neural Information Processing Systems, pp. 4547–4557 (2018)
4. Choi, E., et al.: Graph convolutional transformer: learning the graphical structure of electronic health records. arXiv preprint arXiv:1906.04716 (2019)
5. Dong, Y., Chawla, N.V., Swami, A.: metapath2vec: scalable representation learning for heterogeneous networks. In: Proceedings of the 23rd ACM SIGKDD International Conference on Knowledge Discovery and Data Mining, pp. 135–144 (2017)
6. Glicksberg, B.S., Johnson, K.W., Dudley, J.T.: The next generation of precision medicine: observational studies, electronic health records, biobanks and continuous monitoring. Hum. Mol. Genet. **27**(R1), R56–R62 (2018)
7. Glicksberg, B.S., et al.: Automated disease cohort selection using word embeddings from electronic health records. In: World Scientific (2018)
8. Lipton, Z.C., Kale, D.C., Elkan, C., Wetzel, R.: Learning to diagnose with LSTM recurrent neural networks. arXiv preprint arXiv:1511.03677 (2015)

9. Liu, C., Wang, F., Hu, J., Xiong, H.: Temporal phenotyping from longitudinal electronic health records: a graph based framework. In: Proceedings of the 21th ACM SIGKDD International Conference on Knowledge Discovery and Data Mining, pp. 705–714 (2015)

10. Malinchoc, M., Kamath, P.S., Gordon, F.D., Peine, C.J., Rank, J., Ter Borg, P.C.: A model to predict poor survival in patients undergoing transjugular intrahepatic portosystemic shunts. Hepatology **31**(4), 864–871 (2000)

11. Mikolov, T., Sutskever, I., Chen, K., Corrado, G.S., Dean, J.: Distributed representations of words and phrases and their compositionality. In: Advances in Neural Information Processing Systems, pp. 3111–3119 (2013)

12. Miotto, R., Li, L., Kidd, B.A., Dudley, J.T.: Deep patient: an unsupervised representation to predict the future of patients from the electronic health records. Sci. Rep. **6**(1), 1–10 (2016)

13. Shickel, B., Tighe, P.J., Bihorac, A., Rashidi, P.: Deep EHR: a survey of recent advances in deep learning techniques for electronic health record (EHR) analysis. IEEE J. Biomed. Health Inform. **22**(5), 1589–1604 (2017)

14. Zhu, Z., Yin, C., Qian, B., Cheng, Y., Wei, J., Wang, F.: Measuring patient similarities via a deep architecture with medical concept embedding. In: 2016 IEEE 16th International Conference on Data Mining (ICDM), pp. 749–758. IEEE (2016)

Controlling Level of Unconsciousness by Titrating Propofol with Deep Reinforcement Learning

Gabriel Schamberg[1,2(✉)], Marcus Badgeley[2,3], and Emery N. Brown[1,2,3]

[1] Picower Institute for Learning and Memory, Massachusetts Institute
of Technology, Cambridge, MA 02139, USA
gabes@mit.edu
[2] Department of Brain and Cognitive Sciences, Massachusetts Institute
of Technology, Cambridge, MA 02139, USA
[3] Department of Anesthesiology, Critical Care and Pain Medicine,
Massachusetts General Hospital, Boston, MA 02114, USA

Abstract. Reinforcement Learning (RL) can be used to fit a mapping from patient state to a medication regimen. Prior studies have used deterministic and value-based tabular learning to learn a propofol dose from an observed anesthetic state. Deep RL replaces the table with a deep neural network and has been used to learn medication regimens from registry databases. Here we perform the first application of deep RL to closed-loop control of anesthetic dosing in a simulated environment. We use the cross-entropy method to train a deep neural network to map an observed anesthetic state to a probability of infusing a fixed propofol dosage. During testing, we implement a deterministic policy that transforms the probability of infusion to a continuous infusion rate. The model is trained and tested on simulated pharmacokinetic/pharmacodynamic models with randomized parameters to ensure robustness to patient variability. The deep RL agent significantly outperformed a proportional-integral-derivative controller (median absolute performance error 1.7% ± 0.6 and 3.4% ± 1.2). Modeling continuous input variables instead of a table affords more robust pattern recognition and utilizes our prior domain knowledge. Deep RL learned a smooth policy with a natural interpretation to data scientists and anesthesia care providers alike.

Keywords: Anesthesia · Reinforcement learning · Deep learning

1 Introduction

The proliferation of anesthesia in the 1800s is America's greatest contribution to modern medicine and enabled far more complex, invasive, and humane surgical

This work was supported by Picower Postdoctoral Fellowship (to GS) and the National Institutes of Health P01 GM118629 (to ENB).
Schamberg, G and Badgeley, M.—These authors contributed equally to this work.

© Springer Nature Switzerland AG 2020
M. Michalowski and R. Moskovitch (Eds.): AIME 2020, LNAI 12299, pp. 26–36, 2020.
https://doi.org/10.1007/978-3-030-59137-3_3

procedures. Now nearly 60,000 patients receive general anesthesia for surgery daily in the United States [2]. Anesthesia is a reversible drug-induced state characterized by a combination of amnesia, immobility, antinociception, and loss of consciousness [2]. Anesthesia providers are not only responsible for a patient's depth of anesthesia, but also their physiologic stability and oxygen delivery.

Anesthesiologists need to determine the medication regimen a patient receives throughout a surgical procedure. The anesthetic state is managed by providing inhaled vapors or infusing intravenous medication. The medication most studied for controlling the patient's level of unconsciousness is propofol. Propofol affects the brain's cortex and arousal centers to induce loss of consciousness in a dose-dependent manner. Propofol dosage needs to be balanced: patients should be deep enough to avoid intraoperative awareness, but too much anesthesia can cause physiologic instability or cognitive deficits. Currently anesthesiologists can manually calculate and inject each dose, or select the desired concentration of propofol in the brain, and an infusion pump will adjust infusion rates based on how an average patient processes the medication.

Investigational devices and studies have shown that measuring brain activity can provide personalized computer-calculated dosing regimens. Studies of automatic anesthetic administration have three primary components. First, *sensing* involves automatically obtaining a numerical representation of the patient's anesthetic state. Prior studies have primarily focused on controlling the level of unconsciousness (LoU) using a variety of indices, including the bispectral index (BIS) [1,5], WAV_{CNS} [4], and burst suppression probability [15,17]. Second, *modeling* involves the development of pharmacokinetic (PK) and pharmacodynamic (PD) models of how a patient's LoU responds to specified drug dosages. These models are used to derive optimal control laws [15], tune controller parameters [17], and/or develop robust controller parameterizations [4]. Finally, the *controller* determines the mapping from sensed variables to drug infusion rates. Numerous control algorithms have been studied for LoU regulation, including (but not limited to) proportional integral derivative (PID) controllers, model predictive (MP) controllers, and linear quadratic regulator (LQR) controllers. The performance of these algorithms is restricted by linearity assumptions and/or reliance on a nominal patient model for obtaining the control action.

Reinforcement learning (RL) is a form of optimal control which learns by optimizing a flexible reward system. RL can be used to fit a mapping from anesthetic state to a propofol dose. Contrary to MP and LQR controllers, the RL-based controllers can be *model-free* in that the control law is established without any knowledge of the underlying model. Prior studies using tabular RL created a table where each entry represents a discrete propofol dosage that corresponds to a discrete observation [9–11]. Tabular mappings are flexible but do not scale well with larger state spaces and can have non-smooth policies that result from independently determining actions for each of the discretized states. Continuous states have been used by actor-critic methods that use linear function maps [8] and adaptive linear control [12]. Existing studies that train RL agents to administer anesthesia tend to either underconstrain by disregarding continuity or overconstrain by imposing strict linearity assumptions.

The use of deep neural networks as functional maps in RL (called "deep RL") has been used to learn medication regimens from registry databases outside of the anesthesia context [7,13]. Existing deep RL studies use large observation spaces and disregard known PK/PD properties. By using retrospectively collected data, these studies do not permit the RL agent to learn from its own actions and restrict the agent's "teachable moments" to those that are observed in the data.

We perform the first application of deep RL to anesthetic dosing supported with fundamental models from pharmacology. Using data from a simulated PK/PD model, we implement an RL framework for training a neural network to provide a mapping from a continuous valued observation vector to a distribution over actions. The use of a deep neural network allows the number of parameters of the model to scale linearly with the number of inputs to the policy map, avoiding the exponential growth that occurs when expanding the input dimension of tabular policies. The resultant policy can represent nonlinear functions while still yielding a smooth function of the input variables. By training the RL agent on simulated data, we can control the range of patient models that are included in the learning process. As such, the proposed framework allows us to experiment with a variety of policy inputs in order to directly incorporate robustness to patient variability into the training procedure.

2 Methods

In this section we develop a mathematical formalization for learning how to administer propofol to control LoU. We formalize this propofol dosing task as a partially observable Markov decision process (POMDP) and solve it using the RL method cross-entropy. This RL "agent" learns from data generated by simulated interactions with the environment PK/PD state-space model (see Fig. 1).

2.1 Environment Model

The primary component of the environment is the patient model, which dictates the observed LoU given a drug infusion profile and is composed of three sub-models. First, a discrete time 3-compartment pharmacokinetic (PK) model is used to model the mass transfer of infused drug between the central, slow peripheral, and rapid peripheral compartments:

$$\mathbf{x}_{k+1} = \mathbf{A}\mathbf{x}_k + \mathbf{B}a_k \tag{1}$$

where $\mathbf{x}_k = [x_k^{(1)}, x_k^{(2)}, x_k^{(3)}] \in \mathbb{R}_+^3$ represents the 3-compartment model concentrations (where $\mathbb{R}_+$ represents the set of non-negative reals), $a_k \in \mathcal{A} = \{0, 1\}$ represents whether or not drug is infused at time k, $\mathbf{A} \in \mathbb{R}^{3 \times 3}$ gives the mass transfer rates between compartments, and $\mathbf{B} = [\Delta u, 0, 0]$ represents the mass transfer rate resulting from drug infusion. The parameters of $\mathbf{A}$ are determined

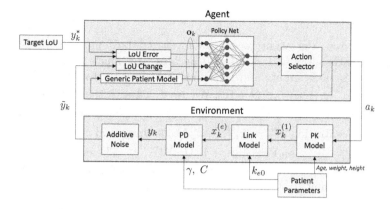

Fig. 1. Block diagram of the proposed RL framework.

by the patient age, height, and weight according to the Schnider model [14]. In the current study we have $\Delta = 5\,\text{s}$ (as in [10]) and $u = 1.67$ mg/s, such that over each five second window is either 0 mg or 8.35 mg of propofol is delivered.

The link function determines the effect site (i.e. brain) concentration from the central compartment concentration:

$$x_{k+1}^{(e)} = \alpha x_k^{(e)} + \beta x_k^{(1)} \tag{2}$$

where $\alpha = \exp(-k_{e0}\Delta/60)$ gives the persistence of drug in the effect site and $\beta = (k_{e0}/60)\exp(-k_{e0}\Delta/60)$ gives the mass transfer rate from the plasma compartment to the effect site for a given plasma-brain equilibration constant k_{e0}.

Finally, we compute how a given effect site concentration of propofol affects LoU using a hill function:

$$y_k = h(x_k^{(e)}) = \frac{x_k^{(e)\gamma}}{C^\gamma + x_k^{(e)\gamma}} \tag{3}$$

where $y_k \in [0,1]$ gives the true LoU at time k, $C \in \mathbb{R}_+$ give the effect site concentration corresponding to a LoU of 0.5, and $\gamma \in \mathbb{R}_+$ determines the shape of the non-linear PD response (higher values give rise to more rapid transitions in LoU). Given that this PD model has been used in studies targeting control of BIS [5], WAV_{CNS} [4], and BSP [17], we choose to treat y_k as a general index of LoU in the present simulation study. As such, y_k can be viewed as representing a non-linear continuum from consciousness (0) to brain death (1). The clinical interpretations of intermediate LoUs depend on the specific choice of index.

The observed LoU $\tilde{y}_k$ is obtained by adding Gaussian measurement noise $v_k \sim \mathcal{N}(0, \sigma_v^2)$ to the true LoU y_k. The resulting observed LoU is clipped to be between zero and one. We set measurement noise at $\sigma_v^2 = 0.0003$ [10].

2.2   Agent Model

At each timestep, the agent receives a measured LoU $\tilde{y}_k$ from the environment along with a target LoU y_k^* and decides how much propofol to infuse. The agent uses these inputs and the infusion history to derive an observation vector: $\mathbf{o}_k = [o_k^{(1)}, \ldots, o_k^{(d)}] \in \mathcal{O} = \mathbb{R}^d$. We use $d = 4$ with the following observed variables: measured LoU error $o_k^{(1)} = \tilde{y}_k - y_k^*$, 30-s ahead predicted change in effect site concentration $o_k^{(2)} = \hat{x}_{k+6}^{(e)} - \hat{x}_k^{(e)}$, 30-s historical change in measured LoU $o_k^{(3)} = \tilde{y}_k - \tilde{y}_{k-6}$, and the target LoU $o_k^{(4)} = y_k^*$. The estimated effect site concentration is computed by maintaining a PK model with generic parameterization (according to Table 1) throughout a trial to estimate concentration levels $\hat{x}_k$ and $\hat{x}_k^{(e)}$. At each time, this model is propagated 30 s forward under the assumption that there will be no further infusion to yield $\hat{x}_{k+6}^{(e)}$, which is used to compute the predicted change. All elements of the observation vector can be computed solely from previous actions and measured LoUs, and they do not assume any knowledge of the specific patient variables in the environment. Each of the observation variables is presumed to provide the agent with a unique advantage in selecting an action. For example, the predicted change in effect site concentration is included to account for the lag between drug administration and arrival at the effect site and including the target LoU enables the agent to have different steady-state infusion rates for different target LoUs.

Table 1. Model parameters for the generic patient and range of parameters selected randomly for training and testing.

Sub-model	Parameter	Units	Generic	Minimum	Maximum
PK	Height	cm	170	160	190
PK	Weight	kg	70	50	100
PK	Age	yr	30	18	90
Link	k_{e0}	min^{-1}	0.17	0.128	0.213
PD	γ	-	5	5	9
PD	C	-	2.5	2	6

The agent uses a neural network to map observations to distributions over the action space. This mapping, known as the policy, is represented by the function $\pi(a_k \mid \mathbf{o}_k)$, which assigns a probability to an action a_k given an observation $\mathbf{o}_k$. The network contains a single hidden layer with 128 nodes and ReLU activation functions, with two output nodes passed through a softmax to obtain action probabilities. The network is fully parameterized by the weights $w_\pi \in \mathbb{R}^{898}$ ($(4+1) \times 128 + (128+1) \times 2 = 898$, where the +1 accounts for a constant offset).

To promote exploration during training, the agent randomly selects an action according to its policy. During testing, we employ three action selection modes.

In *stochastic* mode, the agent randomly selects an action according to its policy as it does in training: $a_k^{(s)} \sim \pi(\cdot \mid \mathbf{o}_k)$. In *deterministic* mode, the agent selects the action with the highest probability: $a_k^{(d)} = \underset{a}{\arg\max}\, \pi(a \mid \mathbf{o}_k)$. In *continuous* mode, the agent selects a continuous action corresponding to the policy's probability of infusing: $a_k^{(c)} = \pi(1 \mid \mathbf{o}_k)$. In all cases, the action yields a normalized infusion rate that is multiplied by the maximum dose u in (1).

2.3   Cross-Entropy Training Algorithm

The agent's policy network is trained using the cross-entropy method [3], an importance sampling based algorithm that is popular in training deep reinforcement learning agents [16]. The algorithm performs batches of simulations (referred to as episodes) and at the end of each batch updates the policy based on the episodes where the agent performed best. The agent's performance in a given episode is assessed using a reward function.

For a given episode n in a collection of N episodes, we define the episode reward to be the cumulative negative absolute error: $r_n = \sum_{k=1}^{K} -|y_{n,k}^* - y_{n,k}| \in (-\infty, 0]$, where K gives the fixed duration of an episode and $y_{n,k}^*$ and $y_{n,k}$ give the target and true LoU at time k for episode n, respectively. After simulating N episodes, the set $\mathcal{N}_p$ of episodes in the pth percentile of rewards is identified, and these are used to update the policy net parameters w_π. Letting $a_{n,k}$ be the action taken at time k in episode n and $\mathbf{o}_{n,k}$ be the corresponding observation vector, define the cross-entropy loss for episode n as:

$$\mathcal{L}_n = -\sum_{k=1}^{K} a_{n,k} \log \pi(a_{n,k} \mid \mathbf{o}_{n,k}) + (1 - a_{n,k}) \log(1 - \pi(a_{n,k} \mid \mathbf{o}_{n,k})) \quad (4)$$

Given that $a_{n,k}$ is either zero or one, reducing the cross-entropy loss results in nudging the policy π to assign a higher probability to the action $a_{n,k}$ when $\mathbf{o}_{n,k}$ is observed. This nudge is accomplished by performing stochastic gradient descent (SGD) on the policy net weights with the computed losses.

In our training, we used a cutoff of $p = 70\%$ (i.e. $\mathcal{N}_p$ includes the best 30%) and a batch size of $N = 16$. Between batches, we selected the patient parameters randomly from a uniform distribution over the parameters specified in Table 1. Four LoU targets y^* were sampled uniformly from $[0.25, 0.75]$, with each target being used for 2,500 s before switching to the next, resulting in a total episode duration of 10,000 s, or $K = 2,000$. The patient parameters and targets were kept fixed within a batch to avoid updating the policy based on the "easiest" environments rather than on the best performance of the agent. Further details on the training and reward system are provided in the Appendix.

2.4   PID Controller

Previous studies on RL-based control of propofol infusion tested control performance against a PID controller [10]. The PID control action is determined by a linear combination of the instantaneous error, the error integral, and the error derivative. We use a discrete implementation of the PID controller:

$$a_k^{(pid)} = K_P e_k + K_I \sum_{i=0}^{k} e_k + K_D \frac{e_k - e_{k-6}}{6} \tag{5}$$

where K_P, K_I, and K_D are the controller parameters and $e_k = y_k^* - \tilde{y}_k$ is the error. The error derivative is approximated using a 30-s lag to avoid being dominated by noise. To be consistent with the RL agent implementation, the PID control action is normalized ($a_k^{(pid)} \in [0,1]$) and multiplied by the maximum infusion rate u in (1). To avoid reset windup during induction and target changes, we implemented clamping on the integral term. The PID parameters were tuned using the Ziegler and Nichols method [18] on simulations using the generic patient model (see Table 1), resulting in $K_P = 9$, $K_I = 0.9$, and $K_D = 22.5$.

3   Results

Performance was evaluated on cases that had different patient demographics and LoU targets. Figure 2A shows sample trajectories of the true and target LoU for the cases with the worst, median, and best performance. The worst case for each controller exhibits similar increases of oscillatory behavior, especially pronounced at set-point escalations. Continuous RL used less propofol than PID during induction (186 mg ± 57 and 210 mg ± 55) and throughout a whole case (2430 mg ± 763 and 2457 mg ± 760), but more during maintenance (15 mg/min ± 5.6 and 14 mg/min ± 5.3). In the state-space view of these trajectories (Fig. 2B), it is apparent that all trajectories involve a nearly linear decision threshold with an intercept near the origin $(o^{(1)}, o^{(2)}) = (0, 0)$.

The controller performances were evaluated using the per-episode median absolute performance error (MAPE) and median performance error (MPE), where performance error is defined as $PE_k = 100 \frac{y_k - y_k^*}{y_k^*}$. All RL test modes outperformed the PID controller (Fig. 3A). Among the RL test modes, the continuous action mode had the best performance. Notably, the continuous mode had a median (across episodes) MPE near zero, suggesting that the ability to select continuous infusion rates helped reduce the controller's bias. On the contrary, the PID controller had nearly equivalent MAPE and MPE, suggesting that its MAPE was limited by maintaining LoU at values slightly above the target. Adjusted 2-sided paired t-tests showed that all controllers had significantly different mean MPE and MAPE (p < 0.05). The continuous RL controller was robust to variation in patient age and height, but sensitive to differences in mass and PD parameters, in particular to C (Fig. 3B). While this initially seems to suggest that our model performs better on patients with a higher drug requirements, it is important to note that *both* γ and C affect the shape of the Hill function. As such, for the range of γ indicated in Table 1, low C values correspond to steeper PD responses than high C values. Sampling γ and C from a

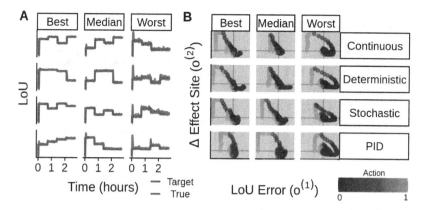

Fig. 2. A: True and target LoU for typical/extreme cases for each controller. **B**: State-subspace trajectory for typical/extreme cases for each controller. Each point indicates a single step in a case's trajectory. The normalized propofol dosage administered at that step is indicated by color. (Color figure online)

joint distribution may reduce the apparent effect of C on performance. Finally, the continuous RL controller had a duration out-of-bounds error (percentage of time at 5% or more off target) of 6.0% as compared with 12.4% for the PID controller.

Figure 4 shows two-dimensional cross sections of the learned policy. We see that the agent learns to transition sharply between the non-infusing and infusing actions. While the decision boundary is essentially linear in the measured error and predicted effect site concentration change, this boundary shifts to promote more infusion when the LoU has been increasing to approach the target.

4   Discussion

Our experiments show that the proposed RL controllers significantly outperform a PID controller. We attribute RL's superior performance to the fact that its observation provides a much richer representation of the latent state of the system under control than PID (which only observes the error). It is worth emphasizing that we used a heuristic tuning method to optimize PID parameters on a generic patient model, and it is possible that alternative tuning methods could improve PID performance in our experiment. Nevertheless, other tuning methods would also involve heuristics, and the ability to incorporate robustness considerations directly into the RL training paradigm yields a considerable benefit.

The behavior of the model can be interpreted by inspecting the policy. In Fig. 4 we see expected increases in propofol administration when the patient is further below a target concentration and when the projected effect site concentration is more rapidly falling. The tendency to administer more propofol when the LoU has been rising may be related to behavior learned during set-point

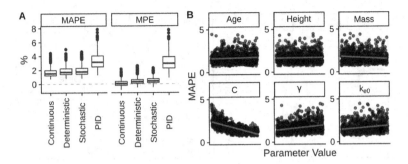

Fig. 3. A: Median absolute performance error (MAPE) and median performance error (MPE) across 1,000 test parameterizations for each of the four controllers. **B**: Association between MAPE and PK/PD parameters for continuous action mode. Each point represents a test-episode, positioned by that episode's performance and a PK/PD parameter. Overlaid blue lines represent linear trend. (Color figure online)

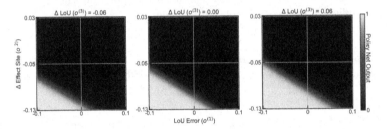

Fig. 4. Policy maps show the policy net outputs ($\pi(1 \mid \mathbf{o}_k)$) for three different 30 s LoU changes ($o^{(3)}$) and a fixed target LoU ($o^{(4)}$) of 0.5.

transitions or an encoding of the patient sensitivity to a given dosage history of propofol. Given that the agent's internal generic patient model encodes the previously administered drugs, these policy maps suggest that the agent has learned the interaction between change in LoU and predicted effect site concentration change to determine at which error level drug should be administered.

The ideal way to test this algorithm would be conducting a closely monitored prospective clinical trial. Reinforcement learning algorithms are notoriously difficult to evaluate. Usually there is not an opportunity to collect prospective data according to the agent's policy, and policies are instead evaluated on retrospective data collected according to a different policy [6]. In this study we changed the propofol dosage exactly every 5 s, whereas in standard practice dosages are changed sporadically with individual infusion rates lasting minutes to hours. Reasonable approaches to evaluating this algorithm prospectively include non-human studies with standard dose-safety limits or clinically with a human-in-the-loop study where the agent acts as a recommender system.

Appendix

The algorithm described in Sect. 2.3 is provided in detail in Algorithm 1. Training will terminate once either a maximum number of batches i_{max} is executed or the desired batch reward $\bar{r}_{min}$ is obtained.

Algorithm 1: Cross-entropy Training

Input: p, N,i_{max},$\bar{r}_{min}$
Output: $\pi : \mathcal{O} \to \mathcal{A}$
1   randomly initialize policy net weights w_π ;
2   set $i = 0$, $\bar{r} = -\infty$;
3   **while** $i < i_{max}$ and $\bar{r} < \bar{r}_{min}$ **do**
4 | sample model parameters and targets;
5 | simulate N episodes and compute rewards $\{r_n\}_{n=1,...,N}$;
6 | select episodes with top p percentile of rewards;
7 | compute cross-entropy loss between actions performed in the top episodes and associated policy net outputs $\mathcal{L}$;
8 | perform stochastic gradient descent step to reduce the cross-entropy loss with respect to the policy net parameters w_π;
9 | set $i = i + 1$, $\bar{r} = \frac{1}{N}\sum_n r_n$;
10   **end**

The per-batch mean reward $\bar{r}$ and policy network loss $\mathcal{L}$ associated with the training of our model are shown in Fig. 5, where the policy network loss is found by summing the loss over the best performing cases in the batch: $\mathcal{L} = \sum_{n \in \mathcal{N}_p} \mathcal{L}_n$. We set the maximum number of iterations to 4,000 and visually confirmed that both the reward and loss converged. Given that the reward is represented as the negative of an absolute value, the maximum possible reward is zero, which is obtained only when the true LoU exactly matches the target LoU for the entirety of a case. Due to the inherent limitations of the environment model (for example the delay between infusion and change in effect site concentration), we expect some non-negligible error to occur at induction and target change points.

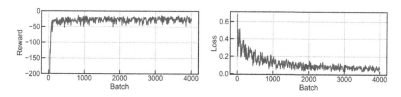

Fig. 5. Round mean reward and policy network loss for each batch (corresponding to the iteration index i in Algorithm 1).

References

1. Absalom, A.R., Sutcliffe, N., Kenny, G.N.: Closed-loop control of anesthesia using bispectral index: performance assessment in patients undergoing major orthopedic surgery under combined general and regional anesthesia. Anesthesiol. J. Am. Soc. Anesthesiologists **96**(1), 67–73 (2002)
2. Brown, E.N., Lydic, R., Schiff, N.D.: General anesthesia, sleep, and coma. New Engl. J. Med. **363**(27), 2638–2650 (2010)
3. De Boer, P.T., Kroese, D.P., Mannor, S., Rubinstein, R.Y.: A tutorial on the cross-entropy method. Ann. Oper. Res. **134**(1), 19–67 (2005)
4. Dumont, G.A., Martinez, A., Ansermino, J.M.: Robust control of depth of anesthesia. Int. J. Adapt. Control Signal Process. **23**(5), 435–454 (2009)
5. Gentilini, A., et al.: Modeling and closed-loop control of hypnosis by means of bispectral index (BIS) with isoflurane. IEEE Trans. Biomed. Eng. **48**(8), 874–889 (2001)
6. Gottesman, O., et al.: Evaluating reinforcement learning algorithms in observational health settings. arXiv preprint (2018). arXiv:1805.12298
7. Lopez-Martinez, D., Eschenfeldt, P., Ostvar, S., Ingram, M., Hur, C., Picard, R.: Deep reinforcement learning for optimal critical care pain management with morphine using dueling double-deep Q networks. In: 2019 41st Annual International Conference of the IEEE Engineering in Medicine and Biology Society (EMBC), pp. 3960–3963. IEEE (2019)
8. Lowery, C., Faisal, A.A.: Towards efficient, personalized anesthesia using continuous reinforcement learning for propofol infusion control. In: 2013 6th International IEEE/EMBS Conference on Neural Engineering (NER), pp. 1414–1417. IEEE (2013)
9. Moore, B.L., Pyeatt, L.D., Kulkarni, V., Panousis, P., Padrez, K., Doufas, A.G.: Reinforcement learning for closed-loop propofol anesthesia: a study in human volunteers. J. Mach. Learn. Res. **15**(1), 655–696 (2014)
10. Moore, B.L., Quasny, T.M., Doufas, A.G.: Reinforcement learning versus proportional-integral-derivative control of hypnosis in a simulated intraoperative patient. Anesth. Analg. **112**(2), 350–359 (2011)
11. Padmanabhan, R., Meskin, N., Haddad, W.M.: Closed-loop control of anesthesia and mean arterial pressure using reinforcement learning. Biomed. Signal Process. Control **22**, 54–64 (2015)
12. Padmanabhan, R., Meskin, N., Haddad, W.M.: Optimal adaptive control of drug dosing using integral reinforcement learning. Math. Biosci. **309**, 131–142 (2019)
13. Prasad, N., Cheng, L.F., Chivers, C., Draugelis, M., Engelhardt, B.E.: A reinforcement learning approach to weaning of mechanical ventilation in intensive care units. arXiv preprint (2017). arXiv:1704.06300
14. Schnider, T.W., et al.: The influence of method of administration and covariates on the pharmacokinetics of propofol in adult volunteers. Anesthesiol. J. Am. Soc. Anesthesiologists **88**(5), 1170–1182 (1998)
15. Shanechi, M.M., Chemali, J.J., Liberman, M., Solt, K., Brown, E.N.: A brain-machine interface for control of medically-induced coma. PLoS Comput. Biol. **9**(10), e1003284 (2013)
16. Szita, I., Lörincz, A.: Learning tetris using the noisy cross-entropy method. Neural Comput. **18**(12), 2936–2941 (2006)
17. Westover, M.B., Kim, S.E., Ching, S., Purdon, P.L., Brown, E.N.: Robust control of burst suppression for medical coma. J. Neural Eng. **12**(4), 046004 (2015)
18. Ziegler, J.G., Nichols, N.B.: Optimum settings for automatic controllers. Trans. ASME **64**(11), 759–765 (1942)

SMOOTH-GAN: Towards Sharp and Smooth Synthetic EHR Data Generation

Sina Rashidian[1(✉)], Fusheng Wang[1,2], Richard Moffitt[2], Victor Garcia[2],
Anurag Dutt[1], Wei Chang[1], Vishwam Pandya[1], Janos Hajagos[2], Mary Saltz[2],
and Joel Saltz[2]

[1] Stony Brook University, Stony Brook, NY 11794, USA
{sina.rashidian,fusheng.wang,anurag.dutt,wei.chang,
vishwam.pandya}@stonybrook.edu
[2] Stony Brook Medicine, Stony Brook, NY 11794, USA
{fusheng.wang,richard.moffitt,victor.garcia,janos.hajagos,
mary.saltz,joel.saltz}@stonybrookmedicine.edu

Abstract. Generative adversarial networks (GANs) have been highly successful for generating realistic synthetic data. In healthcare, synthetic data generation can be helpful for producing annotated data and improving data-driven research without worries on data privacy. However, electronic health records (EHRs) are noisy, incomplete and complex, and existing work on EHR data is mainly devoted to generating discrete elements such as diagnosis codes and medications or frequent laboratory values. In this work, we propose SMOOTH-GAN, a novel approach for generating reliable EHR data such as laboratory values and medications given diagnosis codes. SMOOTH-GAN takes advantage of a conditional GAN architecture with WGAN-GP loss, and is able to learn transitions between disease stages with high flexibility over data customization. Our experiments demonstrate the model's effectiveness in terms of both statistical similarity and accuracy on machine learning based prediction. To further demonstrate the usage of our model, we apply counterfactual reasoning and generate data with occurrence of multiple diseases, which can provide unique datasets for artificial intelligence driven healthcare research.

Keywords: Generative adversarial networks · Electronic health records · Synthetic data generation · Counterfactual machine learning

1 Introduction

Electronic health records (EHRs) include rich information to support artificial intelligence (AI) driven healthcare. Analyzing EHR data has many practical applications such as predicting mortality [3], phenotyping diseases [6], detecting missing/missed diagnosis codes [17] and predicting unplanned readmissions [2].

© Springer Nature Switzerland AG 2020
M. Michalowski and R. Moskovitch (Eds.): AIME 2020, LNAI 12299, pp. 37–48, 2020.
https://doi.org/10.1007/978-3-030-59137-3_4

In the meantime, EHR data is difficult to access due to privacy protection. It is also noisy, incomplete and complex, thus difficult for researchers to work with. Generating synthetic EHR datasets can help both AI and medical communities to share datasets for developing new algorithms and comparing results.

Synthetic data generation can provide the opportunity for researchers to share large datasets without privacy concerns and improve the quality of studies with competitive and reproducible experiments. Having a reliable data generator can also be useful for augmentation tasks and building more robust machine learning models that can potentially provide new insights into how models can interpret and capture patterns from EHR data. However, for a various number of reasons including, but not limited to, large dimensions, longitudinal irregularity, missing values, and heterogeneity it is more challenging to provide synthetic data generation for EHR data, compared with other applications such as imaging.

Generative adversarial networks (GANs) are generative models for creating realistic synthetic data based on an adversarial process which are proven to be more effective than their statistical counterparts [10]. GANs have been very successful with image generation, and there are many interesting applications of GANs such as real images augmenting with Invertible Conditional GANs [16]. This success inspired studies to adapt strategies to tabular data [19].

In recent years, the concept of counterfactual reasoning has gained attention within the machine learning community as one of the potential methods for explainable AI and generating never-before-seen patterns [13]. This concept has a lot of potential in AI driven healthcare, where physicians encounter new patterns among diseases and are skeptical about black box models. Such patterns can be potentially uncovered through GAN based methods.

In this paper, we take advantage of GANs for high quality synthetic data generation and data augmentation, and explore how the models can track patients over the course of their disease using EHR data. Instead of a human-based perception of disease progression by a clinical expert, we are interested in understanding how the models can observe and capture these patterns. We believe these observations can help building more robust models and provide essential knowledge for understanding decisions made by neural networks. We will first introduce SMOOTH-GAN (Sharp sMOOTh eHr), a new approach for generating synthetic EHR data, and then, we will provide in-depth analysis of the models generated by defining new metrics and concepts. At the end, we explore an application of counterfactual data generation.

2   Related Work

Recently, generating synthetic EHR data using GANs has become an active research area. However, there is limited work due to several challenges associated with EHR data. One notable project is MedGAN which focuses on generating discrete data elements -medications and diagnosis codes- by adding an additional encoder decoder inside the GAN architecture [7]. Another inspiring work is RCGAN which provides a framework for generating frequent sequences using conditional recurrent GANs designed for medical time series data [8]. Moreover, the SSL-GAN augments medications and diagnosis codes for improving classification tasks with a semi-supervised learning approach [5].

In this study, we design a conditional GAN which generates both medications and laboratory values for given diseases. Our work has the following salient features. First, the generator generates both continuous and binary values and there is no need to have separate generators. Secondly, we created new methodology to have more control over conditions, which can help with generating patients with different stages of disease. Furthermore, conditions in SMOOTH-GAN can be combined together, creating more realistic and diverse encounters.

3   Data

We extracted inpatient encounter data for adults ($\geq$18) from the Cerner Health-Facts database, a large multi-institutional de-identified database derived from EHRs and administrative systems. From the 10 highest volume inpatient facilities, we randomly chose one acute-care facility (143) and extracted encounters with at least one diagnosis code, laboratory value, and medication from 1/1/2016 to 12/31/2017. We used 47,412 encounters that were broken into 80% for the training set and the rest for the test set.

As multiple values for each laboratory test exist for an encounter, we take the median of each test for each encounter. For medications, we consider them binary whether they were ordered or not. After filtering out features with less than 5% occurrence rate to reduce sparsity and noise, 166 features remained. Diagnosis codes for 5 major chronic conditions, hypertension, congestive heart failure (CHF), diabetes mellitus, cardiac arrhythmias, and chronic kidney disease (CKD) were defined according to [18] and used in this study.

4   Methods

We first briefly review the GAN concept and the architecture we are adapting, and then discuss the details of our algorithm and methods.

4.1   Generative Adversarial Networks Concept

A GAN is normally comprised of two neural networks, which compete with each other in a minimax game: a discriminator and generator. The generator's $G(z; \theta_g)$ goal is to generate samples intended to come from the same distribution as the training set, where z is random noise usually from the normal distribution. The discriminator $D(x, \theta_d)$ tries to detect whether the samples generated by the generator are real or fake. Ideally, the data distribution by G (p_g) should be the same as the real data distribution (p_{data}) [9,10]. Conditional GANs are extensions where generators generate data based on some extra information as conditions or labels [14]. The formal optimization formula is:

$$\min_{G} \max_{D} V(\theta_g, \theta_d) = E_{x \sim p_{data}}[log(D(x|y)] + E_{z \sim p_z}[log(1 - D(G(z|y)))]$$

where θ_g, θ_d are parameters for the generator and discriminator, and p_z is the normal distribution. The ideal generator would create authentic samples similar to the training set that force the discriminator to guess randomly. y is the vector condition, which is given to both the generator and the discriminator in *cgan* architecture. We adapted Wasserstein-GAN with gradient penalty (WGAN-GP) as the loss function in this work. It has several advantages including not suffering from the gradient diminishing problem during training and producing more robust results [1,12]. The discriminator becomes critic in this method which assigns a real value score instead of a binary value.

4.2 SMOOTH-GAN

The SMOOTH-GAN is a conditional GAN adapting WGAN-GP for healthcare data. Its main objective is to generate high quality EHRs, including laboratory values and medications, given diagnosis codes as conditions. We refer to diagnosis codes as set C, where $c \in \{0, 1\}^{|C|}$ is a random set of conditions, and the ith dimension c_i shows presence or absence of ith disease in a patient's encounter record. In EHR data, diagnosis codes are recorded as binary values indicating which diseases patients have. Although having a disease is a binary status, reaching the certain threshold to have the disease is in a probabilistic continuous space for most chronic diseases. For instance, patients with a "hemoglobin A1C" of 6.0 and 4.5 are both below the threshold of diabetes, but the first patient is closer to being a positive case and has a higher risk of getting diabetes. However, in EHR data both of these patients are labeled as 0.

A generative model needs to be reliable and adjustable to have practical usage. We observed by generating a GAN model directly with those binary values as conditions, the generated data in many cases was borderline and did not pass the cutoff for that disease. The GAN was learning broad patterns and the control of the output was limited. The outcome was not deterministic by input conditions and was highly dependent on random variables.

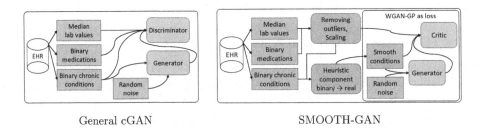

General cGAN   SMOOTH-GAN

Fig. 1. Illustrating how the heuristic function is added to the model.

Based on the issued discussed above, we added a unit which would change the GAN input conditions to smooth labels. Note that assigning an exact probability is a very difficult task, especially when the definition of what is the probability is debatable. Therefore, we are looking for a heuristic function that given binary conditions and input data, can estimate the condition in a continuous space, $H(c, x) = \tilde{c}$ where $\tilde{c} \in [0, 1]$. Although finding the perfect function for assigning probabilities/risk scores to encounters is an active field of research in healthcare, finding a heuristic function simplifies the task and provides a fast solution. The architecture is shown in Fig. 1.

There are different ideas and models to use as the heuristic function. We use random forest (RF) models as the core part of this heuristic function in this work. These models can be trained on the training set and assigns probabilities for each disease accordingly. When the estimated probability is in contrary with the original label, we adjust it to the center (0.5). It is necessary that the model can label the majority of each class correctly. We demonstrate how the model is capable of generating more diverse synthetic data with traceable disease progress by using $\tilde{c}$ instead of c in training the GAN in Sect. 5.3.

Training Details. The generator has two *leakyRelu* hidden layers with $\alpha = 0.2$ each one is followed by a batch normalization layer and *tanh* output layer. The critic has two *leakyRelu* hidden layers with $\alpha = 0.2$ and *linear* output layer. The critic is trained 5 times more than the generator in each epoch. Moreover, the heuristic function is pre-trained in advance (RF models). The model was trained for 600 epochs. Data is scaled to $[-1, 1]$ and outliers with Z-score more than 4 are removed for non-binary features before the median imputation.

5   Results

In this section, we provide in-depth analysis of our GAN method and innovative applications. We used random forest as the prediction model since we needed to know the important features and output probability of inputs for most experiments. To have a reasonable comparison, the synthetic dataset is generated given a set of conditions similar to the training set.

5.1   Statistical Analysis

The first step is to measure how the synthetic data distribution fits to the real training set. We measured the mean absolute error (MAE) for means and standard deviations of columns and element-wise Pearson correlations as shown in Table 1. For medications which are binary values, we calculated MAE for dimension-wise probability. The Loss functions based on Wasserstein distance have robust progress even when data is partially binary. Figure 2 shows heatmaps of Pearson correlations for real and synthetic data.

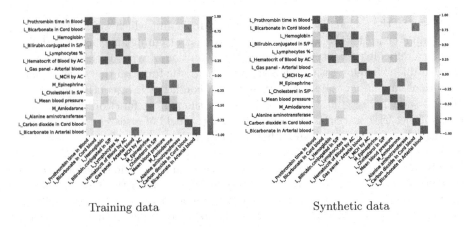

Training data   Synthetic data

Fig. 2. Heatmap for 15 features with highest correlation. MC: Manual Count, AC: Automated Count, S/P: Serum or Plasma, L: lab value, M: medication

Table 1. Mean absolute error for statistics between real and synthetic data

Method name	Laboratory mean	Laboratory std	Medications prob	Correlation
cGAN	248.10	22.92	0.382	NaN
AC-GAN	79.49	14.53	0.068	0.196
WGAN	1.33	5.39	0.007	0.058
WGAN-GP	0.80	1.77	0.003	0.039
SMOOTH-GAN	0.68	2.29	0.003	0.039

5.2   Synthetic Data Prediction Models

One major goal of generating synthetic data is to use it in place of real data when training machine learning models. We are comparing RF models trained on real training data and generated data with the same **real test set** which was untouched in Table 2. This experiment has become the main metric to measure a GAN's success in related publications [7,8,19]. Moreover, it is critical that the synthetic model is making predictions based on similar factors to the real training set. Otherwise, GANs might have altered other features correlated with the input conditions and generated new patterns which is undesired. The last rows of Table 2 show the number of overlapping features in the top 15 most important features of both the synthetic and real trained models. Note that AC-GAN is designed for mutually exclusive input conditions [15].

Table 2. Performance of trained RF models on synthetic and real data measured on **real test set**. "#/15" represents number of common features with top 15 most important features identified by real RF model.

Disease name	Metric	Real	cGAN	WGAN	AC-GAN	WGAN-GP	SMOOTH-GAN
Hypertension	AUROC	.8822	.5896	.8434	.5165	.8515	**.8625**
	AUPRC	.7965	.3929	.7474	.3324	.7562	**.7688**
	#/15	-	2	8	2	8	**9**
Diabetes	AUROC	.9357	.5849	.8412	.5759	.8641	**.8708**
	AUPRC	.8905	.3821	.7702	.3872	.8061	**.8089**
	#/15	-	2	9	2	11	**11**
Congestive heart failure	AUROC	.9000	.5663	.8239	.5795	.8619	**.8633**
	AUPRC	.7471	.2483	.5885	.2708	.6551	**.6577**
	#/15	-	3	8	2	9	**12**
Chronic kidney disease	AUROC	.9544	.6331	.9386	.4240	**.9404**	.9380
	AUPRC	.8705	.3654	.8380	.1740	**.8384**	.8321
	#/15	-	3	9	1	10	**12**
Cardiac arrhythmias	AUROC	.8110	.5191	.7065	.4791	**.7564**	.7512
	AUPRC	.7037	.3609	.5825	.3353	**.6352**	.6144
	#/15	-	3	5	4	7	**8**

5.3   Smooth Conditions, Sharp Synthetic Data

The ideal conditional generator should be capable of generating high quality data according to given conditions. Here we define two terms, sharpness and smoothness. The generator must be *sharp* as generated data reflects attributes of given conditions clearly when expected. For instance, a patient with 100% chance of diabetes must have obvious observations/or medications. Second, it must be *smooth*, which means that it has control over what is generated with a realistic continuous distribution of the data. In other words, it should learn to transit between disease stages, which is natural for chronic diseases. Sharpness is more obvious at the boundaries, aka disease chances are closer to 0 or 1, while smoothness is a characteristic for transitions between stages.

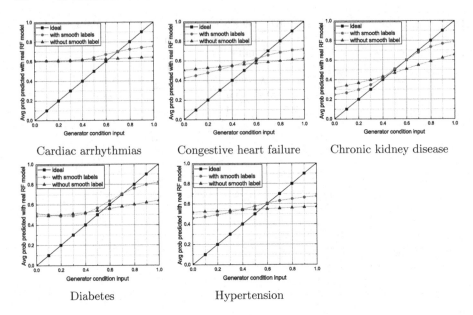

Fig. 3. Gradually increasing input conditions to measure average probability of generated data according to the random forest model trained on real data.

Having a set of conditions $\tilde{C}$, for the ith disease, we changed $\tilde{c}_i$ increasing from 0 to 1 by 0.1 steps gradually while all other $\tilde{c}_j$ (where $j \neq i$) remained the same as conditions passed for training the GAN. This process lead to creation of 11 data groups. For each group, we measured the average probability assigned by the random forest model trained on real data. Results are shown in Fig. 3. The ideal result in this model would be the solid diagonal line, where for given input condition generated data would get similar probability by the model trained on real data. Considering g_i as the average for the ith group and n as the number of steps (here $n = 10$), we define $sharpness = (g_0 - 0) + (1 - g_n)$ and $smoothness = (\sum_{i=0}^{n} |\frac{i}{n} - g_i|))/(n+1)$. In both metrics lower magnitude is better. As a baseline, when probabilities of all groups are 0.50 (horizontal line in middle), the sharpness and the smoothness are 1 and 0.27 respectively.

Training with smooth labels decreased the sharpness and the smoothness from 0.86 and 0.23 to 0.69 and 0.18 average among all diseases. Of note, the other conditions in the input also affect the output of the generator. For this reason, reaching absolute 0 or 1 probability over a reasonable set of conditions is unrealistic. For instance, a passed vector condition with high diabetes and hypertension with exactly 0 % chance of CKD is not possible. This can explain why the curves are bent when they get closer to 0 or 1.

Table 3. Sample synthetic CKD cases generated.

#	Lab name	Initial State	CKD GAN input probability										
			0.0	0.1	0.2	0.3	0.4	0.5	0.6	0.7	0.8	0.9	1.0
1	BUN	54.03	37.82	38.12	40.31	43.99	48.98	53.58	58.59	63.60	67.62	70.79	73.21
	Creatinine	1.53	0.70	0.76	0.88	1.05	1.27	1.50	1.81	2.31	2.92	3.63	4.35
	GFR	27.32	85.49	76.52	64.23	50.61	38.45	28.22	19.47	13.14	8.67	5.82	4.18
2	BUN	33.90	29.60	30.09	31.09	32.90	35.29	38.60	42.07	45.19	47.49	48.97	50.73
	Creatinine	0.58	0.41	0.42	0.46	0.53	0.64	0.78	0.98	1.23	1.52	1.87	2.29
	GFR	84.76	100.85	97.51	93.70	87.63	79.58	69.58	56.57	43.04	31.38	22.59	15.78

In Table 3, we show how samples made by the SMOOTH-GAN change over given CKD conditions for three important features: blood urea nitrogen (BUN), glomerular filtration rate (GFR) and creatinine in serum/plasma. The initial state is what is generated by passing a set of random conditions and random noise to the generator. We set the CKD condition from 0 to 1 to get a spectrum of potential states for this encounter.

5.4 Counterfactual Disease Generation

Generally, counterfactuals are hypothetical "what would happen/have happened if" questions. We designed a very specific experiment to show GANs can also be used for generating special combinations of diseases in healthcare. We removed all cases with both hypertension and diabetes from the training set, and we call this new set the "pruned training set". Then we trained our GAN on this new training set to measure whether the model can produce acceptable encounters having both conditions. We chose these two diseases to have a reasonable amount of data for validating the results as this combination happens often in EHR data. Similarly, we measure machine learning efficacy as the ultimate test. In Table 4, we measure RF performance when trained on 1) real data 2) synthetic data from a GAN model trained on the original training set 3) synthetic data from a GAN model trained on the pruned training set. We observe that while the pruned model does not outperform other models in detecting positive cases, it has captured a significant amount of the existing patterns.

Table 4. Performance for counterfactual disease generation

Disease name	Metric	Real	Original training set	Pruned training set
Hypertension & Diabetes	AUROC	.9106	.8720	.8317
	AUPRC	.7122	.6223	.5252
	# /15	–	10	8

There are several challenges for this type of experiment. First, for disease pairs that usually occur together, there might be very few examples of either disease alone. Thus, the pruned dataset would be inefficient. For instance, 88% of patients with CKD also had hypertension. Secondly, the combination of two diseases might be rare when diseases are less relevant to each other, leaving the validation set very small. Last, it is time consuming to train a GAN model for all permutations. We believe that this approach has high potential and can lead to the discovery of novel patterns, which we will further study with larger datasets in our future work.

6   Conclusion

In this paper, we propose SMOOTH-GAN, a new approach for generating synthetic EHR data based on recent advances in generative adversarial networks. We show it is possible to produce high quality synthetic data that maintains important relations and factors in the original data and can be useful for training competitive machine learning models. We define sharpness and smoothness as vital concepts which are applicable in other domains as well. Furthermore, we demonstrate how to create synthetic EHR data with meaningful clinical implications. By combining this approach and Invertible cGANs it is possible to augment existing patient data, as well as helping to produce more accurate machine learning models. Our approach opens doors to new research opportunities and has high potential for generating unseen combinations to support novel research projects such as counterfactual use cases.

Acknowledgments. Authors wish to thank Aryan Arbabi for his constructive comments.

A   Appendix

A.1   Binary Data Distribution

As GANs were known to struggle with generating binary values, we added Fig. 4 to illustrate dimension-wise probability for medications comparing real versus synthetic data.

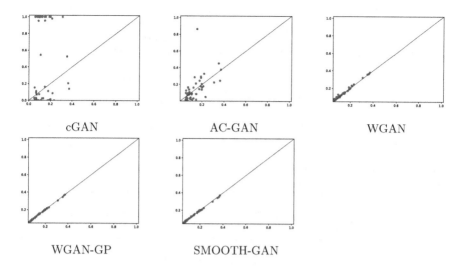

Fig. 4. Dimension-wise probability performance for binary values.

A.2   Is Training Data Memorized by the GAN?

For ensuring privacy and discovering whether the GAN is generating new cases or memorizing the training set, we followed the footsteps of [8] by measuring maximum mean discrepancy (MMD) and applying the three-sample test [4,11]. MMD can answer whether two sets of samples were generated from the same distribution. If the synthetic data is memorized then MMD(synthetic, training) would be *significantly* lower than MMD(synthetic, test). For this reason, we state the null hypothesis as GAN has not memorized the training set, and consequently MMD(synthetic, test) $\leq$ MMD(synthetic, training). We sampled from these three datasets 35 times and calculated MMDs and p-values for the hypothesis. The mean p-value with its standard deviation is 0.26 ± 0.15 which means we cannot reject the null hypothesis and we can establish that GAN did not memorize from the training set.

References

1. Arjovsky, M., Chintala, S., Bottou, L.: Wasserstein generative adversarial networks. In: International Conference on Machine Learning. pp. 214–223 (2017)
2. Ashfaq, A., Sant'Anna, A., Lingman, M., Nowaczyk, S.: Readmission prediction using deep learning on electronic health records. Journal of biomedical informatics 97, 103256 (2019)
3. Avati, A., Jung, K., Harman, S., Downing, L., Ng, A., Shah, N.H.: Improving palliative care with deep learning. BMC medical informatics and decision making 18(4), 122 (2018)
4. Bounliphone, W., Belilovsky, E., Blaschko, M.B., Antonoglou, I., Gretton, A.: A test of relative similarity for model selection in generative models. arXiv preprint arXiv:1511.04581 (2015)

5. Che, Z., Cheng, Y., Zhai, S., Sun, Z., Liu, Y.: Boosting deep learning risk prediction with generative adversarial networks for electronic health records. In: 2017 IEEE International Conference on Data Mining (ICDM). pp. 787–792. IEEE (2017)
6. Choi, E., Bahadori, M.T., Schuetz, A., Stewart, W.F., Sun, J.: Doctor AI: predicting clinical events via recurrent neural networks. In: Machine Learning for Healthcare Conference, pp. 301–318 (2016)
7. Choi, E., Biswal, S., Malin, B., Duke, J., Stewart, W.F., Sun, J.: Generating multi-label discrete patient records using generative adversarial networks. In: Machine Learning for Healthcare Conference, pp. 286–305 (2017)
8. Esteban, C., Hyland, S.L., Rätsch, G.: Real-valued (medical) time series generation with recurrent conditional gans. arXiv preprint arXiv:1706.02633 (2017)
9. Goodfellow, I.: Nips 2016 tutorial: Generative adversarial networks. arXiv preprint arXiv:1701.00160 (2016)
10. Goodfellow, I., et al.: Generative adversarial nets. In: Advances in neural information processing systems. pp. 2672–2680 (2014)
11. Gretton, A., Borgwardt, K., Rasch, M., Schölkopf, B., Smola, A.J.: A kernel method for the two-sample-problem. In: Advances in neural information processing systems, pp. 513–520 (2007)
12. Gulrajani, I., Ahmed, F., Arjovsky, M., Dumoulin, V., Courville, A.C.: Improved training of wasserstein gans. In: Guyon, I., et al. (eds.) Advances in Neural Information Processing Systems 30, pp. 5767–5777. Curran Associates, Inc. (2017), http:// papers.nips.cc/paper/7159-improved-training-of-wasserstein-gans.pdf
13. Liu, S., Kailkhura, B., Loveland, D., Han, Y.: Generative counterfactual introspection for explainable deep learning. arXiv preprint arXiv:1907.03077 (2019)
14. Mirza, M., Osindero, S.: Conditional generative adversarial nets. arXiv preprint arXiv:1411.1784 (2014)
15. Odena, A., Olah, C., Shlens, J.: Conditional image synthesis with auxiliary classifier gans. In: Proceedings of the 34th International Conference on Machine Learning. **70**, pp. 2642–2651. JMLR. org (2017)
16. Perarnau, G., Van De Weijer, J., Raducanu, B., Álvarez, J.M.: Invertible conditional gans for image editing. arXiv preprint arXiv:1611.06355 (2016)
17. Rashidian, S., et al.: Deep learning on electronic health records to improve disease coding accuracy. In: AMIA Summits on Translational Science Proceedings. vol. 2019, p. 620 (2019)
18. Steiner, C.A., Barrett, M.L., Weiss, A.J., Andrews, R.M.: Trends and projections in hospital stays for adults with multiple chronic conditions, 2003–2014: Statistical brief# 183. Healthcare Cost and Utilization Project (HCUP) Statistical Briefs. Rockville: Agency for Health Care Policy and Research (US) (2006)
19. Xu, L., Skoularidou, M., Cuesta-Infante, A., Veeramachaneni, K.: Modeling tabular data using conditional gan. In: Advances in Neural Information Processing Systems, pp. 7333–7343 (2019)

A Multi-task LSTM Framework
for Improved Early Sepsis Prediction

Theodoros Tsiligkaridis$^{(\boxtimes)}$ and Jennifer Sloboda

MIT Lincoln Laboratory, Lexington, MA 02421, USA
{ttsili,jennifer.sloboda}@ll.mit.edu

Abstract. Early detection for sepsis, a high-mortality clinical condition, is important for improving patient outcomes. The performance of conventional deep learning methods degrades quickly as predictions are made several hours prior to the clinical definition. We adopt recurrent neural networks (RNNs) to improve early prediction of the onset of sepsis using times series of physiological measurements. Furthermore, physiological data is often missing and imputation is necessary. Absence of data might arise due to decisions made by clinical professionals which carries information. Using the missing data patterns into the learning process can further guide how much trust to place on imputed values. A new multi-task LSTM model is proposed that takes informative missingness into account during training that effectively attributes trust to temporal measurements. Experimental results demonstrate our method outperforms conventional CNN and LSTM models on the PhysoNet-2019 CiC early sepsis prediction challenge in terms of area under receiver-operating curve and precision-recall curve, and further improves upon calibration of prediction scores.

Keywords: Sepsis prediction · LSTM · Recurrent neural network · Calibration

1 Introduction

Sepsis is a serious medical condition that occurs when a body's dysregulated response to infection causes life-threatening organ dysfunction [18]. Sepsis is a major public health concern: it is estimated to be present in over 30% of U.S. hospitalizations culminating in death and accounted for more than 20 billion (5.2%) of total U.S. hospital costs in 2011. Tentative global estimates – limited by a lack of reporting in middle- and low-income countries – suggest that there

DISTRIBUTION STATEMENT A. Approved for public release. Distribution is unlimited. This material is based upon work supported by the Under Secretary of Defense for Research and Engineering under Air Force Contract No. FA8702-15-D-0001. Any opinions, findings, conclusions or recommendations expressed in this material are those of the author(s) and do not necessarily reflect the views of the Under Secretary of Defense for Research and Engineering.

© Springer Nature Switzerland AG 2020
M. Michalowski and R. Moskovitch (Eds.): AIME 2020, LNAI 12299, pp. 49–58, 2020.
https://doi.org/10.1007/978-3-030-59137-3_5

are over 30 million sepsis cases, resulting in 5 million deaths, annually [4]. As it is a condition with multiple causative organisms and an evolving nature over time, people with sepsis can present various signs and symptoms at different times. This makes diagnosis difficult, even for experienced clinicians [18].

Early detection of sepsis and timely clinical management is important for improving patient outcomes. It is known that each hour of delayed treatment after hypotension onset incurs an approximate 7.6% increase in mortality rates [12], and early treatment with a 3 h boundle of sepsis care and fast antibiotic administration leads to lower in-hospital mortality [17]. While attention around the issue of sepsis diagnosis and treatment has risen in recent years, including a revision of clinical criteria for recognizing and treating sepsis in 2016, there is an outstanding need for early and reliable sepsis identification [18].

Data-driven early warning scores such as the National Early Warning Score (NEWS) has the potential to identify acutely deteriorating patients, such as those with sepsis [19]. This score compares a few physiological variables to their normal ranges to get a single score. While the NEWS and similar early warning scores have been adopted broadly by the National Health Service in England and selectively by individual providers world-wide, they are too simplistic to capture patient-specific data variations and not tailored to a specific condition like sepsis. Due to their broad nature, false alarms occur often. They lack the ability to model complex correlations between physiological variables across time series.

Several machine learning approaches have been proposed for sepsis detection. The work [14] uses a Weibull-Cox proportional hazards model but its semi-parametric form is too restrictive. Other works use simplistic machine-learning models, such as logistic regression [2] and decision-trees [3,13] to predict onset of sepsis, but fail to capture temporal patterns. Further works use recurrent neural networks (RNN) to improve on learning temporal trends [5,11].

In this paper, we make use of a class of recurrent neural networks (RNN) known as Long Short-Term Memory (LSTM) networks [10] to learn powerful sequence representations that can be used to predict whether patients will develop sepsis or not. These networks alleviate the learning difficulties of vanishing gradients that standard RNNs suffer from by using a gating mechanism for information flow. In practice, physiological data may be missing and their absence can be partially attributed to decisions made by clinical professionals. To model this, we incorporate the missing data patterns in our learning framework which can be furthermore used to guide how much to trust imputed measurements. Our training objective is based on a new multi-task LSTM model that takes informative missingness into account during training that effectively attributes trust to temporal measurements. We remark that in contrast to [5] that jointly imputes data and trains a conventional LSTM classifier, we propose a multi-task training loss for LSTMs to improve early prediction while taking into account informative missingness without increasing the computational complexity of the model too much. Our experimental results show our proposed method achieves both improved predictive performance measured by area under receiver-operating curve and precision-recall curve, and better calibrated prediction scores.

2   Multi-task Learning Framework

We formulate the early prediction of sepsis problem as a time series classification problem. Given a patient encounter, the model updates the likelihood of sepsis given all the available measurements up to that time.

Our dataset $\mathcal{D}$ consists of N independent labeled patient encounters. Each encounter $\mathcal{D}_i$ consists of a set of vital measurements $\boldsymbol{v}_o(i)$, lab results $\boldsymbol{l}_o(i)$ and demographics $\boldsymbol{d}_o(i)$ (age, gender, weight). Assuming each time series has T measurements[1], the data may be grouped in the observables matrix $\boldsymbol{X}_o(i) = [\boldsymbol{v}_o(i)^T, \boldsymbol{l}_o(i)^T, \boldsymbol{d}_o(i)^T]^T \in \mathbb{R}^{d \times T}$.

In practice, data is missing and a subset of the matrix $\boldsymbol{X}_o(i)$ is observed. Multivariate imputation is performed to fill in the missing variables using iterative regression (experimental details in Sect. 3.2). The missingness pattern $\boldsymbol{m}(i)$ is appended to the data as well and used in the training process. The augmented data matrix is then formed as $\boldsymbol{X}(i) = [\boldsymbol{v}(i)^T, \boldsymbol{l}(i)^T, \boldsymbol{d}(i)^T, \boldsymbol{m}(i)^T]^T = [\boldsymbol{x}_1, \boldsymbol{x}_2, \dots, \boldsymbol{x}_T] \in \mathbb{R}^{2d \times T}$.

Each labeled encounter has a sepsis label sequence $\boldsymbol{y}(i) = [y_1, y_2, \dots, y_T]$ where $y_t \in \mathcal{Y} = \{0, 1\}$. For non-sepsis patients, $y_t = 0$, and for sepsis patients $y_t = 1$ for $t \geq t_{sepsis} - 6$ and $y_t = 0$ otherwise. The time t_{sepsis} is based on the Sepsis-3 definition; i.e., the earliest time of (1) a two-point change in Sequential Organ Failure Assessment (SOFA) score and (2) clinical suspicion of infection guided by ordering of blood cultures or IV antibiotics [16]. For convenience, we align $t_{sepsis} = T$.

2.1   Long-Short Term Memory Networks

In RNNs, hidden states are used to carry information from the past towards the current step serving the role of memory in neural networks. At each time step, hidden state $\boldsymbol{h}_t$ is updated by taking into account the previous time step's hidden state $\boldsymbol{h}_{t-1}$ and new input vector $\boldsymbol{x}_t$. Long-short term networks (LSTMs) overcome the vanishing gradient limitation prevalent in standard RNNs.

LSTM units contain a carry track in parallel to the information sequence that can be used to store and transport information at a later time step which has the effect of retaining older signals. This is known as the internal cell state, $\boldsymbol{c}_t$, of an LSTM unit. A gating mechanism controls the information flow of the cell. It consists of an input gate, that controls what information enters the cell, a forget gate, that controls what information to keep in the cell, and an output gate that controls what information in the cell will affect the LSTM unit activation or

---

[1] For time series with less than T times steps, zero-padding is performed prior to the measurements.

hidden state. The updates for LSTM units with inputs x_t, h_{t-1} and output h_t are as follows

$$g_t = \text{Tanh}(U_g x_t + V_g h_{t-1} + b_g)$$
$$i_t = \sigma(U_i x_t + V_i h_{t-1} + J_i c_{t-1} + b_i) \qquad \text{(input gate)}$$
$$f_t = \sigma(U_f x_t + V_f h_{t-1} + J_f c_{t-1} + b_f) \qquad \text{(forget gate)}$$
$$c_t = i_t \circ g_t + c_{t-1} \circ f_t \qquad \text{(cell state)}$$
$$o_t = \sigma(U_o x_t + V_o h_{t-1} + J_o c_t + b_o) \qquad \text{(output gate)}$$
$$h_t = \text{Tanh}(c_t) \circ o_t \qquad \text{(hidden state)}$$

where U, V, J are weight matrices and b are bias vectors. Here $\circ$ denotes element-wise multiplication and $\sigma(x) = 1/(1 + e^{-x})$ denotes the sigmoid activation function. We remark that peephole connections [7] are used as the gate layers incorporate the internal cell state.

Given the structure of the augmented data, the LSTM gates take into account the missingness pattern of the data in a sequential fashion as, e.g. for the input gate without loss of generality, we obtain:

$$i_t = \sigma(U_i^v v_t + U_i^l l_t + U_i^d d_t + U_i^m m_t + V_i h_{t-1} + J_i c_{t-1} + b_i)$$

A similar expansion holds for the rest of the gates. As a result, the missingness pattern of the data influences what to input and store in the carry track. This further controls the hidden state evolution and this additional level of control allows the LSTM cell to learn how much trust to place on the imputed data values. A dense layer with sigmoid activation function is used to map the hidden vector to a likelihood score given by

$$\hat{p}_t = \sigma(W_c h_t + b_c)$$

which represents the likelihood of sepsis. The loss function is given by the binary cross-entropy applied to the last hidden vector

$$L(y_T, \hat{p}_T; w) = -y_T \log \hat{p}_T - (1 - y_T) \log(1 - \hat{p}_T) \qquad (1)$$

where y_T corresponds to the sepsis label at the last time step T. These losses are summed across all patient encounters in the training set to form the total training loss.

2.2   Multi-task LSTM Model

We propose to modify the loss to jointly predict sepsis labels for several time steps prior to the clinical sepsis definition as:

$$\tilde{L}(w) = \sum_{t=T-M}^{T} \alpha_t L(y_t, \hat{p}_t; w) \qquad (2)$$

where $L(\cdot, \cdot; \boldsymbol{w})$ is defined in (1) and M is a lag parameter. The weight parameters obey the decay relation $\alpha_j \geq \alpha_k$ for $j > k$ to ensure that losses associated with older time steps do not dominate newer loss contributions. These losses (where i is suppressed in (2)) are summed across all patient encounters, i, in the training set to form the total training loss for the multi-task model.

The decision at time $t \leq T$ is made by thresholding the prediction score, i.e., $\hat{y}_t = 1$ if $\hat{p}_t \geq c$ and 0 otherwise, where c controls the tradeoff between true positive and false positive rates (operating point on receiver-operating characteristic (ROC) curve).

2.3   Performance Metrics

To measure the predictive performance of the algorithm and compare with others, the main metrics of interest include the area under Receiver Operating Characteristic (AUROC) and the area under Precision-Recall curve (AUPRC). The AUROC metric measures the model's discriminative ability between septic and non-septic patients using a statistic that may be interpreted as the probability that the method correctly assigns a random patient encounter with sepsis a higher score than a random patient encounter without sepsis. Furthermore, we measure the balanced false-positive rate (FPR) at 85% balanced true-positive rate (TPR). We monitor these metrics as a function of number of hours prior to the sepsis clinical definition to observe how early one can predict the onset of sepsis with a given performance guarantee.

Of additional interest is the calibration performance of the prediction scores, i.e., how well prediction scores match the actual likelihood of correct predictions, which is measured as follows. Consider a partition of the $[0, 1]$ interval into Q sub-intervals of equal width and define S_q to be the patient patterns whose prediction scores coincide with bin q. The accuracy and confidence of bin S_q is

$$\text{acc}(S_q) = \frac{1}{|S_q|} \sum_{i \in S_q} I(\hat{y}(i) = y(i)), \quad \text{conf}(S_q) = \frac{1}{|S_q|} \sum_{i \in S_q} \bar{p}(i)$$

where $\bar{p}(i) = \max\{\hat{p}(i), 1 - \hat{p}(i)\}$ is the winning score of the i-th pattern. The expected calibration error (ECE) measure [9] is given by

$$\text{ECE} = \sum_{q=1}^{Q} \frac{|S_q|}{n} |\text{acc}(S_q) - \text{conf}(S_q)|$$

and the overconfidence error (OE) measure [8] is

$$\text{OE} = \sum_{q=1}^{Q} \frac{|S_q|}{n} \text{conf}(S_q) \max \{\text{conf}(S_q) - \text{acc}(S_q), 0\}$$

The ECE measures miscalibration by computing a weighted average of the accuracy/confidence difference across the sub-intervals, while OE computes a weighted average of the confidence only when the confidence exceeds accuracy. OE is an appropriate measure for safety-critical applications since confident but incorrect predictions have disastrous consequences. We remark that these calibration metrics where computed on a balanced dataset.

3   Experimental Results

3.1   Data Description

The PhysioNet 19 challenge data [16] was used to evaluate the performance of our algorithm and compare with other baselines. It consists of $40,336$ patient encounters spanning 36 variables that include vitals, labs and demographic information, with data coming from two hospitals, hospital A ($20,336$ patients) and hospital B ($20,000$ patients). The median age is 63 years old, and the male/female proportion is 44/56%. The average length of the time series for septic-patients was 59 h and for non-septic patients 37 h. Approximately 7.3% of the patients developed sepsis.

The physiological variables include 8 vital signs (e.g., heart rate, respiration rate, temperature, systolic blood pressure), 26 laboratory measurements (e.g., bicarbonate, fraction of inspired oxygen, platelets), and 2 demographic variables (age, gender). The sepsis label signal indicates the onset of sepsis according to the Sepsis-3 guidelines [18], where 1 indicates sepsis (labeled 6 h prior to $t_{sepsis} = T$) and 0 no sepsis.

3.2   Experimental Setup

Patients' time series with duration between 8 and 150 h were included in the dataset. The time series with less than 150 h were appropriately zero-padded. This pre-processing step only removed 0.16% of the non-septic population and 8.6% of the septic population, leaving a total of $40,023$ patients for our cohort.

Labs variables were typically recorded on a daily basis thus are sparsely populated and have an average per-patient density of samples (i.e., fraction of non-missing hourly samples per patient) of 5.2%. Vitals variables were typically recorded on an hourly basis thus are densely populated and have an average per-patient density of 82.7%, excluding temperature and end-tidal carbon dioxide (EtCO2), which have average per-patient densities of 33.6% and 2.9%, respectively. Patient records from hospital A tended to have a higher fraction of missing samples as well as a higher variance in non-missing sample density per patient than hospital B. In particular, patients from hospital A have a significantly larger spread in per-patient density of Diastolic Blood Pressure samples. The entry density for the pooled set of patients from hospital A and B is shown in Fig. 1.

Multivariate imputation, using scikit-learn's IterativeImputer class, was used to fill missing values for each patient [15]. A population-based imputation model was learned – using the pooled data from all patients, a regressor is fit for each

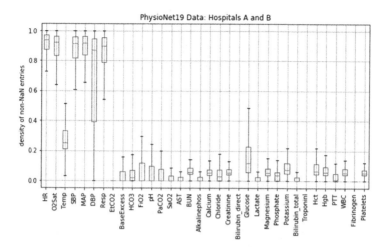

Fig. 1. Density of non-missing entries, excluding outliers, in PhysioNet 19 dataset. It is observed that the lab values have higher missingness rates than the vital sign variables.

variable as a function of the others for known samples of the designated output, in an iterative fashion, and the process is repeated for 50 imputation rounds – and subsequently applied to each patient. We remark that the implementation of IterativeImputer was inspired by the R MICE package [1].

We compare our multi-task LSTM method with single-task CNN-1d classification model without the missingness mask as input, and single-task LSTM classification models without and with the missingness mask as inputs. This setup allows us to measure the additional value of using informative missingness in training, and a multi-task training loss. As a remark, we do not compare with early-warning scores such as NEWS [19] and MEWS [6] since it has been already shown that their prediction performance lags quite a bit behind RNN methods [5]. The CNN-1d architecture was composed of four one-dimensional convolutional and max-pooling layers composed of $16, 32, 64, 128$ filters with kernel sizes $7, 5, 3, 3$ respectively, followed by a dense layer of size 10. A single-layer LSTM architecture was trained with 100 hidden nodes. All models were trained with a small L_2 regularization on the final classification weights, a batch size of 512 and a learning rate of 10^{-4}. All experiments were implemented in Tensorflow with the RMSprop optimizer.

We train our method using 90% of the full dataset and test on the remaining 10%. Sample minority oversampling was used to obtain balance train/test splits, which is critical when training models with random minibatches. Performance metrics were averaged over a total of five train/test splits to account for train/test split variation.

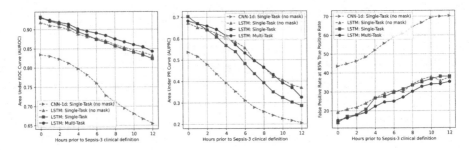

Fig. 2. Area under Receiver-Operating Characteristic (left), Area under Precision-Recall Curve (middle) and False Positive Rate at 85% True Positive Rate (right).

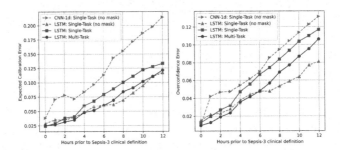

Fig. 3. Expected calibration error (left) and overconfidence error (right) as a function of hours prior to Sepsis-3 clinical definition.

3.3   Results

Overall, the experimental results show that our proposed method offers a consistent performance improvement over the baseline methods for various metrics. Figure 2 shows the AUROC, AUPRC and FPR at 85% TPR results as a function of time prior to the Sepsis-3 clinical definition. On a high level, the performance of all methods degrades as a function of how early the prediction is made. Throughout this time horizon, our method achieves improved prediction performance. We note that LSTMs significantly improve upon the CNN-1d model. Figure 3 displays the calibration performance. We observe that our method aligns the confidence of predictions (measured by maximum probability score) better with the empirical likelihood of correct predictions. We remark that the LSTM models provide better-calibrated scores in comparison to the CNN-1d model.

4   Discussion

In regards to the dataset under study, while the methodology was demonstrated on the PhysioNet-2019 CiC challenge that contains vital measurements, lab results and demographics, predictive performance can benefit from additionally incorporating medication, procedures and underlying health conditions data which may be explored in future work.

Our experimental results show that LSTM networks outperform CNNs with an additional benefit gained from sequentially incorporating the missingness patterns into the learning process and performing multi-task learning over several time points leading up to time t_{sepsis}. This may be attributed to the long-memory aspect of LSTM while CNN-1d performs a multi-scale convolution of with fixed-size kernels. Other ways of training may be compared with the current training method, e.g., train at 6 h prior to t_{sepsis} or use different time lags M, to see if early prediction performance can be further improved without deteriorating performance at later times.

Our proposed model improved calibration, which allows for a more direct interpretation of probability scores as confidence levels. This is particularly important for the high-risk application of sepsis detection; for patients on which the model has low confidence, a physician should intervene and make decisions based on clinical judgments. When our model is deployed, a streaming sequence of temporal measurements is used as inputs to sequentially update the likelihood of a patient becoming septic. When the likelihood crosses a threshold (defined to achieve an acceptable sensitivity-specificity tradeoff), an automated monitoring system may trigger an alarm and physicians can take appropriate action.

Future modeling directions on data imputation include studying the effects of different imputation methods (e.g. multivariate Gaussian processes) and incorporating expected variations in imputed values into the training process.

5   Conclusion

In this paper, we proposed a novel multi-task LSTM model for early sepsis detection based on physiological times series that learns how to make use of informative missingness. We demonstrate the superior performance obtained with our algorithm on the PhysioNet-2019 CiC challenge data by showing absolute improvements 6 h before sepsis in AUROC scores up to ~13%/1.4% and reductions in false-alarm rates by up to ~34%/5.6% for a fixed true-positive rate of 85%, over conventional CNN and LSTM models respectively. As an additional benefit, our method provides better-calibrated prediction scores that allow for improved interpretation of scores as confidence levels. Although our work is on sepsis detection, the proposed framework can be applied to predict other clinical events as well, e.g., cardiac conditions, onset of hemorrhagic shock, and other critical events in ICUs.

References

1. Buuren, S.V., Groothuis-Oudshoorn, K.: MICE: multivariate imputation by chained equations in R. J. Stat. Softw. **45**, 1–67 (2011)
2. Calvert, J.S., et al.: A computational approach to early sepsis detection. Comput. Biol. Med. **74**, 69–73 (2016)
3. Delahanty, R.J., Alvarez, J., Flynn, L.M., Sherwin, R.L., Jones, S.S.: Development and evaluation of a machine learning model for the early identification of patients at risk for sepsis. Ann. Emerg. Med. **73**(4), 334–344 (2019)

4. Fleischmann, C., Scherag, A., Adhikari, N.: Assessment of global incidence and mortality of hospital-treated sepsis current estimates and limitations. Am. J. Respir. Crit. Care Med. **193**, 259–272 (2016)

5. Futoma, J., Hariharan, S., Heller, K.: Learning to detect sepsis with a multitask gaussian process RNN classifier. In: International Conference on Machine Learning (2017)

6. Gardner-Thorpe, J., Love, N., Wrightson, J., Walsh, S., Keeling, N.: The value of modified early warning score (MEWS) in surgical in-patients a prospective observational study. Ann. Roy. Coll. Surg. Engl. **88**(6), 571–575 (2006)

7. Gers, F.A., Schmidhuber, J.: Recurrent nets that time and count. In: International Joint Conference on Neural Networks (2000)

8. Gers, F.A., Schmidhuber, J.: On mixup training: improved calibration and predictive uncertainty for deep neural networks. In: Advances in Neural Information Processing Systems (2019)

9. Guo, C., Pleiss, G., Sun, Y., Weinberger, K.Q.: On calibration of modern neural networks. In: International Conference on Machine Learning (2017)

10. Hochreiter, S., Schmidhuber, J.: Long short-term memory. Neural Comput. **9**(8), 1735–1780 (1997)

11. Kam, H.J., Kim, H.Y.: Learning representations for the early detection of sepsis with deep neural networks. Comput. Biol. Med. **89**, 248–255 (2017)

12. Kumar, A., Roberts, D., Wood, K.E., et al.: Duration of hypotension before initiation of effective antimicrobial therapy is the critical determinant of survival in human septic shock. Crit. Care Med. **34**(6), 1589–1596 (2006)

13. Mao, Q., et al.: Multicentre validation of a sepsis prediction algorithm using only vital sign data in the emergency department, general ward and ICU. BMJ Open. **8**(1), e017833 (2018)

14. Nemati, S., Holder, A., Razmi, F., Stanley, M.D., Clifford, G.D., Buchman, T.G.: An interpretable machine learning model for accurate prediction of sepsis in the ICU. Crit. Care Med. **46**(4), 547–553 (2018)

15. Pedregosa, F., et al.: Scikit-learn: machine learning in Python. J. Mach. Learn. Res. **12**, 2825–2830 (2011)

16. Reyna, M., et al.: Early prediction of sepsis from clinical data - the physionet computing in cardiology challenge 2019 (version 1.0.0). Physionet. (2019) https://doi.org/10.13026/v64v-d857

17. Seymour, C.W., Gesten, F., Prescott, H.C., et al.: Time to treatment and mortality during mandated emergency care for sepsis. New England J. Med. **376**, 2235–2244 (2017)

18. Singer, M., et al.: The third international consensus definitions for sepsis and septic shock (Sepsis-3). JAMA **315**(8), 801–810 (2016)

19. Smith, G.B., Prytherch, D.R., Meredith, P., et al.: The ability of the national early warning score (news) to discriminate patients at risk of early cardiac arrest, unanticipated intensive care unit admission, and death. Resuscitation **84**(4), 465–470 (2013)

Deep Learning Applied to Blood Glucose Prediction from Flash Glucose Monitoring and Fitbit Data

Pietro Bosoni[1(✉)], Marco Meccariello[1], Valeria Calcaterra[1,2],
Cristiana Larizza[1], Lucia Sacchi[1], and Riccardo Bellazzi[1]

[1] University of Pavia, 27100 Pavia, Italy
pietro.bosoni02@universitadipavia.it
[2] IRCCS Policlinico San Matteo, 27100 Pavia, Italy

Abstract. Blood glucose (BG) monitoring devices play an important role in diabetes management, offering real time BG measurements, which can be analyzed to discover new knowledge. In this paper we present a multi-patient and multivariate deep learning approach, based on Long-Short Term Memory (LSTM) artificial neural networks, for building a generalized model to forecast BG levels on a short-time prediction horizon. The proposed framework is evaluated on a clinical dataset of 17 patients, receiving care at the IRCCS Policlinico San Matteo hospital in Pavia, Italy. BG profiles collected by a flash glucose monitoring system were analyzed together with information collected by an activity tracker, including heart rate, sleep, and physical activity. Results suggest that a model with good prediction performance can be obtained and that a combination of HR and lifestyle monitoring signals can help to predict BG levels.

Keywords: Flash glucose monitoring · Diabetes · Time series analysis · Deep learning · Data integration

1 Introduction

Recent developments in blood glucose (BG) monitoring technology [1] have given a great support in the management of diabetes. Devices for continuous glucose monitoring (CGM) and flash glucose monitoring (FGM) allow patients to measure their BG almost in real-time. This is crucial, since a good glycemic control represents a key point to reduce risks of serious and life-threatening complications, including stroke and heart disease [2]. However, the accurate prediction of BG is still a challenge due to the complexity of glycemic dynamics.

Given the considerable amount of data made available by CGM and FGM devices, deep learning algorithms have recently been applied to BG forecasting [3]. Most of these approaches entail the development of personalized models that are based on a specific patient BG profile and clinical data, but not suited to be extended to a different subject. Thus, a general model built on a population of patients could be useful in real

© Springer Nature Switzerland AG 2020
M. Michalowski and R. Moskovitch (Eds.): AIME 2020, LNAI 12299, pp. 59–63, 2020.
https://doi.org/10.1007/978-3-030-59137-3_6

clinical scenarios, where the information for a new patient can be scarce at the beginning of the monitoring period.

In [4], a multi-layer convolutional Neural Network (NN) is implemented to learn a generalized model including also information on meals and insulin dosages recorded by the patients. In [5], a Long-Short Term Memory (LSTM) model is trained using multi-patient BG profiles.

In this paper, we introduce a multi-patient and multivariate deep learning approach to build a generalized BG prediction model, testing whether the combined use of monitoring signals and BG profile could improve predictions compared to the uni-variate scenario. The results herein presented represent an update of a preliminary work [6]. The clinical dataset used to train the models was obtained from the AID-GM (Advanced Intelligent Distant-Glucose Monitoring) project [7]. In this project, a new platform to jointly collect and analyze BG monitoring and Fitbit [8] data was proposed and tested in a pilot study on a group of young patients treated at the Pediatric Diabetology outpatient service of the IRCCS Policlinico San Matteo hospital in Pavia, Italy. The study was approved by the Institutional Review Board (IRB) of the hospital.

2   Methods

2.1   Dataset Description and Preprocessing

In the AID-GM project [7], BG data are collected using the Abbott FreeStyle Libre FGM system [9, 10], while heart rate (HR) profiles are collected using a Fitbit tracker. BG and HR measurements are contextualized within the day thanks to the Fitbit *tag*, which characterizes each measure with one of the following values: sleep, workout, routine, or NA. The routine value is assigned when the patient is not sleeping and is not training, while the NA value is assigned if the Fitbit tracker is not being worn during the measurement.

According to the sampling frequency of the two sensors, BG measurements are recorded every 15 min, whereas HR measurements are taken every minute. We dealt with the different sampling frequency by computing a weighted mean of the HR values in the interval between two consecutive BG measurements. HR measurements closer to the next BG value are assigned a higher weight when computing the mean. Then, the available time-series were split in subseries using frames of 96 timestamps (24 h). The resulting subseries were discarded either if the percentage of missing data was greater than 20%, or if there were more than 8 consecutive missing values (2 h). Quadratic bidirectional interpolation was then applied to deal with the remaining missing data for short gaps.

Fitbit tag information was processed using a one-hot encoding, adding a new binary variable for each unique tag value. The preprocessed dataset resulted into 57888 observations, related to 17 patients.

2.2 Model Architecture

The core of the proposed deep learning framework is represented by a LSTM layer, surrounded by a set of hidden layers. LSTM networks are a specific subclass of Recurrent Neural Networks, and they were herein selected because of their suitability for time-series forecasting and support of multivariate inputs [11].

The input layer of our framework is designed to work in three different scenarios: (i) univariate, with only the BG time-series as input, (ii) multivariate, using BG and HR time-series, (iii) multivariate, using a combination of BG, HR, and Fitbit information. Sliding windows of various sizes (15, 30, and 45 min) were used to control the volume of historical data used for BG predictions on a forecasting horizon of 15 min. Besides the input layer, the overall architecture of the proposed NN includes the following layers: hidden layer (16 neurons), hidden layer (32 neurons), LSTM layer with 64 LSTM cell, hidden layer (32 neurons), hidden layer (16 neurons), and output layer (single neuron).

The architecture was implemented using the high-level neural network API Keras version 2.1.6 in Python 3.6.7 environment with TensorFlow backend. A rectified linear unit (ReLU) activation function was used in all but the output layer, which exploits a linear activation function. Weight matrices and bias vectors were randomly initialized at the beginning of the NN training procedure, and then updated using the Truncated Backward Propagation Trough Time method [12]. A RMSprop optimizer with a learning rate of 0.001 was adopted to minimize Mean Absolute Error (MAE) as the cost function. We considered 250 training epochs and a batch size equal to the subseries length.

2.3 Evaluation

The preprocessed dataset was divided into a training set including 13 randomly selected patients, and a test set of 4 patients to evaluate prediction performances. This procedure was repeated 10 times varying the composition of training and test sets. At the end, an average of the prediction performances is computed for every scenario.

The forecasting accuracy was evaluated both in analytical and clinical terms. For the analytical assessment, we used the Root-Mean Square Error (RMSE), which returns a quantitative measure of the forecast error on the same unit scale as the data, i.e. mg/dL.

To provide an indication of the consequences of prediction errors on treatments decisions, we used the Clarke Error Grid analysis [13], a non-parametric graphical method to interpret the mapping between the BG measurements and the corresponding predictions in terms of severity of the potential harm caused by the prediction error. As shown in Fig. 1, the grid is divided into five zones. Zone A includes the area where the difference between reference and predicted BG values is less than 20%, leading to correct clinical decisions based on the prediction. In zone B, the resulting clinical decision is not correct, but still uncritical. In zone C, BG prediction errors can lead to inappropriate treatment but without dangerous consequences for the patient, whereas in zone D the necessary corrections are not triggered, both in case of hypoglycemia and hyperglycemia. Prediction errors in zone E are the most dangerous because they lead to treat hypoglycemia instead of hyperglycemia and vice versa.

3 Results

Table 1 shows a comparison between the average prediction errors over 10 repetitions in the three scenarios, considering sliding windows of different sizes. With a short sliding window (15 min), the model that relies only on the BG time-series performs slightly better than the others. Increasing the volume of historical data allows obtaining more accurate predictions. Moreover, for a longer sliding window (45 min), feeding the model with a combination of BG, HR and Fitbit time-series provides on average a lower prediction error. Interestingly, such values are comparable to those obtained by other generalized models trained in similar conditions [4, 5].

Table 1. RMSE mean ± standard deviation [mg/dL] over 10 repetitions by window size

Model	Window = 15 min	Window = 30 min	Window = 45 min
BG	15.82 ± 1.79	11.62 ± 0.85	11.05 ± 1.01
BG-HR	15.88 ± 1.64	11.45 ± 1.16	10.92 ± 1.27
BG-HR-Fitbit	15.88 ± 1.75	11.50 ± 1.36	10.82 ± 1.02

Figure 1 displays the Clarke Error Grid for a test patient using the BR-HR-Fitbit model with a sliding window of 45 min. The majority of the points (97.65% on average over 10 repetitions) are in the A zone, with just a few points in zone B and no points in the dangerous areas. A positive clinical assessment is confirmed also for the univariate and BG-HR scenarios, with less than 0.2% points on average outside zones A and B.

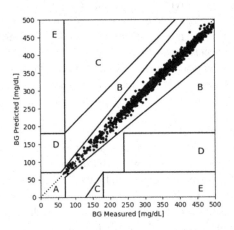

Fig. 1. Clarke Error Grid for a test patient (BG-HR-Fitbit model, window = 45 min)

4 Conclusion

In this paper we presented a multi-patient and multivariate deep learning approach to develop a generalized model for BG prediction. The model is trained on a set of diabetes patients' data, collected using Freestyle Libre FGM system and Fitbit activity tracker. Although a limited number of patients was considered, the resulting prediction performances are encouraging, and comparable with the literature. Interestingly, it was shown that, when using longer sliding windows for prediction, a combination of BG, HR and Fitbit time-series can help in BG forecasting, providing a higher accuracy compared to the univariate scenario. Future developments include an extension of the proposed model to perform a multinomial classification discriminating between hypoglycemia, normoglycemia and hyperglycemia episodes. Moreover, a deeper architecture will be investigated.

References

1. Rodbard, D.: Continuous glucose monitoring: a review of successes, challenges, and opportunities. Diabetes Technol. Ther. **180**(S2), S3–S13 (2016)
2. Pickup, J., Freeman, S., Sutton, A.: Glycaemic control in type 1 diabetes during real time continuous glucose monitoring compared with self monitoring of blood glucose: meta-analysis of randomised controlled trials using individual patient data. BMJ **343**, d3805 (2011)
3. Oviedo, S., Vehí, J., Calm, R., Armengol, J.: A review of personalized blood glucose prediction strategies for T1DM patients. Int. J. Numer. Meth. Biomed. Eng. **33**(6), e2833 (2016)
4. Li, K., Liu, C., Zhu, T., Herrero, P., Georgiou, P.: GluNet: a deep learning framework for accurate glucose forecasting. IEEE J. Biomed. Health Inform. **24**(2), 414–423 (2020)
5. Aliberti, A., et al.: A multi-patient data-driven approach to blood glucose prediction. IEEE Access **7**, 69311–69325 (2019)
6. Bosoni, P., Meccariello, M., Calcaterra, V., Larizza, C., Sacchi, L., Bellazzi, R.: Blood glucose prediction from flash glucose monitoring and fitbit data: a deep learning approach. In: Proceedings of the 7th Congress of the National Group of Bioengineering (GNB), Patron (2020)
7. Salvi, E., et al.: Patient-generated health data integration and advanced analytics for diabetes management: the AID-GM platform. Sensors **20**(1), 128 (2019)
8. Fitbit tracker homepage. https://www.fitbit.com/us/home/. Accessed 30 Apr 2020
9. FreeStyle Libre system Homepage. https://www.freestylelibre.us/. Accessed 30 Apr 2020
10. Massa, G., et al.: Evaluation of the FreeStyle® Libre flash glucose monitoring system in children and adolescents with type 1 diabetes. Horm. Res. Paediatr. **89**(3), 189–199 (2018)
11. Hochreiter, S., Schmidhuber, J.: Long short-term memory. Neural Comput. **9**, 1735–1780 (1997)
12. Werbos, P.: Backpropagation through time: what it does and how to do it. Proc. IEEE **78** (10), 1550–1560 (1990)
13. Clarke, W.: The original Clarke Error Grid Analysis (EGA). Diabetes Technol. Ther. **7**(5), 776–779 (2005)

Natural Language Processing

Comparing NLP Systems to Extract Entities of Eligibility Criteria in Dietary Supplements Clinical Trials Using NLP-ADAPT

Anusha Bompelli[1] , Greg Silverman[2] , Raymond Finzel[3] ,
Jake Vasilakes[1,3] , Benjamin Knoll[1] , Serguei Pakhomov[3] ,
and Rui Zhang[1,3(✉)]

[1] Institute for Health Informatics, University of Minnesota,
Minneapolis, MN, USA
zhan1386@umn.edu
[2] Department of Surgery, University of Minnesota, Minneapolis, MN, USA
[3] Department of Pharmaceutical Care and Health Systems,
University of Minnesota, Minneapolis, MN, USA

Abstract. Natural Language Processing (NLP) techniques have been used extensively to extract concepts from unstructured clinical trial eligibility criteria. Recruiting patients whose information in Electronic Health Records matches clinical trial eligibility criteria can potentially facilitate and accelerate the clinical trial recruitment process. However, a significant obstacle is identifying an efficient Named Entity Recognition (NER) system to parse the clinical trial eligibility criteria. In this study, we used NLP-ADAPT (Artifact Discovery and Preparation Toolkit) to compare existing biomedical NLP systems (BiomedI-CUS, CLAMP, cTAKES and MetaMap) and their Boolean ensemble to identify entities of the eligibility criteria of 150 randomly selected Dietary Supplement (DS) clinical trials. We created a custom mapping of the gold standard annotated entities to UMLS semantic types to align with annotations from each system. All systems in NLP-ADAPT used their default pipelines to extract entities based on our custom mappings. The systems performed reasonably well in extracting UMLS concepts belonging to the semantic types *Disorders* and *Chemicals and Drugs*. Among all systems, cTAKES was the highest performing system for *Chemicals and Drugs* and *Disorders* semantic groups and BioMedICUS was the highest performing system for *Procedures, Living Beings, Concepts and Ideas,* and *Devices*. Whereas, the Boolean ensemble outperformed individual systems. This study sets a baseline that can be potentially improved with modifications to the NLP-ADAPT pipeline.

Keywords: Natural Language Processing · Named Entity Recognition · Clinical trial eligibility

---

A. Bompelli and G. Silverman—Equal-contribution first authors

M. Michalowski and R. Moskovitch (Eds.): AIME 2020, LNAI 12299, pp. 67–77, 2020.
https://doi.org/10.1007/978-3-030-59137-3_7

1 Introduction

Extraction of information from unstructured clinical trial eligibility criteria using Natural Language Processing (NLP) techniques is essential to support clinical trial recruitment process [1, 2]. Many NLP tools including MedLEE [3], the Clinical Language Annotation, Modeling, and Processing Toolkit (CLAMP) [4]; the Clinical Text Analysis and Knowledge Extraction System (cTAKES) [5]; etc. have been developed to extract information related to anatomical location, signs and symptoms, diseases, procedures, laboratory tests and medications [6, 7]. In the clinical domain, extraction of clinical information or concepts is not adequate since the concepts are significantly affected by attributes such as negation modifier, temporal information and qualifiers which describe condition status or severity [8, 9]. NLP tools are generally trained using a specific dataset and are suitable to extract certain concepts. For example, CLAMP's pipeline was trained on the 2010 VA challenge i2b2 corpus to recognize problems, drugs, treatments and lab tests [10]. Thus, it is challenging to find an NLP tool capable of extracting diverse concepts, modifiers and attributes.

The objective of this study was to address this issue by examining the performance of standard open-source clinical NLP systems for the task of Named Entity Recognition (NER) for a corpus outside of the domain for which these systems were developed. We examined a particular strategy for combining annotations generated from out-of-the-box clinical NLP systems into ensembles using NLP systems provided by the NLP Artifact Discovery and Preparation Toolkit (NLP-ADAPT) [11]. NLP-Ensemble-Explorer [12] integrates output from NLP-ADAPT, and through use of a custom mapping to UMLS concepts, allowed us to investigate performance of individual systems and their ensembles for the task of NER for several semantic groupings of UMLS concepts on a novel corpus. NLP-ADAPT and NLP-Ensemble-Explorer were both developed as a complete pipeline for clinical researchers to help improve the experience of the exploration phase of NLP and Information Extraction (IE) projects using individual NLP systems and their ensembles.

2 Background

2.1 Unified Medical Language Systems (UMLS) System

The Unified Medical Language Systems (UMLS) developed and maintained by the National Library of Medicine (NLM) provides a unified global biomedical terminology [13]. The UMLS semantic network organizes many concepts and groups them according to the semantic types [14]. There are 15 semantic groups, 133 semantic types and 54 semantic relationships [15]. The semantic network has been widely used in information extraction, clinical annotation, and knowledge representation [16].

2.2 NLP Systems

Parsing clinical notes is a critical task for information extraction (IE) as it leverages information from narrative text to support clinical and translational research [6].

Clinical NLP systems such as the BioMedical Information Collection and Understanding System (BioMedICUS) [17]; the Clinical Language Annotation, Modeling, and Processing Toolkit (CLAMP) [18]; the Clinical Text Analysis and Knowledge Extraction System (cTAKES) [19]; and MetaMap [20] have been developed to perform Named Entity Recognition (NER) and IE tasks on free text clinical notes or biomedical literature. Because many of these systems were developed to extract specific types of information, adopting these systems for use beyond their original purpose without customization of each system's statistical models and dictionaries can potentially result in reduced performance [4].

3   Methods

3.1   Overview of the Study

This study compares performances of different NLP systems and their ensembles for the task of concept extraction from unstructured dietary supplements (DSs) clinical trial eligibility criteria. The study was performed following these steps: (1) obtain the clinical trial eligibility criteria of DS clinical trials from ClinicalTrials.gov; (2) develop gold standard annotations; (3) map entities to UMLS semantic types; (4) Apply NLP-ADAPT to extract entities mapped to semantic types; and (5) use NLP-Ensemble-Explorer to create ensembles and compare the performance of individual and ensembled annotator systems against the gold standard annotations (see Appendix, Fig. 2).

3.2   Corpus and Annotation

We obtained the dietary supplements clinical trial data corpus from ClinicalTrials.gov, which is an online repository developed by the National Library of Medicine (NLM) and the National Institutes of Health (NIH). We randomly selected 150 clinical trials from the Behaviors & Mental Disorders and Nervous System Diseases categories and parsed the clinical trial XML files to obtain the eligibility criteria. We annotated the eligibility criteria by following the annotation guidelines[1] developed in our previous unpublished study. Three annotators independently annotated 5 randomly selected clinical trials by understanding the first iteration of the guidelines. The team compared the annotation results, discussed the difference of opinions and revised the annotation guidelines. The team then annotated another set until a reasonable interrater agreement is reached and until no discrepancy among annotators. Later, Inter-annotator agreement among three annotators was computed over 10 trials, revealing a kappa of 0.94. The annotated entities and attributes include: *Demographics, Observation, Condition, Procedure, Device, Drug, Dietary Supplement, Negation Modifier, Qualifier, Measurement,* and *Temporal Measurement.* While annotating we observed that certain criteria can be easily computable to extract from EHR data while the rest are either difficult or impossible to compute. Examples of such criteria are given below: criteria referring to willingness or unwillingness to use or decline certain medications or

---

[1] https://z.umn.edu/annotation_guidelines.

methods (contraception); requiring consent; ability/inability and compliance of the individual, and caregivers or study partners; criteria about participant's enrollment in any other clinical trial. The eligibility criteria whose corresponding data cannot be found in the EHR were not annotated even if certain terms in the criteria qualify for one of the entities or attributes as this information is not computable.

3.3   Mapping to UMLS Semantic Groups Across NLP Systems

To compare the performance of individual NLP systems against our gold standard annotations, the entities and attributes present in the gold standard annotation were mapped to UMLS semantic types/groups and annotation types available in individual NLP systems. We observed that some entities can be mapped to one or more semantic groups. For example, *Condition* and *Observation* were mapped to *Disorders* (see Appendix, Fig. 3). Figure 3 also illustrates that CLAMP and cTAKES don't annotate four and two semantic groups respectively.

3.4   NLP-ADAPT

Text notes representing eligibility criteria were processed using the version of NLP-ADAPT for Kubernetes (NLP-ADAPT-kube), which includes the following NLP systems that are compatible with the Unstructured Information Management Architecture (UIMA) [21]: BioMedICUS, CLAMP, cTAKES, and MetaMap (with UIMA adapter). All NLP systems in NLP-ADAPT utilized their default pipelines to extract entities mapped to UMLS semantic types. To minimize false positives, as determined by our prior experience with these systems, we used 800 as the threshold for Meta-Map's evaluation score. MetaMap outputs all entity mapping candidates with corresponding mapping scores in a range of 0-1000 (where 1000 indicates a complete mapping) [22]. Annotations produced by NLP-ADAPT-kube were extracted using dkpro-cassis, a software library developed by the Technische Universität Darmstadt [23]. The following versions of the UMLS, by system, were used: 2019AB by MetaMap; 2016AB by cTAKES; 2016AA by BioMedICUS; 2014AB by CLAMP.

NLP-Ensemble-Explorer [12] was used to create ensembles and evaluate individual systems and their ensembles on the task of Named Entity Recognition (NER) of text spans representing UMLS concepts across the DS clinical trial corpus. Individual systems and their ensembles were evaluated using standard performance measures of precision, recall and F1-score. NLP-Ensemble-Explorer takes comprehensive lists of all permutations for NLP systems as input and transforms these into an exhaustive set of Boolean combinations using the logical $\lor$ operator - to represent a UNION set operation (or $\cup$); and the logical $\land$ operator - to represent an INTERSECTION set operation (or $\cap$). NLP-Ensemble-Explorer then evaluates Boolean combinations by creating a merged set of system annotations to assess performance against gold standard annotations. Once a Boolean expression is generated it is stored and evaluated as a binary tree using the parse tree algorithms provided by Miller and Ranum [24]. NLP-Ensemble-Explorer uses character-level binary i-o classification on the positive label (labeled as 1 and 0, respectively) to determine whether there is overlap between an annotated span in the system, merged span set and gold standard span set for each document [25]. We used a character-level partial matching scheme to adjust the weight based on the length of the match, in order to appropriately weight matches to the number of characters in overlap.

4   Results

4.1   Entities and Attributes in DS Clinical Trials

The distribution of entities and attributes in the DS clinical trials is shown in Fig. 1. Out of the annotated 150 trials, *Condition* entity (mapped to *Disorders* in Fig. 1.) was the largest (1,832 terms) followed by *Qualifier* (1,137 terms), *Drug* (890 terms) and *Observation* (868 terms) while *Device* was the smallest (37 terms).

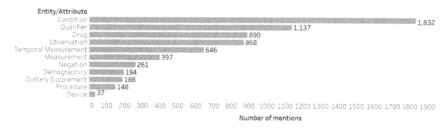

Fig. 1.   Distribution of entities and attributes in DS clinical trials

4.2   Performances of Individual NLP Systems and Boolean Ensemble

As individual NLP systems were trained on different datasets and had distinct strengths and weaknesses, the top performers for one semantic group often struggled in other areas. Among the individual NLP systems, cTAKES was the highest performing system for *Chemicals and Drugs* and *Disorders* semantic groups and BioMedICUS was the highest performing system for *Procedures, Living Beings, Concepts and Ideas,* and *Devices.* We evaluated Boolean combinations of all 4 systems on the DS corpus for all semantic groups in our mapping with the exceptions. For *Concepts and Ideas,* and *Phenomena,* only 3 systems were evaluated, because CLAMP does not annotate for these semantic groups, whereas for *Living Beings,* and *Devices,* only 2 systems were evaluated, because CLAMP and cTAKES does not annotate for these groups. The total number of potential Boolean combinations when combining 4 systems was 238; while for 3 systems it was 28; and for 2 systems it was 4 [26]. The highest performing Boolean combination in each semantic group were shown in Table 1. For example, Boolean combinations like ((BioMedICUS ∧ cTAKES) ∨ CLAMP) and (((BioMedICUS ∨ CLAMP) ∧ MetaMap) ∨ cTAKES) have higher F1-scores for the *Chemicals and Drugs* and *Disorders* semantic groups, respectively than any single system.

Table 1. NLP-ADAPT individual system and Boolean combinations performance for NER; measures are **p** = precision, **r** = recall, **F1** = F1-score; systems are A: BioMedICUS, B: CLAMP, C: cTAKES and D: MetaMap. The first part of the table contains individual NLP system performance and the second part of the table contains only the highest performing Boolean combination in each semantic group

Individual NLP system performance for NER											
Corpus	BioMedICUS			CLAMP			cTAKES			MetaMap		
Semantic Group	p	r	F1	p	r	F1	p	r	F1	p	r	F1
Chemicals and Drugs	0.542	0.528	0.535	0.638	0.476	0.545	0.518	0.591	**0.552**	0.464	0.345	0.396
Disorders	0.641	0.532	0.582	0.617	0.603	0.610	0.687	0.565	**0.620**	0.619	0.429	0.507
Procedures	0.292	0.267	**0.279**	0.237	0.328	0.275	0.293	0.184	0.226	0.209	0.172	0.188
Living Beings	0.033	0.051	**0.040**	-	-	-	-	-	-	0.021	0.040	0.027
Concepts and Ideas	0.605	0.229	**0.332**	-	-	-	0.277	0.001	0.001	0.306	0.186	0.231
Devices	0.278	0.483	**0.353**	-	-	-	-	-	-	0.247	0.386	0.301
Phenomena	0.109	0.006	0.011	-	-	-	0.377	0.019	0.036	0.100	0.007	**0.014**
Micro-average	0.528	0.354	0.424	0.508	0.520	0.514	0.578	0.315	0.408	0.426	0.270	0.331

Boolean combination performance for NER												
Corpus	Highest F1-score				Highest precision				Highest recall			
Semantic Group	combination	p	r	F1	combination	p	r	F1	combination	p	r	F1
Chemicals and Drugs	((A∧C)∨B)	0.583	0.576	0.579	(((A∧B)∧C)∧D)	0.704	0.186	0.294	(((A∨B)∨C)∨D)	0.457	0.662	0.540
Disorders	(((A∨B)∧D)∨C)	0.669	0.608	0.637	(((A∨D)∧B)∧C)	0.841	0.249	0.384	(((A∨B)∨C)∨D)	0.523	0.746	0.615
Procedures	(((C∧D)∨A)∨B)	0.228	0.399	0.290	(((C∧D)∨A)∧B)	0.370	0.135	0.198	(((A∨B)∨C)∨D)	0.209	0.422	0.279
Living Beings	(A∧D)	0.041	0.040	0.041	(A∧D)	0.041	0.040	0.041	(A∨D)	0.021	0.051	0.029
Concepts and Ideas	((A∨C)∨D)	0.404	0.355	0.378	((C∧D)∨A)	0.605	0.229	0.332	((A∨C)∨D)	0.404	0.355	0.378
Devices	(A∨D)	0.238	0.524	0.328	(A∧D)	0.294	0.307	0.301	(A∨D)	0.238	0.524	0.328
Phenomena	((A∨C)∨D)	0.188	0.029	0.050	((A∧D)∨C)	0.377	0.019	0.037	((A∨C)∨D)	0.188	0.029	0.050

5   Discussion

The four NLP annotator systems used in this study were developed with different datasets. Both BioMedICUS and cTAKES utilize pipelines that were developed on the MiPACQ corpus, which consists of fully anonymized outpatient clinical narratives [27, 28]. CLAMP's pipeline has elements that were trained on the 2010 VA challenge i2b2 corpus, which consists of hospital discharge summaries and progress notes taken from multiple independent institutions [10]. MetaMap was designed for extracting text from biomedical literature [29].

Because of this, each individual system has its own internal strengths and weaknesses but may improve performance for particular tasks when ensembled with systems that have complementary strengths and weaknesses, as discussed by Derczynski [30]. Thus, ensemble performance of systems would provide increased performance over any one individual system. The DS clinical trial eligibility criteria corpus used in this study is significantly different from any corpora used to develop these annotator systems. The corpus covers criteria related to patient characteristics, disease characteristics, laboratory tests, lifestyle and concurrent therapies which can further be classified according to several dimensions that characterize the content of corresponding eligibility criteria, including but not limited to temporal status, time independent status, constraint types and subject.

BioMedICUS uses a tiered scoring technique for matching UMLS concepts to phrases by first performing direct dictionary phrase matches, second by lower-cased dictionary phrase matches, and lastly using a discontinuous bag of SPECIALIST normalized terms matches. cTAKES matches UMLS concepts to phrases, by each phrase's lexical and non-lexical permutations and variations against concepts in a dictionary and a list of maintained terms [5]. CLAMP matches UMLS concepts to phrases using the BM25 algorithm for UMLS lookup to find candidate concepts from the UMLS and then applies RankSVM to rank those candidates, from which the top ranked concept is selected. MetaMap uses a shallow parser to generate candidate phrases then, for each candidate phrase, many lexical variations are generated; finally, each phrase is then assigned a score based on its distance to concepts in the UMLS [31]. For this study, we did not use word sense disambiguation functionality from these systems.

We tested the above-mentioned NLP systems and ensemble methods on clinical trial corpus to examine the differences between systems. We saw improved performance when systems were combined. The performance improvement from particular Boolean ensembles confirms the complementary nature of the individual NLP systems, which we suspect exists due to the previously mentioned differences in development.

In order to evaluate the system annotations, we compared the FPs and FNs for each system against the gold standard annotations. Out of 556 FNs, 28 belong to *Chemicals and Drugs*, *Disorders* and *Procedures* semantic groups whereas 515 belong to *Living Beings*, *Concepts* and *Ideas*, *Devices*, and *Phenomena*. Among the 8622 FPs, 3984, 941, 485 and 3222 belong to BioMedICUS, CLAMP, cTAKES and MetaMap, respectively. We investigated the reasons and found that the systems annotated the text from the sentences which were not manually annotated resulting in high FPs and affecting NER performance. According to our annotation guidelines, we omitted annotating the sentences which in our view cannot be converted into computable queries. For example, any sentence which is focused on the willingness of the patient, informed consent or in the investigator's opinion.

This study has some limitations. As this was our first attempt at defining the mapping to UMLS semantic types across NLP systems, our choice of mappings might not be optimal. Additionally, since the NLP systems used in this study are by default each configured with different versions of UMLS, it is possible different versions can supplement or hinder the system performance. However, in our study, we observed that the older version of UMLS outperformed the systems which used newer versions in some tasks (e.g., BioMedICUS). We believe inclusion of different versions of the UMLS may be beneficial, since, for example, developers of SemRep continue to use

2006AA, as opposed to newer versions of the UMLS, because there are fewer concepts/synonyms which decreases ambiguity [32]. Furthermore, since systems within NLP-ADAPT are configured with different versions of the UMLS, we believe this exploits complementarity between systems (as discussed by Derczynski), with potential for increased ensemble performance. Thus, complementarity due to UMLS version differences warrants further research. Lastly, we only examined Boolean combination ensembles in this study. NLP-Ensemble = Explorer has an option for generating all combinations of majority-vote ensembles, and will be explored in future work.

All systems used in NLP-ADAPT were based on the UIMA architecture, and as such this could be a potential source of bias in our results. Thus, use of non-UIMA based NLP/IE systems, such as QuickUMLS and SpaCy for use in extraction of UMLS concepts based on our mapping would be worth exploring. Also, customization of the systems in NLP-ADAPT to use customized dictionaries such as iDISK [33] would be worth exploring, which demonstrated better performance to identify supplement entities compared to UMLS [34].

Our results indicate that currently publicly available traditional biomedical NLP systems do not seem to generalize well beyond the tasks for which they were originally designed. These findings are consistent with other previously published results of applying standard NLP tools and their combinations in the domain of pre-hospital trauma notes which also showed limited generalizability [35, 36]. It is possible that the new generation of biomedical NLP tools based on neural network models may help overcome some of these issues; however, given the results obtained so far we believe that heavy domain adaptation is necessary in order to realize the full potential of general-purpose biomedical NLP tools.

6 Conclusion

We used NLP-ADAPT which is configured with NLP systems and ensemble methods to extract data elements from the unstructured DS clinical trial eligibility criteria. Results indicated that the ensemble of NLP systems can improve NER performance of each individual system, thus setting a baseline that can be potentially improved with modifications to the NLP-ADAPT pipeline.

Acknowledgements. This work was partially supported by the NIH's National Center for Complementary and Integrative Health and the Office of Dietary Supplements under grant number R01AT009457 (Zhang); and supported by the National Center for Advancing Translational Sciences under grant number UL1TR002494 and U01TR002062.

Appendix

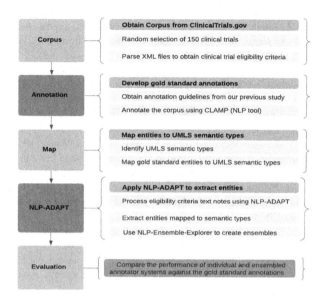

Fig. 2. Overview of the study

UMLS Semantic Groups	BioMedICUS	CLAMP	cTAKES	MetaMap	Gold Standard Entities
Chemicals and Drugs	T130, T121, T195, T192, T123, T131, T103, T196, T125, T116, T120, T129, T114, T122, T104, T197, T109, T127, T200, T126	• drug	• MedicationMention	irda, phsu, antb, rcpt, bacs, hops, chem, elii, horm, aapp, chvf, imft, nnon, bodm, chvs, inch, orch, vita, clnd, enzy	Drug Dietary Supplement
Disorders	T049, T033, T047, T019, T048, T184, T191, T050, T020, T190, T037, T046	• problem • lab value	• DiseaseDisorderMention • SignSymptomMention	comd, fndg, dsyn, cgab, mobd, sosy, neop, emod, acab, anab, inpo, patf	Condition Observation
Procedures	T060, T059, T058, T063, T062, T065, T061	• test • treatment	• ProcedureMention	diap, lbpr, hlca, mbrt, resa, edac, topp	Procedures Observation
Living Beings	T100, T098	-	-	aggp, popg	Demographics
Device	T203, T074, T075	-	-	drdd, medd, resd	Device
Concepts and Ideas	T080, T081, T079	-	DateAnnotation	qlco, qnco, tmco	Temporal Measurement Measurement Qualifier
Phenomena	T034	-	MeasurementAnnotation	lbtr	Measurement Qualifier

Fig. 3. Mapping to UMLS semantic groups across NLP systems

References

1. Kuo, T-T., et al.: Ensembles of NLP tools for data element extraction from clinical notes. In: AMIA Annual Symposium Proceedings, vol. 2016, pp. 1880–1889 (2017)
2. Kang, N., Afzal, Z., Singh, B., van Mulligen, E.M., Kors, J.A.: Using an ensemble system to improve concept extraction from clinical records. J. Biomed. Inform. **45**, 423–428 (2012). https://doi.org/10.1016/j.jbi.2011.12.009
3. Friedman, C.: Towards a comprehensive medical language processing system: methods and issues. In: Proceedings AMIA Annual Fall Symposium, pp. 595–599 (1997)
4. Soysal, E., et al.: CLAMP - a toolkit for efficiently building customized clinical natural language processing pipelines. J. Am. Med. Inform. Assoc. **25**, 331–336 (2018). https://doi.org/10.1093/jamia/ocx132
5. Savova, G.K., et al.: Mayo clinical Text Analysis and Knowledge Extraction System (cTAKES): architecture, component evaluation and applications. J. Am. Med. Inform. Assoc. **17**, 507–513 (2010). https://doi.org/10.1136/jamia.2009.001560
6. Conway, M., et al.: Moonstone: a novel natural language processing system for inferring social risk from clinical narratives. J Biomed. Seman. **10**, 1–10 (2018). https://doi.org/10.1186/s13326-019-0198-0
7. Wang, Y., et al.: Clinical information extraction applications: a literature review. J. Biomed. Inform. **77**, 34–49 (2018). https://doi.org/10.1016/j.jbi.2017.11.011
8. Friedman, C., Shagina, L., Lussier, Y., Hripcsak, G.: Automated encoding of clinical documents based on natural language processing. J. Am. Med. Inform. Assoc. **11**, 392–402 (2004). https://doi.org/10.1197/jamia.M1552
9. ten Teije, A., et al.: Knowledge Engineering and Knowledge Management: 18th International Conference, EKAW 2012, Galway City, Ireland, October 8-12, 2012. Proceedings. Springer, Heidelberg (2012). https://doi.org/10.1007/978-3-642-33876-2
10. Uzuner, Ö., South, B.R., Shen, S., DuVall, S.L.: 2010 i2b2/VA challenge on concepts, assertions, and relations in clinical text. J. Am. Med. Inform. Assoc. **18**, 552–556 (2011). https://doi.org/10.1136/amiajnl-2011-000203
11. University of Minnesota, NLP/IE. nlp-adapt-kube (2019). https://github.com/nlpie/nlp-adapt-kube. Accessed 06 Jan 2020
12. University of Minnesota, NLP/IE, nlp-ensemble-explorer, UMN NLPIE (2020). https://github.com/nlpie/ensemble-explorer. Accessed 06 Jan 2020
13. Azam, S.S., Raju, M., Pagidimarri, V., Kasivajjala, V.: Q-Map: clinical concept mining from clinical documents. arXiv:1804.11149 (2018)
14. McCray, A.T., Burgun, A., Bodenreider, O.: Aggregating UMLS semantic types for reducing conceptual complexity. Stud. Health Technol. Inform. **84**, 216–220 (2001)
15. Semantic types and groups. https://metamap.nlm.nih.gov/SemanticTypesAndGroups.shtml. Accessed 05 May 2020
16. He, Z., Perl, Y., Elhanan, G., Chen, Y., Geller, J., Bian, J.: Auditing the assignments of top-level semantic types in the UMLS semantic network to UMLS concepts. In: Proceedings (IEEE International Conference Bioinformatics and Biomedicine), vol. 2017, pp. 1262–1269 (2017). https://doi.org/10.1109/BIBM.2017.8217840
17. University of Minnesota N, biomedicus (2019). https://github.com/nlpie/biomedicus. Accessed 06 Jan 2020
18. University of Texas, UT health, CLAMP (2020). https://clamp.uth.edu. Accessed 06 Jan 2020
19. Apache software foundation, cTAKES. https://ctakes.apache.org. Accessed 06 Jan 2020

20. The National Institutes of Health, MetaMap (2019). https://metamap.nlm.nih.gov. Accessed 06 Jan 2020
21. Apache foundation. UIMA project (2013). https://uima.apache.org. Accessed 08 Feb 2020
22. Aronson, A.R.: MetaMap evaluation (2001). https://ii.nlm.nih.gov/Publications/Papers/mm. evaluation.pdf
23. Technische Universität Darmstadt, ubiquitous knowledge processing lab, dkpro-cassis (2019). https://github.com/dkpro/dkpro-cassis. Accessed 06 Jan 2020
24. Miller, B.N., Ranum, D.L.: Parse tree. In: Problem Solving with Algorithms and Data Structures using Python. Section 7.6. https://runestone.academy/runestone/books/published/pythonds/Trees/ParseTree.html. Accessed 06 Jan 2020
25. Sang, E.F.T.K., Veenstra, J.: Representing text chunks. In: Proceedings of the 9th Conference on European Chapter of the Association for Computational Linguistics, Bergen, Norway, pp. 173–179. Association for Computational Linguistics (1999). https://doi.org/10. 3115/977035.977059
26. University of Minnesota, NLP/IE. expected_number_boolean_combinations_n_eq_5.py. expected_number_boolean_combinations_n_eq_5.py (2020). https://gist.github.com/ GregSilverman/3e09cb6b7c7bf664b4df14d309192bb3. Accessed 07 Feb 2020
27. Knoll, B.C., Melton, G.B., Liu, H., Xu, H., Pakhomov, S.V.S.: Using synthetic clinical data to train an HMM-based POS tagger. In: 2016 IEEE-EMBS International Conference on Biomedical and Health Informatics (BHI), pp. 252–255 (2016). https://doi.org/10.1109/BHI. 2016.7455882
28. Albright, D., et al.: Towards comprehensive syntactic and semantic annotations of the clinical narrative. J. Am. Med. Inform. Assoc. **20**, 922–930 (2013). https://doi.org/10.1136/ amiajnl-2012-001317
29. Aronson, A.R.: Effective mapping of biomedical text to the UMLS Metathesaurus: the MetaMap program. In: Proceeding AMIA Symposium, pp. 17–21 (2001)
30. Derczynski, L.: Complementarity, F-score, and NLP evaluation. In: Proceedings of the 10th International Conference on Language Resources and Evaluation (LREC 2016), Portorož, Slovenia, pp. 261–266. European Language Resources Association (ELRA) (2016)
31. Aronson, A.R., Lang, F.-M.: An overview of MetaMap: historical perspective and recent advances. J. Am. Med. Inform. Assoc. **17**, 229–236 (2010). https://doi.org/10.1136/jamia. 2009.002733
32. Kilicoglu, H., Rosemblat, G., Fiszman, M., Shin, D.: Broad-coverage biomedical relation extraction with SemRep. BMC Bioinform. **21**, 1–28 (2020). https://doi.org/10.1186/s12859-020-3517 7
33. Rizvi, R.F., et al.: iDISK: the integrated dietary supplements knowledge base. J. Am. Med. Inform. Assoc. **27**, 539–548 (2020). https://doi.org/10.1093/jamia/ocz216
34. Vasilakes, J., Bompelli, A., Bishop, J., Adam, T., Bodenreider, O., Zhang, R.: Assessing the enrichment of dietary supplement coverage in the UMLS. J. Am. Med. Informa. Assoc. (2020, in press)
35. Silverman, G.M., et al.: Named entity recognition in prehospital trauma care. Stud. Health Technol. Inform. **264**, 1586–1587 (2019). https://doi.org/10.3233/SHTI190547
36. Tignanelli, C.J., et al.: Natural language processing of prehospital emergency medical services trauma records allows for automated characterization of treatment appropriateness. J. Trauma Acute Care Surg. **88**, 607–614 (2020). https://doi.org/10.1097/TA.00000000 00002598

Ontology-Guided Data Augmentation for Medical Document Classification

Mahdi Abdollahi[1]([✉]) [iD], Xiaoying Gao[1] [iD], Yi Mei[1] [iD], Shameek Ghosh[2], and Jinyan Li[3] [iD]

[1] Victoria University of Wellington, Wellington, New Zealand
{mahdi.abdollahi,xiaoying.gao,yi.mei}@ecs.vuw.ac.nz
[2] Medius Health, Sydney, Australia
shameek.ghosh@mediushealth.org
[3] University of Technology Sydney, Sydney, Australia
Jinyan.Li@uts.edu.au

Abstract. Extracting meaningful features from unstructured text is one of the most challenging tasks in medical document classification. The various domain specific expressions and synonyms in the clinical discharge notes make it more challenging to analyse them. The case becomes worse for short texts such as abstract documents. These challenges can lead to poor classification accuracy. As the medical input data is often not enough in the real world, in this work a novel ontology-guided method is proposed for data augmentation to enrich input data. Then, three different deep learning methods are employed to analyse the performance of the suggested approach for classification. The experimental results show that the suggested approach achieved substantial improvement in the targeted medical documents classification.

Keywords: Ontology · Data augmentation · Medical document classification

1 Introduction

Medical document classification is different from the commonly considered document classification in terms of text terminology and their repetitiveness. In medical document classification, the content explains a set of medical events in a discharge note, with the objective of providing a clarification as accurately and comprehensively as conceivable when explaining the health condition of a patient. Mainly, such text massively uses domain-specific vocabulary and acronyms, making medical note analysis significantly different from commonly considered document classification. In addition, different combinations of domain-specific clinical events in a medical discharge note can explain a patient's health status completely differently. Hence, extracting important information to analyze clinical documents is exceptionally imperative.

One of the important factors which has effect on the classification accuracy is the size of the data set for training the model. Generally, there is a lack of

© Springer Nature Switzerland AG 2020
M. Michalowski and R. Moskovitch (Eds.): AIME 2020, LNAI 12299, pp. 78–88, 2020.
https://doi.org/10.1007/978-3-030-59137-3_8

adequate data in medical area [1]. When the training data set is not big enough, the trained classification model has not sufficient instances to learn. Hence, the prediction of the classifier will not be satisfactory. This issue can be worse when the data set has not enough text inside of the documents such as document abstracts. One possible solution to address the issue is to augment data for training the model.

Data augmentation is a methodology that empowers experts to fundamentally build the assorted variety of data accessible for training models, without really gathering new data. Data augmentation has many applications in image classification, sound and speech classification [2]. But there is not much work for text. In terms of text, it is not appropriate to augment the text by utilizing signal transformations as commonly used in image or speech classification. Because the order of words in text is important and may has semantic meaning. Hence, the best approach for doing data augmentation is to paraphrase the sentences in the documents by human. But this is very expensive due to the large size of instances in the data set. Replacing words and expressions with their synonyms can be a reasonable choice in data augmentation [3]. However, these methods are using normal dictionaries for augmentation and some domain specific terms or acronyms do not have synonyms in normal dictionaries.

As there are domain-specific vocabulary and acronyms in medical discharge notes, finding synonyms is not trival and this requires domain knowledge. In this paper, an ontology-based method is introduced for data augmentation by targeting concepts of words and expressions in the documents. This method will replace all of the words and phrases with their scientific names if they belong to a concept in medical field. This paper plans to study the following research questions:

1. Whether the ontology-guided approach can produce new discriminative instances from the original document set; and
2. Whether the proposed method can improve the classification accuracy in the targeted medical documents classification task.

2   Related Work

2.1   Data Augmentation in Classification

Data augmentation is a technique to deal with data scarcity in training models for different tasks such as classification. There are some common methods such as adding spelling errors, paraphrasing by utilizing syntax trees or regular expressions, adding textual noise and replacing with synonyms. Among these methods, synonyms replacement is one of the common approaches in textual data augmentation.

Zhang et al. [3] applied data augmentation in Convolutional Neural Network (CNN) for text classification by utilizing English thesaurus obtained from Word-Net. They replaced the words and expressions with their synonyms in the text to make new text based on the main data set. Rosario [2] introduced a method to data augmentation for short texts classification by producing similar words for each short texts to make a longer text by considering a semantic space. Quijas has investigated the effect of data augmentation in training CNNs and RNNs for text classification [4]. Kobayashi has suggested "contextual augmentation" method which produces counterparts of words by using a bidirectional language model and replaces words with their counterparts in sentences. They examined the method on different data set and showed improvements [5]. Coulombe in [6] has introduced another textual data augmentation by applying different methods including paraphrase generation, spelling errors, textual noise, back-translation and synonyms replacement. The methods were tested on different neural network architectures. Jungiewicz has proposed an approach to textual data augmentation for training CNNs by applying on sentence classification task. The researcher transformed sentences by keeping their lengths the same as their original lengths. The author has employed a thesaurus which belongs to Princeton University's WordNet [7]. However, these methods are using normal dictionaries for augmentation and some domain specific terms or acronyms do not have synonyms in normal dictionaries.

2.2 Feature Extraction in Medical Document Classification

Shah and Patel have used statistical approaches from features distribution in document classification to rank features [8]. The introduced methods used information gain (IG), mutual information, word frequency and term frequency-inverse document frequency (tf-idf) metrics for textual feature extraction. Nevertheless, these methods weight each feature separately without considering the relationship between features. Ontology-based classification methods is introduced in [9]. Dollah and Aono have introduced ontology-based classification approaches for biomedical abstract text classification [9]. Authors in [10–13] utilize different ontologies such as Unified Medical Language System (UMLS), Systematized Nomenclature of Medicine (SNOMED) and Medical Subject Headings (MeSH) to increase text classification accuracy.

Medical documents have been utilized in different tasks such as analyzing Framingham risk score (FRF), assessing risk factors in diabetic patients, discriminating heart disease risk factors, and finding risk factors for heart disease patients [14]. In this paper, we employ ontology as a feature extraction approach to detect meaningful words and expressions for augmenting documents.

3 Our Ontology-Based Method

In this section, we illustrate a novel data augmentation method and the utilized tools for extracting concepts of words and expressions for producing new documents. The suggested approach targets concepts of words and expressions to replace them with their scientific names. Figure 1 shows the flowchart of the suggested ontology-based approach for data augmentation.

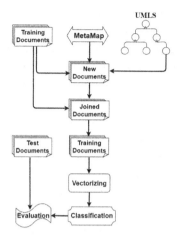

Fig. 1. The proposed data augmentation for medical document classification

The input of the suggested system is a set of clinical documents. Firstly, the method parses each document and tokenize the context based on sentences. Then, MetaMap tool [15] is employed to detect the meaningful phrases and their concepts in each sentence from the Unified Medical Language System (UMLS). After finding the phrases with a concept, the scientific name of the detected words or expressions are used to replace their corresponding phrases in the sentence. All of the new documents are created by applying the method. Next, all of the features are extracted from the original data set and the new created data set. Then, three different neural network approaches including Convolutional Neural Network (CNN), Recurrent Neural Network (RNN) and Hierarchical Attention Network (HAN) are employed for classification. The output predicts the label of a document.

It is expected that the suggested approach which produces meaningful documents and keeps their class name based on the original documents, can enhance the classification accuracy.

3.1 Data Augmentation Method

There are many domain-specific words and expressions in medical document and data augmentation requires domain knowledge. In this section, an ontology-guided approach is introduced for short text augmentation as a preprocessing stage.

UMLS is a domain-specific dictionary in the biomedical field. It provides an ontology structure of medical terminology concepts. In the suggested approach ("SciName"), each document in the data set (D) is analyzed independently. Firstly, the xth document (D_x) is tokenized based on the sentences (S). Then, the ith sentence (S_i) is sent to the UMLS by using the MetaMap tool. MetaMap extracts all of the concepts of the detected meaningful expressions in S_i from the UMLS. Next, all of the detected phrases are replaced with their extracted scientific names from the UMLS. Finally, S_i is updated in D_x. This process is repeated on all of the sentences of documents to make new documents.

A document segment is given below to illustrate how MetaMap works on the input medical documents and what output it returns in the data augmentation process. The following is a sample of a clinical note.

"Early resistance to pathogens requires a swift response from nk cells. In largeint giorgio trinchieri identified an nk growth factor and activator later called interleukin 12 il 12. This discovery helped reveal the regulatory link between innate and adaptive immunity."

```
     ----------------------------------------------------------
 1   Phrase: Early resistance to pathogens
 2   >>>>> Phrase
 3   early resistance to pathogens
 4   <<<<< Phrase
 5   >>>>> Mappings
 6   Meta Mapping (834):
 7       570   Pathogen (Pathogenic organism) [Organism]
 8   <<<<< Mappings
     ----------------------------------------------------------
 9   Phrase: a swift response from nk cells.
10   >>>>> Phrase
11   a swift response from nk cells
12   <<<<< Phrase
13   >>>>> Mappings
14   Meta Mapping (719):
15       586   Swift (Family Apodidae) [Bird]
16       753   Response (Response process) [Organism Attribute]
17       623   NK Cells (Natural Killer Cells) [Cell]
18   <<<<< Mappings
     ----------------------------------------------------------
19   Phrase: an nk growth factor
20   >>>>> Phrase
21   nk growth factor
22   <<<<< Phrase
23   >>>>> Mappings
24   Meta Mapping (901):
25       660   NK (Natural Killer Cells) [Cell]
26   <<<<< Mappings
     ----------------------------------------------------------
27   Phrase: activator later called interleukin 12 il 12.
28   >>>>> Phrase
29   activator later called interleukin 12 il 12
30   <<<<< Phrase
31   >>>>> Mappings
32   Meta Mapping (745):
33       595   Call (Decision) [Mental Process]
34       795   IL (Illinois (geographic location)) [Geographic Area]
35   <<<<< Mappings
     ----------------------------------------------------------
36   Phrase: helped
37   >>>>> Phrase
38   helped
39   <<<<< Phrase
40   >>>>> Mappings
41   Meta Mapping (966):
42       966   Help (Assisted (qualifier value)) [Qualitative Concept]
43   <<<<< Mappings
     ----------------------------------------------------------
44   Phrase: the regulatory link between innate
45   >>>>> Phrase
46   the regulatory link between innate
47   <<<<< Phrase
48   >>>>> Mappings
49   Meta Mapping (695):
50       593   regulatory [Regulation or Law]
51       760   Link (Links List) [Intellectual Product]
52   <<<<< Mappings
     ----------------------------------------------------------
```

Fig. 2. A segment of returned results of extracted concepts using MetaMap

Figure 2 shows the output of the MetaMap for the sample document. Table 1 presents the detected expressions with their concepts and scientific names for each phrase of the sample document. The concepts and scientific names of each detected phrase in the table is extracted by analyzing the lines 7, 15, 16, 17 for the first sentence, lines 25, 33, 34 for the second sentence and lines 42, 50 and 51 for the third sentence (in Fig. 2). Firstly, the phrase appeared in square brackets is extracted as a concept of the detected expression in the sentence. Then, the phrase appeared within the round parentheses at the same line is extracted as a scientific name of the detected expression. Finally, the extracted scientific name is used to replace the original expression in the sentence. This process is applied on all of the three sentences of the sample note. Below is the final output of the proposed method for the example clinical note.

Table 1. The detected phrases of the example notes using MetaMap.

Sentences	Detected phrases	Extracted concepts	Replaced phrases
First sentence	pathogens	[Organism]	Pathogenic organism
	swift	[Bird]	Family Apodidae
	response	[Organism Attribute]	Response process
	nk cells	[Cell]	Natural Killer Cells
Second sentence	nk	[Cell]	Natural Killer Cells
	call	[Mental Process]	Decision
	il	[Geographic Area]	Illinois (geographic location)
Third sentence	help	[Qualitative Concept]	Assisted (qualifier value)
	regulatory	[Regulation or Law]	regulatory
	link	[Intellectual Product]	Links List

"Early resistance to pathogenic organisms requires a family apodidae response process from natural killer cells. In largeint giorgio trinchieri identified an natural killer cells growth factor and activator later decisioned interleukin 12 illinois (geographic location) 12. This discovery assisted (qualifier value) reveal the regulator links list between innate and adaptive immunity."

The length of the output is longer than the input due to the more specific knowledge provided by the UMLS. For example, the acronym *"nk"* is changed to *"natural killer"* and the acronym *"il"* is replaced with *"illinois (geographic location)"*. The proposed method can easily provide more informative knowledge by using the UMLS. Finally, the new produced documents are used for the training stage together with the original documents to improve the performance of medical document classification.

4   Experiment Design

4.1   Classification Methods

In the ontology-based data augmentation, the new documents are made and mixed with the original documents to use for classification. We use three deep learning (DL) models, including a convolutional neural network (CNN), a recurrent neural network (RNN), and a hierarchical attention network (HAN) [16]. The performance is calculated by evaluating macro F1-measure metric for all of the used ML methods:

$$F1 \ measure = \frac{1}{N} \sum_{i=1}^{N} 2 * \frac{(precision * recall)}{(precision + recall)} \tag{1}$$

where N indicates the number of classes. Word2Vec word embedding is used to represent word tokens into numerical vectors. Word embedding represents the semantic meaning of each word in a numerical vector form. Word2Vec makes word embedding by utilizing a feed-forward neural network to anticipate the vicinage words for an input word. The word embedding is trained on all documents and transformed each word to its corresponding embedding. Then, the learned word embedding is used to generate the input for CNN, RNN and HAN. The size of word embedding is 350.

4.2   Dataset and Preprocessing

The performance of the suggested ontology-guided data augmentation is evaluated on the 2010 Informatics for Integrating Biology and the Bedside (i2b2), CAD (Coronary Artery Disease) task of the 2008 Informatics for Integrating Biology and the Bedside (i2b2) and the PubMed data set. The labels of i2b2(2008) and i2b2(2010) data set are CAD and non-CAD that form an imbalanced binary classification task. The numbers of CAD instances for i2b2(2008) training and testing sets are 391 and 272 documents, respectively. The numbers of CAD instances for i2b2(2010) training and testing sets are 25 and 48 documents, respectively. The labels of the PubMed data set are metabolism, physiology, genetics, chemistry, pathology, surgery, psychology, and diagnosis. The data set includes 8000 documents, each class has 1000 documents with 70% of documents for training and 30% for testing. The size of the training sets of all data sets will double by adding the new produced documents to the original ones.

The i2b2(2008) data set has 20701 different terms. It contains 1113 documents which 656 documents for training and 457 documents for testing. The i2b2(2010) data set has 7481 different terms. It contains 426 documents which 170 documents for training and 256 documents for testing. The PubMed data set has 30178 various terms. The number of the train documents is increased to 16000 in PubMed, 2226 in i2b2(2008) and 852 in i2b2(2010) by adding the new augmented documents. The 2223 input documents contain 27248 various terms. The 852 input documents contain 14920 various terms. The 16000 input documents include 59151 different terms.

4.3   Parameter Settings

Three different ML methods are used to evaluate the proposed idea. The suggested parameters in [16] are used for the employed neural network approaches.

The early stopping approach by considering the validation accuracy (three epochs without any change) is used to terminate the training step.

The used CNN architecture has 3 parallel convolutional layers with 100 channels for each one. The window of layers are 3, 4, and 5 words, respectively. The output of this architecture for each input document is 300 channels × number of words. The applied dropout rate is 50% [16].

HAN [16] is a deep learning model developed for document classification. It contains two hierarchies. The lower hierarchy analyzes a line in word level and it feeds with a word embedding. Then, it uses a bidirectional GRU to apply an attention mechanism to find more important words. The output is a line embedding which is feed to the upper hierarchy to analyze a document in line level. A dropout is applied on the produced document embedding and finally, a softmax function is employed to predict a label of each document.

The RNN architecture uses an attention mechanism (which is similar to a single hierarchy of HAN method). A bidirectional GRU with attention and 200 number of neurons is utilized with dropout and softmax. The used optimizer is Adam with learning rate of 0.0002. The applied dropout rate is 50%.

5   Results and Discussion

The performance of the methods are evaluated based on macro F1-measure and accuracy metrics for the PubMed data set and macro F1-measure metric for the i2b2(2010) data set.

Three different approaches are applied on the original documents in each data set. In first approach (SynName), we used WordNet dictionary to extract all of the synonyms of the main word appeared inside of a document. Then, the most similar synonym is found by using the GloVe pre-trained model (from the GloVe website http://nlp.stanford.edu/data/glove.6B.zip) and used to replace the main word to augment new documents. The used GloVe model provides 100-dimensional vector which is trained on Wikipedia data with 6 billion tokens and a 400,000 word vocabulary. In second approach (our proposed method), UMLS is employed to find scientific names of the appeared phrases in the documents based on their concepts to replace with the original phrase in the document to augment a new document. In third approach, we combined the augmented document from the two introduced approaches with the original data set. The proposed augmentation method (SciName) is applied on the training set only. Then, experimental results are calculated using 30 independent runs on the original test set.

Tables 2, 3, 4 and 5 compare the statistical results for three methods. The average and standard deviation of accuracies and F1-measure are provided for each ML method and the significance test is done utilizing the experiment results of the 30 runs to compute the three approaches. The Wilcoxon signed ranks test with significance level of 0.05 is used to assess whether the suggested approach has made significant difference in classification performance. In Tables 2, 3, 4 and 5, "T" column indicates the significance test of the best approach (the third combined approach) against the other methods, where "+" indicates the suggested method is significantly better, "=" mentions no significant difference, and "−" points significantly less accurate. The best results are highlighted in the tables.

By analyzing Tables 2 and 3, it is clear that neural network methods are improved in accuracy and F1-measure by using combination of the original data set with the obtained augmented data sets from the two introduced augmentation methods. The highest accuracy and F1-measure in Tables 2 and 3 belong to RNN. Tables 4 and 5 provide the statistical results for i2b2(2010) and i2b2(2008) data sets, respectively. In Table 4, CNN, RNN and HAN show high F1-measure in the combination approach (SynName+SciName). The highest F1-measure in both data sets belongs to RNN with 96.43% and 995.61%, respectively. In Tables 2, 3, 4 and 5, the suggested ontology-based approach (SciName) shows better performance in comparison with the SynName method [3].

Interestingly, although some of the resulting documents look gibberish, the generated documents improve the classifier performance. This might be because our approach works at word level which can tolerate non-sense sentences as long as they contain meaningful words. Our method not only works well on clinical notes (such as i2b2(2008) and i2b2(2010)), but also it shows promising results on related biomedical notes (PubMed set).

5.1 The Value of the Work

As indicated in this paper, in numerous practical works on modeling, data augmentation is extremely important. This is a situation that we encounter when in practical settings, real life patient cases are unavailable to feed data-hungry models (a rare disease is an example where available cases are few). In fact, synthetic data synthesis and augmentation has strong advantages with respect to advancing healthcare models research by protecting patient confidentiality, and is a promising tool for situations where real world data is difficult to obtain or unnecessary. At that time, in combination with data augmentation we can also perform simulations to generate digital patient cases.

Table 2. Comparison of classification accuracy and standard deviation averages using 30 independent runs for PubMed data set. The significant test is for the combined approach against others (Wilcoxon Test, $\alpha = 0.05$)

Methods	Original		SynName		SciName		SynName+SciName
Classifiers	Accuracy		Accuracy		Accuracy		Accuracy
	Ave ± Std (Best)	T	Ave ± Std (Best)	T	Ave ± Std (Best)	T	Ave ± Std (Best)
CNN	71.64 ± 0.55 (72.67)	+	80.64 ± 0.63 (81.84)	+	80.80 ± 0.51 (82.08)	+	**84.16 ± 0.66 (85.42)**
RNN	71.53 ± 1.03 (73.50)	+	84.42 ± 0.90 (86.38)	+	85.57 ± 0.62 (86.75)	+	**90.80 ± 0.45 (91.63)**
HAN	71.29 ± 0.69 (72.88)	+	84.29 ± 0.85 (85.38)	+	85.00 ± 0.94 (86.92)	+	**90.75 ± 0.49 (91.79)**

Table 3. Comparison of classification F1-measure and standard deviation averages using 30 independent runs for PubMed data set. The significant test is for the combined approach against others (Wilcoxon Test, $\alpha = 0.05$)

Methods	Original		SynName		SciName		SynName+SciName
Classifiers	F1-measure		F1-measure		F1-measure		F1-measure
	Ave ± Std (Best)	T	Ave ± Std (Best)	T	Ave ± Std (Best)	T	Ave ± Std (Best)
CNN	71.54 ± 0.76 (72.64)	+	80.42 ± 0.69 (81.42)	+	80.48 ± 0.71 (81.81)	+	**84.34 ± 0.54 (85.36)**
RNN	71.62 ± 0.73 (72.97)	+	84.07 ± 0.88 (85.64)	+	85.37 ± 0.90 (87.00)	+	**90.91 ± 0.58 (91.98)**
HAN	70.96 ± 0.82 (72.21)	+	84.21 ± 1.23 (86.16)	+	84.95 ± 0.73 (86.36)	+	**90.85 ± 0.79 (91.99)**

Table 4. Comparison of classification F1-measure and standard deviation averages using 30 independent runs for i2b2 2010 data set. The significant test is for the combined approach against others (Wilcoxon Test, $\alpha = 0.05$)

Methods	Original		SynName		SciName		SynName+SciName
Classifiers	F1-measure		F1-measure		F1-measure		F1-measure
	Ave ± Std (Best)	T	Ave ± Std (Best)	T	Ave ± Std (Best)	T	Ave ± Std (Best)
CNN	77.66 ± 13.16 (90.22)	+	92.40 ± 1.95 (95.62)	+	92.72 ± 0.90 (95.10)	+	**94.15 ± 1.12 (97.44)**
RNN	85.51 ± 8.00 (91.37)	+	91.30 ± 2.74 (97.51)	+	94.66 ± 1.32 (96.92)	+	**96.43 ± 0.68 (98.12)**
HAN	56.11 ± 18.78 (90.35)	+	86.90 ± 7.23 (97.48)	+	93.75 ± 2.55 (96.87)	+	**96.27 ± 0.73 (97.51)**

Table 5. Comparison of classification F1-measure and standard deviation averages using 30 independent runs for i2b2 2008 data set (CAD Task). The significant test is for the combined approach against others (Wilcoxon Test, $\alpha = 0.05$)

Methods	Original		SynName		SciName		SynName+SciName
Classifiers	F1-measure		F1-measure		F1-measure		F1-measure
	Ave ± Std (Best)	T	Ave ± Std (Best)	T	Ave ± Std (Best)	T	Ave ± Std (Best)
CNN	90.44 ± 1.40 (92.93)	+	90.53 ± 2.79 (93.13)	+	90.80 ± 0.84 (92.19)	+	**92.09 ± 0.69 (93.64)**
RNN	95.52 ± 1.13 (97.28)	=	95.27 ± 0.64 (96.40)	=	**95.61 ± 0.44 (96.62)**	=	95.25 ± 0.53 (96.37)
HAN	95.16 ± 1.53 (97.29)	=	94.16 ± 1.74 (96.61)	+	95.47 ± 0.34 (96.14)	=	**95.60 ± 0.51 (96.61)**

6   Conclusions and Future Work

This paper proposes a new ontology-based data augmentation method by replacing meaningful expressions with their scientific names to deal with the data shortage issue in medical document classification. The introduced approach is able to improve the precision of classification in the neural network models. Experimental results for accuracy and f1-measure show that the suggested method can increase the performance of the CNN, RNN and HAN models by using the suggested ontology-based approach to provide more samples in the training phase. This paper shows promise in utilizing an ontology-guided data augmentation approach in clinical document classification, however, it is still necessary to do more research to improve the classification performance. We will explore other ways to do data augmentation for medical discharge notes. Meanwhile, we will investigate to employ available domain-specific dictionaries instead of UMLS to enhance the classification precision.

References

1. Sánchez, D., Batet, M., Viejo, A.: Utility-preserving privacy protection of textual healthcare documents. J. Biomed. Inform. **52**, 189–198 (2014)
2. Rosario, R.R.: A data augmentation approach to short text classification. Ph.D. thesis, UCLA (2017)
3. Zhang, X., Zhao, J., LeCun, Y.: Character-level convolutional networks for text classification. In: Advances in Neural Information Processing Systems, pp. 649–657 (2015)
4. Quijas, J.K.: Analysing the effects of data augmentation and free parameters for text classification with recurrent convolutional neural networks. The University of Texas at El Paso (2017)
5. Kobayashi, S.: Contextual augmentation: data augmentation by words with paradigmatic relations. arXiv preprint arXiv:1805.06201 (2018)
6. Coulombe, C.: Text data augmentation made simple by leveraging NLP cloud APIs. arXiv preprint arXiv:1812.04718 (2018)
7. Jungiewicz, M., Smywiński-Pohl, A.: Towards textual data augmentation for neural networks: synonyms and maximum loss. Comput. Sci. **20** (2019)
8. Shah, F.P., Patel, V.: A review on feature selection and feature extraction for text classification. In: 2016 International Conference on Wireless Communications, Signal Processing and Networking (WiSPNET), pp. 2264–2268 (2016)
9. Dollah, R.B., Aono, M.: Ontology based approach for classifying biomedical text abstracts. Int. J. Data Eng. **2**, 1–15 (2011)
10. Buchan, K., Filannino, M., Uzuner, Ö.: Automatic prediction of coronary artery disease from clinical narratives. J. Biomed. Inform. **72**, 23–32 (2017)
11. Abdollahi, M., Gao, X., Mei, Y., Ghosh, S., Li, J.: Uncovering discriminative knowledge-guided medical concepts for classifying coronary artery disease notes. In: Mitrovic, T., Xue, B., Li, X. (eds.) AI 2018. LNCS (LNAI), vol. 11320, pp. 104–110. Springer, Cham (2018). https://doi.org/10.1007/978-3-030-03991-2_11
12. Abdollahi, M., Gao, X., Mei, Y., Ghosh, S., Li, J.: An ontology-based two-stage approach to medical text classification with feature selection by particle swarm optimisation. In: 2019 IEEE Congress on Evolutionary Computation (CEC), pp. 1–8 (2019)
13. Abdollahi, M., Gao, X., Mei, Y., Ghosh, S., Li, J.: Stratifying risk of coronary artery disease using discriminative knowledge-guided medical concept pairings from clinical notes. In: Nayak, A.C., Sharma, A. (eds.) PRICAI 2019. LNCS (LNAI), vol. 11672, pp. 457–473. Springer, Cham (2019). https://doi.org/10.1007/978-3-030-29894-4_37
14. Shivade, C., Malewadkar, P., Fosler-Lussier, E., Lai, A.M.: Comparison of UMLS terminologies to identify risk of heart disease using clinical notes. J. Biomed. Inform. **58**, S103–S110 (2015)
15. Aronson, A.R., Lang, F.-M.: An overview of MetaMap: historical perspective and recent advances. J. Am. Med. Inform. Assoc. **17**, 229–236 (2010)
16. Gao, S., et al.: Hierarchical attention networks for information extraction from cancer pathology reports. J. Am. Med. Inform. Assoc. **25**, 321–330 (2017)

Divide to Better Classify

Yves Mercadier[1(✉)], Jérôme Azé[1], and Sandra Bringay[1,2]

[1] LIRMM UMR 5506, Université de Montpellier, CNRS, Montpellier, France
{yves.mercadier,jerome.aze,sandra.bringay}@lirmm.fr,
http://www.lirmm.fr/
[2] Université Paul-Valéry Montpellier 3, Montpellier, France

Abstract. Medical information is present in various text-based resources such as electronic medical records, biomedical literature, social media, etc. Using all these sources to extract useful information is a real challenge. In this context, the single-label classification of texts is an important task. Recently, in-depth classifiers have shown their ability to achieve very good results. However, their results generally depend on the amount of data used during the training phase. In this article, we propose a new approach to increase text data. We have compared this approach for 5 real data sets with the main approaches in the literature and our proposal outperforms in all configurations.

Keywords: Natural language processing · Document classification · Textual data augmentation

1 Introduction

Medical information is present in various text-based resources such as electronic medical records, biomedical literature, social media, etc. Using all these sources to extract useful information is a real challenge. In this context, the single-label classification of texts is an important task. Recently, in-depth classifiers have shown their ability to achieve very good results. However, their results generally depend on the amount of data used during the training phase.

In this article, we focus on data augmentation methods that can be effective on small data sets. Data augmentation uses limited amounts of data and transforms existing samples to create new ones. Then, an important challenge is to generate new data that retain the same label. More precisely, it is a matter of injecting knowledge by taking into account the invariant properties of the data after particular transformations. The augmented data can thus cover unexplored input space and improve the generalization of the model.

This technique has proven to be very effective for image classification tasks, especially when the training database is limited. For example, for image recognition, it is well known that minor changes due to scaling, cropping, distortion, rotation, etc. do not change the data labels because these changes can occur in real-world observations. However, transformations that preserve text data labels are not as obvious and intuitive.

© Springer Nature Switzerland AG 2020
M. Michalowski and R. Moskovitch (Eds.): AIME 2020, LNAI 12299, pp. 89–99, 2020.
https://doi.org/10.1007/978-3-030-59137-3_9

In this article, we present a new technique for augmenting textual data, called DAIA (**D**ata **A**ugmentation and **I**nference **A**ugmentation). This method is simple to implement because it does not require any semantic resources or a long training phase. We will evaluate DAIA for various medical data sets of different nature and show an improvement over the state of the art, especially on small data sets. To realize our proposition we use the python module Mantéïa. The code is public and accessible at[1].

2   State of the Art

Data augmentation has been used successfully in the field of image analysis [15]. For example, Perez et al. [11] compared several simple techniques such as cropping, rotating and flipping images. They also used more advanced techniques such as GAN (Generative Adversarial Network) to generate images of different styles. The neural network learns which type of augmentation improves the classifier the most. While many solutions exist for image analysis, augmentation methods have been much less studied in the field of text analysis. There are four main approaches that we will describe below:

Approaches using semantic resources were first proposed. For example, Zhang et al. [19] used a thesaurus to replace words with their synonyms in order to create an augmented dataset used for text classification. This increase proved to be inefficient and in some cases even reduced performance. The authors explain that when large amounts of real data are available, models are easily generalized and are not improved by increasing data.

Approaches inspired by the distortions that can be added to images have also been applied to texts. For the classification of texts, [16], in the EDA (Easy Data Augmentation) method, the number of samples is increased by deleting, swapping a word or replacing it with a synonym. Some approaches focused on the choice of words to be changed. [7] analysed the context of the word to find the one to be swapped. The context is defined by the training of an LSTM-type neural network. This approach improved accuracy by 0.5% over five datasets. In the UDA (Unsupervised Data Augmentation) method, [17] replaced words with low information content, identified with a low TF-IDF, with their synonyms while retaining those with high TF-IDF values representing keywords. This heuristic has been tested on six datasets, and the authors have shown that it is possible to reduce the classification error.

Generative approaches have also been explored. [6] has formed GAN models on small datasets and used them to augment the data to improve the generalization of a sentiment classifier. The results improve the accuracy by 1% on two of the datasets.

A final approach is based on increasing the textual data using backtranslation [4]. This involves translating an example into a language and then translating the resulting translation into the original language. Shleifer et al [14]

---

[1] https://github.com/ym001/Manteia/blob/master/notebook/notebook_Manteia_classification_augmentation_run_in_colab.ipynb.

has shown that the back-translation technique has hardly improved with modern classifiers such as UMLfit.

In this article, we will focus on a new approach, simple to implement, which does not require resources such as semantic approaches, or large amounts of computation such as generative approaches, or even access to external resources such as reverse translation approaches. A limitation of distortion approaches found in the literature is that they do not preserve the order of words in sentences. The examples generated are different from those we might find in the real world and do not allow for effective embedding. In the proposed DAIA approach, we will describe three approaches to divide sentences during the learning phase into several sequences, which will be used to augment the textual data. The same approach will be used during the testing phase on the examples to be classified by applying a soft voting technique.

We will show in the experiments that DAIA approach improves the results of text classification over deep classifiers [3], RoBERTa [9], Albert [8], DistilBert [13] or ScienceBert [1]. We will also compare the DAIA proposal to the UDA [17] and EDA [16] distortion approaches, the TextGen generative approach [6] and the back translation [14].

3   DAIA

The DAIA method is structured in two parts: the first is related to training (DA: Data Augmentation) and the second is related to the test phase (IA: Inference Augmentation).

Data Augmentation: In the training phase, we increase the amount of data by dividing the initial text of each sample. We seek to produce new samples without modifying the order between words to the initial sequence. The objective is to not decrease the learning quality of the description of word embeddings. We have conducted preliminary experiments based on different types of divisions, which for lack of space, are not described in this article. We present below only the three methods that we have combined to form the pyramidal division. These three approaches split the initial sentence into n sequences of words which will be associated with the same label as the initial sentence.

- Symmetrical Division 1: We divide the text symmetrically by removing $x\%$ from the text at both ends to generate a new text sequence. For each initial text, we thus obtain an additional text.
- Division 2 by sliding window: We cut the sentence by applying a sliding window of size l which moves m words to the end of the initial text. The number of generated sequences depends on the length of the initial sentence.
- Division 3 into equal parts: The division is done in i equal parts. The results is i new documents plus the original document.

After data augmentation, from each text in the initial training data set, we generated a set of new texts, associated with the same label as the initial text, and are used as input into the learning model. We tested the advantages and disadvantages of these different types of division and finally proposed a pyramidal division that combines divisions 1 and 3 and is described in Fig. 1. This new division is based on n levels. The increase for level 1 is done by the symmetrical division 1. The increase for level 2 is done by dividing the text into two equal parts and adding the increase obtained in level 1. The increase for level i is done by dividing the text into i equal parts and adding the increase at level $i - 1$. For each text, this results in $\frac{n \times (n+1)}{2}$ new labeled segments.

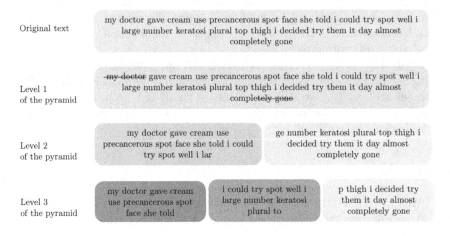

Fig. 1. Description of the pyramidal division. The figure shows the division of the original text into 6 new documents according to the tree levels.

Inference Augmentation: The test phase consists in predicting the labels for new texts from the model learned during the learning phase. We divide the text to be classified according to the same protocol as described for the training phase. Then we give the classifier all the generated sequences. For each sequence, the classifier returns a prediction which is aggregated for the initial text by soft-voting (sum of the values for each predicted class of each element of the set). Thus, for each initial text, one prediction per class is obtained.

4   Presentation of the Experiments

4.1   Data Sets

In order to show the generalisation of our approach, we selected five data sets in the medical field, described in Table 1. The first two data sets correspond to medical publications. The other three data sets correspond to texts written by patients. It is important to note that these data sets are unbalanced.

Table 1. Description of the textual data sets according to the number of classes and documents, the length of the documents in term of words and classification tasks.

Data sets	Classes	Documents	Length	Task
PubMed 200k RCT	5	2 211 861	26.22	Text analysis
WHO COVID-19	4	26 909	166	Provenance analysis
Drugs.com	10	53 766	85.58	Sentiment analysis
eR anorexie	2	84 834	38.24	Sentiment analysis
eR depression	2	531 394	36.76	Sentiment analysis

PubMed 200k RCT[2] Dernoncourt et al. [2] is a database containing more than 200,000 abstracts of articles dealing with randomized controlled trials, with more than 2 million sentences. Each sentence is labeled according to its meaning in the abstract (context, objective, method, result and conclusion).

WHO COVID-19[3], was proposed by the Allen for AI Institute. This data set is composed of more than 29,000 articles about COVID-19 and the corona virus family. In this study we only used articles with an abstract. Each text is labeled according to its source: CZI, PMC, biorxiv, medrxiv.

Drugs.com[4] Gräßer et al. [5] corresponds to patients opinions about drugs. Data were obtained by analyzing online pharmaceutical sites. Each text is labeled with a score from 1 to 10 corresponding to patients' satisfaction.

The eR Depression and eR Anorexia data sets were produced for the CLEF eRisk 2018 challenge[5]. The texts correspond to messages from users in the social network Reddit[6]. [12]. Each text is labeled according to the depression/non-depression and anorexia/non-anorexia classes.

4.2 Data Pre-processing

For each dataset, we applied the following pre-processing: removal of punctuation, special characters, stop words, shift from upper to lower case and lemmatization. For lemmatization, we use the NLTK python module associated with the Wordnet dictionary. Each text can be associated with zero, one or more classes depending on the dataset, used as prediction output.

4.3 Comparison to Other State of the Art Approaches

We compare our proposal to three state-of-the-art methods, which are described in detail below.

---

[2] https://github.com/Franck-Dernoncourt/pubmed-rct.

[3] https://pages.semanticscholar.org/coronavirus-research.

[4] https://archive.ics.uci.edu/ml/datasets/Drug+Review+Dataset+%28Drugs.com %29.

[5] https://early.irlab.org/2018/index.html.

[6] https://www.reddit.com/.

For semantic word distortion approaches, we considered the EDA and UDA approaches. For EDA, we implement the four simple data augmentation techniques described by Wei et al. [16]: replacement by synonyms, random insertion, random exchange, random deletion. For this, we used Wordnet thesaurus of NLTK[7]. For UDA, in order to identify the words with the most informational content, we proceed as Xie et al. [17] who identify these words as being those negatively correlated with their TF-IDF score. For this, they define the probability $min(p(C - TFIDF(x_i))/Z, 1)$, where p is a hyper-parameter controlling the variation of augmentation, C is the maximum TF-IDF score for words x_i of a text x and $Z = \sum_i (C - TFIDF(x_i))/|x|$. For our experiments, we have chosen $p = 0.9$. Xie et al. [18] do not change the keywords, identified by their frequency. Spotted words which are not key words are replaced by one of the non essential words of the corpus.

As the state-of-the-art text generators used for data augmentation have been exceeded in semantic quality by the generator GPT2 [6], we used the latter according to the following protocol. For each text, an augmented text is constructed by concatenating two parts. An initial part, the seed, corresponding to half of the original text and a second part obtained using the GPT2 generator having taken the seed as input.

Data augmentation by back translation [14] consists of producing paraphrases which globally preserve the semantics of the initial sentence. We use the translation web service Yandex[8] in order to produce these paraphrases. We first translate the texts into Japanese and then translate them back into English and then label them with the original label of the text.

4.4 Assessment Protocol

We proceed as follows for each data set described in Sect. 4.1. We extract 5,000 texts while respecting the weighting of the classes. We then separate the data into two sets: a first set of 1,250 texts (i.e. 25%) is used for the test phase and a second set of 3,750 texts (i.e. 75%) is used for the learning phase while respecting the class stratification. To estimate the quality of learning, we use the accuracy metric, which is calculated as the ratio of the number of labels correctly assigned by the classifier to the total number of labels. All the values produced in this study were calculated on the average of a four-fold cross-validation.

4.5 Hyper-parameters and Training

All hyper-parameters in the networks are set for all data sets. The learning rate is set to 0.00001. All our neural networks are trained by a back-propagation process using a cross-entropy error (*loss*) function. The gradient calculation optimizers are of the *Adam weight* type. In addition, we use a linear learning rate update with warm-up. The experiments are performed on two GeForce RTX-type GPUs.

---

[7] https://www.nltk.org/howto/wordnet.html.

[8] https://yandex.com/.

5   Results of the Experiments

We have made available some preliminary experiments at this URL[9] in order to compare the tree divisions that allow us to define the pyramidal division that is evaluated in the rest of this section.

5.1   Impact of the Classifier and Comparison to the State of the Art

We're working on the drugs.com dataset. The baseline corresponds to the use of a single classifier without increasing the data. The classifiers compared are the most efficient in the literature: Bert [3], RoBERTa [9], Albert [8], DistilBert [13] or ScienceBert [1]. The DAIA method is applied here for data augmentation during the learning phase and the test phase as detailed in Sect. 3.

The results are presented in the Table 2. The pyramidal division during the learning phase outperforms the other divisions, regardless of the classifier. It is also superior to other approaches in the literature such as EDA and UDA or even reverse translation, for Roberta and Xlnet. Combined with increasing inference during the test phase, DAIA outperforms for all classifiers. The Roberta network over-performs in terms of accuracy.

Table 2. DAIA impact on the accuracy for 6 classifiers and comparison with other data augmentation approaches.

Classifier	Roberta	Bert	Xlnet	Albert	Distilbert	Scibert
Baseline without DA	0.3661	0.3637	0.3760	0.3200	0.3601	0.3392
Distorsion EDA	0.38	0.3669	0.3712	0.3226	0.3689	0.3510
Distorsion UDA	0.3811	0.3632	0.3525	0.3084	0.3660	0.3327
Text generation GPT2	0.3384	0.3252	0.3425	0.3122	0.3204	0.3078
Back translation	0.3798	0.3462	0.3728	0.3426	0.3584	0.3361
DA - symmetrical division	0.3769	0.3491	0.3568	0.3154	0.3359	0.3498
DA - sliding division	0.3494	0.3399	0.3657	0.3266	0.3527	0.3156
DA - pyramid division	0.3877	0.3660	0.3762	0.334	0.3688	0.3590
DAIA	**0.3931**	**0.3755**	**0.3786**	**0.3444**	**0.3693**	**0.3625**

5.2   Impact of Number of Classes

In Table 3, we study the performance of the DAIA according to the number of categories for the five corpora in the study. We selected Roberta as a classifier with a sample of 500 texts. We notice a greater improvement in DAIA performance for datasets with a high number of classes such as Drugs.com (10).

---

[9] https://github.com/ym001/DAIA/blob/master/Preliminary%20experiment.pdf.

Table 3. Impact of the number of classes on the accuracy for text classification. We used the best classifier Roberta alone and combined with DAIA.

Classifier	Drugs.com	PubMed 200k RCT	WHO COVID-19	eR Depression	eR anoxeria
Roberta	0.2846	0.6413	0.9319	0.8926	0.9079
DAIA	**0.316**	**0.6513**	**0.938**	**0.9066**	**0.9153**

5.3   Impact of the Amount of Data on the Learning Phase

In this experiment, we used the Roberta classifier as a baseline without increasing the data. We show in Fig. 2 that DAIA provides an improvement for all sizes of the eR anorexia (left) and Drugs.com (right) datasets. In particular, from the 500 text datasets, we observe an improvement of 2.7%. The impact is reduced when the dataset reaches the size of 5,000 texts.

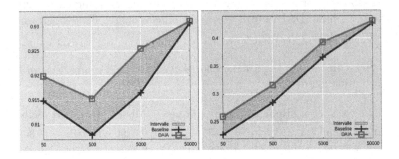

Fig. 2. Impact of DAIA on the accuracy of the text classification according to the size of the corpus on eR Anorexia (left) and Drugs.com (right)

5.4   Impact of the Selected Sequence Size

In Fig. 3, we study the variation in DAIA according to the size of the divisions obtained with approach 3 and the length of the texts in input for the data set drugs.com. Too small a division doesn't make a difference. We got our best result for a level of three combined with a sequence length of 128 words. The size of the input text has little impact in the range studied with a very slight maximum for the 128 word value. We will therefore advise those who wish to test the implementation of the DAIA a level three of the pyramid division with a maximum sequence size of 128. Similar results have been obtained with other data sets.

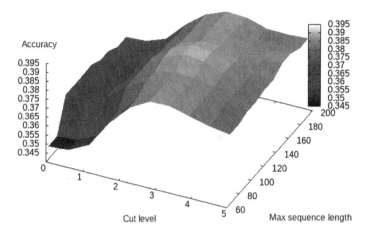

Fig. 3. Accuracy according to the level of pyramidal division and the maximum length of the sequences used as the input of the network.

6   Conclusion

In this study, we proposed a new method to increase textual data, which consists of dividing texts into several segments to increase the variety of training examples while preserving the quality of the learning words embedding. This method, called Data Augmentation and Inference Augmentation, is a distortion approach that does not require any semantic resources, nor any very important training phase, and no external resources. Our DAIA approach has proven to be effective on 5 medical datasets, for 5 classifiers and has been successfully compared to the main approaches in the literature. Comparison with approaches not based on Bert must be conducted to confirm other results. As differences in performance are small, statistical significance tests must also be performed. The impact of the choice of language on the retro-translation must be assessed. Furthermore, it is possible to keep the meaning without keeping the order of words. We could then imagine new experiments consisting in preserving certain levels of syntactic articulation rather than the order of words in order to generate more variability while preserving the semantics of the sentence. In the future, we plan to use multimodal data sets composed of text, sound and images for classification purposes and we will also focus on more complex tasks such as multi-label classification. Finally, DAIA will also be applied to deep active learning heuristics in the medical field [10].

References

1. Beltagy, I., Lo, K., Cohan, A.: SciBERT: a pretrained language model for scientific text. In: Inui, K., Jiang, J., Ng, V., Wan, X. (eds.) EMNLP-IJCNLP 2019, Hong Kong, China, 2019. pp. 3613–3618. Association for Computational Linguistics (2019). https://doi.org/10.18653/v1/D19-1371

2. Dernoncourt, F., Lee, J.Y.: Pubmed 200k RCT: a dataset for sequential sentence classification in medical abstracts. In: Kondrak, G., Watanabe, T. (eds.) IJCNLP 2017. Volume 2: Short Papers, Taipei, Taiwan, 2017, pp. 308–313. Asian Federation of Natural Language Processing (2017)

3. Devlin, J., Chang, M., Lee, K., Toutanova, K.: BERT: pre-training of deep bidirectional transformers for language understanding. In: Burstein, J., Doran, C., Solorio, T. (eds.) NAACL-HLT 2019. Volume 1 (Long and Short Papers), Minneapolis, MN, USA, 2019, pp. 4171–4186, Association for Computational Linguistics (2019). https://doi.org/10.18653/v1/n19-1423

4. Edunov, S., Ott, M., Auli, M., Grangier, D.: Understanding back-translation at scale. CoRR abs/1808.09381 (2018)

5. Gräßer, F., Kallumadi, S., Malberg, H., Zaunseder, S.: Aspect-based sentiment analysis of drug reviews applying cross-domain and cross-data learning. In: Kostkova, P., Grasso, F., Castillo, C., Mejova, Y., Bosman, A., Edelstein, M. (eds.) Digital Health, DH 2018, pp. 121–125. ACM, Lyon (2018). https://doi.org/10.1145/3194658.3194677

6. Gupta, R.: Data augmentation for low resource sentiment analysis using generative adversarial networks. CoRR abs/1902.06818 (2019)

7. Kobayashi, S.: Contextual augmentation: Data augmentation by words with paradigmatic relations. In: Walker, M.A., Ji, H., Stent, A. (eds.) NAACL-HLT, Volume 2 (Short Papers), New Orleans, Louisiana, USA, 1–6 June 2018, pp. 452–457. Association for Computational Linguistics (2018). https://doi.org/10.18653/v1/n18-2072

8. Lan, Z.Z., Chen, M., Goodman, S., Gimpel, K., Sharma, P., Soricut, R.: Albert: A lite BERT for self-supervised learning of language representations. http://arxiv.org/abs/1909.11942 (2019)

9. Liu, Y., et al.: RoBERTa: a robustly optimized BERT pretraining approach. CoRR abs/1907.11692 (2019)

10. Maldonado, R., Harabagiu, S.M.: Active deep learning for the identification of concepts and relations in electroencephalography reports. J. Biomed. Inform. **98**, 103265 (2019). https://doi.org/10.1016/j.jbi.2019.103265

11. Perez, L., Wang, J.: The effectiveness of data augmentation in image classification using deep learning. CoRR abs/1712.04621 (2017)

12. Ragheb, W., Moulahi, B., Azé, J., Bringay, S., Servajean, M.: Temporal mood variation: at the CLEF eRisk-2018 tasks for early risk detection on the Internet. In: Cappellato, L., Ferro, N., Nie, J., Soulier, L. (eds.) Working Notes of CLEF 2018, Avignon, France, 10–14 September 2018, vol. 2125. CEUR-WS.org (2018)

13. Sanh, V., Debut, L., Chaumond, J., Wolf, T.: DistilBERT, a distilled version of BERT: smaller, faster, cheaper and lighter https://arxiv.org/abs/1910.01108 (2019)

14. Shleifer, S.: Low resource text classification with ULMFit and backtranslation. CoRR abs/1903.09244 (2019)

15. Shorten, C., Khoshgoftaar, T.M.: A survey on image data augmentation for deep learning. J. Big Data **6**(1), 1–48 (2019). https://doi.org/10.1186/s40537-019-0197-0

16. Wei, J.W., Zou, K.: EDA: easy data augmentation techniques for boosting performance on text classification tasks. CoRR abs/1901.11196 (2019)
17. Xie, Q., Dai, Z., Hovy, E.H., Luong, M., Le, Q.V.: Unsupervised data augmentation. CoRR abs/1904.12848 (2019)
18. Xie, Z., et al.: Data noising as smoothing in neural network language models. In: ICLR Proceedings of the Conference on Track 2017, OpenReview.net, Toulon (2017)
19. Zhang, X., LeCun, Y.: Text understanding from scratch. CoRR abs/1502.01710 (2015)

Automatic Breast Cancer Cohort Detection from Social Media for Studying Factors Affecting Patient-Centered Outcomes

Mohammed Ali Al-Garadi[1(✉)], Yuan-Chi Yang[1], Sahithi Lakamana[1], Jie Lin[3], Sabrina Li[3], Angel Xie[3], Whitney Hogg-Bremer[1], Mylin Torres[2], Imon Banerjee[1,4], and Abeed Sarker[1]

[1] Department of Biomedical Informatics, School of Medicine, Emory University, Atlanta, GA 30322, USA
{m.a.al-garadi,yuan-chi.yang,slakama,whitney.hogg,imon.banerjee, abeed.sarker}@emory.edu
[2] Department of Radiation Oncology, School of Medicine, Emory University, Atlanta, GA 30322, USA
matorre@emory.edu
[3] Department of Computer Science, College of Arts and Sciences, Emory University, Atlanta, GA 30322, USA
{jie.lin,linyi.li,angel.xie}@emory.edu
[4] Department of Radiology, School of Medicine, Emory University, Atlanta, GA 30322, USA

Abstract. Breast cancer patients often discontinue their long-term treatments, such as hormone therapy, increasing the risk of cancer recurrence. These discontinuations may be caused by adverse patient-centered outcomes (PCOs) due to hormonal drug side effects or other factors. PCOs are not detectable through laboratory tests, and are sparsely documented in electronic health records. Thus, there is a need to explore complementary sources of information for PCOs associated with breast cancer treatments. Social media is a promising resource, but extracting true PCOs from it first requires the accurate detection of real breast cancer patients. We describe a natural language processing (NLP) pipeline for automatically detecting breast cancer patients from Twitter based on their self-reports. The pipeline uses breast cancer-related keywords to collect streaming data from Twitter, applies NLP patterns to filter out noisy posts, and then employs a machine learning classifier trained using manually-annotated data (n = 5,019) for distinguishing firsthand self-reports of breast cancer from other tweets. A classifier based on bidirectional encoder representations from transformers (BERT) showed human-like performance and achieved F_1-score of 0.857 (inter-annotator agreement: 0.845; *Cohen's kappa*) for the positive class, considerably outperforming the next best classifier—a recurrent neural network with bidirectional long short-term memory (F_1-score: 0.670). Qualitative analyses of posts from automatically-detected users revealed discussions about side effects, non-adherence and mental health conditions, illustrating the

© Springer Nature Switzerland AG 2020
M. Michalowski and R. Moskovitch (Eds.): AIME 2020, LNAI 12299, pp. 100–110, 2020.
https://doi.org/10.1007/978-3-030-59137-3_10

feasibility of our social media-based approach for studying breast cancer related PCOs from a large population.

Keywords: Breast cancer · Social media · Natural language processing

1 Introduction

1.1 Background

Women with breast cancer comprise one of the largest groups of surviving cancer patients in high-income countries such as the United States, particularly due to the availability of advanced treatments (*e.g.*, hormone therapy) that have significantly reduced mortality rates. Due to the treatment-driven increased life expectancies of breast cancer patients, their physical and psychological well-being are regarded as important patient-centered outcomes (PCOs), specifically among younger patients. Breast cancer patients often suffer from various treatment-related side effects and other negative outcomes, which range from short-term pain, nausea and fatigue, to lingering psychological dysfunctions such as depression, anxiety, and suicidal tendency. Consequently, up to one-third to half of young breast cancer patients discontinue their treatments, such as endocrine therapy, thus increasing the risk of cancer recurrence and even death [6,7]. In addition, non-adherence to prescribed therapy is associated with poor quality of life, more physician visits and hospitalizations, and longer hospital stays [8].

PCOs, including treatment-related side-effects, are not captured in laboratory or diagnostic tests, but are gathered through patient communications. Sometimes these outcomes are captured as free text in clinical narratives written by caregivers. PCOs documented in this manner, however, are often subject to biases and incompleteness of data in the Electronic Health Records (EHRs). In many cases, PCOs are not documented at all. We demonstrated the under-documentation of PCOs of oncology patients in EHRs in a recent study [2]. Specifically, with the approval of Stanford Institutional Review Board (IRB), we deployed a simple rule-based NLP pipeline for breast cancer, which searched for documentation of physical and mental PCOs affecting patient well-being in EHRs. Physical PCOs (type 1 PCOs) consisted of pain, nausea, hot flush, fatigue, while mental PCOs (type 2) included anxiety, depression and suicidal tendency. On 100 randomly selected clinical notes of breast cancer patients, the model achieved 0.90 F_1-score when validated against manually-labeled ground truth. We applied the validated model on the Stanford breast cancer dataset (Oncoshare), which contains an assortment of clinical notes (*e.g.*, progress notes, oncology notes, discharge summaries, nursing notes) associated with 8,956 women diagnosed with breast cancer from 2008 to 2018. As depicted in Table 1, only 8% of clinical notes and 12% of progress notes contained any documentation (affirm/negation) of PCOs. Importantly, for as many as 30% of breast cancer patients, there were no documented PCOs at any time point at all.

Table 1. Results of patient-centered outcome extraction from clinic notes of Stanford Breast Cancer Cohort (2008–2018).

Data	Total counts	Documentation of type 1 PCO (lymphedema, nausea, fatigue)	Documentation of type 2 PCO (anxiety, depression, suicidal)
SHC (2008–2018) breast cancer patients			
Number of patients	9755	6970	6726
Number of clinical notes	1003210	85039 (8.47%)	82466 (8.22%)
Outpatient progress notes	240486	30219 (12.56%)	29701 (12.35%)
Inpatient progress notes	153915	18714 (12.16%)	15754 (10.23%)
History and physical	21475	3216 (14.97%)	2531 (11.78%)
Consultation note	25557	3824 (14.96%)	3979 (15.57%)
Nursing note	58859	3690 (6.26%)	2404 (4.08%)
Discharge summary	10334	1126 (10.89%)	2404 (23.26%)
Other notes (ED, letters etc.)	492584	24250 (4.92%)	25693 (5.21%)

The under-documentation of PCOs acts as a limiting factor to study the long-term treatment outcomes of young breast cancer patients. Most of the past studies focusing on PCOs have either relied on only small populations of clinical trial patients or analyzed short-term side effects collected during frequent clinic visit periods. Another important limiting factor to understanding the outcomes that matter to patients is that studies focusing on EHRs only capture clinical information, not other relevant factors and patient characteristics that influence their long- and short-term outcomes. Some studies have investigated the feasibility of monitoring patient-reported outcomes (PROs)[1] among oncology patients using sources other than EHRs, such as web portals, mobile applications and automated telephone calls, and their findings suggest that monitoring PROs outside of clinic visits may be more effective and reduce adverse outcomes. However, engaging oncology patients in such routine monitoring activities is extremely resource intensive (expensive) and they only enable the collection of limited information from homogeneous cohorts. Given the under-documentation in EHRs and the expenses associated with conducting patient surveys, there is a need to identify complementary sources of information for PCOs associated with breast cancer patients, and to develop new strategies for capturing diverse patient-level and population-level health-related outcomes.

One promising, albeit challenging, source of information for population-level breast cancer PCOs/PROs is social media. Several studies, including our own, have utilized social media to identify large cohorts of users with common health-related conditions, and then mine relevant longitudinal information about the cohorts using NLP methods. For example, in our past research, we proposed carefully-designed NLP pipelines to discover targeted cohorts, such as pregnant women [10] or nonmedical users of prescription drugs [11] from social media. We have also showed that once many cohort members are detected, it is possible to mine important cohort-specific information from their social media posts (*e.g.*,

---

[1] A major different between PCOs and PROs is that the former may depend on the interpretation of the caregiver, while the latter is not.

medication usage and recovery strategies). For cancer, studies have investigated the role of social media platforms for tasks such as spreading breast cancer awareness, health promotion, and cancer prevention [1,3]. However, to the best of our knowledge, no past research has attempted to accurately detect cancer patients from social media—to build a large cohort and then study long-term cohort-specific information at scale by mining their public posts.

1.2   Objectives

We had the following 3 specific objectives for this study, each dependent on the previous one:

(a) Assess if breast cancer patients discuss personal health-related information on Twitter, including the self-reporting of their positive breast cancer diagnosis/status;
(b) Develop a social media mining pipeline for detecting self-reports of breast cancer using NLP and machine learning methods from Twitter (the primary aim of the paper); and
(c) Gather longitudinal information from the profiles of the automatically-detected users, and qualitatively analyze the information to ascertain if long-term research can be conducted on this cohort.

We refer to the group of breast cancer positive users as a *cohort* although we do not focus on conducting a typical clinical study, with specific start and end dates, in this paper. The purpose of automatically building this cohort is to enable targeted future studies based on the entire set of users or specialized subsets of them.

2   Materials and Methods

2.1   Data and Annotation

We collected data from Twitter using keywords and hashtags via the public streaming application programming interface (API). We used four keywords: (i) cancer, (ii) breastcancer (one word), (iii) tamoxifen, (iv) survivor, and their hashtag equivalents. An inspection of Twitter data retrieved by these keywords showed that while there are many health-related posts from real breast cancer patients, they were hidden within large amounts of noise. Table 2 shows examples of tweets mentioning these keywords, including breast cancer self-reports (category: **S**), and tweets that were not relevant (category: **NR**). We filtered out most of the irrelevant tweets by employing several simple rule- and pattern-matching methods, only keeping tweets that matched the patterns, which were as follows:

– Tweet contains [#]breast & [#]cancer & [#]survivor; OR
– Tweet contains [#]breastcancer & #survivor; OR
– Tweet contains [#]tamoxifen AND ([#]cancer OR [#]survivor)

- Tweet contains a personal pronoun (*e.g.*, 'my', 'I', 'me', 'us') AND [#]breast & [#]cancer

These patterns were developed via a brief manual analysis of Twitter chatter using the website (*i.e.*, the search option). From Table 2, we see that the pattern-based filter does not remove all irrelevant tweets. To fully automate the detection and collection of a Twitter breast cancer cohort, it is necessary to detect self-reports with higher accuracy. Therefore, we employed supervised classification, similar to our past research focusing on Twitter and a pregnancy cohort [10]. We chose a random sample of the pre-filtered tweets for manual annotations. We excluded duplicate tweets, retweets and tweets shorter than 50 characters. Four annotators performed the annotation of tweets, with a random number of overlapping tweets between each pair of annotators. Each tweet was labeled as one of three classes–(i) self-report of breast cancer (**S**), (ii) report of breast cancer of a family member or friend (**F**), or (iii) not relevant (**NR**). We computed pair-wise inter-annotator agreements using Cohen's κ [4]. Since we were only interested in first person self-reports of breast cancer for this study, we combined the classes F and NR for the supervised machine learning experiments.[2]

Table 2. Sample tweets from keyword-based retrieval of data from Twitter. Tweets have been modified to preserve anonymity. '*' - tweet filtered by pattern-matching; '**' - tweet not filtered by pattern-matching (requiring supervised classification).

Tweet	Pattern/keyword match	Category
I am blessed. I know this. As one of the lucky ones, my breast cancer was caught early on. Almost five years ago. @USERNAME URL #survivor #amwriting #writingcommunity #writerlift screenwriters	Breast & cancer & survivor	S
It's damn hard to fight cancer when you cold, hungry & live with constant financial stress	Cancer*	NR
Check out Shelby J's latest single regarding her recent struggle with breast cancer and what sustained her throughout. #Survivor #EarlyDetectionSavesLives #MusicMonday	Breast & cancer & survivor**	NR
Im officially a 16 year breast cancer survivor, mammogram came back all clear no evidence of recurring disease. So grateful	Breast & cancer & survivor	S

2.2   Supervised Classification

We experimented with multiple supervised classification approaches and compared their performances on the same dataset. These approaches were naïve

---

[2] We intend to use information from tweets labeled as **F** in our future studies.

Bayes (NB), random forest (RF), support vector machine (SVM), neural network (NN with solver = lbfgs, hidden layer sizes = (32, 16, 8)), recurrent neural network with bidirectional long short-term memory (BLSTM), and a classifier based on bidirectional encoder representations from transformers (BERT). For the NB, RF, and SVM classifiers, we pre-processed by lowercasing, stemming, removing URLs, usernames, and non-English characters. Following the pre-processing, we converted the text into features: n-grams (contiguous sequences of n words ranging from 1 to 3), and word clusters (a generalized representations of words learned from medication-related chatter collected from Twitter) [12]. For these classifiers, we used *count vector* representations—each tweet is represented as a sparse vector whose length is the size of the entire feature-set/vocabulary and each vector position represents the number of times a specific feature (*e.g.*, a word or bi-gram) appears in the tweet. In addition to being sparse (*i.e.*, most of the vector positions are 0), these count-based representations do not capture word meanings or their similarities. For instance, the terms 'bad' and 'worst' are represented by orthogonal vectors despite being very semantically similar. For the NN and BLSTM (parameters: unit = 100, dropout = 0.2, recurrent dropout = 0.2, optimizer = Adam, epoch = 40) classifiers, we used word embedding based representations (GloVe [9]), which overcome the limitations of n-gram vectors and are able to capture the meanings of words.

Transformer-based approaches, such as BERT, encode contextual semantics at the sentence or word-sequence level, and have vastly improved the state-of-the-art in many NLP tasks [5]. BERT-based classifiers had not been previously used for health cohort detection from Twitter, and in this study, we used the *BERT large* model [5], which consists of 24 layers (transformer blocks), 1024 output dimensions, and 16 attention heads with total of 340M parameters. The tweets are converted into the BERT representations, which capture contextual meanings of character sequences. Following vectorization, a neural network (dense layer) with a softmax activation is used to predict the class of the tweet (S or NR).

2.3   Post-classification Analyses

Following the classification experiments, we conducted manual analyses to (i) study the causes of classification errors, (ii) analyze the associations between training set sizes and classification performances for all classifiers, and (iii) verify if the users detected by the classification approach actually discussed factors that influenced their cancer-related PCOs/PROs on Twitter. For (i) we manually reviewed a sample of the misclassified tweets to identify potential patterns. For (ii), our objective was to assess if the number of tweets required to obtain acceptable classification performance was practical and feasible. We drew stratified samples of the training set consisting of 20%, 40%, 60% and 80% of the set, and computed the F_1-scores over the same test set. For (iii), we collected, via the API, the past posts of a subset of automatically-detected breast cancer positive users, and then qualitatively analyzed them. We used simple string-matching to identify potentially relevant tweets. We discuss the results of these analyses and all the methods described above in the next section.

3 Results

3.1 Annotation and Supervised Classification Results

We annotated a total of 5,019 unique tweets (training: 3,513; validation: 302; evaluation: 1,204). 3,736 (74%) tweets belonged to the NR class (training: 2,615; validation: 225; test: 896) and 1,283 (26%) belonged to the S class (training: 898; validation: 77; test: 308). Micro-average of the pair-wise agreements among all annotators was 0.845 (Cohen's κ) [4], which represents almost perfect agreement [13]. The agreements ranged between 0.806 and 0.907, with a low standard deviation (0.038), which suggests that the annotation/labeling task is relatively straightforward and is not typically subjective to the annotators' interpretations. Table 3 presents the performances of learning models in terms of class-specific recall, precision, F_1-scores, and overall accuracy.

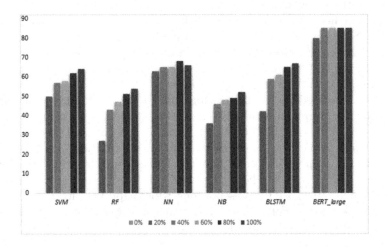

Fig. 1. Classifier performances (F_1-score) at different training set sizes for the different classifiers.

Table 3. Performances of learning models in terms of class-specific recall, precision, F_1-score, and overall accuracy. Best F_1-score on the S class is shown in bold. tables.

Classifiers	Precision (NR)	Precision (S)	Recall (NR)	Recall(S)	F_1-score (NR)	F_1-score (S)	Accuracy
SVM	0.861	0.767	0.941	0.55	0.899	0.646	0.843
RF	0.826	0.849	0.975	0.402	0894	0.546	0.828
NN	0.877	0.701	0.907	0.633	0.892	0.665	0.837
NB	0.953	0.361	0.430	0.938	0.593	0.522	0.560
BLSTM	0.870	0.751	0.930	0.610	0.901	0.670	0.882
BERT large	0.945	0.877	0.959	0.837	0.952	**0.857**	0.928

3.2 Post Classification Analyses Results

Classification Error Analyses: As per our analysis, the possible reasons for misclassifications could be attributed to factors that are common with social media data, primarily the lack of context, ambiguous references, and the use of colloquial language. The following tweets were labeled by the annotator as S, but were misclassified by our best-performing classifier:

Tweet-1: *"we are sisters in this breast cancer club we never wanted to join. bless you my friend. you are an inspiration to all of us."*
Tweet-2: *"when the breast cancer center calls and asks you to donate for the patients' medication and you're just like "i can barely afford my own""*

Classification Performance at Different Training Data Sizes: Figure 1 shows the classifier performances with increments of 20% of the full training set size. From the figure, we see that the BERT-based classifier shows remarkable performance even at small training set sizes. However, the performance of this classifier does not improve further as more training data is added, suggesting that further improving performance will require the incorporation of other strategies.

Content Exploration: We found many informative tweets that covered a wide variety of health-related, and potentially cancer-related, information. Table 4 presents some examples of tweets that were posted by the detected users, and were potentially relevant to the users' PCOs. A number of users reported that they suffered from anxiety/depression, although it was not immediately clear if or how their mental health conditions were related their cancer diagnoses and treatments. Similarly, users reported experiencing or worrying about the side effects of prescribed medications, including Tamoxifen, and their intentions to not adhere to the treatment. These tweets could provide important information about how these patients cope with their treatment and medications, and may complement the information in their EHRs.

4 Discussion

The capability to detect self-reports of breast cancer very accurately is a necessary condition for utilizing Twitter to study PCOs associated with treatments, and our approach has produced promising results. The transformer-based classifier (BERT) is capable of producing performances that far outperform traditional approaches. Thus, our study demonstrates that it is indeed possible to build a large breast cancer cohort from Twitter via an automatic NLP pipeline.

Manual annotation of data is a very time-consuming task and the need to annotate large numbers of samples for supervised classification often act as a barrier to practical deployment. Our experiments show that the BERT-based

Table 4. Sample posts that are relevant to the users' health conditions, collected from the timelines of automatically-detected users. The posts were manually curated and categorized. URLs and emoji's have been removed; usernames have been anonymized.

#	Tweet	Comment
# 1	Sooooo..... the doc put me on an anxiety/anti-depression med the other day (cuz cancer is still a b*tch). She told me to take in the morning. Uh no. I've been asleep for 2 days almost. Taking that joint at night	Mental health issues
# 2	my #mentalhealth suffered unnecessarily and drastically due to #thyroid medications that didn't work for me, for my body. Even when #hypothyroid (on paper) is treated it can make you feel even more unwell. Keep asking for help from new medical professionals until one listens	Mental health issues
# 3	Here we are at my Oncology follow up appointment. I didnt really get on with the tablets prescribed for hot flushes. They made me so sleepy I felt like a zombie and a lower mood than usual so I stopped them. Hopefully get echocardiogram results today too	Side effects, nonadherence intention
# 4	I'm learning something new every day about my #breastcancer. While seeing the oncologist yesterday, I said I know if I stay on my 5 year hormone therapy plan, there is a 9% chance of recurrence. So I asked what if I stop taking the medicine so I no longer have joint pain	Side effects
# 5	New drug today Docetaxel. Not got my usual anti sickness prescribed so I'm feeling quite nervous about how it's going to take me I was vomiting on the EC treatment. But on the positive this is number 5 of 8. #breastcancer #chemotherapy	Side effects
# 6	And Im having a mentally poor day. For all its benefits in preventing #breastcancer recurrence, I think I am going to have to stop taking #Tamoxifen I have a review at the hospital shortly to discuss. Yes, I am grateful that this drug is available but the quality of life is poor	Side effects, nonadherence intention
# 7	Another night, another with lack of sleep. How Im supposed to continue getting by on 3–4 sleep every night is beyond me and definitely contributing to my emotional state of mind. I havent had one night since pre #breastcancer where Ive slept all night #mentalhealth #tamoxifen	Mental health issues, side effects
# 8	The prize for finishing chemo is taking a drug that can cause uterine cancer. #oneroundleft #breastcancer #tamoxifen	Side effects

model overcomes this obstacle, making full automation feasible. We also discovered that it is difficult to raise the performance of this classifier simply by annotating more data. However, including more keywords and patterns for the data collection component of the full pipeline is likely to require additional annotations. Despite the context-incorporating sentence vectors that are used for BERT, the model still lacks the ability to infer meanings that are typically evident to humans. Also, our annotators benefited from implicit knowledge of the topic and additional contextual cues (*e.g.*, from the users' profiles), which the transformer-based model is not able to capture. In the future, it will be important to study how such implicit information may be encoded in numeric vectors.

5   Conclusion

We investigated the potential of using Twitter as a resource for studying PCOs associated with breast cancer treatment by studying information posted directly by patients. We particularly focused on (i) assessing if breast cancer patients discuss health-related information on Twitter, including the self-reporting of their positive breast cancer status; (ii) developing a NLP-based social media mining pipeline for detecting self-reports via supervised classification; and (iii) analyzing health-related longitudinal information of automatically-detected users. The best-performing classifier achieves human-like performance with an F_1-score of 0.857 on the positive class. Qualitative analyses of the tweets retrieved from the users' profiles revealed that they contain information relevant to PCOs, such as mental health issues, side effects of medications, and medication adherence. These findings verify the potential value of social media for studying PCOs that are rarely captured in EHRs. Our future work will focus on collecting large samples of breast cancer patients from Twitter using the methods described, and then implementing further NLP-based methods for studying breast cancer related PCOs from a large cohort.

References

1. Attai, D.J., Cowher, M.S., Al-Hamadani, M., Schoger, J.M., Staley, A.C., Landercasper, J.: Twitter social media is an effective tool for breast cancer patient education and support: patient-reported outcomes by survey. J. Med. Internet Res. **17**(7), e188 (2015)
2. Banerjee, I., Bozkurt, S., Caswell-Jin, J.L., Kurian, A.W., Rubin, D.L.: Natural language processing approaches to detect the timeline of metastatic recurrence of breast cancer. JCO Clin. Cancer Inform. **3**, 1–12 (2019)
3. Bottorff, J.L., Struik, L.L., Bissell, L.J., Graham, R., Stevens, J., Richardson, C.G.: A social media approach to inform youth about breast cancer and smoking: an exploratory descriptive study. Collegian **21**(2), 159–168 (2014)
4. Cohen, J.: A coefficient of agreement for nominal scales. Educ. Psychol. Measur. **20**(1), 37–46 (1960)
5. Devlin, J., Chang, M.W., Lee, K., Toutanova, K.: BERT: pre-training of deep bidirectional transformers for language understanding. arXiv preprint arXiv:1810.04805 (2018)
6. van Herk-Sukel, M.P.P., van de Poll-Franse, L.V., Voogd, A.C., Nieuwenhuijzen, G.A.P., Coebergh, J.W.W., Herings, R.M.C.: Half of breast cancer patients discontinue tamoxifen and any endocrine treatment before the end of the recommended treatment period of 5 years: a population-based analysis. Breast Cancer Res. Treat. **122**(3), 843–851 (2010). https://doi.org/10.1007/s10549-009-0724-3
7. McCowan, C., et al.: Cohort study examining tamoxifen adherence and its relationship to mortality in women with breast cancer. British J. Cancer **99**(11), 1763–1768 (2008). https://doi.org/10.1038/sj.bjc.6604758
8. Milata, J.L., Otte, J.L., Carpenter, J.S.: Oral endocrine therapy nonadherence, adverse effects, decisional support, and decisional needs in women with breast cancer. Cancer Nurs. **41**(1), E9–E18 (2018). https://doi.org/10.1097/NCC. 0000000000000430

9. Pennington, J., Socher, R., Manning, C.D.: Glove: global vectors for word representation. In: Proceedings of the 2014 Conference on Empirical Methods in Natural Language Processing (EMNLP), pp. 1532–1543 (2014)
10. Sarker, A., Chandrashekar, P., Magge, A., Cai, H., Klein, A., Gonzalez, G.: Discovering cohorts of pregnant women from social media for safety surveillance and analysis. J. Med. Internet Res. **19**(10), e361 (2017)
11. Sarker, A., DeRoos, A., Perrone, J.: Mining social media for prescription medication abuse monitoring: a review and proposal for a data-centric framework. J. Am. Med. Inform. Assoc. **27**(2), 315–329 (2019). https://doi.org/10.1093/jamia/ocz162
12. Sarker, A., Gonzalez, G.: A corpus for mining drug-related knowledge from Twitter chatter: language models and their utilities. Data Brief **10**, 122–131 (2017)
13. Viera, A.J., Garrett, J.M., et al.: Understanding interobserver agreement: the kappa statistic. Fam. Med. **37**(5), 360–363 (2005)

Predicting Clinical Diagnosis from Patients Electronic Health Records Using BERT-Based Neural Networks

Pavel Blinov[1](✉), Manvel Avetisian[1], Vladimir Kokh[1], Dmitry Umerenkov[1], and Alexander Tuzhilin[2]

[1] Sberbank Artificial Intelligence Laboratory, Moscow, Russia
{Blinov.P.D,Avetisyan.M.S,Kokh.V.N,Umerenkov.D.E}@sberbank.ru
[2] New York University, New York, USA
atuzhili@stern.nyu.edu

Abstract. In this paper we study the problem of predicting clinical diagnoses from textual Electronic Health Records (EHR) data. We show the importance of this problem in medical community and present comprehensive historical review of the problem and proposed methods. As the main scientific contributions we present a modification of Bidirectional Encoder Representations from Transformers (BERT) model for sequence classification that implements a novel way of Fully-Connected (FC) layer composition and a BERT model pretrained only on domain data. To empirically validate our model, we use a large-scale Russian EHR dataset consisting of about 4 million unique patient visits. This is the largest such study for the Russian language and one of the largest globally. We performed a number of comparative experiments with other text representation models on the task of multiclass classification for 265 disease subset of ICD-10. The experiments demonstrate improved performance of our models compared to other baselines, including a fine-tuned Russian BERT (RuBERT) variant. We also show comparable performance of our model with a panel of experienced medical experts. This allows us to hope that implementation of this system will reduce misdiagnosis.

Keywords: Electronic Health Records · EHR · ICD-10 · Multiclass classification · Natural language processing · Text embedding · BERT

1 Introduction

The process of digital transformation in medicine has been going for a while, providing faster and better treatment results in many cases through the use of modern computer science and Artificial Intelligence (AI) methods [1]. Digitization and subsequent analysis of medical records constitutes one such area of digital transformation that aims to collect broad types of medical information about a patient in the form of EHR, including digital measurements (laboratory

© Springer Nature Switzerland AG 2020
M. Michalowski and R. Moskovitch (Eds.): AIME 2020, LNAI 12299, pp. 111–121, 2020.
https://doi.org/10.1007/978-3-030-59137-3_11

results), verbal descriptions (symptoms and notes, life and disease anamnesis), images (X-Ray, CT and MRI scans, etc.) and document the treatment process of a patient.

In this paper, we focus on the analysis of EHR with the purpose of providing clinical decision support by predicting most probable diagnoses during a patient's visit to a doctor. This problem is complicated by abundance of large volumes of structured and unstructured medical information stored across multiple systems in different data formats that are often incompatible across these systems. Although there exists an emerging FHIR standard (Fast Healthcare Interoperability Resources) for the EHR data [2] the goal of which is to unify the process of storing and exchanging medical information, unfortunately, very few existing Hospital Information Systems (HISes) support it. All this complicates the task of diagnosis prediction based on the EHRs since many of them contain extensive amounts of unstructured, poorly organized and "dirty" data that is less amenable to the analysis using the AI-based methods, unless this data is cleaned and preprocessed appropriately.

Providing clinical decision support in diagnosis prediction during a patient's visit to a doctor is important because many patient's visits, in fact up to 30% in the US, are misdiagnosed [3]. This is also true in some other countries [4]. We formulate the aforementioned clinical decision support problem as a multi-label text classification of clinical notes (anamnesis and stated symptoms) during a patient visit, where the classification is performed for a wide range of diagnosis codes represented by the International Statistical Classification of Diseases (ICD-10) [5].

In this paper, we make the following contributions. First, we propose a novel BERT-based model for classification of textual clinical notes, called *RuPool-BERT*, that differs from the previously proposed models by the way of the FC-layer composition that is described in Sect. 4. Second, we compare the performance of our method with various baselines across different text representation techniques and classification models. Third, we compare the performance of the BERT model pretrained on a large corpus of out-of-domain data [6] with the BERT model pretrained exclusively on in-domain data and using an in-domain tokenizer. Finally, we demonstrate the advantage of the proposed models and their comparable results with a human baseline in Sect. 5.

It is important to note that the clinical decision support system described in the paper will *not* serve as a doctor's replacement but, rather, constitutes an unbiased intelligent diagnosis generator and, therefore, should only *assist* the doctors in their diagnostic decisions.

2   Related Work

There have been many approaches to the analysis of the EHR-data and predicting the diagnosis codes (ICDs) proposed in the literature that are related to our work. In particular, papers [7,8] address the task of diagnoses prediction by using the entire patient history. In our study we explicitly do not take history into account because at the current stage of our project we focus on isolated visits; however, the history is partially accessible for the model due to the avail- ability of the anamnesis field. The level of granularity for ICD-code in paper [7] is quite similar with ours. We also mainly operate on the second level of ICD- 10 classification code hierarchy, but intentionally restrict the number of classes up to 265 (see details in the Data Sect. 3).

The authors of [9] developed a Deep Neural Network-based (DNN) diagnosis prediction algorithm, but only for the pediatric EHRs. In contrast to this, we do not restrict ourselves to any age specifications. Instead, our training set fully covers all the age groups, including having about 20% of the training data visits being children under the age of 14 years. The papers [8–10] proposed methods based on the DNN (mostly its recurrent variants), while in this paper we focus on more recent state-of-the-art transformer-based neural architectures.

In the recent couple of years a new class of neural architectures called trans- formers were developed, which allow to significantly improve performance on the whole range of NLP tasks (e. q. question answering, named entity recognition, sentiment classification etc.). The most known member of this family is the lan- guage representation model called Bidirectional Encoder Representations from Transformers (BERT) [11]. Note that the BERT-like models have been applied to the EHR-data before [12,13]. In particular, [12] presents a system of assign- ing ICD-10 codes to non-technical textual summaries of medical experiments. The original experiment dataset (about 8,000 samples) was in German; but the authors achieved significant performance improvement (more than 6% for the F1-measure) by translating it into English and then applying the BERT model to the translated EHRs. This shows, among other things, that each language has its own linguistic and cultural idiosyncrasies and that there are 'easier' vs. 'harder' languages for machine learning models. In this paper we do not use such a "translation trick" and work directly with the original language.

Fei Li et al. [13] investigated the problem of BERT model fine-tuning for biomedical and clinical entity normalization. The training EHR notes used in that paper have millions of entries and in that way are similar to our study in its ability to handle large EHR datasets. Another significant advantage of this paper is the comparative analysis between pretrained general and biological domain BERT models. In our work, we also use a general Russian pretrained BERT model [6]. Since the lexical and the syntactic structure of a special domain languages can be very different from the general one, we experimented with an in-domain tokenizer and the model trained from scratch on the Russian EHR data.

Most of the EHR research focuses on the English language based elec- tronic health records. Even [12] translated their health records from German into English to leverage the power of the previously conducted English-based research. In contrast to this, we think that it is crucial to successfully apply artificial intelligence and NLP-based methods to other languages, which is done only occasionally. For example, the amount of the EHR-related studies is very limited for the Russian language. In fact we could find only one such recent paper [14], where the authors studied a related classification problem, but only for four ICD codes (D50, E11, E74, E78) and on a small dataset having about 8,000 cases where they used a gradient boosting algorithm on a laboratory tests data as input.

In this paper, we follow this principle and analyze electronic health records of medical patients written in Russian. Unlike several other studies conducted on medium-size EHR datasets, such as the Medical Information Mart for Intensive Care (MIMIC) dataset [15] containing about 60,000 intensive care unit admissions [8,10,16], we work with a large dataset of about 4,000,000 patients' visits to various clinics in Russia.

Finally, the authors in [13] and [6] use the conventional BERT method (and use the output of the classification token (CLS) for the upstream tasks). In this paper we show that this is not an optimal solution since the performance of the BERT model can be further improved by the extension of this layer. Therefore, we venture beyond the conventional BERT architecture and propose a modification to it that we call RuPool-BERT. Furthermore, we show that this extension outperforms the conventional BERT approaches for our classification problem.

3   Data

In this project, we worked with three real-world anonymized datasets containing information about patients' visits to the networks of clinics in Russia. The first two datasets (we call them DataN and DataM in the paper) pertain to two large private networks of clinics and the third one (we call it DataT) pertains to the network of public clinics. We used only the symptom and anamnesis fields for each patient visit since all other fields where substantially different across the datasets. All the relevant data was concatenated into a single textual content field (since it was initially stored in different formats across different datasets). We did not apply any special preprocessing to this field, and therefore it was presented to the model as a raw text including typos, abbreviations and misspellings. The main statistics for each dataset are summarized in Table 1.

Table 1. Statistics of the datasets.

Dataset name	Split	# of patients	# of visits	Avg # of visits per patient	From	To
DataN	Train/Validation	251,763	1,685,253	6.69	2005-01	2018-12
DataM	Train/Validation	177,715	563,106	3.17	2013-01	2019-06
DataT	Test	694,063	1,728,529	2.5	2014-01	2019-10

As Table 1 demonstrates, the three datasets collectively have 3,976,888 visits of 1,123,541 patients over the period of almost 15-years. To the best of our knowledge, this is the largest such study (in terms of the number of cases and time duration) for the Russian language and one of the largest globally[1].

Patient visits were split into train, validation and test sets. The whole DataT part was assigned to the test set, and the union of DataN and DataM sets was randomly split in the 80/20 proportion to make the train and the validation sets. The final cardinalities of the train, validation and test sets were 1,798,687 (45.23%), 449,672 (11.3%) and 1,728,529 (43.47%) respectively. The validation set was used exclusively to fine-tune hyperparameters for the baseline BERT model. The test set was used to compare different baselines with the proposed models. The reason we decided to keep the entire DataT only as the test set lies in the more reliable nature of this data. More specifically, for each visit in DataT, we have a confirmed diagnosis.

The full spectrum of ICD-10 consists of 71,932 codes arranged in a hierarchical manner. A single code can be represented by 3 to 7 characters depending on the level of disease specification. Internally, an ICD code has 3 part structure. The first part, up to the dot, represents a distinct disease. For example, D30 is the code for the neoplasm of urinary organs. After the dot follows potential specifying elements (e. g. D30.0 represents neoplasm of a kidney, D30.01 – neoplasm of the right kidney, D30.02 – neoplasm of the left kidney, etc.).

We selected $K = 265$ categories of codes for this study because such number of codes (diseases) is enough to cover up to 95% of all the cases in the training set. As with many other natural distributions, distribution of diagnoses is very skewed, i.e., the first 19 codes accounting for 50% of all the cases in the training set (with J06 – 11.5%, I11 – 7.4%, E11 – 4.3%, M42 – 3.9% and so on up to D72 – 0.03% and L40 – 0.03%). We have also experimented with $K = 1,000$ codes. This selection of codes were obtained from the same 265 codes by extending them with available second ICD parts (e.g. J06.0, J06.8, J06.9, etc.) and selecting the most frequent 1,000 of them. This second case of $K = 1000$ codes significantly complicates the classification task since it requires to predict more fine-grained diagnoses.

4   Proposed Models

In this section we present our classification model that predicts the 265 ICD codes based only on textual information of the patient visits to the clinics. We solve the classification problem by using a transformer-based BERT model [11] architecture presented in Fig. 1 (referred as RuPool-BERT hereafter). The inputs to the model constitute text from the symptoms and the anamnesis fields of the patients visit to a doctor that are concatenated into a single text sequence. Most of the transformer-based models use the approach proposed in [17] to represent raw input sequences in terms of sub-words (tokens) and keep balance between character and word information. That allows to naturally process out of vocabulary words (typos, misspellings, etc.). The distribution of sequence

---

[1] Although the data does not contain any personal information, we cannot publicly release it due to certain legal restrictions.

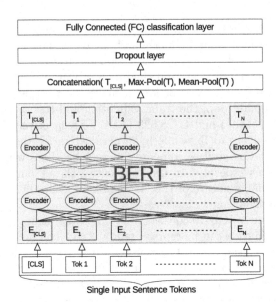

Fig. 1. Architecture of RuPool-BERT model.

lengths (for the train set) showed that the mean number of tokens is around 79 and median value is about 57. Therefore, we decided to allow some margin by limiting each sequence to $N = 128$ tokens.

Each input token is represented by learnable H-dimensional vector called *embedding* (which in turn is the sum of three embeddings: token, segment and positional). Therefore, the text tokenizer and the trained model are linked together. As a base tokenizer (with the vocabulary of approximately 120k tokens) and a model, we used RuBERT (architecturally the same as base BERT model for the English: 12 transformer block layers, hidden size $H = 768$ and 12 self-attention heads) [6] because it significantly outperformed the multilingual variant of BERT, as was shown in [6].

The gray part of the model presented in Fig. 1 is the same as in [11], which methodically explains this part of the network architecture and the process of pretraining. The authors of [11] specifically designed the CLS token as classification one, and the linear fully-connected layer is added to the last hidden state C: $T_{[CLS]} \in R^H$. We propose to concatenate this state with two additional parts, namely max and mean pooling over the whole last encoder states T_i along the sequence dimension. Both operations also return embeddings from R^H. With a new hidden state vector $C \in R^{3H}$ and fully-connected classification layer with weights $W \in R^{K3H}$, where $K = 265$ – number of ICD codes, the diseases probabilities after applying softmax function is then $P = softmax(CW^T)$. To prevent overfitting, the dropout operation was applied to the C layer.

To fully leverage the difference in vocabulary between general texts and medical records, we trained a tokenizer with a vocabulary of 40k tokens on all the texts from DataN dataset. The tokenizer is identical to the one used in the original BERT model with the difference coming from the training data word distribution. Our expectation was that such a medical-domain tokenizer will allow the model to capture a wider range of medical linguistics phenomena. Using this tokenizer, we pretrained the BERT model with masked language modelling task on data from DataN dataset containing about 1.7 million records of patients visits, which took us about 2 weeks on a Tesla K40 GPU. This pretrained model with a standard pooling scheme (referred RuEHR-BERT hereafter) was fine-tuned for disease classification. RuEHR-BERT can be directly compared with the RuBERT model which has the same architecture but a tokenizer trained on general Russian language texts.

5   Experiments

To evaluate the performance of the proposed model, we compared it with the following baselines: an RNN model (with GRU units), the FastText model [18] and the multilingual Universal Sentence Encoder (USE) [19]. We focused on these baselines because it was feasible to do the direct comparisons with them, while comparing our method with several others was either impossible or very hard to achieve due to accessibility, language incompatibility and other practical reasons.

To remove the effect of hyperparameter tuning and compare all the BERT models under the same conditions, hyperparameters were kept the same across all these models. First, we find the best set of parameters (with respect to the validation data) for the baseline RuBERT model and used them thereafter. We report our results after 5 training epochs with $batch\ size = 128$, optimizer being AdamW and the starting learning rate of 3×10^{-5} with the Binary Cross Entropy loss. By fine-tuning the hyperparameters for our proposed models on the validation set the results could be improved further.

We considered the following performance metrics in our study: the macro and weighted variants of the F1-measure [20] and such ranking measures as Mean Reciprocal Rank (MRR) and Hit@k [20] (i.e., Hit@1, Hit@3, Hit@5 and Hit@10). In cooperation with medical experts we empirically selected Hit@3 as our primary metric.

Table 2 summarizes the performance results of all the considered models on the test set. Note that a large performance gap between the $F1_{macro}$ and $F1_{weighted}$ measures across all the baselines reflects great disbalance in the class distribution (that was discussed in Sect. 3). The significant increase in $F1_{macro}$ for the proposed models (4%) combined with increase in $F1_{weighted}$ (1%) compared to baseline RuBERT model show that our proposed models performance gains are due to better classification of less frequent disease classes.

Also, we tried to train the 1,000 class ICD code classifier for our RuPool-BERT model (denoted as "RuPool-BERT 1k" in Table 2). Note that, although the number of classes increased almost by the factor of four for the RuPool-BERT 1k model, the performance measured by the Hit@3 metric degrades significantly slower (42.97 vs. 70.14). As Table 2 demonstrates, RuPool-BERT 265 and RuEHR-BERT 265 show better performance than other tested models in terms of Hit@k and other measures.

Table 2. Models performance (%) on the test set.

Model name	$F1_{macro}$	$F1_{weighted}$	MRR	Hit@1	Hit@3	Hit@5	Hit@10
USE	19.54	39.59	54.69	41.25	62.62	70.66	79.89
RNN	24.33	43.34	58.02	44.17	66.97	74.79	83.37
fastText	24.49	44.19	59.00	45.27	67.81	75.84	84.54
RNN+FastText	25.44	44.00	59.11	45.05	68.43	76.22	84.71
RuBERT 265	25.78	46.34	60.60	47.29	69.50	76.96	84.79
RuPool-BERT 265	29.83	47.13	61.04	47.54	70.14	77.53	85.49
RuEHR-BERT 265	28.61	46.84	61.01	47.51	70.00	77.61	87.76
RuPool-BERT 1k	8.95	24.03	37.13	25.94	42.97	50.51	59.46

Furthermore, we studied the dependence between the input text (symptoms and anamnesis) lengths and the Hit@3 metric by computing the metric for clinical notes with different input lengths. The results are presented in Fig. 2, where the text length is plotted on the x-axis and the number of test samples on the y-axis (for the black solid curve). Moreover, the red dashed and the blue dotted curves in Fig. 2 show how the Hit@3 metric depends on the text length for the 265- and 1k-classification models respectively.

We can conclude from Fig. 2 that the most reliable results of the model can be obtained for the cases where the input text sequence has at least 20 tokens, which constitutes 72.25% of all the test visits. This is an important observation because it indicates when the results of our model can be shared with the doctors in practical clinical settings in the form of the "second opinion", i.e. it can be done only for the more extensively documented cases having at least 20 tokens.

Additionally, we compared the performance of our proposed models with experienced physicians in the diagnosis task described in Sect. 1. In particular, we filtered out test set according to above observation and randomly selected 530 visits ($K * 2 = 265 * 2$) for the further markup process. We invited a panel of 7 medical experts (each with more than 10 years of practical experience) to participate in this study. Each of them was presented with these 530 visits and asked to point out up to 3 appropriate ICD codes (from $K = 265$) for each case or reject it if the information to make a decision is insufficient. For the sake of a fair performance comparison, we asked the clinicians to make the decisions exclusively based on the EHR data (text from anamnesis and symptom fields) without any personal communications with the patients or any other parties. To assess reliability of agreement between the raters we computed Fleiss' kappa coefficient (among 1st place answers) $k = 0.37$ which corresponds to fair strength of agreement [21]. There are only 6 cases with maximum disagreement (7 experts put a different answer on the first place). We invited an independent physician to

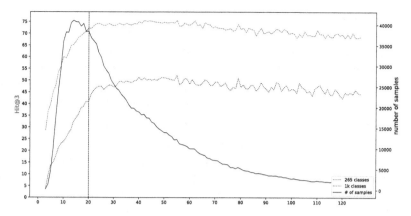

Fig. 2. Distribution of number of samples in the test set (solid black) and RuPool-BERT model Hit@3 metric (%) (dashed red for 265 and dotted blue for 1k classes). (Color figure online)

analyze these cases for the cause of such inconsistency. The physician concluded that all the answers can be treated as relevant given the provided information. The only difference come from diagnosis ordering: each expert select reasonable diagnosis set, but order them based on their primary specialization, experience and other subjective factors that influence the decision making. We assessed Hit@3 metric for each of the 7 experts by inferring the ground truth diagnoses from the panel of 6 experts' answers and comparing them with the remaining expert answers (thus deploying one-against-all labelling strategy [22]). It turned out that the Hit@3 metric varied quite a bit across the clinicians. The worst performing expert reached the value of 57.89% and the best one reached the value of 72.52%, with the mean Hit@3 value for the experts being 68.16% and standard deviation of 4.82%. Our best performing DL-based model (RuPool-BERT 265) against the same 7 panels achieved mean Hit@3 value of 69.15% with standard deviation of 8.52%. To test whether the difference between Hit@3 performance metrics of the models and the experts are significant, we applied the Mann–Whitney U test. Our null hypothesis is that the expert and the model metrics are the same, and the alternative is that there is a significant difference in the metrics. The computed empirical value is $U_{emp} = 22$ and the critical value for the sample size $n = 7$ at the 0.05 significance level is $U_{crit} = 11$. Thus, we fail to reject the null hypothesis and cannot claim that our model is better than humans. Further, based on the model and expert mean metrics, we can conclude that our proposed model shows the performance results comparable to the performance of the medical experts.

6   Conclusions

In this paper we described the challenging and important problem of diagnoses prediction from unstructured real-world clinical text data based on a very large Russian EHR dataset containing about 4 million doctor's visits of over 1 million patients. To provide the diagnosis, we proposed a novel BERT-based model for classification of textual clinical notes, called RuPool-BERT, that differs from the others BERT-based approaches by introducing a novel way of the FC-layer composition. Our experiments of applying the developed prediction model to the practical task of classifying 265 diseases showed the advantage of this model compared to the fine-tuned RuBERT base analog and other text representation models. We also showed that using a BERT model with a vocabulary and pretraining dataset tailored to the medical texts representation (RuEHR-BERT) improves performance on the classification task, specially on less frequent diseases. This improvement is achieved at a small fraction of pretraining time compared to the general Russian language model (2 weeks of Tesla K40 for RuEHR-BERT vs 8 weeks of Tesla P100 for RuBERT).

Comparison of our model with a panel of medical experts showed that the results of our model were similar to the results of experts in terms of the Hit@3 performance measure. Furthermore, we showed that the most reliable performance of our system is achieved on those samples having longer textual inputs, i.e. text sequences having at least 20 input tokens. All this allows us to conclude that our model and system has a strong potential to help doctors with disease diagnosis by providing the "second opinions" to them.

Our partners in the medical community identified one issue with the proposed method: they maintain that the proposed approach would benefit greatly from clear explanations of how our method arrived at each particular diagnoses. This is the topic of future research on which we plan to focus in the immediate future.

References

1. Atasoy, H., Greenwood, B.N., McCullough, J.S.: The digitization of patient care: a review of the effects of electronic health records on health care quality and utilization. Ann. Rev. Public Health **40**(5), 487–500 (2019)
2. Fast Healthcare Interoperability Resources (FHIR). https://www.hl7.org/fhir/. Accessed 20 Apr 2020
3. Liberman, A.L., Newman-Toker, D.E.: Symptom-disease pair analysis of diagnostic error (SPADE): a conceptual framework and methodological approach for unearthing misdiagnosis-related harms using big data. BMJ Qual. Saf. **27**(7), 557–566 (2018)
4. Vardanyan, G.J., et al.: Medical errors: modern condition of the problem. Med. Sci. Armenia **59**(4), 105–120 (2019)
5. World Health Organization: International Statistical Classification of Diseases and Related Health Problems. 10th revision, Fifth edition (2016)
6. Kuratov, Y., Arkhipov, M.: Adaptation of deep bidirectional multilingual transformers for Russian language. arXiv preprint arXiv:1905.07213 (2019)

7. Vasiljeva, I., Arandjelovic, O.: Diagnosis prediction from electronic health records (EHR) using the binary diagnosis history vector representation. J. Comput. Biol. **24**(8), 767–786 (2017)
8. Ma, F., Chitta, R., Zhou, J., You, Q., Sun, T., Gao, J.: Dipole: diagnosis prediction in healthcare via attention-based bidirectional recurrent neural networks. In: Proceedings of the 23rd ACM SIGKDD International Conference on Knowledge Discovery and Data Mining, pp. 1903–1911 (2017). https://doi.org/10.1145/3097983.3098088
9. Shi, J., Fan, X., Wu, J., et al.: DeepDiagnosis: DNN-based diagnosis prediction from pediatric big healthcare data. In: Sixth International Conference on Advanced Cloud and Big Data (CBD), pp. 287–292. IEEE (2018)
10. Qiao, Z., Wu, X., Ge, S., Fan, W.: MNN: multimodal attentional neural networks for diagnosis prediction. In: Proceedings of the 28th International Joint Conference on Artifcial Intelligence, pp. 5937–5943. AAAI Press (2019)
11. Devlin, J., Chang, M.-W., Lee, K., et al.: Bert: pretraining of deep bidirectional transformers for language understanding. arXiv preprint arXiv:1810.04805 (2018)
12. Amin, S., Neumann, G., Dunfield, K., Vechkaeva, A., Chapman, K., Wixted, M.: MLT-DFKI at CLEF eHealth 2019: multi-label classification of ICD-10 codes with BERT. In: Conference and Labs of the Evaluation Forum (Working Notes) (2019)
13. Li, F., Jin, Y., Liu, W., et al.: Fine-tuning bidirectional encoder representations from transformers (BERT)-based models on large-scale electronic health record notes: an empirical study. JMIR Med. Inform. **7**(3), e14830 (2019). https://doi.org/10.2196/14830
14. Sakhibgareeva, M.V., Zaozersky, A.Y.: Developing an artificial intelligence-based system for medical prediction. Bull. Russ. State Med. Univ. **6**, 42–46 (2017). https://doi.org/10.24075/brsmu.2017-06-07
15. Johnson, A.E.W., et al.: MIMIC-III, a freely accessible critical care database. Sci. Data **3**, 160035 (2016). https://doi.org/10.1038/sdata.2016.35
16. Malakouti, S., Hauskrecht, M.: Predicting patient's diagnoses and diagnostic categories from clinical-events in EHR data. In: Riaño, D., Wilk, S., ten Teije, A. (eds.) AIME 2019. LNCS (LNAI), vol. 11526, pp. 125–130. Springer, Cham (2019). https://doi.org/10.1007/978-3-030-21642-9_17
17. Wu, Y., Schuster, M., Chen, Z., et al.: Google's neural machine translation system: bridging the gap between human and machine translation. arXiv preprint arXiv:1609.08144 (2016)
18. Bojanowski, P., Grave, E., Joulin, A., Mikolov, T.: Enriching word vectors with subword information. Trans. Assoc. Comput. Linguist. **5**, 135–146 (2017)
19. Cer, D., Yang, Y., et al.: Universal sentence encoder, CoRR. arXiv preprint arXiv:1803.11175 (2018)
20. Manning, C., Raghavan, P., Schutze, H.: Introduction to Information Retrieval. Cambridge University Press, Cambridge (2008)
21. Landis, J.R., Koch, G.G.: The measurement of observer agreement for categorical data. Biometrics **33**(1), 159–174 (1977)
22. Bishop, C.M.: Pattern Recognition and Machine Learning. Springer, Heidelberg (2006)

Drug-Drug Interaction Prediction on a Biomedical Literature Knowledge Graph

Konstantinos Bougiatiotis[1,2]([⊠]), Fotis Aisopos[1], Anastasios Nentidis[1,3],
Anastasia Krithara[1], and Georgios Paliouras[1]

[1] Institute of Informatics and Telecommunications,
National Center Scientific Research Demokritos, Athens, Greece
kbogas@di.uoa.gr,
{fotis.aisopos,tasosnent,akrithara,paliourg}@iit.demokritos.gr
[2] Department of Informatics and Telecommunications,
National and Kapodistrian University of Athens, Athens, Greece
[3] School of Informatics, Aristotle University of Thessaloniki, Thessaloniki, Greece

Abstract. Knowledge Graphs provide insights from data extracted in various domains. In this paper, we present an approach discovering probable drug-to-drug interactions, through the generation of a Knowledge Graph from disease-specific literature. The Graph is generated using natural language processing and semantic indexing of biomedical publications and open resources. The semantic paths connecting different drugs in the Graph are extracted and aggregated into feature vectors representing drug pairs. A classifier is trained on known interactions, extracted from a manually curated drug database used as a golden standard, and discovers new possible interacting pairs. We evaluate this approach on two use cases, Alzheimer's Disease and Lung Cancer. Our system is shown to outperform competing graph embedding approaches, while also identifying new drug-drug interactions that are validated retrospectively.

Keywords: Literature mining · Knowledge graph · Path analysis · Knowledge discovery · Drug-drug interactions

1 Introduction

Drug-Drug Interactions (DDIs) often occur in cases of simultaneous administration of multiple drugs. This results in high risks for patient safety, seriously affecting the biological action of implicated drugs and may cause various adverse effects. The extent of the problem becomes more evident given that in the United States alone, DDIs are responsible for up to 195,000 hospital admissions [11].

However, the prognosis of such effects can pose a serious challenge due to the absence of sufficient clinical data and knowledge. Thus, automated software solutions that discover potential drug interactions can be valuable tools to improve health care and help pharmacovigilance. Many approaches try to address this

© Springer Nature Switzerland AG 2020
M. Michalowski and R. Moskovitch (Eds.): AIME 2020, LNAI 12299, pp. 122–132, 2020.
https://doi.org/10.1007/978-3-030-59137-3_12

by assessing structural or other kinds of drug similarities, based mainly on targets, pathways and transporters [14]. However, most of these approaches fail to capture and combine information from heterogeneous sources of data which are important to address the complexity of the task.

The current paper proposes a holistic framework towards DDI prediction, based on a Biomedical Literature Knowledge Graph (DDI-BLKG)[1]. In the proposed framework, we extract knowledge items from biomedical publications and manually curated databases, using automated Natural Language Processing (NLP) tools. The results are integrated in a disease-specific Knowledge Graph (KG). Then, a human-curated drug database is used to train a classifier that identifies patterns of interactions between drug pairs. As features for the patterns, the classifier uses the semantic relations in the paths connecting interacting drugs. We showcase the usefulness of our approach by testing it on drug interactions for two prevalent diseases: Alzheimer's Disease (AD) and Lung Cancer (LC). Our experiments show that the proposed approach achieves better results than other graph embedding techniques on the same task. Moreover, through a small-scale qualitative evaluation we showcase the predictive potential of the method and its usefulness in providing novel DDIs.

Overall, the main contributions of this work correspond to the following:

- We present an automated DDI prediction approach, utilizing a disease-specific biomedical literature Knowledge Graph.
- We propose the use of the semantic relations connecting different drugs in the literature, as features for the DDIs.
- We make available for further experimentation two real-world disease-specific KGs, related to Alzheimer's Disease (AD) and Lung Cancer (LC) respectively, alongside the probable DDIs predicted by our model.

2   Related Work

Various existing approaches aim to extract associations and identify relations between biomedical entities directly from text [1,12]. However, in order to extend over the narrow scope of a sentence that rarely contains all the information needed, one needs to combine multiple sources of information. Such an example is the method proposed in [5], which builds a heterogeneous network and performs link prediction to construct an integrative model of drug efficacy.

Most relevant to our approach is the work in [2,17], presenting drug discovery methods, based on biomedical knowledge graphs. The former method focuses on treatment and causative relations exploiting connections of biomedical entities as found in literature. The latter publication presents SemaTyP, a method for discovering drug-disease relations based on a literature Knowledge Graph. Its successor, GrEDeL [16] extends the previous model by employing graph embedding techniques and deep learning approaches.

---

[1] https://github.com/kbogas/DDI_BLKG.

During the last years, there have been many other approaches to utilize graph embeddings for DDIs. Authors in [18] present KMR, a procedure for similarity computation based on chemical structure and side effects. Shtar et al. [19] employ adjacency matrix factorization to embed the drugs based on their interactivity as derived by DrugBank. Finally, authors in [6] also construct a biomedical Knowledge Graph from structured data (i.e. DrugBank, PharmGKB and KEGG databases), but with a limited set of drug-related nodes and relations.

Our work is mainly compared against competing methods employing knowledge graph embeddings, rather than existing biochemical or traditional DDI approaches, as it combines multiple sources of information with the latest scientific knowledge entailed in a literature knowledge graph, to provide a fast and automated DDI prediction solution. We aim at illustrating that a disease-specific approach, analysing all available meta-information related to drugs in a multi-type knowledge graph, could provide significant benefits.

3   Approach

3.1   General Workflow

Our approach to the creation of the biomedical literature Knowledge Graph and the development of a link prediction model consists of a sequence of distinct steps as shown in Fig. 1. In the following sections, we will discuss each one in detail, leading to the predictive model to be validated.

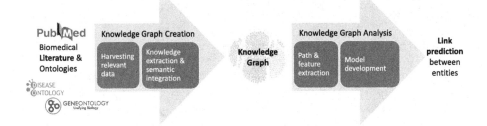

Fig. 1. Overview of the steps for the knowledge graph creation and link prediction.

3.2   Knowledge Graph Creation

First, a disease-specific KG is created based on disease-related biomedical literature and structured resources. To this end, we utilize the project iASiS Open Data Graph [8] generation pipeline. According to this, biomedical articles related to the disease of interest are fetched from PubMed. This is done by filtering the available publications through disease-related topic tags (MeSH terms). The articles are analyzed using SemRep [13], a UMLS-based [3] tool that extracts biomedical predications i.e. semantic triples in the form of subject-predicate-object. Edges denoting co-occurrence of terms and topic annotations of the articles are also added to the KG. Finally, structured ontologies (e.g. Gene Ontology, Disease Ontology) are also indexed under the UMLS schema, allowing for connections with the pre-existing entities from literature. The KG is built using the capabilities of the graph database Neo4j[2].

3.3 Knowledge Graph Analysis

The result of the previous step is a multi-relational directed KG, as seen in Fig. 2 (a). The next step is to identify probable interactions between biomedical entities. Given a pair of drugs, the drug-drug interaction problem can be formulated as a link prediction problem on the KG. In our setting, we model this task as a supervised learning task where data samples are generated from different drug pairs found in the KG and the goal is to find which drug pairs interact.

In order to generate the data samples, we use DrugBank as a source of known interactions (ground truth) and map the drugs to the corresponding UMLS entities. For example, *Bivalirudin* (DB00006), an antithrombin drug, is mapped to two entities under the UMLS schema, *Hirulog* (C0210057) and *Bivalirudin* (C0168273). We assume that when two drugs interact according to DrugBank (e.g. Bivalirudin and Lacosamide), then their corresponding UMLS entities also interact with each other. This process is depicted in Fig. 2(b).

Then, we aggregate all possible paths in the KG between the examined pair of drug nodes. Let $E = \{e_1, e_2, ..., e_M\}$ denote all the relations (i.e. the edges) found in the graph. Also, let d_1, d_2 be two drug nodes and let π^l be a path of length l connecting d_1 and d_2. This path π consists of a series of relations starting from node d_1 and ending in node d_2 in the form of: $d_1 e_0 e_1 e_2 ... e_{l-1} d_2$. In this work, we limit $l \leq 3$ after observing that longer paths were of lower quality due to the high interconnectedness of the graph (i.e. with $l \geq 4$ almost all nodes were within reach from any other node). Thus, the representation of each path between a pair of drugs becomes: $\pi = e_0 e_1 e_2$.

Given two drugs d_1 and d_2 let $\Pi = \{\pi_1^l, \pi_2^l, ..., \pi_{N_{d_1, d_2}}^l\}$ be the set of all possible paths between them. An illustrated example of two such paths, as found between two drugs in the KG, can be seen in Fig. 2(c). Once the paths are retrieved, each one is processed in order to extract a feature representation of it.

We use 35 unique relation types from the UMLS Semantic Network, after merging the semantically similar ones and downsizing them from 55. We use one-hot encoding of these 35 relations as features on every possible hop. Therefore with $l = 3$ hops, the feature vector will have: $3 \times 35 = 105$ features. Each feature value in this vector will be either zero or one, denoting whether the specific relation was found in the specific hop.

Then, for a specific path $\pi_{d1, d2}^l$ between the drugs d_1, d_2 the corresponding feature vector will be: $x_i = [c_{r_1}, c_{r_2}, ..., c_{r_{105}}]$ where c_{r_j} is either 0 or 1 as

---

[2] https://neo4j.com/.

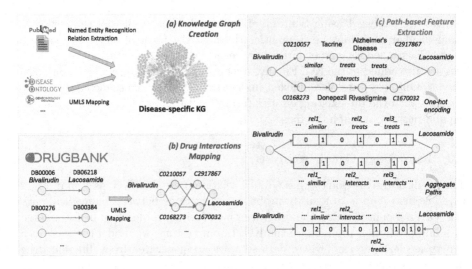

Fig. 2. Overview of the process forming the final feature representations of drug pairs.

mentioned before. For paths $\pi_{d1,d2}^{m}$ with length $m < l = 3$ the last $(l - m) \times 35$ elements of the vector are set to zero. The result of this one-hot encoding process is illustrated in the middle-right part of Fig. 2.

Eventually, we want to aggregate the information of all the paths into a single feature vector for each drug pair. Thus, we combine the feature vectors of the individual paths leading from d_1 to d_2 by summing their corresponding feature vectors. Hence, the feature vector for a drug pair with N_{d_1,d_2} paths will be :

$$x_{d_1,d_2} = \sum_{i=1}^{N_{d_1,d_2}} x_i = \sum_{i=1}^{N_{d_1,d_2}} [c_{r_1}, c_{r_2}, ..., c_{r_{105}}]_i = [\sum_i c_{r_1}^i, \sum_i c_{r_2}^i, ..., \sum_i c_{r_{105}}^i] \quad (1)$$

where $c_{r_j}^i$ denotes the feature c_{r_j} of path x_i. The final outcome is shown in the bottom-right part of Fig. 2. These features will be used downstream to train a classifier and predict new probable DDIs.

4   Evaluation and Results

In this section, we first introduce the details of the generated Knowledge Graph and the corresponding dataset for the experiments. Then, the competing methods are outlined and the evaluation task and procedure are described. Finally, the results are presented and discussed.

4.1   Knowledge Graph Creation

The pipeline described above was configured with the MeSH descriptors "Dementia" (D003704) and "Lung Neoplasms" (D008175) to create two disease-specific

KGs related to two prevalent diseases with high societal impact. As a result, more than 250,000 documents were analysed yielding more than 27 million semantic triples for both diseases. Moreover, hierarchical relations were integrated from the Disease Ontology, the Gene Ontology and the MeSH hierarchy adding more than 250,000 semantic triples in each KG. More details about the composition of the data and the type of nodes and edges as ingested by different data sources can be found in the repository of the proposed method.

4.2   Drug-Drug Interaction

Focusing on the task of predicting interactions between drugs, we used Drug-Bank 5.0.3[3] taking into account only AD and LC related drugs. There were 94 and 68 drugs respectively, according to their textual description. Using these drugs and their corresponding interactions with all other drugs in DrugBank we found 1326 and 4494 interacting drug pairs existing in the two KGs.

Additionally, we generated the (implicitly) negative pairs. We opted for a *corrupted sampling* procedure adapted to our setting. Specifically, let D^+ be the set of positive pairs. Then, for each positive pair $(d_1, d_2 \in D^+)$ a negative pair $(d_1, d' \notin D^+)$ was formed. This way of creating the dataset allows for the most "interacting" drugs to be represented equally in the two classes, marginalizing the factor of the different "interactivity" levels among the drugs. Moreover, to mimic the real world where actual DDIs are rare, in comparison to all possible drug pairs, we retrieved as many negative pairs could be found in the KGs, essentially oversampling the negative class. Details on the final datasets can be seen on Table 1.

Table 1. Drug pairs for each use case to be used for training and validation.

	Positive	Negative
AD	1,326	6,554
LC	4,494	28,752

4.3   Competing Methods

In order to evaluate the semantic path approach, as a way for creating expressive features for the drug pairs, we compared our method against several graph embedding approaches that are popular for link prediction tasks over KGs. To this end, we generated entity and relation embeddings, by training the respective methods on the triples that make up each KG. We experimented with methods from different families of embeddings (i.e. translational, semantic matching,

---

[3] https://www.drugbank.ca/releases/5-0-3.

etc.), focusing heavily on tensor decomposition methods, due to their robustness in link prediction tasks [15].

The details and references on the competing methods can be seen in Table 2. The symbol ; symbolizes the concatenation of the embeddings, D_1 and D_2 the embedding of the first and second drug of the pair respectively and $R_{INTERACTS}$ the embedding of the relation $INTERACTS$.

Table 2. Feature representations of the drug pairs for each competing method.

Method	Feature vector from embeddings	Size
TransE [4]	$[D_1; R_{INTERACTS}; D_2]$	300
RESCAL [10]	$[(D_1 \times R_{INTERACTS}); D_2]$	200
HolE [9]	$[D_1; R_{INTERACTS}; D_2]$	300
DistMult [20]	$[D_1; R_{INTERACTS}; D_2]$	300

It is worth noting that we also tried using directly the score for each drug pair as generated by the corresponding graph embedding model. However, the results were much worse compared to using the embeddings generated from each model as feature vectors for a classifer.

4.4   Experimental Setup

Having generated the various feature presentations, the effectiveness of each one is measured using an extensive cross-validation (cv) procedure. Specifically, a nested cv scheme with an outer 10-fold cv to estimate the performance of the model and an inner 5-fold cv to tune the hyperparameters of the classifier was used. For all the different feature representations a *Random Forest* was used as the final predictor. The hyperparameters of each forest (e.g. number of trees, maximum depth, etc.) were optimized independently for each model. Performance is calculated using the *area under the receiver-operating characteristic* (AUROC), while also the F_1-score and the *area under the precision-recall curve* (AUPRC) for the positive class are calculated. Higher values always indicate better performance for all measures.

Finally, regarding the graph embedding procedures, the *TorchKGE*[4] library was used to train the models and generate the embeddings. All methods were allowed to train for a maximum of 100 epochs while early stopping was used, utilizing 10% of the data for validation purposes. For each model, 100-sized embeddings were used, as they seemed to converge faster and increasing the embedding size did not provide better results.

---

[4] https://torchkge.readthedocs.io.

4.5   Results

The results of the evaluation can be seen in Fig. 3. We can clearly see that the *DDI-BLKG* method outperforms the competing approaches on both disease-specific KGs. Although all of the models seem to be doing well when focusing on their ROC-AUC scores, our model surpasses the competition by far when focusing on the positive class. This indicates that the proposed methodology, captures important information regarding the DDIs and the predicted interactions are precise despite the class imbalance.

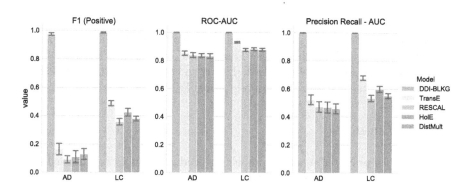

Fig. 3. Results of the 10-fold validation for the both use-cases.

To evaluate the predictive capabilities of our method, we also performed a small-scale qualitative analysis of its top predictions. Taking the top-10 scoring false positives (i.e. DDIs not present in DrugBank 5.0.3), we evaluated them using a newer version (5.3.1). The results were promising as 5/10 drug pairs for the AD, and 7/10 for the LC graph, were actual DDIs in the more recent version. The predicted drug-pairs that were validated on the new database can be seen in Table 3. As an example of the AD drugs, *Estradiol* is a form of estrogen used to treat menopause symptoms and its relation with *Memantine* (medication for mild-to-severe AD) has been thoroughly studied [7]. Taking into account the interaction with *Rivastigmine* (another medication for moderate AD), our model captured the possible DDIs that emerge with hormone therapy trials after menopause and the adverse effects they can cause in Alzheimer's patients.

Regarding the LC drugs, *Cisplatin* (a chemotherapy drug) is commonly provided in conjunction with the predicted interacting drugs to combat the heavy side-effects of chemotherapy (e.g. *Flunitrazepam* as an anti-emetic treatment and *Mangesium hydroxide* to combat cisplatin-induced hypomagnesemia). However, adverse effects may arise, as in these DDIs, with unwanted changes in the metabolism and the serum concentration of the drugs. It is also worth noting that the predicted DDIs are related to a small set of drugs, namely *Estradiol* and *Trientine* for the AD use case and *Cisplatin* and *Gemcitabine* for the LC use case. It is interesting to look further into the predicted DDIs, taking into account

Table 3. Most probable DDIs that have been retrospectively validated in DrugBank 5.3.1.

Disease	Drug	Interacting drugs
AD	Estradiol	Memantine, Rivastigmine, Trientine
AD	Trientine	Leuprolide, Rivastigmine
LC	Cisplatin	Flunitrazepam, Magnesium hydroxide Sparfloxacin
LC	Gemcitabine	Abacavir, Anagrelide, Isosorbide Mononitrate
LC	Docetaxel	Pranlukast

the interactivity of the drugs, their popularity (e.g. in how many publications they were mentioned) and their type. Overall, 60% of the predicted DDIs are present in the new version of DrugBank. This is an indication of the usefulness of our method and its capability to propose probable DDIs for further validation.

5   Conclusion

In this work, we proposed a new approach for predicting DDIs as a downstream task in a semantically-rich Knowledge Graph. We utilized a KG creation workflow to integrate knowledge extracted from biomedical literature and structured databases into a common semantic graph representation. Then, we extracted expressive features for the drug pairs based on the semantic paths that connect those. The experimental results validated the basic premise of our research regarding the latent knowledge of the extracted paths and that our method can be used to effectively discover interacting drugs. Source code for the methods and pre-processed datasets to replicate the results have also been made available.

As next steps, further experiments on the graph motifs indicating a DDI should be conducted, employing reasoning techniques as well. Also, a qualitative analysis of the resulting KG should be conducted, to gain insights on errors generated by the automatic extraction tools. As a final note, it is important to stress that the presented methodology can be used in other tasks involving different node pairs, such as drug-side effect, gene-disease, etc. This stems from the fact that the feature extraction methodology was use case agnostic, depending only on the topological and relational aspects of paths between different node pairs.

Acknowledgments. This work is supported by European Union's Horizon 2020 research and innovation programme under grant agreement No. 727658, project iASiS (http://project-iasis.eu/) (Integration and analysis of heterogeneous big data for precision medicine and suggested treatments for different types of patients). We would also like to acknowledge the help we received from the iASiS consortium for the completion of this work.

References

1. Arnold, P., Rahm, E.: Semrep: a repository for semantic mapping. In: Daten-banksysteme für Business, Technologie und Web (BTW 2015) (2015)
2. Bakal, G., Talari, P., Kakani, E.V., Kavuluru, R.: Exploiting semantic patterns over biomedical knowledge graphs for predicting treatment and causative relations. J. Biomed. Inform. **82**, 189–199 (2018)
3. Bodenreider, O.: The unified medical language system (UMLS): integrating biomedical terminology. Nucleic Acids Res. **32**(suppl_1), D267–D270 (2004)
4. Bordes, A., Usunier, N., Garcia-Duran, A., Weston, J., Yakhnenko, O.: Translating embeddings for modeling multi-relational data. In: Burges, C.J.C., Bottou, L., Welling, M., Ghahramani, Z., Weinberger, K.Q. (eds.) Advances in Neural Information Processing Systems, vol. 26, pp. 2787–2795. Curran Associates, Inc. (2013)
5. Himmelstein, D.S., et al.: Systematic integration of biomedical knowledge prioritizes drugs for repurposing. Elife **6**, e26726 (2017)
6. Karim, M.R., Cochez, M., Jares, J.B., Uddin, M., Beyan, O., Decker, S.: Drug-drug interaction prediction based on knowledge graph embeddings and convolutional-LSTM network. In: Proceedings of the 10th ACM International Conference on Bioinformatics, Computational Biology and Health Informatics, pp. 113–123 (2019)
7. Lamprecht, M.R., Morrison III, B.: A combination therapy of 17β-estradiol and memantine is more neuroprotective than monotherapies in an organotypic brain slice culture model of traumatic brain injury. J. Neurotrauma **32**(17), 1361–1368 (2015)
8. Nentidis, A., Bougiatiotis, K., Krithara, A., Paliouras, G.: Semantic integration of disease-specific knowledge. In: IEEE 33rd International Symposium on Computer Based Medical Systems (CBMS) (2020, to appear)
9. Nickel, M., Rosasco, L., Poggio, T.: Holographic embeddings of knowledge graphs. In: Thirtieth AAAI Conference on Artificial Intelligence (2016)
10. Nickel, M., Tresp, V., Kriegel, H.P.: A three-way model for collective learning on multi-relational data. In: ICML, vol. 11, pp. 809–816 (2011)
11. Percha, B., Altman, R.B.: Informatics confronts drug-drug interactions. Trends Pharmacol. Sci. **34**(3), 178–184 (2013)
12. Percha, B., Altman, R.B.: A global network of biomedical relationships derived from text. Bioinformatics **34**(15), 2614–2624 (2018)
13. Rindflesch, T.C., Fiszman, M.: The interaction of domain knowledge and linguistic structure in natural language processing: interpreting hypernymic propositions in biomedical text. J. Biomed. Inform. **36**(6), 462–477 (2003)
14. Rohani, N., Eslahchi, C.: Drug-drug interaction predicting by neural network using integrated similarity. Sci. Rep. **9**(1), 1–11 (2019)
15. Rossi, A., Firmani, D., Matinata, A., Merialdo, P., Barbosa, D.: Knowledge graph embedding for link prediction: a comparative analysis (2020)
16. Sang, S., et al.: GrEDeL: a knowledge graph embedding based method for drug discovery from biomedical literatures. IEEE Access **7**, 8404–8415 (2018)
17. Sang, S., Yang, Z., Wang, L., Liu, X., Lin, H., Wang, J.: SemaTyP: a knowledge graph based literature mining method for drug discovery. BMC Bioinform. **19**(1), 193 (2018). https://doi.org/10.1186/s12859-018-2167-5
18. Shen, Y., et al.: KMR: knowledge-oriented medicine representation learning for drug-drug interaction and similarity computation. J. Cheminform. **11**(1), 22 (2019). https://doi.org/10.1186/s13321-019-0342-y

19. Shtar, G., Rokach, L., Shapira, B.: Detecting drug-drug interactions using artificial neural networks and classic graph similarity measures. PloS One **14**(8), e0219796 (2019)
20. Yang, B., Yih, W., He, X., Gao, J., Deng, L.: Embedding entities and relations for learning and inference in knowledge bases. arXiv preprint arXiv:1412.6575 (2014)

AI Medical School Tutor: Modelling and Implementation

Shazia Afzal[1], Tejas Indulal Dhamecha[1], Paul Gagnon[2], Akash Nayak[1],
Ayush Shah[1], Jan Carlstedt-Duke[2], Smriti Pathak[3], Sneha Mondal[1],
Akshay Gugnani[1], Nabil Zary[2], and Malolan Chetlur[1(✉)]

[1] IBM Research, New Delhi, India
{shaafzal,tidhamecha,anayak09,snemonda,aksgug22,mchetlur}@in.ibm.com
[2] Lee Kong Chian School of Medicine,
Nanyang Technological University, Singapore, Singapore
{pgagnon,jan.carlstedt-duke,nzary}@ntu.edu.sg
[3] Imperial College, London, UK
smriti.pathak@kcl.ac.uk

Abstract. In this paper we present our experience in the design, modelling, implementation and evaluation of a conversational medical school tutor (MST), employing AI on the cloud. MST combines case-based tutoring with competency based curriculum review, using a natural language interface to enable an adaptive and rich learning experience. It is designed both to engage and tutor medical students through Digital Virtual Patient (DVP) interactions built around clinical reasoning activities and their application of foundational knowledge. DVPs in MST are realistic clinical cases authored by subject matter experts in natural language text. The context of each clinical case is modelled as a set of complex concepts with their associated attributes and synonyms using the UMLS ontology. The MST conversational engine understands the intent of the user's natural language inputs by training Watson Assistant service and drives a meaningful dialogue relevant to the clinical case under investigation. The curriculum content is analysed using NLP techniques and represented as a related and cohesive graph with concepts as its nodes. The runtime application is modelled as a dynamic and adaptive flow between the case and student characteristics. We describe in detail the various challenges encountered in the design and implementation of this intelligent tutor and also present evaluation of the tutor through two field trials with third and fourth year students comprising of 90 medical students.

Keywords: Case-based tutoring · Digital virtual patient · Natural language interface

---

A. Shah—Work done while working at IBM.

M. Michalowski and R. Moskovitch (Eds.): AIME 2020, LNAI 12299, pp. 133–145, 2020.
https://doi.org/10.1007/978-3-030-59137-3_13

1   Introduction

Virtual case based learning (CBL) exposes students to a variety of clinical scenarios, allowing practice and explicit training in the stages and skills associated with expert clinical reasoning like collecting relevant data, formulating a better abstraction of the patient problem using semantic qualifiers, and the ability to store and recall 'illness scripts' that aid in rapid hypothesis generation and verification leading to a focused comparison of differentials [3]. This implies that the meaningful organization and structuring of knowledge is more important for its contextual retrieval and building of a knowledge-base rather than just the acquisition of foundational science. In fact, the majority of diagnostic errors are known to be caused by cognitive errors that are not related to knowledge deficiency (3%) but to flaws in data collection (14%), data integration (50%) and data verification (33%) [17]. This explains the considerable importance given to clinical exposure early on in medical education. However, traditional modes of imparting such clinical experience through clinical rotations fall short in providing in-depth personalised learning encounters because of coursework schedules and limited availability of expert clinicians. An internal survey conducted by LKC School of Medicine, NTU Singapore found that students often wish to get more time and dedicated one-on-one learning support for clinical cases. Anticipating this gap and thanks to an e-learning grant, LKC School of Medicine, NTU Singapore, partnered with IBM Research to envision an AI driven medical school tutor (MST) based on Digital Virtual Patient (DVP). The aim was to prepare students in the transition from academic study to clinical application in an engaging manner. Real life clinical cases that were culturally and contextually relevant for the target students were selected by subject matter experts (SMEs). Pedagogical strategies for promoting diagnostic skills as recommended in medical literature were chosen to deliver these cases. A conversational interface was deemed appropriate to support interaction in natural language. The clinical cases were authored and annotated in a manner that permitted dialog-based interaction marked with key references to coursework. The medical school curriculum that was organised in the form of learning outcomes (LOs) was linked and mapped to serve as the back-end for driving recommendations in the tutor. Assessment data of the students was retrieved from the school's data warehouse to build mastery profiles of students and seed the learner model. Finally, MST was optimised for mobile device usage in order to maximise value to students who had expressed the need of being able to access it anytime, often on the go. The following sections describe the challenges faced during modelling of different components that eventually led to the implementation and cloud based delivery of MST.

2   Related Work

One of the earliest attempts in the medical tutoring space was the knowledge based system GUIDON [4] for training in infectious diseases. Its rule base was subsequently reconfigured to create GUIDON 2 that reasoned more like human experts and provided better explanations [5]. Later, other intelligent tutoring systems like MR Tutor [19], SlideTutor [6], CIRCSIM [8], COMET [20], and SIAS [16] were introduced. These catered to very specific topics, but as a proof-of-concept they did report better motivation and engagement in addition to showing significant learning gains. More popular commercial medical education apps like MedScape [9], Prognosis[1] and Human Dx[2] - amongst others, are also worth mentioning because of their wide user base and ease of use. Finally, even though MST has a conceptual resemblance with the clinical reasoning tool described in [10] it differs significantly with respect to being tutor-driven interaction design, use of natural language, resource recommendation, continuous evaluation based just-in-time feedback and other features as described in the following sections. These features we believe can overcome the lower adoption rates and student engagement found in [10].

MST differs from existing tutoring systems on several key aspects. Firstly, it provides a holistic learning experience where medical cases are used to train students in diagnostic reasoning skills in relation to their foundational curriculum. So the focus is not entirely on diagnostic accuracy but rather the diagnostic ability of students to integrate and apply the relevant knowledge from their curriculum in the context of each medical case. Secondly, MST employs a set of diagnostic activities for each case and the assessment of students on each of these is represented in an open learner model (OLM). This assessment is done automatically and dynamically, and can eventually support the tracking of student's diagnostic ability in an evidence-based manner. Thirdly, the underlying knowledge-base built as the foundation of MST has resulted in an entire medical school curriculum being automatically linked and mapped across the entirety of learning outcomes and their associated resources. Finally, each medical case is structured and authored by SMEs to drive an engaging interaction with anchors to suggested reading at strategic points. The outcome is a rich DVP schema that can drive an intelligent conversation with students by replicating a real clinical encounter in terms of case presentation, information flow and knowledge interrogation. The authoring effort is expected to reduce drastically by using semi-automated methods using advanced NLP/AI technology. To sum, MST offers a comprehensive learning experience by leveraging foundational curriculum knowledge with established educational strategies to provide medical students with anytime, anywhere access to relevant medical cases for the ongoing development and refinement of their clinical reasoning skills and competency. The novelty in MST is to build a complete tutoring system that understands the natural language conversations interspersed with clinical terms and connects the background knowledge with curriculum while responding with meaningful interactions to provide a rich learning experience.

---

[1] https://www.medicaljoyworks.com/prognosis-your-diagnosis.

[2] https://www.humandx.org/.

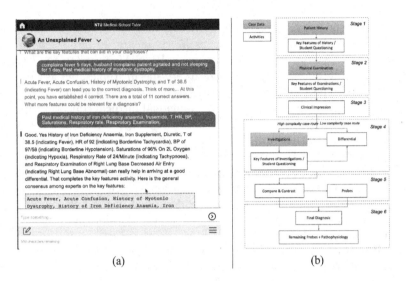

(a)   (b)

Fig. 1. (a) Sample MST interaction for key features activity on a clinical case. (b) Interaction flow indicating sequence of activities for a typical interaction around a DVP.

3   Case Representation and Modelling

CBL is an educational paradigm closely related to Problem based Learning (PBL) in which real-life cases form an authentic context for learning activities meant to promote active problem solving and foster deeper knowledge [13,22]. A DVP is an instantiation of a clinical case that is often used for CBL. DVPs can have varying levels of realism ranging from text based descriptions to high fidelity simulations [12]. While the granularity and design of DVPs depends on their proposed usage, their development and authoring costs remain high making it challenging to scale these for new or unseen cases. MST also has its unique requirement for a DVP: clinical cases need to be authored and annotated in a way that enables a conversational interaction scenario while fulfilling the pedagogical goals of knowledge acquisition, application and reinforcement. The challenge is to not just annotate each case in an elaborate manner but to also ensure that it conforms to the unfolding of a real clinical encounter. The annotation needed to be objectively done so that students performance could be evaluated on each activity. SMEs from LKC School of Medicine carefully selected relevant clinical cases based on real-life patient cases. Together with computer scientists, an annotation protocol was devised and a DVP schema was generated that captured the essential knowledge elements as well as the information flow and related curriculum linkages. This knowledge modelling evolved from custom spreadsheets into DVP JSON (JavaScript Object Notation) objects - a lightweight machine readable data format. This was a non-trivial and time consuming task that involved several rounds of iterations and consultations until a final structure was agreed upon. The resulting DVP schema was a collaborative effort between expert clin-

Table 1. Activities driving interaction with a clinical case.

Activity name	Description
Key features	It trains students in recognizing key clinical findings (e.g. high body temperature) potentially relevant to their differential diagnoses. It aligns with the first stage of data acquisition, mental abstraction and promotes systematic case-reading for distilling pertinent information to formulate possible hypotheses
Clinical impression (CI)	CI is a concise description of patients' key clinical issue(s) with relevant history and examination including functional impairment in clinical terminology (e.g. fever). CI is similar to Problem Representation [21] The purpose is to check students' conceptual understanding of the problem and encourage use of clinical vocabulary that can facilitate retrieval of relevant content from students clinical knowledge/experience
Student questioning	An important part of history taking is the ability to follow a clear line of enquiry and interrogate the patient. To be able to ask the right questions ensures focused data collection which increases the likelihood of potential diagnoses. This activity guides students to ask targeted questions of the patient given the case information presented so far
Knowledge/skills recall	Relating the case data and the path to a possible diagnosis tests medical knowledge and its application. Therefore, probes in the form of multiple answer questions are interspersed in the MST interaction to help students recall and apply the knowledge acquired in the foundational academic years. MST also has provision for open text probes for flexible testing should the SME require
Differential	This involves listing of the three most likely diagnosis in an attempt to identify diseases that may cause patients presentation. The differential diagnosis is the first attempt to distinctly pronounce the potential hypotheses
Compare & contrast	Rather than listing out the features of possible diagnoses, expert clinicians compare and contrast the evidence against possible diagnoses which makes the focus on the discriminating features for each differential. MST uses this method to teach students to differentiate between diagnostic possibilities by reflecting on the relationship among the key clinical features. Contrasting key clinical features against multiple diagnoses leads to prioritizing a final diagnosis and can help in building of appropriate knowledge networks for future application
Final diagnosis	The final diagnosis is the identification of the disease that is most likely the cause of the patients illness. The process of concluding the final diagnosis is as important as arriving at the decision itself so all other activities enable and support this decision making and help students make a conclusion

icians and computer scientists that laid the foundation of the MST. We realised that to make the project scalable and viable in the long run the authoring had to be complemented with proper tooling to make it easy for SMEs to create and review cases. Eventually a complete case authoring tool was developed to semi-automate the process using advanced AI/NLP techniques.

4   Interaction Modeling

MST takes Bowens' [3] elements of the diagnostic reasoning process as the core framework to design and steer learning activities that target student training and practice in: data acquisition skills; problem summarizing using clinical vocabulary; analysing competing hypotheses; prioritizing of diagnostic possibilities; creation of illness scripts by integrating contextual case knowledge; understanding prototypical presentations of diseases; retrieval and recall of acquired curriculum knowledge; encouraging targeted and timely reading; and finally reinforcing the foundational clinical knowledge with clinical practice. The following sections describe the activities that enable development of these skills and how they are interleaved with the DVP to create an interactive seamless flow.

4.1   Learning Activities and Case Flow

Figure 1b illustrates the stages within MST and how the various clinical reasoning activities are surfaced. The stages correspond to blocks or gates that end with learning points and takeaways. This sequence template tries to replicate the stages of a real clinical encounter and was created in consultation with SMEs. Any variations on this case flow are defined by SMEs as part of the authoring process. Table 1 provides a brief description of the specific activities. In order to make the interaction crisp and engaging, students are given only two attempts to specify their answers. Exception to this are the probing questions that are delivered in a multiple choice question answer style of interrogation. The expert answers with explanations are provided at the end of each activity. These help students introspect and validate their understanding.

4.2   Response Generation

MST models conversation with students using the above-mentioned case flow to achieve the tutor goals of case presentation, activity sequencing, activity evaluation, resource recommendation and simultaneously building of a student model. A major challenge is to meet the pedagogical goals while maintaining conversational efficiency and providing constructive remediation to the student. MST uses a state space model design to generate tutor responses depending on the state and location in the dialog. A library of responses collectively compiled by experts are used to seed the tutor responses in order to ensure culturally and professionally appropriate language. A dialog is enabled by alternate turn-taking where student responses are evaluated in the context of ongoing task and recognised intent. Expectation setting occurs at the beginning of each activity when students are given activity specific instructions. An appropriate response is accordingly generated which consists of an acknowledgment, an evaluation statement and a dialog advancer to keep the conversation going. At the end of each activity the tutor response also includes an evaluation component to show the score attained. Refer to Fig. 1a for an example of how MST responds to student input by acknowledging the response, asserting correct information and also taking the dialog forward. The response generation has scope to include a students performance as well as engagement parameters so that tutor responses are specifically aligned to the actual state of a student.

5   Context and Domain Modeling

Understanding the students' natural language input in the context of a clinical case and its medical vocabulary is one of the main challenges for MST. The conversational agent within MST has to understand the intention of the student and has to respond appropriately to retain engagement. Responses that do not meet student's expectations or are incorrect will result in disengagement. The tutor's response has to be relevant and valuable with respect to the immediate learning activity. Thus, understanding the student's intent is essential for ensuring smooth interaction and superior experience.

In MST, students' input are modelled as a bag of intents. Intents are small group of one-to-four words uttered during interaction with the tutor. There may be any number of intents within a student response. MST relies on the Watson Assistant service [1] to identify an intent given a group of words. MST identifies all the different intents (bag of intents) within a student response and their appropriateness in the ongoing conversation. The bag of intents are returned to tutoring engine for further processing by Input Curator module within MST. This machine learning module relies on Unified Medical Language System [2] knowledge-base for training and understanding medical terms. For example, when a student writes `high temperature` or `fever` or `pyrexia`, it should all be understood as the same intent. Figure 1a shows the correctly expressed intents are also part of system response in addition to unexpressed intents (shown in bold face). This is to further inquire the student to provide missing parts of the answer.

6   Content Modeling and Curriculum Linkage

Appropriate content recommendation customized to each learner is a crucial feature in MST. This is achieved by linking each clinical case with the underlying curriculum content in the form of Learning Resources (LRs) via Learning Objectives (LOs). MST uses advanced NLP techniques to extract all related LOs (LO-LO Relatedness prediction) and identifies pages of relevant LRs (LO-LR Relevance prediction) to relate concepts from the case with those in the curriculum. The detailed modeling, experiments and results of LO-LO relatedness prediction and LO-LR relevance prediction can be found in [15, 18] respectively.

6.1   LO-LO Relatedness

Predicting LO-LO relationship requires looking into the semantic content of disparate LOs, in addition to their relatedness in the curriculum hierarchy [15]. This is formulated as a three-class classification task where given a pair of LOs, a classifier is trained to categorize them as being either strongly related, weakly related, or unrelated based on annotations obtained from Subject Matter Experts (SMEs). The representation for a pair of LOs is the concatenation of their curriculum and semantic features, that encode their relative position in the curriculum hierarchy and similarity in meaning, respectively. This feature-set is used with a random forest classifier, that learns to classify the relationship between pairs of LOs. This LO relationship extraction system is then applied to uncover LOs relevant to clinical cases (DVP) in MST so that useful LRs can be recommended.

6.2   LO-LR Relevance Prediction

Identifying relevance of an LR page with an LO has three key aspects–(1) Lexical, (2) Semantic, and (3) Spatial [18]. The overlap between the terms in the LO and those in the page can be used to identify some of the relevant pages. Similarly, semantic overlap between the LO and the page can identify relevant pages that are not identified by lexical matches. Additional pages that are adjacent to highly relevant pages (either lexical or semantic) are also likely to be relevant. Exploiting this spatial aspect identifies additional relevant pages to the existing set of pages. This problem is formulated as a binary classification task, where given an LO and a page of an LR, a machine learning model predicts its relevance. It would be ideal to train a joint model that is able to utilize all three aspects (lexical, semantic, spatial) to make relevance predictions. However, obtaining annotated training data for this task is expensive as the SME has to go through the entire LR for annotating each LO. Thus, a pipelined approach consisting of separate models capturing each of the aspects is used to convert the alignment problem into a page relevance classification problem.

7   Learner Modeling

The student model for MST tracks knowledge and behaviour including estimation of mastery level from prior assessment data, performance on the various reasoning activities, and behaviour based implicit metrics of engagement. The idea was to build a student profile that could enable open learner model visualisation, provide adaptive learning pathways, allow personalised feedback generation, and tailored sequencing and recommendation of content - eventually leading to customising interactions and optimising the tutor strategy overtime.

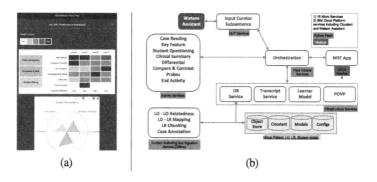

Fig. 2. (a) Learner model visualisation (b) Different components of MST and their interaction.

7.1 Representation

The knowledge and behaviour models in MST have two components to separate the prior student profile from the MST based profile. A combination of overlay and stereotype modelling is used for prior mastery estimation by conducting IRT [11] analysis on the historical performance data from the medical schools' formative assessments called IRA (Individual Readiness Assessment). Students are categorised into five levels of mastery ranging from beginner to expert using overall, cohort-level, year-level, block-level, and LO level IRT estimates. The prior behaviour model is captured in the form of various metrics like login frequency, dwell time, media preferences and resource access to get an idea of the interaction style of students.

MST tracks detailed activity level performance at case level and an overall global estimation across all cases for each student. The activity level assessments form the basis of an evaluation model that bins students into five mastery levels. Activity level scores are computed by comparing student answers against the corresponding expert answers. Additionally, an overall reasoning ability is approximated using dimensions of Pattern Recognition, Knowledge and Skills, and Decision-Making using scores across cases. Key features and Compare & Contrast scores give an estimation of the Pattern Recognition skills of the student. Differential and Final Diagnosis give an indication of the Decision-Making skills while the Probes showcase Knowledge & Skills. Taken together, these three higher order measures highlight the various components of diagnostic reasoning ability and help in identifying areas of strengths and weaknesses. The MST learner model can be instrumental in getting first evidence on automatic assessment of diagnostic reasoning.

7.2 Visualisation

In the spirit of OLMs [7], MST provides students with a powerful visualisation of their student model in order to motivate and track their learning and help them reflect on their performance Two data visualisation techniques are used in the learner dashboard. The first one is based on the Mastery Grid [14] concept wherein the student mastery levels are depicted using colour gradients in a grid format. The second visualisation is a spider diagram that provides a summary view of higher order skills of pattern recognition, knowledge and skills, and decision-making. The arcs represent the different activities and their gradations correspond to the mastery levels. The scores for this visualization are aggregated across all the cases completed by the student so that it gives an overview of reasoning ability in an intuitive manner. Figure 2a shows how the dashboard looks for a dummy user who has gone through four cases and completed one out of them.

Fig. 3. Feedback survey results

8   Implementation Details

MST is implemented in a cloud-native microservice-based architecture on IBM Cloud. Figure 2b shows its key components and how they interact. The Input Curator service analyses each student input to identify the intent and answer. This service uses advanced NLP techniques and relies on IBM's Watson Assistant service [1]. The identified intents and current state of student in the interaction flow 1b is passed back to the Orchestration which in turn delegates control to the appropriate activity specific microservices. The Learner Model service updates the performance and engagement parameters after completion of individual activities in the interaction flow - refer to Fig. 1b. Additional offline components that complement the tutor system to support case authoring, LO-LO relation extraction, LO-LR mapping, LR chunking and training of Watson Assistant are shown separately in the figure.

9   Experimental Validation

We conducted two field trials with medical students to validate the goals of MST and evaluate its efficacy. The purpose of these trials was not to compare these two user groups but to assess the functioning of the system and get direct feedback on its perceived value from end-users. The trials were conducted a few weeks apart so that feedback taken from students in the first field trial was incorporated into MST before conducting the second trial. The first field trial was conducted with 55 third year medical students while the second trial was done with 35 fourth year medical students. All students voluntarily participated and were given a brief introduction about the purpose of the experiment. More than 75% of students used laptops while less than 10 % of students used their smartphones during the trials. All students worked on the same dengue case in all sessions. Student's participation was monitored through a muted video conference session to avoid distraction. All students appeared to be considerably engaged while working through the cases in MST. A feedback survey questionnaire was completed by all participants in both trials. It had 14 Likert questions aligned to the dimensions of usability, learning gains and content engagement. Additionally, there were two open style questions asking students about what they liked in MST and what they wished to see in future improvements. The results of the survey comparing feedback between the two groups are depicted in Fig. 3 with scores normalized on 0–1 scale. All scores for usability, learning and content engagement are well above 0.6 with the exception of one score of 0.52 given in the first field trial on ease of finding information on the app. The change of this score to over 0.7 in the second field trial shows that the feedback was taken constructively and utilised to improve the app. Overall, all students rated MST over 0.65 on usability and experience. On the dimension of learning, students gave a score of more than 0.7 vindicating the value of MST as a supplement to learning. Students found themselves in control of their learning, and expressed their interest to explore more cases. Students also found the cases both challenging and engaging. They found the pacing of activities appropriate and were generally happy with the time it took to complete a case. Students comments from the open style questions gave interesting insights into their experience and views on MST. Figure 4a presents the word cloud of the notable features liked by students. For improvements, students mainly expected clarity in instructions, better response classification and an ability to ask counter questions about the systems evaluation or explanation of answers. We aim to follow up on these findings and suggestions prior to conducting a longitudinal study with students. We would especially focus on using the learner model to compare learning gains and changes in diagnostic skills overtime.

Both field trials collectively gave us about 50 h of interaction data. Students spent an average of 33 min on the dengue case with the least time spent being 10 min and the maximum 50 min. Considering time spent as a proxy measure of student engagement, we can perhaps conclude that students were motivated and all of them completed the case till the end. Figure 4b shows the distribution of overall time spent on the case as well as in individual learning activities. Students appear to have spent more time in key features, clinical impression and student questioning activities.

To sum, the survey results and data from these initial evaluations are positive and encouraging. Usability, learning gains and engagement seems to be high among the students from the field trials. Students also proposed areas of improvement in MST specifically regarding instructions not being clear and not

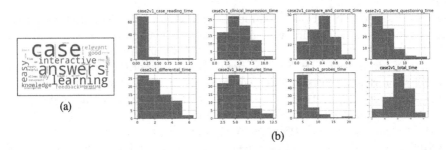

Fig. 4. (a) Word-cloud of what students liked (b) Distribution of time-spent in MST (in mins)

knowing the response expected by the tutor. Students also cited instances where the tutor was not able to understand their responses. Understanding nuances in natural language and medical vocabulary is a huge NLP challenge and this is being currently worked at to improve the experience. However, the reported limitations do not come in the way of achieving a superior learning experience.

10   Summary

We described the Medical School Tutor (MST) as a holistic learning tool driven by AI techniques to prepare medical students in their transition from academic study to clinical application. The interaction design modeled on CBL pedagogy engaged students with realistic clinical cases. The content and context understanding enabled using NLP techniques allows seamless introduction of curriculum resources and generation of appropriate feedback. The survey results show high scores on usability, learning gains and engagement; and validate the purpose and efficacy of conversational tutoring in complex medical domain. The combination of clinical cases with medication foundation curriculum through conversational interface seems to be a novel and an enriching experience for students. In future, MST aims to provide a truly personalized tutoring experience featuring responses adapted to student's learner model.

References

1. IBM Watson Assistant. https://www.ibm.com/cloud/watson-assistant/
2. Bodenreider, O.: The unified medical language system (UMLS): integrating biomedical terminology. Nucleic acids Res. **32**, D267–D270 (2004)
3. Bowen, J.L.: Educational strategies to promote clinical diagnostic reasoning. New Engl. J. Med. **355**(21), 2217–2225 (2006)
4. Clancey, W.J.: Guidon. J. Comput.-Based Instr. (1983)
5. Clancey, W.J., Letsinger, R.: NEOMYCIN: Reconfiguring a rule-based expert system for application to teaching. Stanford University Stanford, Department of Computer Science (1982)

6. Crowley, R.S., Medvedeva, O.: An intelligent tutoring system for visual classification problem solving. Artif. Intell. Med. **36**(1), 85–117 (2006)
7. Desmarais, M.C., Baker, R.S.: A review of recent advances in learner and skill modeling in intelligent learning environments. User Model. User-Adap. Inter. **22**(1–2), 9–38 (2012)
8. Evens, M.W., et al.: Circsim-tutor: an intelligent tutoring system using natural language dialogue. In: Conference on Applied Natural Language Processing: Descriptions of System Demonstrations and Videos (1997)
9. Frishauf, P.: Medscape-the first 5 years. Medscape General Med. **7**(2), 5 (2005)
10. Hege, I., Kononowicz, A.A., Adler, M.: A clinical reasoning tool for virtual patients: design-based research study. JMIR Med. Educ. **3**(2), e21 (2017)
11. Johns, J., Mahadevan, S., Woolf, B.: Estimating student proficiency using an item response theory model. In: International Conference on Intelligent Tutoring Systems, pp. 473–480 (2006)
12. Kononowicz, A.A., Zary, N., Edelbring, S., Corral, J., Hege, I.: Virtual patients-what are we talking about? A framework to classify the meanings of the term in healthcare education. BMC Med. Educ. **15**(1), 11 (2015)
13. Lajoie, S.P., Faremo, S., Wiseman, J.: Tutoring strategies for effective instruction in internal medicine. Int. J. Artif. Intell. Educ. **12**(3), 293–309 (2001)
14. Loboda, T.D., Guerra, J., Hosseini, R., Brusilovsky, P.: Mastery grids: an open source social educational progress visualization. In: European Conference on Technology Enhanced Learning (2014)
15. Mondal, S., et al.: Learning outcomes and their relatedness in a medical curriculum. In: Innovative Use of NLP for Building Educational Applications, pp. 402–411 (2019)
16. Muñoz, D.C., Ortiz, A., González, C., López, D.M., Blobel, B.: Effective e-learning for health professional and medical students: the experience with sias-intelligent tutoring system. Stud. Health Technol. Inform. **156**, 89–102 (2010)
17. Nendaz, M., Perrier, A.: Diagnostic errors and flaws in clinical reasoning: mechanisms and prevention in practice. Swiss Med. Weekly **142**(4344), (2012)
18. Saha, S., Chetlur, M., Dhamecha, T.I.: Aligning learning outcomes to learning resources: a lexico-semantic spatial approach. In: IJCAI (2019)
19. Sharples, M., Du Boulay, J., Teather, B., Teather, D., Jeffery, N., Du Boulay, G.: The MR tutor: computer-based training and professional practice. In: Cognitive Science Research Paper (1995)
20. Suebnukarn, S., Haddawy, P.: Comet: a collaborative tutoring system for medical problem-based learning. IEEE Intell. Syst. **22**(4), 70–77 (2007)
21. Ten Cate, O., Custers, E.J., Durning, S.J.: Principles and Practice of Case-Based Clinical Reasoning Education: A Method for Preclinical Students, vol. 15. Springer, Heidelberg (2017). https://doi.org/10.1007/978-3-319-64828-6
22. Williams, B.: Case based learning—a review of the literature: is there scope for this educational paradigm in prehospital education? Emerg. Med. J. **22**(8), 577–581 (2005)

Predictive Modeling

Lung Cancer Survival Prediction Using Instance-Specific Bayesian Networks

Fattaneh Jabbari[1]([⊠]), Liza C. Villaruz[2], Mike Davis[3],
and Gregory F. Cooper[1,3]

[1] Intelligent Systems Program, University of Pittsburgh, Pittsburgh, USA
{faj5,gfc}@pitt.edu
[2] UPMC Hillman Cancer Center, Pittsburgh, USA
villaruzl@upmc.edu
[3] Department of Biomedical Informatics, University of Pittsburgh, Pittsburgh, USA
midavis@pitt.edu

Abstract. Lung cancer is the most common cause of cancer-related death in men worldwide and the second most common cause in women. Accurate prediction of lung cancer outcomes can help guide patient care and decision making. The amount and variety of available data on lung cancer cases continues to increase, which provides an opportunity to apply machine learning methods to predict lung cancer outcomes. Traditional population-wide machine learning methods for predicting clinical outcomes involve constructing a single model from training data and applying it to predict the outcomes for each future patient case. In contrast, instance-specific methods construct a model that is optimized to predict well for a given patient case. In this paper, we first describe an instance-specific method for learning Bayesian networks that we developed. We then use the Markov blanket of the outcome variable to predict 1-year survival in a cohort of 261 lung cancer patient cases containing clinical and omics variables. We report the results using AUROC as the evaluation measure. In leave-one-out testing, the instance-specific Bayesian network method achieved higher AUROC on average, compared to the population-wide Bayesian network method.

Keywords: Lung cancer survival prediction · Instance-specific modeling · Population-wide modeling · Bayesian network classifiers

1 Introduction

Lung cancer is the most frequent cause of cancer-related death in men worldwide and the second most common cause in women [4], despite significant improvements in diagnosis and treatment during the past decade. Lung cancer is typically divided into two major subtypes: small cell lung cancer (SCLC) and non-small cell lung cancer (NSCLC), where the latter is more prevalent. The overall 5-year survival rate for lung cancer is 19% but it can be increased to 57% if diagnosis occurs at a localized stage of the disease [2], which is not often the case. Accurate

© Springer Nature Switzerland AG 2020
M. Michalowski and R. Moskovitch (Eds.): AIME 2020, LNAI 12299, pp. 149–159, 2020.
https://doi.org/10.1007/978-3-030-59137-3_14

prediction of lung cancer outcomes is important, because it can facilitate patient care and clinical decision making. With rapid advancements in technology, large amounts of lung cancer data with clinical and molecular information are being collected and made available. This trend provides an opportunity for researchers to apply machine learning techniques to predict outcomes of lung cancer.

Most machine learning methods for predicting clinical outcomes construct a single model M from training data. M is then applied to predict outcomes in future instances. We refer to such a model as a *population-wide model* because it predicts outcomes for a future population of instances. It may be difficult for population-wide models to perform well in domains in which instances are highly heterogeneous. Studies have revealed that heterogeneity exists both within individual lung cancer tumors and between patients [20]. In such domains, a reasonable approach is to learn a model that is tailored to a particular instance (e.g., a patient), which we refer to as an *instance-specific model*. An instance-specific approach builds a model M_T for a given instance T from the features that we know about T (e.g., clinical and molecular features) and from a training set of data on many other instances. It then uses M_T to predict the outcome for T. This procedure repeats for each instance that is to be predicted.

In this paper, we use Bayesian network (BN) classifiers to predict patient survival in a cohort of 261 lung cancer patients with clinical and omics information. More specifically, we apply a score-based instance-specific BN learning method, called IGES, which we introduced in [14]. The IGES algorithm searches the space of BNs to learn a model that is specific to an instance T by guiding the search based on T's features (i.e., variable-value pairs). We also apply a state-of-the-art score-based population-wide BN learning method, called GES [14], as a control method. These methods are summarized in Sect. 4. The main goal of this paper is to investigate the effectiveness of instance-specific modeling in predicting survival outcomes for lung cancer patients.

2   Related Work

Various population-wide methods such as neural networks, support vector machines, random forests, and Bayesian models have been applied in cancer research to predict cancer survival and reported AUROCs of ∼0.80 [1,8,19]. A review of such methods is provided in [16]. None of these methods learn instance-specific models. Additionally, they use different sets of predictors and different patient cohorts, so a direct comparison to the results in the current paper is not possible.

Several machine learning methods have been developed to learn instance-specific models. Zheng and Webb [22] introduced a lazy Bayesian rule learning (LBR) method to construct a model that is specific to a test instance. In an LBR rule, the antecedent is a conjunction of the variable-value pairs that are present in the test instance and the consequent is a local naïve Bayes classifier in which the target variable is the parent of the variables that do not appear in the antecedent. Visweswaran and Cooper [21] developed an instance-specific Markov Blanket

(ISMB) algorithm that searches over the space of Markov blankets (MBs) of the target variable by utilizing the features of a given test instance. ISMB first uses a greedy hill-climbing search to find a set of MBs that best fit the training data. Then, it greedily adds single edges to the MB structures from the previous step, if doing so improves the prediction of a given test instance. Ferreira et al. [10] implemented two patient-specific decision path (PSDP) algorithms using two variable selection criteria: balanced accuracy and information gain. A decision path is a conjunction of features that are present in a given test instance and a leaf node that contains the probability distribution of the target variable.

Recently, Lengerich et al. [17] introduced a personalized linear regression model that learns a specific set of parameters for each test case based on the idea that the similarity between personalized parameters is related to the similarity between the features. Accordingly, a regularizer is learned to match the pairwise distance between the features (i.e., variable-value pairs) and the personalized regression parameters. In other related work, Cai et al. [5] developed a method to learn tumor-specific BN models from data; this method uses bipartite BNs and makes other assumptions that restrict its generality.

The IGES method [14] that we use in this paper is different from the methods mentioned above. IGES differs from a population-wide method in a principled way; it learns a model for each instance T that is optimized for T, while a population-wide model is designed to learn a single model that is optimized for a population of instances. Also, the IGES method learns a more general model than does LBR [22], ISMB [21], PSDP [10], and tumor-specific BN models [5], because it learns Bayesian network models, which can be used for both prediction and causal discovery [13–15].

3   Preliminaries

A Bayesian network (BN) encodes probabilistic relationships among a set of variables. A BN consists of a pair $(\mathcal{G}, \boldsymbol{\Theta})$, where $\mathcal{G}$ is a directed acyclic graph (DAG) and $\boldsymbol{\Theta}$ is a set of parameters for $\mathcal{G}$. A DAG $\mathcal{G}$ is given as a pair $(\boldsymbol{X}, \boldsymbol{E})$, where $\boldsymbol{X} = \{X_1, X_2, ..., X_n\}$ denotes a set of nodes that correspond to domain variables and $\boldsymbol{E}$ is a set of directed edges between variables. The presence of an edge $X_i \rightarrow X_j$ (X_i is called the parent and X_j is called the child) denotes that these variables are probabilistically dependent. The absence of an edge denotes that X_i and X_j are conditionally probabilistically independent. A set of DAGs that encode the same independence and dependence relationships are statistically indistinguishable from observational data; such DAGs are called Markov equivalent. The second component, $\boldsymbol{\Theta}$, is a set of parameters that encodes the joint probability distribution over $\boldsymbol{X}$, which can be efficiently factored based on the parent-child relationships in $\mathcal{G}$.

As mentioned above, the edges present and absent in a DAG represent conditional dependence and independence relationships between variables, respectively. Any such relationship should hold for all combinations of values of the variables in the BN. There is a more refined form of conditional independence that holds only in a specific context, which is known as *context-specific inde-pendence* (CSI) [3]. Figure 1 shows a BN that includes two CSI relationships: $X_4 \perp\!\!\!\perp \{X_2, X_3\}|X_1 = 0$ and $X_4 \perp\!\!\!\perp X_3|\{X_1 = 1, X_2 = 1\}$. The IGES method models such local independence structures for each instance T, which results in a more expressive BN structure and a more efficient BN parameterization.

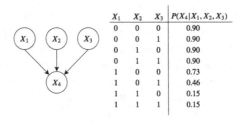

X_1	X_2	X_3	$P(X_4|X_1, X_2, X_3)$
0	0	0	0.90
0	0	1	0.90
0	1	0	0.90
0	1	1	0.90
1	0	0	0.73
1	0	1	0.46
1	1	0	0.15
1	1	1	0.15

Fig. 1. An example Bayesian network with two context-specific independencies (CSIs): $X_4 \perp\!\!\!\perp \{X_2, X_3\}|X_1 = 0$ and $X_4 \perp\!\!\!\perp X_3|\{X_1 = 1, X_2 = 1\}$.

4   Methodology

4.1   Population-Wide Greedy Equivalence Search (GES)

Score-based methods are one of the main approaches to learn BN structure from data that involve (1) a scoring function to measure how well the data (and optional background knowledge) support a given DAG and (2) a search strategy to explore the space of possible DAGs. Since recovering the data-generating BN is an NP-hard problem [6], these methods often utilize a greedy search strategy. GES [7] is a state-of-the-art score-based BN learning algorithm that searches over the Markov equivalence classes of DAGs using two local graph operations: single edge addition and single edge removal. First, it greedily adds single edges to the current graph as long as doing so leads to score improvement. It then greedily removes single edges as long as doing so results in a higher score. Under reasonable assumptions, the GES algorithm converges to the data-generating BN or one Markov equivalent to it [7].

The Bayesian Dirichlet (BD) scoring function [12] is used to score a BN with discrete variables and is calculated as follows:

$$\mathrm{BD}(\mathcal{G}, D) = P(\mathcal{G}) \cdot \prod_{i=1}^{n} \prod_{j=1}^{q_i} \frac{\Gamma(\alpha_{ij})}{\Gamma(\alpha_{ij} + N_{ij})} \cdot \prod_{k=1}^{r_i} \frac{\Gamma(\alpha_{ijk} + N_{ijk})}{\Gamma(\alpha_{ijk})}, \tag{1}$$

where $P(\mathcal{G})$ is the prior structure probability of $\mathcal{G}$. The first product term is over all n variables, the second product is over the q_i parent instantiations of variable X_i, and the third product is over all r_i values of variable X_i. The term N_{ijk} is the number of cases in the data in which variable $X_i = k$ and its parents

$\boldsymbol{Pa}(X_i) = j$; also, $N_{ij} = \sum_{k=1}^{r_i} N_{ijk}$. The term α_{ijk} is a Dirichlet prior parameter that may be interpreted as representing "pseudo-counts" and $\alpha_{ij} = \sum_{k=1}^{r_i} \alpha_{ijk}$. The pseudo-counts can be defined to be evenly distributed, in which case Eq. (1) represents the so-called *BDeu* score [11]:

$$\alpha_{ijk} = \frac{\alpha}{r_i \cdot q_i}, \tag{2}$$

where α is a positive constant called the prior equivalent sample size (PESS).

4.2 Instance-Specific Greedy Equivalence Search (IGES)

We introduced an instance-specific version of GES called IGES in [14]. Similar to GES, IGES is a two-stage greedy BN learning algorithm that uses an instance-specific score (IS-Score). First, it runs GES using the training data D and the BDeu score to learn a population-wide BN model $\mathcal{G}_{PW}$; this is the population-wide BN for all test instances. In the second stage, IGES uses the data D, the population-wide model $\mathcal{G}_{PW}$, and a single test instance T to learn an instance-specific model for T, which is denoted as $\mathcal{G}_{IS}$. To do so, IGES starts with $\mathcal{G}_{PW}$ and runs an adapted version of GES with a specialized IS-Score to maximize the marginal likelihood of the data given both BN models $\mathcal{G}_{PW}$ and $\mathcal{G}_{IS}$.

For each variable X_i, the IS-Score is calculated at the parent-child level and is composed of two components: (1) the instance-specific structure that includes X_i's parents in $\mathcal{G}_{IS}$, which we denote by $\boldsymbol{Pa}_{IS}(X_i)$ and (2) the population-wide structure that includes X_i's parents in $\mathcal{G}_{PW}$, which we denote by $\boldsymbol{Pa}_{PW}(X_i)$. In order to score the instance-specific structure $\boldsymbol{Pa}_{IS}(X_i) \rightarrow X_i$, we use the instances in data that are similar to the current test instance T in terms of the values of the variables in $\boldsymbol{Pa}_{IS}(X_i)$. These instances are selected based on the values of $\boldsymbol{Pa}_{IS}(X_i)$ in T; let $D_{\boldsymbol{Pa}_{IS}(X_i)=j}$ be the instances that are similar to T assuming $\boldsymbol{Pa}_{IS}(X_i) = j$. Then the instance-specific score for $\boldsymbol{Pa}_{IS}(X_i) \rightarrow X_i$ given data $D_{\boldsymbol{Pa}_{IS}(X_i)=j}$ is as follows:

$$P(D_{\boldsymbol{Pa}_{IS}(X_i)=j} | \boldsymbol{Pa}_{IS}(X_i) \rightarrow X_i) = \frac{\Gamma(\alpha_{ij})}{\Gamma(\alpha_{ij} + N_{ij})} \cdot \prod_{k=1}^{r_i} \frac{\Gamma(\alpha_{ijk} + N_{ijk})}{\Gamma(\alpha_{ijk})}, \tag{3}$$

where r_i denotes all values of X_i, N_{ijk} is the number of instances in $D_{\boldsymbol{Pa}_{IS}(X_i)=j}$ in which $X_i = k$, and $N_{ij} = \sum_{k=1}^{r_i} N_{ijk}$. The terms α_{ijk} and $\alpha_{ij} = \sum_{k=1}^{r_i} \alpha_{ijk}$ are the corresponding Dirichlet priors.

Since the instances in $D_{\boldsymbol{Pa}_{IS}(X_i)=j}$ are being used to score the instance-specific structure, they should no longer be used to also score the population-wide structure; therefore, the score for $\boldsymbol{Pa}_{PW}(X_i) \rightarrow X_i$ must be adjusted accordingly. We re-score $\boldsymbol{Pa}_{PW}(X_i) \rightarrow X_i$ using the remaining instances in $D_{\boldsymbol{Pa}_{IS}(X_i) \neq j}$ as follows:

$$P(D_{\boldsymbol{Pa}_{IS}(X_i) \neq j} | \boldsymbol{Pa}_{PW}(X_i) \rightarrow X_i) = \prod_{j'=1}^{q_i} \frac{\Gamma(\alpha_{ij'})}{\Gamma(\alpha_{ij'} + N_{ij'})} \cdot \prod_{k=1}^{r_i} \frac{\Gamma(\alpha_{ij'k} + N_{ij'k})}{\Gamma(\alpha_{ij'k})}, \tag{4}$$

where r_i and q_i are the number of possible values of X_i and $\boldsymbol{Pa}_{PW}(X_i)$, respectively. $N_{ij'k}$ is the number of instances in $D_{Pa_{IS}(X_i)\neq j}$ for which $X_i = k$ and $\boldsymbol{Pa}_{PW}(X_i) = j'$, and $N_{ij'} = \sum_{k=1}^{r_i} N_{ij'k}$. The terms $\alpha_{ij'k}$ and $\alpha_{ij'} = \sum_{k=1}^{r_i} \alpha_{ij'k}$ are the corresponding Dirichlet priors.

Finally, to obtain the overall score for variable X_i using all instances in D, we multiply these two scores. In essence, this method searches for the most probable context-specific BN in light of how well (1) the instance-specific model predicts data for instances like the current test instance, and (2) the population-wide model predicts data for the remaining instances. See [14] for a detailed description of the IS-Score and the IGES method.

5   Experimental Results

This section describes a comparison of the performance of an instance-specific machine learning method to its population-wide counterpart when predicting 1-year survival in lung cancer patients, using clinical and molecular data, which are described in Sect. 5.1. We used Bayesian network classifiers as machine learning models to predict the target variable. More specifically, we applied IGES and GES methods to learn instance-specific and population-wide BN classifiers.

To predict the target variable, we first ran the IGES and GES methods to construct a BN structure over all variables (i.e., the predictors and target). Then, we obtained the Markov blanket (MB) of the target variable that includes the variable's parents, children, and its children's parents. Finally, we calculated the probability distribution of the target variable given its MB and output the most probable outcome as the prediction. We report evaluation criteria to measure the effectiveness of the instance-specific BN model versus the population-wide BN model. In particular, as a measure of discrimination, we report the area under the ROC curve (AUROC) when predicting 1-year survival. We also report the differences between the variables in the MB of the target variable found by the instance-specific models compared to the population-wide model.

5.1   Data Description

This was a retrospective analysis of banked tumor specimens that were collected from patients with lung cancer at the University of Pittsburgh Medical Center (UPMC) in 2016. Baseline demographics, smoking history, staging, treatment, and survival data were collected through the UPMC Network Cancer Registry. We replaced the missing values of the predictor variables with a new category called "missing" and removed the cases for which the value of the outcome variable was not known. Demographic and clinical characteristics of the 261 patients are summarized in Table 1. DNA sequencing was performed using the Ion AmpliSeq^{TM} Cancer Panel (Ion Torrent, Life Technologies, Fisher Scientific). Gene rearrangements of ALK, ROS1, and RET, and MET amplification were detected using FISH. PD-L1 SP263 and PD-L1 22C2 assays were performed on lung cancer samples to determine the PD-L1 tumor proportion score

Table 1. One-year survival given demographic and clinical characteristics. A 95% confidence interval is included for each sub-group of patients.

Variable name	Variable value	Greater than 1 year	
		# Patients (Total)	% Patients (Confidence Interval)
Age	22–62	54 (84)	64.29 (54.04, 74.54)
	63–72	41 (88)	46.59 (36.17, 57.01)
	73–88	45 (89)	50.56 (40.17, 60.95)
Sex	Female	80 (135)	59.26 (50.97, 67.55)
	Male	60 (126)	47.62 (38.9, 56.34)
Race	White	119 (224)	53.13 (46.58, 59.66)
	Black	16 (31)	51.61 ((34.02, 69.2)
	Other	5 (6)	83.33 (53.33, 113.33)
Tobacco history	Cigar/pipe smoker	0 (1)	0
	Cigarette smoker	42 (85)	49.41 (38.78, 60.04)
	Never used	22 (32)	68.75 (52.69, 84.81)
	Previous tobacco use	76 (142)	53.52 (45.32, 61.72)
	Snuff/chew/smokeless	0 (1)	0
Diagnosis	Adenocarcinoma	53 (89)	59.55 (49.35, 69.75)
	Squamous	3 (7)	42.86 (45.3, 62.06)
	Other	11 (29)	37.93 (6.20, 79.52)
	NA	73 (136)	53.68 (20.27, 55.59)

(TPS). Table 2 provides information about the type, name, and description of the variables that are included in the lung cancer dataset. The outcome-prediction research reported here was performed under the auspices of study protocol number PRO15070164 from the University of Pittsburgh Institutional Review Board (IRB).

5.2   Results

We performed leave-one-out cross-validation on the lung cancer dataset. For each instance T, we used T as a test instance and all the remaining instances as the training set D. We applied IGES search using T, D, and the IS-Score to learn an instance-specific BN model for T, which is called $\mathcal{G}_T$ and is used to predict the outcome for T. We also applied GES search using D and the BDeu score to learn a population-wide BN model for T; this BN model is denoted by $\mathcal{G}_{PW}$ and is used to predict the outcome for T. We repeated this procedure for every instance in the dataset. Note that this leave-one-out cross-validation does not involve any hyperparameter tuning; therefore, we do not expect it to

Table 2. Type, name, and description of the variables of the lung cancer dataset.

Variable type	Variable name	Variable description
Demographics	Sex, Race, Age, Tobacco History	
Clinical	Site	Location of tumor
	Surgical Procedure	Type of surgical resection or biopsy
	Diagnosis	Lung cancer type (Adenocarcinoma, Squamous, Other, NA)
	Mets at Dx-Brain, Mets at Dx-Bone, Mets at Dx-Distant Lymph Nodes, Mets at Dx-Lung, Mets at Dx-Liver, Mets at Dx-Other	Location of metastasis at diagnosis (Dx), if any
	Histo Behavior ICD-O-3	Histological classification
	cT, cN, cM, cStage Group	Clinical staging
	pT, pN, pM, pStage Group, Pathologic Stage Descriptor	Pathologic staging
Molecular	PD-L1 IHC, PD-L1 Comment	PD-L1 immunohistochemistry measures the amount of PD-L1 staining on tumor cells
	MET, KRAS, EGFR-summary, EGFR-Exon-18, EGFR-Exon-19, EGFR-Exon-20, EGFR-Exon-21, BRAF, PIK3CA, ALK Mutation	Status of gene mutations
	ALK IHC	ALK gene immunohistochemistry
	ALK Trans	ALK gene translocation
	ROS Trans	ROS gene translocation
	RET Trans	RET gene translocation
	cMET Ratio	Measurement of cMET gene amplification

cause overfitting. We used an efficient implementation of GES, called FGES [18], which is available in the Tetrad system[1]. We used prior equivalence sample sizes of PESS = $\{0.1, 1.0, 10.0\}$ for both IGES and GES methods. IGES also has a parameter that penalizes the structural difference between the population-wide and instance-specific BN models, called κ $(0.0 < \kappa \leq 1.0)$, where a lower value

---

[1] https://github.com/cmu-phil/tetrad.

indicates more penalty for differences (see [14] for details). We report the results of using multiple values of κ.

Table 3 shows the AUROC results on the lung cancer dataset, using both GES and IGES searches; boldface indicates that the results are statistically significantly better, based on DeLong's non-parametric test [9] at a 0.001 significance level. The results indicate that with PESS = 1.0 and κ = 1.0, the instance-specific search resulted in the highest AUROC; also, for almost all values of κ, IGES performs better. Table 3 also suggests that it is important to define PESS properly when applying a Bayesian method on a dataset with small to moderate sample size, which is the case in this paper.

Table 3. AUROC of the GES and IGES methods on the lung cancer dataset. Boldface indicates statistically significantly better results.

Method	GES	IGES									
PESS	–	$\kappa = 0.1$	$\kappa = 0.2$	$\kappa = 0.3$	$\kappa = 0.4$	$\kappa = 0.5$	$\kappa = 0.6$	$\kappa = 0.7$	$\kappa = 0.8$	$\kappa = 0.9$	$\kappa = 1.0$
0.1	0.68	0.67	0.70	0.70	0.70	0.70	0.71	0.71	0.70	0.71	0.72
1.0	0.68	0.68	**0.75**	**0.75**	**0.75**	**0.75**	**0.75**	**0.74**	**0.75**	**0.75**	**0.81**
10.0	0.65	**0.75**	**0.75**	**0.77**	**0.76**	**0.75**	**0.72**	**0.73**	**0.73**	0.69	0.70

Table 4. Comparison of the target variable's Markov blanket (MB) in the instance-specific BNs vs. the population-wide BN for PESS = 1.0 and κ = 1.0.

(a) Structural differences of the variables in the MBs in instance-specific BNs vs. the population-wide BN.

# Added	# Deleted	# Reoriented	% Patients
0	0	0	16.9
5	0	0	10.7
4	0	0	7.7
1	0	0	6.9
3	0	2	4.2
0	0	2	4.2
6	0	0	3.8
other	other	other	45.6

(b) Percentage of top-7 variables of the MBs of instance-specific BNs. The MB of the population-wide BN includes the first two variables denoted by *.

Variable name	% Occurrence in patients
EGFR-Exon-19*	98.1
Mets at Dx-Other*	92.8
Race	37.9
EGFR-Exon-18	35.6
EGFR-Exon-20	31.8
cM	26.8
cStage Group	24.5

Table 4a shows the results of comparing the target variable's MB in the instance-specific models versus the population-wide models with PESS = 1.0 and κ = 1.0. It indicates that in 16.9% of the patient cases, the MB of the target variable was exactly the same in instance-specific and population-wide BNs. Also, in 10.7% of the cases, the MB of the target variable had 5 additional variables in instance-specific models compared to the population-wide model. Table 4b also shows the percentage of the 7 variables that occurred the most in the instance-specific MBs. Table 4 supports that instance-specific structures exist for the lung cancer cases in the dataset we used.

6   Discussion and Conclusions

In this paper, we evaluated the performance of an instance-specific BN classifier, which uses the IGES algorithm, to predict 1-year survival for 261 lung cancer patient cases. We compared IGES results to its population-wide counterpart, GES. We compared IGES to GES method for two reasons. First, the goal of this study was to evaluate the effectiveness of instance-specific modeling in predicting lung-cancer survival; therefore, we wanted the only difference between the two methods to be instance-specific versus population-wide modeling, while keeping the type of machine learning classifier the same (i.e., Bayesian networks). Since to date we have only implemented instance-specific and population-wide algorithms for learning BNs, we compared these two methods. Additionally, BNs continue to be an important machine learning method for clinical outcome prediction because they generally perform well and provide interpretable models that clinicians can understand relatively well. We compared the predictive performance using AUROC and the structural differences between the instance-specific and population-wide BNs. The results provide support that the instance-specific models are often different and have better predictive performance than the population-wide ones. Future extensions include (1) tuning hyperparameters of the methods such as PESS and κ, and (2) implementing instance-specific versions of other machine learning classification methods and comparing them to their population-wide counterparts.

Acknowledgement. The research reported in this paper was supported by grant #4100070287 from the Pennsylvania Department of Health (DOH), grant U54HG008540 from the National Human Genome Research Institute of the National Institutes of Health (NIH), and grant R01LM012095 from the National Library of Medicine of the NIH. The content of the paper is solely the responsibility of the authors and does not necessarily represent the official views of the Pennsylvania DOH or the NIH.

References

1. Agrawal, A., Misra, S., Narayanan, R., et al.: Lung cancer survival prediction using ensemble data mining on SEER data. Sci. Program. **20**(1), 29–42 (2012)
2. American Cancer Society: Cancer Facts & Figures 2020. American Cancer Society, Atlanta (2020)
3. Boutilier, C., Friedman, N., Goldszmidt, M., Koller, D.: Context-specific independence in Bayesian networks. In: Proceedings of Uncertainty in Artificial Intelligence (UAI), pp. 115–123 (1996)
4. Bray, F., Ferlay, J., Soerjomataram, I., et al.: Global cancer statistics 2018: GLOBOCAN estimates of incidence and mortality worldwide for 36 cancers in 185 countries. CA Cancer J. Clin. **68**(6), 394–424 (2018)
5. Cai, C., Cooper, G.F., Lu, K.N., et al.: Systematic discovery of the functional impact of somatic genome alterations in individual tumors through tumor-specific causal inference. PLoS Comput. Biol. **15**(7), e1007088 (2019)

6. Chickering, D.M.: Learning Bayesian networks is NP-complete. In: Learning from Data, pp. 121–130. Springer (1996). https://doi.org/10.1007/978-1-4612-2404-4_12

7. Chickering, D.M.: Optimal structure identification with greedy search. J. Mach. Learn. Res. **3**, 507–554 (2003)

8. Deist, T.M., Dankers, F.J., Valdes, G., et al.: Machine learning algorithms for outcome prediction in (chemo) radiotherapy: An empirical comparison of classifiers. Med. Phys. **45**(7), 3449–3459 (2018)

9. DeLong, E.R., DeLong, D.M., Clarke-Pearson, D.L.: Comparing the areas under two or more correlated receiver operating characteristic curves: A nonparametric approach. Biometrics **44**(3), 837–845 (1988)

10. Ferreira, A., Cooper, G.F., Visweswaran, S.: Decision path models for patient-specific modeling of patient outcomes. In: AMIA Annual Symposium Proceedings, pp. 413–421 (2013)

11. Heckerman, D.: A tutorial on learning with Bayesian networks. In: Learning in Graphical Models, pp. 301–354. Springer (1998)

12. Heckerman, D., Geiger, D., Chickering, D.M.: Learning Bayesian networks: The combination of knowledge and statistical data. Mach. Learn. **20**(3), 197–243 (1995)

13. Jabbari, F., Cooper, G.F.: An instance-specific algorithm for learning the structure of causal Bayesian networks containing latent variables. In: Proceedings of the SIAM International Conference on Data Mining (SDM), pp. 433–441 (2020)

14. Jabbari, F., Visweswaran, S., Cooper, G.F.: Instance-specific Bayesian network structure learning. Proc. Mach. Learn. Res. **72**, 169–180 (2018)

15. Jabbari, F., Visweswaran, S. and Cooper, G.F.: An empirical investigation of instance-specific causal Bayesian network learning. In: IEEE International Conference on Bioinformatics and Biomedicine (BIBM), pp. 2582–2585. IEEE (2019)

16. Kourou, K., Exarchos, T.P., Exarchos, K.P., et al.: Machine learning applications in cancer prognosis and prediction. Comput. Struct. Biotechnol. J. **13**, 8–17 (2015)

17. Lengerich, B.J., Aragam, B., Xing, E.P.: Personalized regression enables sample-specific pan-cancer analysis. Bioinformatics **34**(13), i178–i186 (2018)

18. Ramsey, J.D.: Scaling up greedy equivalence search for continuous variables. arXiv preprint arXiv:1507.07749 (2015)

19. Sesen, M.B., Kadir, T., Alcantara, R.B., et al.: Survival prediction and treatment recommendation with Bayesian techniques in lung cancer. In: AMIA Annual Symposium Proceedings, pp. 838–847 (2012)

20. Travis, W.D., Brambilla, E., Noguchi, M., et al.: International association for the study of lung cancer/American Thoracic Society/European Respiratory Society international multidisciplinary classification of lung adenocarcinoma. J. Thorac. Oncol. **6**(2), 244–285 (2011)

21. Visweswaran, S., Cooper, G.F., Chickering, M.: Learning instance-specific predictive models. J. Mach. Learn. Res. **11**(Dec), 3333–3369 (2010)

22. Zheng, Z., Webb, G.I.: Lazy learning of Bayesian rules. Mach. Learn. **41**(1), 53–84 (2000)

Development and Preliminary Evaluation of a Method for Passive, Privacy-Aware Home Care Monitoring Based on 2D LiDAR Data

Paolo Fraccaro[1]($\boxtimes$), Xenophon Evangelopoulos[1,2],
and Blair Edwards[1]

[1] Hartree Centre, IBM Research Europe, Daresbury, UK
paolo.fraccaro@ibm.com
[2] Computer Science Department, University of Liverpool, Liverpool, UK

Abstract. With an ageing population, the healthcare sector struggles to cope worldwide. Home monitoring using technology is increasingly used to release the pressure on healthcare professionals and keep elderly at home for longer. However, current solutions present technical (i.e. low resolution and convenience) and ethical issues. In this paper, we used 2D LiDAR, as a sensor that can provide significant information on patient's activities, whilst still ensuring their privacy (i.e. 2D LiDAR only produces anonymous point clouds). Particularly, we developed an algorithm that uses clustering on the raw 2D LiDAR data, object tracking on cluster centroids to identify a user in a room, and semantic enrichment using metadata about the room (i.e. areas of interest and furniture position) to associate stationary and non-stationary points with every day activities (e.g. relaxing on the couch, working at the desk, standing by the window, and walking). We tested our method across different users (N = 3) and two rooms for a total 60 randomly ordered activity sequences, with five activities per sequence and each activity performed for 30 s. We obtained an overall accuracy in identifying the activities of 0.88 (standard deviation [SD], 0.06). Walking was the activity with the highest F1 score, with values of 0.97 (SD, 0.04) and 1.00 (SD, 0.00). As expected, activities where occlusion from pieces of furniture might be in the way had worse performance with an F1 score of 0.81 (SD, 0.24). Although performed on a limited sample, our paper shows potential for 2D LiDAR to be used for remote monitoring of mobility and daily activities of elderly in their home.

Keywords: Ambient assisted living · Home monitoring · LiDAR · Clustering · Daily activities

1 Introduction

With an ageing population and stretched resources, the public and private healthcare sector is failing to cope worldwide, and this can only get worse, with the number of people older than 60 raising dramatically by 2050 [1]. Therefore, keeping elderly in their homes safely for as long as possible must be a priority [2]. However, home care still requires a vast amount of resources and costs (i.e. carers and nurses to visit elderly

M. Michalowski and R. Moskovitch (Eds.): AIME 2020, LNAI 12299, pp. 160–169, 2020.
https://doi.org/10.1007/978-3-030-59137-3_15

at home), with potential safety risks when elderlies are by themselves (i.e. missing falls or behaviors indicators of cognitive decline or impairment).

Home care monitoring using technology is becoming an increasingly popular solution to overcome the above-mentioned issues faced by the healthcare sector [2]. However, despite the steady release of new technology on the market, current solutions still present challenges [2, 3]. Particularly, wearable solutions are often not ideal when caring for elderly patients, as they often do not like them and have to remember to wear and charge them. Looking at totally passive sensors, the spectrum of possible solutions available on the market ranges between two possible approaches. On the one side, there are simpler systems that use more established type of sensors (e.g. infrared motion sensors, smart plugs, and ambient sensors) [3]. Nevertheless, these solutions are less intrusive (i.e. only sensor's event name and timestamp is recorded) and do not require data to be trained, they often do not provide enough details on the actual behaviors of a user, unless many sensors are spread around the home. On the other side, there are more intrusive solutions that use cameras or microphones with advanced Artificial Intelligence methods to derive patient's behaviors and status. These solutions provide a great level of detail and allow for real-time monitoring of patients (i.e. watch/listen to live stream). However, they require vast amounts of training data and, more importantly, present significant privacy and ethical concerns [4]. "Middle-ground" solutions that provide enough information to analyze and track patient's behavior, with much smaller privacy implications, are also under development (e.g. WiFi to track people behaviors indoor [3, 5]). However, for the most part these technologies are not mature enough or too expensive.

Another of such "middle-ground" technology that has been used for years in geospatial surveying and robotics, and lately has received substantial attention in the context of self-driving cars, is Light Detection And Ranging (LiDAR) technology [6]. By generating laser pulses while spinning and measuring the reflected signal, LiDAR is able to accurately detect obstacles and surroundings as completely anonymous point clouds. Particularly, 2D LiDAR, which are becoming more and more affordable and smaller in size, have recently been used to detect and track people [7, 8], as well as derive their activities [9]. However, these studies relay on approaches that are based on trained machine learning models (i.e. specific to the room the sensor was deployed in) or require the sensor to be placed at a specific height to work. These make them difficult to be used elsewhere.

In this paper, we present an approach that overcomes these drawbacks. Particularly, our method identifies and tracks a patient by clustering point clouds 2D LiDAR data, and then derives daily activities by using metadata about the room the sensor has been deployed in. We tested our approach across different rooms, users, and heterogeneous set of activities.

2   Methods

2.1   2D LiDAR Sensor

For this project, we used a 2D LiDAR (RPLiDAR A2 from SlamTech [10]) with a 12 m range and a sampling frequency of 2 Hz (i.e. two full scans in a second). Each full scan, which we will call *time frame* in the remainder of this paper, produces 2D point clouds in polar coordinates, representing the reflected signal from the surroundings. The sensor was powered and controlled via a Docker container on a Raspberry Pi 3B + (https://www.raspberrypi.org/), which saved the data in batches every 60 s and securely transferred it to IBM Cloud Object Storage [11].

2.2   2D LiDAR Data Processing

Figure 1 shows our method to derive daily activities from 2D LiDAR data in its three main steps. Below, details on each step are reported.

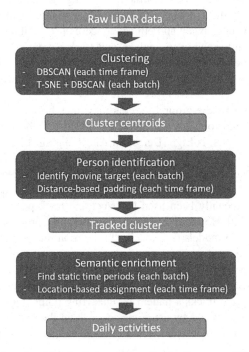

Fig. 1. Data processing algorithm to derive daily activities from raw 2D LiDAR data. Smaller light blue rectangles represent data that flows through the algorithm, while bigger dark blue rectangles represent the methodological steps applied. (Color figure online)

2.2.1   Clustering

The aim of the first step is to identify and separate the different 2D point clouds present in each time frame, which represent not only the person moving in the room but also the room's furniture and room walls. First, we convert the sensor readings from polar to cartesian XY coordinates. Then, we apply DBSCAN [12], a density-based clustering method that is particularly suited for point cloud data, to the points in each time frame, recording the centroid for each cluster. The centroid is calculated by finding the point within the cluster with the minimum cumulative distance from all the other points in the same cluster.

We then divide the data in different batches (i.e. group of consecutive time frames) where we stack the centroids in time. Heuristically, we identified that 60-s long time windows were appropriate for our data. To this 3D data (i.e. centroids x and y coordinates as well as time) we apply t-SNE [13] that allows the further separation of clusters across time and space. By applying DBSCAN to the t-SNE transformed data, we are able to track clusters across consecutive scans.

2.2.2   Person Identification

Currently, our algorithm makes the assumption that there is always only one moving cluster in the room (i.e. it does not track multiple people). Therefore, the first stage in identifying a moving person, is to find moving clusters among the ones tracked in the previous step. To identify moving clusters within a batch, we use linear regression on each tracked cluster with x and y against time. This is done separately for each axis. The idea is that clusters that do not move (e.g. walls or furniture) will be represented by a horizontal line over consecutive time frames in x and y, while the moving ones will have a slope or curve. Heuristically, we found that a slope coefficient of >0.6 cm/s either on x or y was indicative of a moving cluster.

Out of all identified moving clusters we disregard the ones on the boundaries (e.g. walls and peripheral furniture), as we make the assumption that it would be really unlikely for a person to be performing an activity in that area. To identify the room boundaries, we use the occupancy grid mapping approach [14]. Particularly, we calculate a spatial histogram with a bin size of 20 cm for each time frame of the first 10 min of our data, and label as boundaries those bins that have a count greater than 0 at least 50% of the times. From these bins the boundaries of the room are derived by using Orthogonal Convex Hull [15], using an internal buffer of 20 cm.

Occasionally, more than one moving cluster can be found at the same time. This might happen because the person moving in the room changes the way the laser hits objects and therefore some static surfaces can be seen as "moving" by the algorithm. To deal with this, we use a distance-based approach to identify which one among the moving clusters is our target (i.e. the person in the room). Particularly, we calculate the distance between the moving clusters and last known target position, with the closest moving cluster within 50 cm being considered as our new target location. If no moving clusters are found, we attempt to match a stationary cluster to the last known target position based on their distance, with a threshold of 50 cm. If no cluster is found, the target location is padded from the last known location of the target over the previous time frames.

At the end of this stage, we obtain for each time frame, an estimated location for the person in the room (i.e. target).

2.2.3   Semantic Enrichment

The final stage identifies consecutive static periods and assigns an activity to them. The former, is done by looking at consecutive target locations within 5 s and calculating the distance of each point to the others. If all distances are within a certain threshold (i.e. 50 cm), those points represent a static period. If not, the first time point in the 5 s sequence is labelled as 'walking' and the algorithm moves on focusing on the next 5 s.

Finally, an activity is assigned to each static period by applying semantic enrichment [16]. This uses the metadata about the room, such as position of furniture or specific areas of interest. To each of those an activity is associated, and if the target coordinates fall within that area, that activity is inferred. For example, relaxing on the sofa is associated to the area where the sofa is, or watering the plants is associated to the area where the plants are. At the end of this stage, we obtain a list of time-stamped (start and finish) activities that we can use to assess overall behaviors.

2.3   Evaluation

2.3.1   Experimental Procedure

We deployed our 2D LiDAR sensor on a table at a height of 1.1 m in two rooms at the Hartree Centre (Warrington, UK). These rooms contained furniture typical of an office environment (e.g. desks, office chairs, or boards) and living room environment (e.g. sofa and armchairs). This included furniture to which we associated sitting activities and different areas where "standing activities" could be performed. Figure 2 shows the rooms overall layout, while the related areas of interest dimensions and associated activity are shown in Table 1.

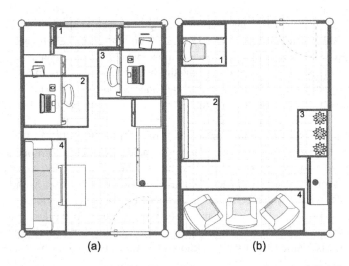

(a)   (b)

Fig. 2. Layout of room 1 (a) and 2 (b). The red dot represents where the LiDAR was placed. Blue rectangles represent the different areas on interest with their associated number in Table 1. (Color figure online)

Table 1. Activities and areas of interest across room 1 and 2. Width and height are in meters.

Room (W × H)	Area (W × H)	Furniture/Location	Activity
1 (3.7 × 5.0)	1 (1.6 × 0.8)	Window	Standing by window
	2 (1.5 × 1.0)	Left desk & chair	Working at desk 1
	3 (1.5 × 1.0)	Right Desk & chair	Working at desk 2
	4 (0.9 × 2.1)	Sofa	Relaxing on the sofa
2 (3.5 × 5.5)	1 (1.1 × 1.0)	Chair and bookcase	Sitting corner chair
	2 (1.0 × 1.4)	Board	Writing on the board
	3 (0.7 × 1.5)	Plants	Watering plants
	4 (2.5 × 1.0)	Armchairs	Relaxing on armchair

To test our approach, we collected data among healthy volunteers part of the research team (N = 3) whilst performing the activities listed in Table 1. For each room, we also included "walking" among the tested activities. In order to evaluate how our approach would perform in real life, users were encouraged to interact as naturally as possible with objects and space around them while performing activities. For example, when "working at the desk" they were told to use the available laptop and phone. Furthermore, for "relaxing on the sofa" participants were also instructed to perform the activity using a range of different postures, including laying and sitting.

A Python programming language script randomized the activities under study in different sequences, where for each sequence an activity would be done once, and instructed the user on which activity to perform, while labelling the data for performance evaluation. Following the approached used by [9], each activity lasted for 30 s and then the user was asked to move to the next activity. Overall, each user performed 10 activity sequences in each room, for a total of 60 sequences.

2.3.2   Outcome Measures

We evaluated the overall performance in terms of the accuracy with which our algorithm identified the activities that the user was performing (i.e. the presence of the user in a specific area of interest or walking). We calculated this as the proportion of correct activity identification over all the activities performed based on the labelled data. To identify which activity was identified by our algorithm, we calculated the most frequent identified activity (i.e. modal activity) over each 30 s time slot. To have an idea of how stable the activity identification was over that period of time, we also calculated mean and standard deviation (SD) of the proportion of time the modal activity was identified over all 30 s time slots.

We also evaluated activity specific performance in terms of precision, recall, and F1 score. Precision was the proportion of time an identified activity was correct over all the times that activity was identified. Recall was the proportion of times an activity was correctly identified over all the times an activity was actually performed. F1 score was the harmonic mean among the two.

For each room, overall and activity specific performance was calculated for each user, and then averaged.

3   Results

3.1   Qualitative Analysis

Figure 3a shows the spatial distribution of the 2D LiDAR data over the tests in Room 1. Brighter colour represents areas with higher concentration of 2D LiDAR data points, and the LiDAR is located in the origin of the axis. From Fig. 3a it is possible to visually identify the boundaries of the room, which we identified algorithmically with the occupancy mapping grid approach, as well as the areas of interest where users performed the different activities. Another feature that can be spotted, are the areas where no data was collected (i.e. no user visit) because of the presence of furniture. A clear example is the coffee table in front of the sofa in Room 1 (see Fig. 2), which is represented by a hole in the middle of the spatial histogram in Fig. 3a.

Figure 3b shows an example of results from applying our algorithm for a time frame in Room 1. Again, the LiDAR is located at the axis origin. In the figure, the black points were the ones that were not considered as room boundaries by the occupancy mapping grid algorithm. In addition to pieces of the sofa (bottom left), these points represent the office chairs (left and right between 1000 and 2000 y coordinates), the drawer in the top left corner, and the person in the room (bottom right). From the point cloud representing the person, it is also possible to spot the person's legs as two separate blobs part of the same point cloud. Finally, the red shaded markers are indicative of the current and previous coordinate of the person in the room over time. This allows us to calculate distance covered, as well as walking speed statistics.

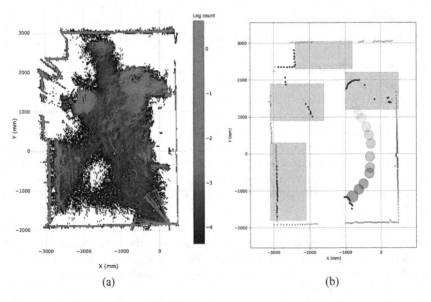

(a)   (b)

Fig. 3. (a): Spatial histogram of 2D LiDAR data over tests in room 1. Bins have dimensions 2 × 2 cm, and counts are shown in log scale. (b): Example of algorithm results. Grey points and red markers represent boundaries (e.g. walls) and current/previous user location, respectively. Yellow rectangles represent the areas on interest. (Color figure online)

3.2   Algorithm Performance

Overall accuracy in correctly identifying user's activities (i.e. presence in a specific area or walking) was 0.88 (SD, 0.06). This broke down as 0.84 (SD, 0.07) for Room 1 and 0.92 (SD, 0.03) for Room 2, and ranged from 0.80 to 0.92 in Room 1 and from 0.90 to 0.96 in Room 2. The mean proportion of time for each modal activity during each 30 s time slot was 0.93 (SD, 0.12) for Room 1 and 0.93 (SD, 0.03) for Room 2.

Table 2 shows activity specific performance. In both rooms, walking was consistently identified by our algorithm, with F1 scores of 0.97 (SD, 0.04) and 1.00 (SD, 0.00), respectively. In terms of the static activities, the best performance was for those where the user was standing, with the exception of standing by the window in room 1. As a matter of fact, this activity got the worst F1-score across all activities and rooms. This was primarily due to low recall, driven by a recall score of 0.38 for one of the users.

Table 2. Precision, recall and F1 score across different rooms and activities.

Room	Area of interest	Activity	Precision (SD)	Recall (SD)	F1 score (SD)
1	Window	Standing by window	1.00 (0.00)	0.72 (0.32)	0.81 (0.24)
	Left desk & chair	Working at the desk 1	0.85 (0.07)	0.92 (0.07)	0.88 (0.02)
	Right Desk & chair	Working at the desk 2	0.75 (0.05)	0.89 (0.10)	0.82 (0.08)
	Sofa	Relaxing on the sofa	1.00 (0.04)	0.78 (0.17)	0.87 (0.01)
	/	Walking	1.00 (0.00)	0.95 (0.08)	0.97 (0.04)
2	Chair and bookcase	Sitting corner chair	0.93 (0.07)	0.83 (0.15)	0.87 (0.09)
	Board	Writing on the board	0.89 (0.12)	0.97 (0.06)	0.92 (0.07)
	Plants	Watering plants	1.00 (0.00)	1.00 (0.00)	1.00 (0.00)
	Armchairs	Relaxing on armchair	0.93 (0.13)	0.80 (0.10)	0.87 (0.09)
	/	Walking	1.00 (0.00)	1.00 (0.00)	1.00 (0.00)

4   Discussion

We developed an approach that uses clustering, object tracking and semantic enrichment to infer daily activities from 2D LiDAR data. Preliminary evaluation of our approach across different rooms and users, obtained an overall accuracy of 0.88 (SD, 0.06). The activity specific performance showed good results in tracking non-static activities such walking, as well as difficulties to track activities where occlusions were present. This was clear in Room 1, where the window was occluded by the office chairs.

To the best of our knowledge, this study is the first one to use a "model-free" approach and in principle could be deployed elsewhere with no or minor changes, provided the room metadata are available (e.g. areas of interest position and related activities). In fact, although a previous study [9] followed a similar approach to ours with 2D LiDAR data (i.e. associating activities based on the location of the user in the room), they trained a recursive neural network to recognize the different activities in the room. This makes it specific to the room the model was developed in, and new training would be needed to deploy the sensor somewhere else. This is also another advantage of our method, which would require only metadata about the room, rather than data to train the model. Another study used 2D LiDAR data to identify people in the room [7, 8], however they did not contextualize the person position on any activity the person might be performing, making it difficult to monitor behaviours. Nevertheless, this approach could be used in combination with our approach, to reduce uncertainty on the cluster identified as the target actually being the person in the room.

Our results show potential for 2D LiDAR data to be used for remotely monitor elderly activities at home. Particularly, the high performance obtained in tracking walking seems to indicate that this approach could be particularly suited to track mobility. Good performance in tracking other more "static" activities could create new streams of data that will allow anomaly detection in longitudinal behavioral patterns as well as signs of early decline. Although our evaluation does not explicitly consider it, our method could also, in principle, be used to identify falls, which are one of main issues affecting the elderly [17]. In fact, if a person is lost while being tracked in a location where no known areas of interest are present for a certain amount of time and with no subsequent movement detected around the house, it is reasonable to assume that the person has fallen below the 2D LiDAR plane and is laying on the floor.

Despite the clear innovation on the current landscape of passively deriving daily activities for home remote monitoring, our paper has some limitations. First, we tested our approach in a small study population. However, we believe we provided enough variability in our experiments to show the potential of our method. Second, some activities we tested (e.g. working at the desk) were not directly relatable to elderly behaviours. However, these activities are very similar to others that elderly would perform (e.g. eating at the dining table). Furthermore, our main objective was to demonstrate the potential of our approach in tracking a person across different areas of interest in a room, which is directly translatable to elderlies remote monitoring. Finally, by processing 60 s batches, our approach cannot be used for totally synchronous real-time monitoring. However, we believe that 60 s lag is an acceptable delay, which would not impact on usefulness of the information provided.

For the future, we plan to tackle multiple occupancy households by integrating multitarget hypothesis tracking [18] in our person detection step. We will also explore ways to improve detection in presence of occlusions by potentially integrating the 2D LiDAR data with simpler sensors' data (e.g. infrared motion sensor) placed in occluded areas. Furthermore, we plan to improve our method by exploring opportunities to automatically detect landmarks in a room, through the identification of areas of high static occupancy (similarly to Fig. 3a), for easier deployment and further privacy preservation. Finally, we plan to expand our method for real-time use by using less computationally intensive clustering techniques such as Point-Distance-Based Segmentation [19]. All these improvements will further enhance usefulness in the real-world, where we plan to validate our method with elderly patients in a pilot study in the near future.

Acknowledgements. This work was supported by the STFC Hartree Centre's Innovation Return on Research programme, funded by the UK Department for Business, Energy & Industrial Strategy. XE was supported by the IBM Research internship program during summer 2019.

References

1. World Report on Ageing and Health. World Health Organisation, Geneva (2015)
2. Majumder, S., et al.: Smart homes for elderly healthcare–recent advances and research challenges. Sensors **17**(11), 2496 (2017). https://doi.org/10.3390/s17112496
3. Uddin, M.Z., Khaksar, W., Torresen, J.: Ambient sensors for elderly care and independent living: a survey. Sensors **18**(7), 2027 (2018)
4. Marikyan, D., Papagiannidis, S., Alamanos, E.: A systematic review of the smart home literature: a user perspective. Technol. Forecast. Soc. Change **138**, 139–154 (2019). https://doi.org/10.1016/j.techfore.2018.08.015
5. Zhao, M., et al.: Through-wall human pose estimation using radio signals. In: Proceedings of the IEEE Conference on Computer Vision and Pattern Recogniton, pp. 7356–7365 (2018)
6. Dong, P., Chen, Q.: LiDAR Remote Sensing and Applications. CRC Press, New York (2017)
7. Guerrero-Higueras, Á.M., et al.: Tracking people in a mobile robot from 2D LIDAR scans using full convolutional neural networks for security in cluttered environments. Front. Neurorobotics **12**, 85 (2019). https://doi.org/10.3389/fnbot.2018.00085
8. Álvarez-Aparicio, C., Guerrero-Higueras, Á.M., Rodríguez-Lera, F.J., Ginés Clavero, J., Martín Rico, F., Matellán, V.: People detection and tracking using LIDAR sensors. Robotics **8**(3), 75 (2019)
9. Bodanese, E., Ma, Z., Bigham, J., et al.: Device-free daily life (ADL) recognition for smart home healthcare using a low-cost (2D) lidar. In: 2018 IEEE Global Communications Conference (GLOBECOM) (2018). https://ieeexplore.ieee.org/document/8647251
10. SlamTech. https://www.slamtec.com/en/Lidar/A2, RPLiDAR A2
11. IBM Cloud Object Storage. https://www.ibm.com/uk-en/cloud/object-storage
12. Ester, M., Kriegel, H.-P., Sander, J., Xu, X.: A density-based algorithm for discovering clusters in large spatial databases with noise. In: KDD 1996 (1996). https://dl.acm.org/doi/10.5555/3001460.3001507
13. van der Maaten, L., Hinton, G.: Visualizing data using t-SNE. J. Mach. Learn. Res. **9**, 2579–2605 (2008)
14. Colleens, T., Colleens, J.: Occupancy grid mapping: an empirical evaluation. In: 2007 Mediterranean Conference on Control & Automation, pp. 1–6 (2007). https://ieeexplore.ieee.org/document/4433772
15. Wikipedia, Orthogonal Conven Hull. https://en.wikipedia.org/wiki/Orthogonal_convex_hull
16. Fraccaro, P., Lavery-Blackie, S., Van der Veer, S.N., Peek, N.: Behavioural phenotyping of daily activities relevant to social functioning based on smartphone-collected geolocation data. Stud. Health Technol. Inform. **264**, 945–949 (2019)
17. Duque, G.: Age-related physical and physiologic changes and comorbidities in older people: association with falls. In: Huang, A.R., Mallet, L. (eds.) Medication-Related Falls in Older People, pp. 67–73. Springer, Cham (2016). https://doi.org/10.1007/978-3-319-32304-6_6
18. Blackman, S.S.: Multiple hypothesis tracking for multiple target tracking. IEEE Aerosp. Electron. Syst. Mag. **19**(1), 5–18 (2004)
19. Premebida, C.: Segmentation and geometric primitives extraction from 2d laser range data for mobile robot applications. Robotica **2005** (2005). https://www.semanticscholar.org/paper/Segmentation-and-Geometric-Primitives-Extraction-2D-Premebida/b0017df5fe86501e8338ff737077c13777d55a9d

Innovative Method to Build Robust Prediction Models When Gold-Standard Outcomes Are Scarce

Ying Zhu[1], Roshan Tourani[1], Adam Sheka[2], Elizabeth Wick[3],
Genevieve B. Melton[1,2], and Gyorgy Simon[1,4(✉)]

[1] Institute for Health Informatics, University of Minnesota, Minneapolis,
MN 55455, USA
simo0342@umn.edu
[2] Department of Surgery, University of Minnesota, Minneapolis,
MN 55455, USA
[3] Department of Surgery, University of California San Francisco,
San Francisco, CA 94158, USA
[4] Department of Medicine, University of Minnesota, Minneapolis,
MN 55455, USA

Abstract. Postsurgical hospital acquired infections (HAIs) are viewed as a quality benchmark in healthcare due to their association with morbidity, mortality and high cost. The prediction of HAIs allows for implementing prevention strategies at early stages to reduce postsurgical complications. In the United States, the National Surgical Quality Improvement Program (NSQIP) maintains a registry considered a "gold-standard" for HAIs outcome reporting, but it relies heavily on costly manual chart review and therefore only includes a small percent of surgery cases from each participating site. Most HAI prediction models rely on a wide range of weak risk factors, which are combined into models with many parameters and require larger sample sizes than available from NSQIP at a single health system. In this study, we propose an alternative approach to develop a robust prediction model, using the few NSQIP cases efficiently. Rather than training the HAIs prediction models directly on the small number of NSQIP patients, we leverage a simple detection model which detects HAIs after the fact on postoperative data and use this detection model to label a large non-NSQIP perioperative dataset on which prediction models are constructed. Detection models rely on strong signals requiring fewer samples to learn. We evaluate this approach in a single academic health system with 115,202 surgeries (10,354 in NSQIP). The prediction models were evaluated on the NSQIP "gold-standard" labels. While organ-space surgical site infection showed comparable performance, the proposed model demonstrated better performance for prediction of superficial surgical site infection, sepsis or septic shock, pneumonia, and urinary tract infection.

Keywords: Artificial intelligence · Medicine · Hospital acquired infection · Machine learning · Predictive modeling

© Springer Nature Switzerland AG 2020
M. Michalowski and R. Moskovitch (Eds.): AIME 2020, LNAI 12299, pp. 170–180, 2020.
https://doi.org/10.1007/978-3-030-59137-3_16

1 Introduction

Infections acquired in a hospital or other healthcare facility, or hospital-acquired infections (HAIs), are associated with significant morbidity, mortality, and prolonged hospital length of stay and are expensive to treat [1–4]. Among surgical adverse events, HAI rates are a quality benchmark and are increasingly emphasized by payers and the public due to their clinical severity and associated costs. Thus, prevention and reduction of HAIs are top priorities for hospitals nationwide. For surgical patients, it is imperative to develop ongoing interventions and tailor treatment plans aimed to reduce the chance of postsurgical infections including identifying modifiable risk factors associated with HAIs at early stage of patient care.

A growing body of literature has explored risk factors predictive of postoperative complications [5–8]. One common resource used for postoperative complications research is the ACS National Surgical Quality Improvement Program (NSQIP) registry database, which provides high-fidelity data and serves as a "gold-standard" for HAIs outcome reporting. NSQIP data is manually extracted and reviewed by trained nurse reviewers and includes outcomes occurring within 30 days of the index surgical case. Due to the cost associated with this manual review process, most centers abstract only a subset of the surgical cases from each participating institution. Effective machine learned predictive models, however, typically require larger labeled datasets; data with "gold-standard" NSQIP labels may prove to be of insufficient size.

In this study, we propose a semi-supervised approach leveraging larger quantities of unlabeled data (data not in the NSQIP sample) to build robust predictive models. Instead of building the predictive models directly on the NSQIP cases, we first apply a detection model on postoperative data using a protocol we previously developed [9] and then use the detection model to assign a "silver-standard" label to the non-NSQIP perioperative dataset. We then train our predictive model on the large non-NSQIP perioperative data with the "silver standard" labels.

Intuitively, this approach works for two reasons. First, we incorporate information external to the prediction problem. The postoperative data used to learn the "silver standard" labels is not part of the original prediction problem (which primarily uses perioperative data); hence, it contains information that the perioperative data does not have. The postoperative data is closer to the outcome, so it likely contains more information about it. Our approach stands in contrast to traditional semi-supervised learning approaches that typically rely solely on the same perioperative data and try to exploit its structure to extract additional information about the outcome for unlabeled instances. Second, because the postoperative data is closer to the outcome, building the detection models is a far simpler task than learning the prediction models. Detection models contain fewer (but stronger) predictors, requiring fewer samples to learn, thus the NSQIP sample size suffices. These stronger postoperative features include diagnosis codes, imaging and lab test orders and procedures, and microbiological results. On the other hand, none of these strong features are available to the prediction model, which must instead rely on a larger number of weaker features, requiring a larger sample size to train than what is available in the NSQIP sample.

We evaluated the proposed approach on data from 115,202 surgeries (of which 10,354 are included in NSQIP) performed at Fairview Health Services (FHS) between 2011 and 2019 for five HAIs: superficial and organ-space surgical site infection, sepsis/septic shock, pneumonia and urinary tract infection.

2   Materials and Methods

Figure 1 presents the overview of the modeling and evaluation approach used in this study.

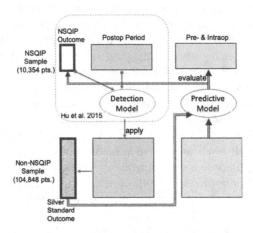

Fig. 1. Overview of model development and evaluation process

2.1   Study Design

This study was performed at M Health Fairview, a health system comprised of a flagship academic hospital, the University of Minnesota Medical Center (UMMC), and 11 community hospitals located in Minneapolis, Minnesota. Institutional review board approval (IRB) was obtained and informed consent waived for this minimal risk study.

The study population includes 115,202 adult surgical cases recorded in the clinical data repository (CDR) of the University of Minnesota between years 2011 and 2019. Data used for this study were collected from two major sources: the CDR and NSQIP outcomes. We collected structured EHR data from the CDR as model variables and the reported postsurgical complications from the NSQIP registry as "gold-standard" labels. The patients' medical record number and date of surgery are used to link the CDR data with the NSQIP outcome. Since only a small set of surgical cases is selected in the NSQIP registry, the study population can be further divided into two groups: NSQIP group (10,354 cases) with "gold-standard" HAIs labels and non-NSQIP group (104,848 cases) without such labels.

Variables. The independent variables include demographic, pre-, intra-, and postoperative data. We limit the demographics variables to those generally available (e.g., age at surgery and gender). Preoperative variables include information related to medical history, such as problem list, procedures, medications and preoperative indication for surgery up to 30 days prior to surgery. Preoperative laboratory results and vital sign measurements are used to establish a baseline. Intraoperative variables include orders, medications, and high-resolution vital signs and lab values recorded during the operation. Continuous variables are aggregated to mean values as described in Tourani et al. [5]. Postoperative variables are associated with events occurring during the postoperative window (from day 3 to 30 after surgery), including relevant diagnosis codes, orders, procedures, microbiology and lab test results and vital sign values. Due to the difference in temporal occurrence of the variables, preoperative and intraoperative features are mainly used for prediction model training and testing purposes, while the postoperative features are used for detection model development and application purposes.

Outcomes. We consider five types of HAIs outcomes: pneumonia (PNA), urinary tract infection (UTI), occurrence of sepsis or septic shock (SE|SS), superficial surgical site infections (Superficial SSI), and organ-space SSI. Outcome information collected from the NSQIP registry source is considered a "gold-standard" since NSQIP data is abstracted through a manual review process established by the NSQIP program. We do not have outcome information for non-NSQIP cases; therefore, we generated labels by applying the detection model onto the postoperative variables as our "silver-standard". Details regarding detection model development and application are discussed in the section below.

2.2   Analysis

Construct Detection Model. The detection models for each outcome were built independently on the NSQIP dataset with postoperative features using "gold-standard" outcomes using the algorithms we previously developed [9]. The 95% confidence interval of area under the ROC curve (AUC) score for internal validation are calculated through 1,000 bootstrap iterations.

Label Non-NSQIP Cases. The probability of each HAI for non-NSQIP cases was calculated by applying the detection models to the postoperative features of the non-NSQIP cases. We adapt a three-way labeling system to assign the cases into three categories: positive (complication is present), negative (complication is unlikely) and unknown. In this system, two thresholds are determined by weighing the specificity, sensitivity, positive and negative predictive values (PPV and NPV). The cases with probabilities larger than or equal to the higher threshold are assigned a positive outcome; and the cases with probabilities smaller than the lower threshold are assigned a negative outcome. For each outcome, the cases with unknown labels were excluded from predictive model training. We refer to this labeling as the "silver standard" label.

Train Predictive Model. For each outcome, the predictive model was trained on non-NSQIP dataset with pre- and intraoperative features using the "silver-standard" labels. Logistic regression model was used. Missing values were imputed using median imputation. Causal variables screening process [10] through the PC-Simple algorithm [11] followed by backward elimination was used as feature selection. For additional details, such as aggregating the intraoperative features, the reader is referred to Tourani et al. [5]. Prediction models were evaluated on the NSQIP dataset with "gold-standard" labels using the AUC score.

Comparison Model. To assess the performance of the proposed approach, we built a comparison model using the pre- and intraoperative features directly on the NSQIP outcome data using the "gold-standard" NSQIP labels. This model is also a logistic regression model, using the same imputation and features selection methodology as the proposed predictive models. Model performance was measured as the AUC score through 10-fold cross validation.

Comparison of the Comparison and Proposed Models. Bootstrap estimation with 20 replications was conducted to compare the model performances. For each bootstrap iteration, the NSQIP and non-NSQIP cases were sampled accordingly and used for models. We measured the difference in model performance as the mean difference in AUC between the two modeling approaches. The statistical significance in model performance was calculated through a paired t-test.

3   Results

3.1   Cohort Description

Table 1 presents a summary of the two cohorts (10,354 NSQIP and 104,848 non-NSQIP surgeries) included in this study with demographics, HAIs outcome, and selected perioperative features. Binary variables are reported as counts with percentages in parenthesis; the continuous variables are reported as median with interquartile range in parenthesis. For the NSQIP surgeries, we report the "gold-standard" outcomes as reported by the registry; and for the non-NSQIP surgeries, we report the "silver-standard" outcomes determined by the detection model applied to the postoperative data. The non-NSQIP cohort has overall higher percentage of outcomes than the NSQIP cohort. Due to the large number of predictors, only features among the top five standardized coefficients of each predictive model are reported.

Table 1. Cohort description

	NSQIP	Non-NSQIP
Total surgical cases	10,354	104,848
Demographics		
Age	54 (42, 65)	55 (44, 67)
Gender (male)	4,382 (42.3%)	50,005 (47.7%)
Outcomes		
PNA	226 (2.2%)	2,926 (2.8%)
SE|SS	230 (2.2%)	4,491 (4.3%)
UTI	245 (2.4%)	3,248 (3.1%)
Superficial SSI	231 (2.2%)	4,262 (4.1%)
Organ-space SSI	231 (2.2%)	5,215 (5.0%)
Predictive Features		
Asthma (HX)	38 (0.4%)	577 (0.6%)
Bacteremia (HX)	50 (0.5%)	1,332 (1.3%)
Metastatic disease (HX)	162 (1.6%)	2,066 (2.0%)
Leukopenia (HX)	13 (0.1%)	596 (0.6%)
Malnutrition (HX)	115 (1.1%)	2,928 (2.8%)
SSI related diagnosis (HX)	93 (0.9%)	1,118 (1.1%)
Opport. Inf. (HX)	73 (0.7%)	1,514 (1.4%)
PNA (HX)	133 (1.3%)	3,844 (3.7%)
SSI (HX)	38 (0.4%)	1,076 (1.0%)
UTI (HX)	207 (2.0%)	2,698 (2.6%)
Lymphoma (HX)	78 (0.8%)	1,592 (1.5%)
Ascites (HX)	177 (1.7%)	4,136 (3.9%)
Gangrene (HX)	42 (0.4%)	430(0.4%)
Paraplegia & Paraparesis (HX)	23 (0.2%)	334 (0.3%)
Severe sepsis (HX)	33 (0.3%)	758 (0.7%)
Stroke (HX)	61 (0.6%)	1,195 (1.1%)
Immunosup. Med (Preop)	328 (3.2%)	5,568 (5.3%)
PNA Vac. (Preop)	46 (0.4%)	474 (0.5%)
Steroid (Intraop)	3,134 (30.3%)	23,576 (22.5%)

* Abbreviation: PNA: Pneumonia; SE|SS: Sepsis or septic shock; UTI: Urinary tract infection; SSI: Surgical site infection; HX: History of complications; Opport. Inf.: Opportunistic infections; Immunosup. Med: Immunosuppressants medicine; PNA Vac.: Pneumococcal vaccine

3.2   Model Performance and Variable Selection

First, for each outcome, we constructed detection models built on NSQIP data. In Table 2, we report the performance of these detection models as the mean and empirical 95% confidence interval of the AUC measure estimated from 1,000 bootstrap iterations.

Table 2. Detection model performance through internal validation

Outcome	AUC (95% CI)
PNA	0.945 (0.917, 0.966)
SESS	0.953 (0.933, 0.973)
UTI	0.936 (0.912, 0.958)
Superficial SSI	0.879 (0.844, 0.911)
Organ-space SSI	0.935 (0.902, 0.960)

* 95% confidence interval (CI) of AUC values are calculated based on internal validations using 1,000 times bootstrap iterations.

Figure 2 and Table 3 compare the performance of the proposed predictive model with the comparison model (that uses only the NSQIP data) estimated through 20 bootstrap replications. Figure 2 depicts the 20 AUC values as a boxplot for each outcome. Table 3 shows the mean of AUC and standard deviation in parenthesis; the last column shows the difference in mean AUC between the proposed and comparison modeling approaches and p-value from the paired t-test in parenthesis. The differences are statistically significant at 0.05 confidence level for the prediction of all outcomes except for the Organ-Space SSI.

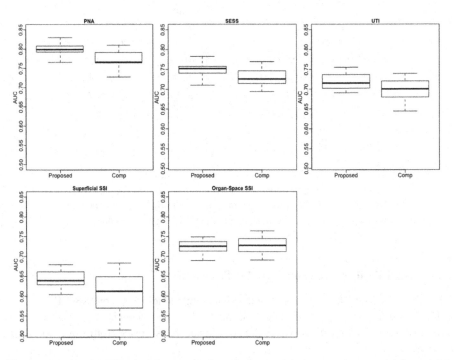

Fig. 2. Boxplot of AUC values for proposed predictive model vs comparison model through 20 times bootstrap iterations

Table 3. Comparison of AUC from proposed and comparison models

Outcome	Proposed	Comparison	Difference
PNA	0.799 (±.016)	0.773 (±.022)	0.026 (6.53e−6)
SE|SS	0.748 (±.018)	0.732 (±.021)	0.016 (2.27e−3)
UTI	0.719 (±.021)	0.700 (±.027)	0.019 (1.84e−3)
Superficial SSI	0.643 (±.022)	0.608 (±.049)	0.035 (1.80e−3)
Organ-space SSI	0.725 (±.016)	0.727 (±.021)	−0.002 (6.42e−1)

To illustrate the difference between a detection and a prediction model, Table 4 presents the coefficients of both the detection and prediction for the SE|SS (sepsis or septic shock) outcome. The detection model built on NSQIP dataset contains only three postoperative factors with strong signals. Compared to the comparison model, the proposed predictive model captures more risk factors associated with the SE|SS prediction.

Table 4. Selected variables and modeling coefficients of SE|SS detection and predictive models

	DDetection	Predictive		
			Proposed	Comparison
(Intercept)	-9.586	(Intercept)	−18.664	−2.466
Sepsis Dx (Postop)	0.503	Gender (Female)	−0.405	−0.687
Blood Cult. (Postop)	0.879	Bacterial infection (HX)	0.211	
Max. Temp. (Postop)	0.055	Bacteremia (HX)	0.999	
		Metastatic disease (HX)	0.405	
		Other infections (HX)	0.273	
		Leukopenia (HX)	1.143	
		Malnutrition (HX)	0.416	
		SSI related Dx (HX)	–	1.302
		Opport. Inf. (HX)	0.555	
		PNA (HX)	0.369	
		Sepsis (HX)	0.653	1.134
		UTI (HX)	0.376	
		Drug abuse (HX)	0.411	
		Lymphoma (HX)	0.582	
		Neuro. Disorders (HX)	0.262	
		CVA w/o Neuro. (HX)	0.336	
		Insulin (Prior 1 year)	0.245	1.054
		Antibiotics (Prior 30 days)	0.553	0.561
		Antibiotics (Preop)	0.367	
		Immunosup. Med (Preop)	0.390	

(continued)

Table 4. (*continued*)

	DDetection	Predictive		
		PTT (Prior 30 days)	0.006	
		Respiratory Rate (Preop)	0.022	
		Temperature (Preop)	0.441	
		Pulse (Preop)	0.016	
		GCS (Preop)	−0.078	
		DBP (Preop)	−0.018	
		Bilirubin (Intraop)	0.056	
		Calcium (Intraop)	−0.184	
		Glucose (Intraop)	0.003	
		Hemoglobin (Intraop)	−0.147	−0.166
		Lactate (Intraop)	0.052	0.190
		pO2 Arterial (Intraop)	0.003	
		Pulse (Intraop)	0.012	
		PIP (Intraop)	0.045	
		FiO2 (Intraop)	−0.008	
		CVP (Intraop)	0.032	

* Abbreviation: Sepsis Dx: Sepsis diagnosis; Blood Cult.: Blood culture; Max. Temp.: Maximum temperature; SSI Related Dx: SSI related diagnosis; Neuro. Disorders: Neurological disorders; CVA w/o Neuro.: CVA/Stroke with no neurological deficit; GCS: Glasgow coma score; DBP: Diastolic blood Pressure; pO2: Partial pressure of oxygen; PIP: Peak inspiratory pressure; FiO2: Fraction of inspired oxygen; CVP: Central venous pressure

4 Discussion

In this study, we propose an innovative approach to build robust prediction models when the "gold-standard" outcomes from NSQIP are scarce. This method uses the few NSQIP cases efficiently by first building a simple detection model on postoperative data (after the fact) and using these detection models to assign "silver-standard" labels to a much larger non-NSQIP perioperative dataset, on which the predictive models are then built.

The proposed method shows significantly better performance to predict four out of five HAIs investigated (superficial surgical site infection, sepsis or septic shock, pneumonia, and urinary tract infection) than the comparison model which is built directly on the NSQIP cases. As the first step, the detection models show high performances, which provides reliable "silver-standard" labels for the larger non-NSQIP cohort (Table 2). Compared to pre- and intraoperative features, the postoperative features used for detection models have strong signals indicative of the HAIs occurrence. For example, the detection model for sepsis or septic shock (Table 4) contains three features: the postoperative discharge diagnosis relevant to sepsis, blood culture, and maximum temperature during day 3 to day 30 after the surgery. With only three features, the 95% confidence interval of the AUC for the constructed detection model is

0.933 to 0.973 (Table 2). Even though the number of positive cases in the NSQIP data set is low, this is sufficient to estimate so few variables. Additionally, by labeling the non-NSQIP dataset, we are able to build the predictive models with a much larger dataset, allowing the model to capture more details and therefore show better predictive performance.

For predicting the Organ-space SSI, the proposed model does not show better performance than the comparison model but has comparable performance measures (Table 3). The proposed model appears to select more (noise) features than the comparison model and has slightly overfit the training data. A possible explanation is that the "silver-standard" outcome generated by the detection model introduces systematic false positives. A model that is generalizable within the non-NSQIP cohort will appear as overfit in the NSQIP cohort, which does not have these false positives. Another possible explanation is that the NSQIP sample is not a random sample: NSQIP over-samples difficult and rare cases thus the NSQIP population is different from (contains more serious cases) the non-NSQIP surgical population.

The proposed multi-step modeling approach can be generalized for predicting other postsurgical complication as long as the NSQIP and non-NSQIP samples are not too different in terms of this outcome or more broadly to consider other datasets with smaller numbers of outcomes. The main limitation of this study is that the NSQIP population is potentially different than the overall surgical populations which may produce false positive outcomes on the (less severe) non-NSQIP dataset.

5 Conclusion

This study investigated the feasibility of applying an alternative modeling approach to build predictive model when the "gold-standard" labels are not enough. The multi-step modeling approach was proposed and validated. The proposed model shows better performance for predicting four out of five HAIs outcomes than the comparison model.

Acknowledgements. This work was supported in part by NIGMS award R01 GM 120079, AHRQ award R01 HS024532, and the NCATS University of Minnesota CTSA UL1 TR002494. The views expressed in this manuscript are those of the authors and do not necessarily reflect the views of the funding agencies.

References

1. Khan, N.A., Quan, H., Bugar, J.M., Lemaire, J.B., Brant, R., Ghali, W.A.: Association of postoperative complications with hospital costs and length of stay in a tertiary care center. J. Gen. Intern. Med. **21**(2), 177–180 (2006). https://doi.org/10.1111/j.1525-1497.2006. 00319.x
2. Lawson, E.H., et al.: Association between occurrence of a postoperative complication and readmission. Ann. Surg. **258**(1), 10–18 (2013). https://doi.org/10.1097/SLA.0b013e3182-8e3ac3

3. Horan, T.C., Andrus, M., Dudeck, M.A.: CDC/NHSN surveillance definition of health care–associated infection and criteria for specific types of infections in the acute care setting. Am. J. Infect. Control **36**(5), 309–332 (2008). https://doi.org/10.1016/j.ajic.2008.03.002

4. Dimick, J.B., Pronovost, P.J., Cowan, J.A., Lipsett, P.A., Stanley, J.C., Upchurch, G.R.: Variation in postoperative complication rates after high-risk surgery in the United States. Surgery **134**(4), 534–540 (2003). https://doi.org/10.1016/S0039-6060(03)00273-3

5. Tourani, R., Murphree, D.H., Melton-Meaux, G., Wick, E., Kor, D.J., Simon, G.J.: The value of aggregated high-resolution intraoperative data for predicting post-surgical infectious complications at two independent sites. Stud. Health Technol. Inform. **264**, 398–402 (2019). https://doi.org/10.3233/SHTI190251

6. Leaper, D., Ousey, K.: Evidence update on prevention of surgical site infection. Curr. Opin. Infect. Dis. **28**(2), 158–163 (2015). https://doi.org/10.1097/QCO.0000000000000144

7. Seamon, M.J., Wobb, J., Gaughan, J.P., Kulp, H., Kamel, I., Dempsey, D.T.: The effects of intraoperative hypothermia on surgical site infection. Ann. Surg. **255**(4), 789–795 (2012). https://doi.org/10.1097/SLA.0b013e31824b7e35

8. Chang, Y.-J., et al.: Predicting hospital-acquired infections by scoring system with simple parameters. PLoS One **6**(8), e23137 (2011). https://doi.org/10.1371/journal.pone.0023137

9. Hu, Z., Simon, G.J., Arsoniadis, E.G., Wang, Y., Kwaan, M.R., Melton, G.B.: Automated detection of postoperative surgical site infections using supervised methods with electronic health record data. Stud. Health Technol. Inform. **216**, 706–710 (2015). https://doi.org/10.3233/978-1-61499-564-7-706

10. Ma, S., Statnikov, A.: Methods for computational causal discovery in biomedicine. Behaviormetrika **44**(1), 165–191 (2017). https://doi.org/10.1007/s41237-016-0013-5

11. Kalisch, M., Mächler, M., Colombo, D., Maathuis, M.H., Bühlmann, P.: Causal inference using graphical models with the R package pcalg. J. Stat. Softw. **47**(11), 1–26 (2012). https://doi.org/10.18637/jss.v047.i11

Consensus Modeling: A Transfer Learning Approach for Small Health Systems

Roshan Tourani[1], Dennis H. Murphree[2], Ying Zhu[1], Adam Sheka[3], Genevieve B. Melton[1,3], Daryl J. Kor[4], and Gyorgy J. Simon[1,5(✉)]

[1] Institute for Health Informatics, University of Minnesota, Twin Cities, MN, USA
simo0342@umn.edu
[2] Department of Health Sciences Research, Mayo Clinic, Rochester, MN, USA
[3] Department of Surgery, University of Minnesota, Twin Cities, MN, USA
[4] Department of Anesthesia, Mayo Clinic, Rochester, MN, USA
[5] Department of Medicine, University of Minnesota, Twin Cities, MN, USA

Abstract. The adoption of predictive modeling for clinical decision support is accelerating in healthcare, however, the need for large sample sizes puts smaller health systems at a disadvantage. Small health systems have insufficient positive cases to build models are left with three choices. First, they can obtain already trained models, which are often too generic. Second, they can participate in research networks, building a model through a network-wide data set. Since small hospitals can only contribute small amounts of data influencing the resulting shared model minimally, this approach yields only minimal specialization. The third option is transfer learning, where a model previously trained on a large population is refined to the specific population, which carries the danger of over-specializing to the idiosyncrasies of the small data set. In this paper, we present a novel paradigm, consensus modeling, that allows a small health system to collaborate with a larger system to build a model specific to the smaller system without sharing any data instances. The method is similar to transfer learning in that it refines models from the larger system to be specific to the small system, but through iterative refinement, the larger system alleviates the risk of over-specializing to the small system. We evaluated the approach on predicting postoperative complications at two health systems with 9,044 and 38,545 patients. The model obtained from the proposed consensus modeling paradigm achieved a predictive performance on the small system that is as good as the transfer learning approach (AUC 0.71 vs 0.71) but significantly outperformed the transfer learning approach on the large dataset (AUC 0.80 vs 0.65) suggesting significantly reduced over-specializing.

Keywords: Machine learning · Hospital acquired infection · Transfer learning · Predictive modeling

1 Introduction

The adoption of predictive modeling-based artificial intelligence (AI) for clinical decision support is accelerating in healthcare. The promise of these modern AI

© Springer Nature Switzerland AG 2020
M. Michalowski and R. Moskovitch (Eds.): AIME 2020, LNAI 12299, pp. 181–191, 2020.
https://doi.org/10.1007/978-3-030-59137-3_17

technologies is that health systems can use their own data to promote customize care for their specific patient populations. Small health systems, or any health system with small amounts of training data, cannot follow suit. Building models using small amounts of data carries the twin dangers of overfitting and being over-specific. *Overfitting* is a lack of generalizability within the health system: the model is too specific to the training set and does not generalize to a test set from the same health system. There are well-established methods to detect overfitting. In contrast, an *over-specific* model lacks the ability to generalize to a different health system: it captures artifacts of the health system, factors such as care policies or procedures that are not directly related to disease pathophysiology. Being *specific* to a health system by better capturing disease modalities is desirable; however, being *over-specific* is dangerous, because the underlying artifacts can change over time, without warning, and can render the models inaccurate. The inability to build specialized (but not over-specialized) models due to small patient populations puts smaller health systems at a distinct disadvantage.

If a health system with smaller amount of training data wishes to adapt modern machine learned clinical decision support tools, they have only a few choices. They could obtain a model from (say) a commercial vendor and use it as is. This approach does not require any training samples, but also does not offer a model specialized to the population. Implementation of models in this context often result in suboptimal predictions. An alternative approach is to join a practice network, share data with other health systems and collectively build a model. As health systems are often unwilling to share data, federated learning [4,6,17] methods have been developed, which approximate learning a model on shared data without actually sharing data; participants only share aggregates from their data. Although such a model could offer better specialization to the population at hand, the degree of specificity depends on the size of the participant health systems' data. Larger systems with more data exert greater influence in the common model and hence the model produced is more specific to their area. Conversely, small health systems that contribute fewer data points influence the shared model less and, in return, receive a less specific model. Ironically, the small health systems with the greatest need for clinical decision support derived from federated shared data stand to benefit the least. A third approach is transfer learning. In case of transfer learning, the health system would receive a generic model and refine the model to their population at hand. Since refining a model requires fewer samples than building a model from scratch, this approach appears more promising, but the small training set size can still render the model over-specific to the population. We can detect that a model is over-specific by taking the model back to the large data set (where it was originally trained) and see whether its performance dropped significantly on the exact same task as a result of having refined it to the smaller health system.

In this paper, we propose a novel modeling paradigm that we term consensus modeling. Consensus modeling allows a small health system to collaborate with a larger health system to build a model specific to the smaller system without sharing any data instances. The outcome is a model that performs almost as well as a transfer learning model in the smaller health system, yet the involvement of the large health system alleviates the risk of learning idiosyncrasies of the small system and thus becoming over-specific.

We demonstrate our methodology by predicting hospital-acquired postoperative infectious complications. We use data from a health system, Mayo Clinic (MC), with extensive participation in the National Surgical Quality Improvement Project (NSQIP) registry (38,545 surgeries) and a single hospital, University of Minnesota Medical Center (UMMC) with typical participation in NSQIP (9,044 surgeries) from another health system, Fairview Health Services (FHS). Outcomes are ascertained from the NSQIP registry. UMMC is "small" in the sense that it did not participate in full case sampling and has fewer than 100 NSQIP positive cases for some of the outcomes of interest.

2   Materials and Methods

2.1   Setting, Study Design, and Data

Data from two independent Midwestern health systems, Mayo Clinic (MC) and University of Minnesota-affiliated Fairview Health Services are used. Both health systems provide a wide range of surgical services and are members NSQIP registry, with only one site at FHS participating in NSQIP, the University of Minnesota Medical Center (UMMC). NSQIP provides high-quality gold-standard information on surgical outcomes [3]. We include all patients from MC and UMMC between 2010 and 2017 who are part of the NSQIP sample. In our cohort, we have 38,545 patients from MC and 9,044 from UMMC. For these patients, we collected all available information about their NSQIP index surgery and a 30-day history before the index surgery from the respective institutions' EHR repositories.

Outcomes. We are considering seven infectious outcomes: sepsis, septic shock, urinary tract infections (UTI), pneumonia (PNA), and three kinds of surgical site infections (SSI) (superficial, wound or deep tissue, and organ space), as defined by NSQIP. The NSQIP registry collects complications within a 30-day postoperative window.

Independent Variables. We primarily rely on demographics, history of complications, and preoperative observations (vital signs and laboratory results) that are known risk factors of postoperative infections [5,10,11]. Pre-operative diagnosis codes were rolled up into higher-level disease groups partly using the Clinical Classification Software [14] along with domain experts. Pre-operative lab results are preferentially taken from the pre-operative evaluation, or from historic records no more than 30 days before surgery. This resulted in a total of 54 independent variables, defined identically between MC and FHS.

The reader is referred to [13] for a more detailed description of the analytic cohort at the two health systems.

2.2 Baseline or Comparison Methods

To predict outcomes at both sites, we have three baseline approaches. The first approach, in-situ model, constructs a model for each outcome on the larger (MC) cohort and applies it to both the MC and the smaller FHS cohorts. The second approach, transfer learning, constructs a model for each outcome on the larger MC cohort and refines the model to the FHS cohort. The third approach is federated learning, where a model is jointly constructed on the MC and FHS cohorts (without sharing data) and then applied to both.

(a) *Source in-situ model.* For each outcome, a logistic regression model is constructed using the above independent variables. Missing laboratory results and vital signs are imputed by the median of the patient set with no complications on the MC cohort (larger data set). We used causal feature screening [8] (with a maximal condition set size of three [1]), followed by backward elimination for feature selection. The significance level α is adjusted for multiple comparisons incurred during the causal feature screening.

(b) *Transfer Learning: In-situ model as offset.* We start with the model built on MC as described in (a), and transfer it to the target (FHS) site. A logistic regression model is fit at the FHS site, using the score from the source in-situ model as an offset. Imputation and feature selection were identical to (a). Intuitively, using the prediction from the source model as an offset allows the fitting procedure to only estimate the difference between the institutions (rather than the entire mechanism of the disease). We hypothesize that this results in better performance from a smaller sample size than fitting a model from scratch. To check whether the model is over-specific to the target, we transfer the model back to the original site. Even for *homogeneous transfer learning* [16] problems, this is uncommon practice; the sole purpose of re-transferring the model back to source and re-testing it is to check whether it is over-specific to the target. Note, that the purpose of transfer learning in this application significantly differs from that of the typical modern transfer learning setting [16]. Typically, a deep learning model is trained to learn low-level representation of the data (e.g. extract lines, shapes from images) and then is transferred to a specific problem (pathology). In our application, both models are specific to the problem.

(c) *Federated Learning: FedAvg.* Federated learning is the problem of training a shared model between a large number of clients under the constraint that they only share summary information of their data and share no instances. The interested reader is referred to a survey by Xu and Wang [17]. The basic algorithm FedAvg [9] approximates the gradient descent algorithm, by averaging the shared client gradients on a central server. Though simple, this method is shown to perform surprisingly well [9,12]. (Other more developed federated learning methods often have requirements that are not met in this study or focus on improving issues [4] which are not relevant to our study. For example, a federated transfer learning approach is introduced in [7], but it assumes there is shared data between sites and thus not applicable to our problem.) Given the risk of overfitting on the small FHS dataset, we modified the FedAvg slightly. First, instead of starting from random or zero initial parameters, we start with the source in-situ model as the initial parameterization; this speeds convergence by starting closer to the optimal solution, and makes the method more stable by removing the variability in starting point. Second, we use a validation set to choose an optimal stopping point, reducing the risk of overfitting on the training set. Finally, we use FedSGD setting [9] which seems to give better convergence [15].

2.3 Proposed Method: Consensus Modeling

Overview. In the proposed Consensus Modeling, model construction is carried out as an iterative refinement process as described in Fig. 1. An initial model is constructed at the source, which is transferred to the target. The target then sends information about the residual back to the source, which, in turn, constructs a new model, taking the information about the residuals into account. Taking information about the target's residual into account allows the source to build a model that fits the target better; however, the source's own data and residuals alleviate the risk of the resulting model overfitting or becoming over-specific to the target. The resulting new model is transferred back to the target, which sends information about the residuals back to the source, so that the model can be refined again. The process continues until the model cannot be significantly improved or until the residuals at the target are insignificant.

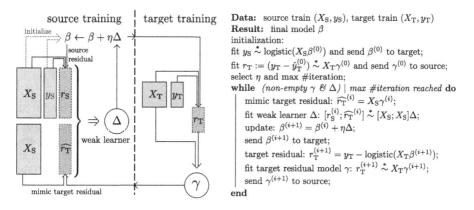

Fig. 1. Algorithm for consensus modeling, where β and γ are the information passed between the source and the target sites. The $*$ on top of fitting $\sim$ notation is used to include the feature selection methods.

Algorithm Details. Let X_S denote the predictor matrix of the (large) source data and X_T the (smaller) target data. Let y_S be the outcomes on the source site and y_T on the target site. Let β denote the coefficients of the model we wish to construct. Consensus Modeling initializes β with the source model, following Sect. 2.2(a), $y_S \overset{*}{\sim} \text{logistic}(X_S\beta)$ (we use star in $\overset{*}{\sim}$ to denote the feature selection). The current model (parameterized by β) is then transferred to the target and the residuals at the target are computed $r_T = y_T - \text{logistic}(X_T\beta)$. To send information about the residuals on the target to the source without sharing data, a residual model (a linear regression model with coefficients γ) is constructed on the target $r_T \overset{*}{\sim} X_T\gamma$. Feature selection follows Sect. 2.2(a) with significance level corrected for multiple comparisons, $\alpha < 0.05/(\#\text{features} \times i)$, i being the iteration number. The residual model (parameterized with γ) is then transferred back to the source. The source then constructs a "joint" data set containing (residual) information from both sites. The predictors of this joint data set are the source data and an approximation of the target data $\widehat{X_T}$ appended row-wise $[X_S; \widehat{X_T}]$. In this application, we approximate the target with the source data $\widehat{X_T} = X_S$, pretending that they are samples from the same population. A joint residual vector is also constructed as $[r_S; \widehat{r_T}]$, by concatenating the source residuals $r_S = y_S - X_S\beta$ and the approximation of the target residuals $\widehat{r_T} = \widehat{X_T}\gamma = X_S\gamma$. A differential model (a weak learner in boosting terminology) with coefficients Δ is fitted to the joint residuals using the joint predictors and the previously described feature selection method. The model is updated $\beta = \beta + \eta\Delta$ (η is learning rate) and transferred to the target. The process continues until either the differential model Δ or the residual model γ has no significant predictors.

2.4   Evaluation

Models are trained and evaluated through repeated leave-out validation. In each of 10 replications, a randomly selected 20% of the data set on both sites is left out for testing and another 20% for validation (FedAvg uses this to evaluate the stopping criterion). The remaining 60% of the data (80% when validation set is not needed) is used to train the models.

Three baseline models (in-situ, transfer learning, and FedAvg) are compared to the proposed Consensus Model on the same test sets in terms of AUC. Significance of the comparisons are assessed via paired t-tests. The comparisons are carried out on both data sets. Note that the performance of the transfer learned model would typically not be evaluated on the source data after refinement. The purpose of the comparison on the source data set is to examine whether the refined model is over-specific to the target.

3   Results

Figure 2 shows the performance of the four competing approaches on both the MC and FHS cohorts. The seven panels correspond to the seven outcomes. The performance is measured by AUC and is estimated from 10 iterations, yielding 10 AUC estimates from each approach. These 10 AUC estimates are plotted as boxplots.

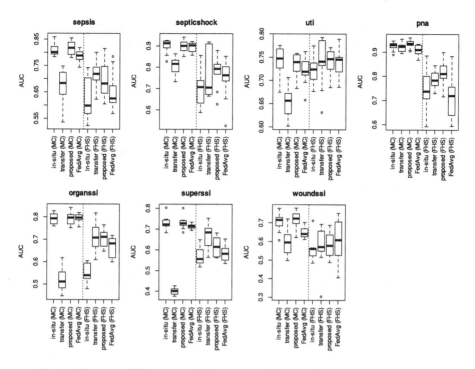

Fig. 2. The test AUC result of different models tested at MC and FHS sites.

To determine the statistical significance of the differences observed in Fig. 2, Table 1 shows specific comparisons described in Sect. 2.4. The top half of the table presents the comparisons on the FHS site; the bottom half on the MC site. Columns correspond to outcomes. On each site, the mean AUC is reported for all four methods and the mean difference in AUC is reported for three comparisons. The significance of the difference is denoted using stars (*** signals < 0.001, ** is < 0.01, and * denotes < 0.05).

Proposed vs in-situ. The proposed method outperformed the in-situ model on all outcomes on the FHS data set except for wound SSI: the p-values in 'proposed - in-situ (FHS)' row are all <.05 except wound SSI. Conversely, there is no detectable degradation in performance on the MC site; the only outcome with a significant difference in performance is wound SSI, where the proposed method outperformed the in-situ model on the source data.

Proposed vs transfer. On the FHS data set, the performance of the transfer learning method and the proposed methods were statistically equivalent with PNA being the only outcome with statistically significant difference favoring the proposed method (surprisingly). However, on the MC data set, as expected, the proposed method outperforms the transfer learned model on all outcomes.

Proposed vs FedAvg. The results show that the proposed method outperforms FedAvg either on the MC or the FHS cohort for each outcome except septic shock (where their performance is statistically equivalent) and that FedAvg never achieves statistically significantly better performance than the proposed method. The other exception is PNA, where the proposed method outperforms FedAvg on both cohorts.

4   Discussion

Health systems strive to optimize patient outcomes at the lowest possible cost, ideally delivering financial benefit through improved patient care. The recent emergence and adoption of artificial intelligence and machine learning technologies offers a vehicle to achieve both of these goals by increasingly tailoring care delivery to the health system's patient population. However, making care delivery increasingly tailored to a particular patient population carries the risk becoming overly specific, leading to a reproducibility crisis [2]. Commonly, machine learned models that perform "optimally" at one health system fail to achieve similar performance at a different health system. Results from our first comparison (in-situ vs proposed) clearly demonstrate this phenomenon. Specifically, comparing the 'in-situ (MC)' and 'in-situ (FHS)' rows in Table 1, shows that postoperative risk models built at a large health system (MC) failed to achieve similar performance at a smaller health system (FHS) for all outcomes.

Surprisingly, the same comparison also suggests that models built on the larger cohort and thought to be "optimal" are in fact occasionally suboptimal. For two (PNA and wound SSI) out of seven outcomes, the in-situ model significantly underperformed the proposed method. This suggests that information transferred from the smaller health system may have captured attributes associated with postoperative complications that exist at the larger health system but were ignored, perhaps as statistical noise. Collaborating with a larger health system clearly benefits the smaller health system (the proposed method outperformed the in-situ model on all outcomes except wound SSI) but may also benefit the larger health system.

As is seen in Table 1, it is clear that for a smaller institution to use a model from a larger one is a suboptimal solution: the received model is not sufficiently specific to the new patient population. Seemingly transfer learning, where a model received from a larger system is refined to the population of the smaller system, offers a solution. Our second comparison (*proposed* vs *transfer*), however, demonstrates that this approach makes the model overly specific to the small health system: the fact that the proposed method outperformed the transfer learned model on the large health system on all outcomes but the transfer learned model never outperformed the proposed model on the small health system supports this point. This means that the proposed model managed to learn a model that offers as high performance on the smaller FHS cohort as the transfer learnt model, without being so specific to the FHS cohort that its performance on the MC cohort would deteriorate; its performance on the MC cohort was as high as that of the in-situ model (which is over-specific to the MC cohort).

A third option that available to a small health system is to join a consortium and participate with its data in the construction of a consortium-wide model.

Table 1. Results of mean AUCs on holdout sets of FHS and MC. Difference between proposed model vs in-situ, transfer, and FedAvg are reported. The paired t-test p-values are translated to symbols: 0< *** < .001 < ** < .01 < * < .05 < .< 0.1.

	sepsis	septicshock	uti	pna	organssi	superssi	woundssi
proposed (FHS)	0.696	0.78	0.74	0.813	0.707	0.617	0.585
in-situ (FHS)	0.625	0.696	0.722	0.749	0.551	0.571	0.559
transer (FHS)	0.716	0.773	0.741	0.788	0.711	0.666	0.572
FedAvg (FHS)	0.653	0.745	0.737	0.718	0.667	0.583	0.599
proposed - in-situ (FHS)	0.070***	0.084***	0.018**	0.064***	0.156***	0.046***	0.026
proposed - transfer (FHS)	−0.021	0.007	−0.001	0.026**	−0.004	−0.048.	0.013
proposed - FedAvg (FHS)	0.042	0.035	0.003	0.095***	0.040**	0.034**	−0.014
proposed (MC)	0.818	0.901	0.733	0.934	0.794	0.73	0.716
in-situ (MC)	0.809	0.9	0.741	0.925	0.792	0.729	0.707
transer (MC)	0.675	0.808	0.65	0.921	0.52	0.401	0.603
FedAvg (MC)	0.786	0.894	0.72	0.909	0.795	0.714	0.649
proposed - in-situ (MC)	0.008.	0.001	−0.008*	0.009**	0.001	0.001	0.010**
proposed - transfer (MC)	0.143***	0.093***	0.083***	0.013*	0.274***	0.329***	0.114**
proposed - FedAvg (MC)	0.032**	0.007	0.013.	0.024**	−0.002	0.016	0.067***

This choice is represented by the FedAvg approach in our analysis. Our third comparison shows that FedAvg builds a model that captures characteristics of both populations to a varying (across outcomes) degree. It captured characteristics of the MC population at least as well as the proposed method for septic shock, organ-space and superficial SSI; and captures the characteristics of the smaller FHS cohort as well as the proposed method for sepsis, septic shock, UTI and wound SSI. However, it never achieved a better performance than the proposed method on any outcome on either cohort. The proposed method achieved significantly better performance on at least one of the two cohorts for all outcomes (except septic shock) and achieved a significantly better performance on both cohorts for PNA.

The central issue addressed in this paper is not the reproducibility crisis, but an issue of inequity, the inability of smaller health systems to tailor care to their specific populations. This puts these health systems and their patients at a pronounced disadvantage with a risk of significant disparity in our artificial intelligence-enabled healthcare future. Modeling approaches of today do not offer a satisfactory solution; using a model developing elsewhere is clearly an inferior solution. Transfer learning makes the model overly specific to the population at hand, while federated learning results in a model that over- or underfits both populations to a varying degree. We propose Consensus Modeling which can capture characteristics of a smaller patient population without sacrificing performance on a larger cohort as a way to address these disparities.

We demonstrated the proposed method in the context of a smaller and a larger health system collaborating without sharing data. The significance of the proposed methods transcends this setup. Even a large system is relatively small when compared to patient populations pooled by a multi-system consortium or by a nation-wide effort; or may become "small" (having only have a few positive cases) when modeling a rare disease. The problems arising from transferring a model (with much higher complexity than those considered in this paper) from a nation-wide data set to a large health system are the same: the health system aims to refine the nation-wide model to be more specific to their patient population, without becoming so specific that the model captures peculiarities rather than disease modalities.

5 Conclusion

The model obtained from the proposed consensus modeling paradigm achieved a predictive performance on the small system that is as good as the transfer learning approach (AUC 0.71 vs 0.71) but significantly outperformed the transfer learning approach on the large set (AUC 0.80 vs 0.65) suggesting significantly reduced over-specialization to the small data set. For each outcome except septic shock, it significantly outperformed federated learning on either the source, the target, or both data sets.

Acknowledgement. This work was supported in part by NIGMS award R01 GM 120079, AHRQ award R01 HS024532, and the NCATS University of Minnesota CTSA UL1 TR002494. The views expressed in this manuscript are those of the authors and do not necessarily reflect the views of the funding agencies.

References

1. Aliferis, C.F., Tsamardinos, I., Statnikov, A.: Hiton: a novel Markov blanket algorithm for optimal variable selection. In: AMIA Annual Symposium Proceedings, vol. 2003, p. 21. American Medical Informatics Association (2003)
2. Beam, A.L., Manrai, A.K., Ghassemi, M.: Challenges to the reproducibility of machine learning models in health care. JAMA (2020)
3. Henderson, W.G., Daley, J.: Design and statistical methodology of the national surgical quality improvement program: why is it what it is? Am. J. Surg. **198**(5), S19–S27 (2009)
4. Kairouz, P., et al.: Advances and open problems in federated learning. arXiv preprint arXiv:1912.04977 (2019)
5. Korol, E., et al.: A systematic review of risk factors associated with surgical site infections among surgical patients. PloS One **8**(12) (2013)
6. Li, Q., Wen, Z., He, B.: Federated learning systems: vision, hype and reality for data privacy and protection. arXiv preprint arXiv:1907.09693 (2019)
7. Liu, Y., Chen, T., Yang, Q.: Secure federated transfer learning. arXiv preprint arXiv:1812.03337 (2018)
8. Ma, S., Statnikov, A.: Methods for computational causal discovery in biomedicine. Behaviormetrika 44(1), 165–191 (2017). https://doi.org/10.1007/s41237-016-0013-5
9. McMahan, H.B., Moore, E., Ramage, D., Hampson, S., et al.: Communication-efficient learning of deep networks from decentralized data. arXiv preprint arXiv:1602.05629 (2016)

10. Pessaux, P., Msika, S., Atalla, D., Hay, J.M., Flamant, Y.: Risk factors for postoperative infectious complications in noncolorectal abdominal surgery: a multivariate analysis based on a prospective multicenter study of 4718 patients. Arch. Surg. **138**(3), 314–324 (2003)
11. Phillips, J., O'Grady, H., Baker, E.: Prevention of surgical site infections. Surgery (Oxford) **32**(9), 468–471 (2014)
12. Reddi, S., et al.: Adaptive federated optimization. arXiv preprint arXiv:2003.00295 (2020)
13. Tourani, R., Murphree, D.H., Melton-Meaux, G., Wick, E., Kor, D.J., Simon, G.J.: The value of aggregated high-resolution intraoperative data for predicting post-surgical infectious complications at two independent sites. In: 17th World Congress on Medical and Health Informatics, MEDINFO 2019, pp. 398–402. IOS Press (2019)
14. U.S. Agency for Healthcare Research and Quality: Clinical Classifications Software (CCS) for ICD-9-CM & ICD-10-CM/PCS (2019). https://www.hcup-us.ahrq.gov/toolssoftware/ccs10/ccs10.jsp
15. Wang, H., Yurochkin, M., Sun, Y., Papailiopoulos, D., Khazaeni, Y.: Federated learning with matched averaging. arXiv preprint arXiv:2002.06440 (2020)
16. Weiss, K., Khoshgoftaar, T.M., Wang, D.D.: A survey of transfer learning. J. Big Data **3**(1), 1–40 (2016). https://doi.org/10.1186/s40537-016-0043-6
17. Xu, J., Wang, F.: Federated learning for healthcare informatics. arXiv preprint arXiv:1911.06270 (2019)

An AI-Driven Predictive Modelling Framework to Analyze and Visualize Blood Product Transactional Data for Reducing Blood Products' Discards

Jaber Rad[1] , Calvino Cheng[2], Jason G. Quinn[2], Samina Abidi[1,3] ,
Robert Liwski[2], and Syed Sibte Raza Abidi[1(✉)]

[1] NICHE Research Group, Faculty of Computer Science, Dalhousie University,
Halifax, Canada
{jaber.rad,samina.abidi,ssrabidi}@dal.ca
[2] Deptartment of Pathology and Laboratory Medicine, Nova Scotia Health
Authority, Halifax, Canada
{calvino.cheng,jason.quinn,robert.liwski}@nshealth.ca
[3] Deptartment of Community Health and Epidemiology, Dalhousie University,
Halifax, Canada

Abstract. Maintaining an equilibrium between shortage and wastage in blood inventories is challenging due to the perishable nature of blood products. Research in blood product inventory management has predominantly focused on reducing wastage due to outdates (i.e. expiry of the blood product), whereas wastage due to discards, related to the lifecycle of a blood product, is not well investigated. In this study, we investigate machine learning methods to analyze blood product transition sequences in a large real-life transactional dataset of Red Blood Cells (RBC) to predict potential blood product discard. Our prediction models are able to predict with 79% accuracy potential discards based on the blood product's current transaction data. We applied advanced data visualizations methods to develop an interactive blood inventory dashboard to help laboratory managers to probe blood units' lifecycles to identify discard causes.

Keywords: Blood Inventory Management · Machine learning · Blood product wastage · Sequence prediction · Data visualization · Visual analytics · Big data

1 Introduction and Background

Blood transfusion is essential, and often, lifesaving treatment that is commonly used to support many therapeutic interventions (surgeries, chemotherapy, etc.). These products include RBCs, platelets, plasma, and other plasma derived products. All major tertiary care hospitals maintain an inventory of blood products within their blood transfusion services (BTS) to ensure timely availability of these products to meet clinical demand. Blood products are perishable—i.e. have an expiry date and require strict storage requirements to ensure vitality—therefore on the one hand a hospital BTS needs to

© Springer Nature Switzerland AG 2020
M. Michalowski and R. Moskovitch (Eds.): AIME 2020, LNAI 12299, pp. 192–202, 2020.
https://doi.org/10.1007/978-3-030-59137-3_18

provide safe and timely blood products, and on the other hand it has to optimally manage limited inventory to ensure adequate supply while minimizing wastage [1].

One of the major factors that affects the efficiency of blood inventory is *wastage* either due to expiry (i.e. outdates) [2] or discard (e.g. due to unsafe temperature, damaged bags, recalls, transfer inefficiencies, etc.) of a blood product. Both these wastage issues affect inventory management; a blood product will expire if it remains unutilized for too long, and it will be discarded if it is handled in an unsafe manner. Minimizing blood product wastage is therefore crucial for optimal inventory management with regard to economical, operational and ethical considerations [2, 3].

Blood Inventory Management has been extensively studied in terms of ordering and issuing policies, distribution scheduling algorithms and inventory modelling using operations research and stochastic dynamic programming methods [3, 4]. Additionally, blood inventory management techniques range from rules of thumb, first-in-first-out, daily standing orders, and reduction in cross-match periods [5]. Recently, simple machine learning methods [6] using k-NN [7] and random forests [8] have also been used. The majority of the research in blood inventory management has focused on reducing wastage by avoiding outdates which is often simply done by prioritizing the transfusion of a blood product that is nearing its expiry. These efforts have resulted in efficient product utilization, where many blood banks have reported outdate rates for Red Blood Cells (RBC) at less than 1%. Anticipating blood product demand to determine optimal inventory levels is an approach implemented by the Nova Scotia Health Authority, Central Zone Blood Transfusion Services (CZBTS), whereby an RBC ordering algorithm determines the probability of RBC demand based on lab characteristics of admitted patients. This strategy reduced overstocking and brought the RBC inventory down from ca. 401 to 309 units and monthly outdates from 19 to 8 [2].

Reducing blood product wastage due to discards, however, is still quite challenging as a discard is not related to the blood product's shelf life. It is related to its transactional lifecycle—i.e. how and when it was collected, handled, stored, transferred and received [9]. The lifecycle of a blood product unit embodies a non-deterministic sequence of transactions until it is either transfused or discarded [9], governed by stipulated processes, actors and measurable outcomes; a unit is deemed to be discarded if there is a safety error at any point in its lifecycle. Given the multifaceted and stochastic nature of a blood units' lifecycle, avoiding discards is far more challenging than reducing outdates. Currently, the most common way to investigate discards is periodic retrospective review of product transactions (e.g. at monthly meetings) by BTS staff; this manual analysis is due to the lack of suitable analytical tools to analyze the blood products' lifecycle to identify underlying *discard patterns* that can help to detect potential discards. The current discard determination process is sub-optimal—our internal data from 2015–2018 for RBCs shows that for every expired unit, 3.2X more are being discarded. Therefore, there is an urgent need for innovative strategies to *predict* whether a blood unit is likely to be discarded so that administrators can proactively act to avoid its discard, and in turn improve blood inventory management.

In this paper, we present our investigations to detect potential blood unit discards to help reduce wastage and optimize blood inventory management. We take a data-driven approach, using machine learning and visual analytics methods, to (a) identify the transactional patterns, within a blood unit's evolving lifecycle, that leads to a discard;

(b) predict a potential blood unit's discard based on analyzing its transaction patterns; and (c) visualize the evolving blood unit's lifecycle to help discern the underlying operational causes leading to discards. We have analyzed the RBC transactional data, for 15 months (June 2017 – August 2018) from CZBTS in Halifax (Nova Scotia, Canada), to develop two separate discard prediction models to predict a potential blood unit discard based on the sequence of transactions within its evolving lifecycle. We used lifecycle-dependent attributes to learn *sequence prediction models* from time-series RBC transactional data (i.e. RBC lifecycle). Our models have shown an accuracy of 79% to predict discard of active RBC units managed by CZBTS. We have also developed a dynamic interactive visualization of the RBC units' lifecycle to provide a high-level overview of the RBC lifecycle within CZBTS. The research outcome—i.e. the visualization dashboard and the predictive model—is intended to empower CZBTS to probe and respond to discard patterns in order to optimize blood product inventory.

2   Blood Product Lifecycle Concept

In the BTS, the lifecycle of a blood unit comprises multiple operational status, denoting its initial testing, transportation, storage, transfer to the transfusion site and finally transfusion or discard. The blood units' lifecycle has a temporal element; the status of the blood unit constantly changes over time (each status change is timestamped and recorded in an event log) until the unit is either transfused or discarded. A blood unit may be discarded due to several reasons, e.g. not passing the screening tests after collection, blood supplier recall after unit is sent to hospital, failure in visual inspection, exceeding the time threshold allowed to be out of refrigerator and so on. Table 1 shows the possible status values a RBC unit can go through in its lifecycle, and Fig. 1 depicts two possible lifecycles for RBC units.

Table 1. All the possible status values that can happen during RBC units' lifecycle

Status name	Description
Received	Product is received in inventory. This is the first status in a unit's lifecycle.
Unconfirmed	ABO confirmation test has not been completed yet.
Confirmed	ABO confirmation test has been completed successfully.
Available	Indicates that the RBC unit is available.
Transferred	RBC unit is transferred to a different inventory area.
Assigned	RBC unit has been allocated to a specific patient.
Issued	RBC unit has left the lab and been dispensed to a patient.
Crossmatched	Necessary tests have been performed on patient serum against the RBC unit to make sure they are fully compatible.
Quarantined	RBC unit is removed from inventory.
Destroyed	RBC unit has been destroyed. This status always comes before being discarded.
Discarded	The RBC unit has been thrown away. If this status happens, it would be the final status in the unit's lifecycle.
Transfused	The RBC unit has been given to the patient. If this status happens, it would be the final status in the unit's lifecycle.

Fig. 1. Examples illustrating potential lifecycles of RBC units

3   Data Description

In this project, we analyzed the RBC transactional dataset, for 15 months (June 2017 – August 2018) from CZBTS in Halifax (Nova Scotia, Canada). This dataset consisted of 174,349 status transitions for 17,108 RBC units with complete lifecycles.

The original dataset is a raw event log dataset with a series of sequential temporal transactions for all active RBC units. We pre-processed the data to develop a temporal lifecycle for individual RBC units in terms of n-grams to represent transiting between two statuses as a *transition step* and the entire lifecycle as a *transition sequence*.

This dataset consisted of 18 transactional attributes divided into two categories: (i) lifecycle-independent attributes: The properties of a blood unit regardless of its lifecycle, e.g. blood type, supplier of the blood unit, etc. (ii) lifecycle-dependent attributes, e.g. status of the unit, location of the units, the actors handling the unit, etc.

4   AI-Driven Predictive Modelling Approach

Our approach is to analyze and model the non-deterministic temporal lifecycle of a blood product—i.e. from being collected to its transfusion or discard—to predict potential discards for new blood units. Our research approach involves three steps:

- *Identifying* the attributes in the dataset which are related to underlying discard patterns. We will perform attribute correlation analysis to determine meaningful relationships between transactional attributes and the outcome of the blood product.
- *Predicting* the future status of an RBC unit based on the trend of its evolving lifecycle. We will learn *sequence prediction models,* using the temporal lifecycles to develop discard prediction models that use *n* previous statuses to predict a discard.
- *Visualizing* the unit's transition patterns to provide a high-level overview of the prevalent transition sequences of blood units, highlighting the sequences that lead to a discard. We will use advanced data visualizations to develop a web-based dynamic interactive visualization of the RBC units' lifecycle to probe the discard patterns.

5   Investigating Correlations Among Lifecycle Attributes

We first performed attribute correlation analysis that helped in the selection of the most informative attributes. Since our dataset comprises both categorical and quantitative attributes, we used three correlation analysis methods for all possible attribute pairs: Pearson's correlation for quantitative-quantitative pairs, Correlation Ratio (η) for quantitative-categorical pairs, and Theil's U for categorical-categorical pairs. Figure 2 shows a heatmap illustrating correlations for all the attribute pairs. Attribute names have a suffix of '(nom for nominal)' or '(con for continuous)' to denote their type. Our results indicate poor correlations between each attribute and a blood unit's outcome (The red rectangle on the heatmap), thus suggesting no relationships, especially for the total handling time of the blood unit, supplier of the blood unit, and the blood type—this is against the general belief held by the blood transfusion staff. To confirm the information in the lifecycle-independent attributes for predicting the target variable, we used Normalized Mutual Information (NMI) [10]. Figure 3 confirms weak correlations between each attribute and a blood unit's outcome as the maximum NMI value is 0.2.

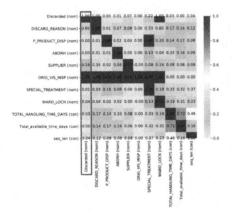

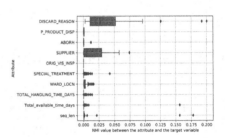

Fig. 2. Heatmap illustrating correlations for all the attribute pairs. (Color figure online)

Fig. 3. NMI criterion between each attribute and the target variable.

6   Predictive Modeling for Blood Unit Status Prediction

We approach the predictive modeling of the targeted discarded status of a blood unit as a sequence prediction problem based on RBC units' lifecycle (i.e. transition sequences). For our purpose, the term sequence prediction refers to predicting the next item(s) of a sequence. The training of a machine learning based prediction model using temporal sequences uses a dataset that takes into account the order of the observations. There are various ways of performing sequence prediction such as LSTMs/RNNs from neural networks area [11, 12] and Markov models from Machine Learning area [13–15]. We investigated two machine learning models to predict a blood unit's next status —i.e. (a) All-K-Order Markov (AKOM) and (b) Compact Prediction Tree (CPT).

The AKOM model consists of all the ith-order Markov models ($i = 1, 2, ..., k$) to address two issues: (i) poor prediction accuracy of 1st-order Markov model since it does not look far into the sequences, and (ii) low coverage and sometimes even worse accuracy of higher order Markov models [15]. Figure 4 shows the transition graph for the 1st-order Markov model on the RBC dataset, which is the basis for building higher order Markov models (the graph is too large to present here). The resultant transition graph highlights the underlying inefficiencies in the transfusion system (e.g. the circled numbers show the conditional probabilities associated with transition steps that are not ideal in the blood supply chain). On the other hand, higher order Markov models may not cover transition sequences shorter than the order of the model.

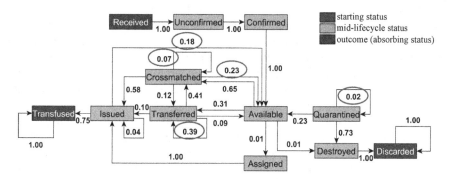

Fig. 4. The transition graph for the 1st-order Markov model. All the edges with conditional probabilities less than 1% are removed from the graph for visualization purposes.

We also investigated a new CPT method which claims to have higher accuracy than the AKOM model, Prediction by Partial Matching, and Dependency Graph [16, 17].

6.1 Training the Prediction Models

For both the prediction modeling approaches, a sequence of k previous statuses (i.e. transition history) is used to predict the k + 1 status. As suggested in [14], working with longer transition history can lead to better prediction accuracy, however the optimal value of k that gives the highest prediction accuracy had to be determined through experimentation. To determine the best performing prediction model, we created multiple AKOM and CPT models, with varying k values, and applied a 10-fold cross validation (90%–10% ratio for training set and test set) for each k value on 90% of the RBC dataset (i.e. validation set). We left the remaining 10% of the dataset untouched for comparing the predictive accuracy of the models (i.e. evaluation set).

In order to maximize our usage from the test set/evaluation set for calculating the accuracy score, each transition sequence in the test set/evaluation set is split into subsequences in a way that each transition step in the sequence is predicted using its previous statuses—i.e. if L is the length of a sequence in the test set/evaluation set, then the sequence contributes $L - 1$ times in the accuracy score. Accordingly, the accuracy score for validation/evaluation of the models is measured as the number of true predictions against the total number of sub-sequences.

6.2   k Parameter Tuning

Through experimentation we determined the optimum value of the k parameter for achieving the highest prediction accuracy—this basically informs us the extent of the transition history that is needed to make a meaningful prediction about a future status of a blood unit. Figure 5 shows the prediction accuracy of the AKOM model over the course of k values ranging from 1–20 (the maximum value of the k parameter is 20 since 97% of the RBC units have less than 20 transition steps). We decided to use $k = 4$ as the optimum k parameter among the k values which are associated with the highest accuracy results ($k = 4, 5, 6, 7, 8$); the result set for $k = 4$ is not significantly different from the others according to pairwise Student's t-test at a 95% confidence level, and it makes the AKOM model relatively smaller as compared to the other models with higher k values yet with no discernable predictive accuracy gains.

Figure 6 illustrates the prediction accuracy trend for different k values as used in the CPT model. By applying a One-way ANOVA test at a 95% confidence level, we note that there is no significant difference among the means of predictive accuracies of all the k values—i.e. increasing the value of k has no influence on the accuracy of the prediction model for this dataset. Therefore, we selected $k = 9$ for the CPT model as it provides the best accuracy results yet not significantly better than other k values. It is worth mentioning that unlike the AKOM model, the size of the k parameter has no influence on the size of the CPT model since it utilizes all the information in the training transition sequences to build the training model regardless of the value of k.

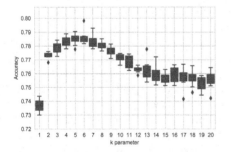

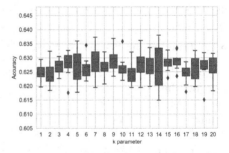

Fig. 5. Influence of the size of k parameter on accuracy of the AKOM model

Fig. 6. Influence of the size of k parameter on accuracy of the CPT algorithm

6.3   Prediction Model Evaluation

Figure 7 shows a comparison between the AKOM model and the CPT model while their k parameters are tuned. To compare these models not only on frequent sequence patterns but also on less frequent patterns, we define a 5% pattern frequency threshold to report their accuracy results.

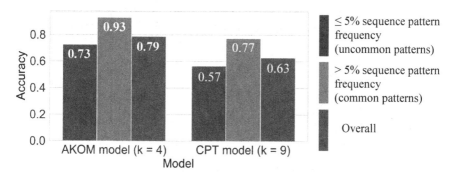

Fig. 7. Comparison between accuracies of the two investigated models

7   Blood Product Lifecycle Visualization

To provide a high-level understanding of the type and frequency of RBC unit transitions, we developed a web-based interactive visualization that allows the researchers to dig into the RBC units' lifecycle whilst they have a big picture of the entire dataset. We have employed a User-Centered Design (UCD) approach and have used React[1]. and D3[2]. as advanced data visualization software to implement the dynamic visualization.

Figure 8 illustrates the dashboard while the user interacting with the visualization. The core of the visualization is the sunburst diagram which aggregates all the transition sequences. The first transition step is the innermost ring, and as we move outward from the center, we approach to the ultimate statuses which can be either Transfused or Discarded. The angle of each segment shows how much frequent the occurrence of the corresponding status is in that specific level. As the user hovers the mouse on the sunburst diagram, the chosen sequence is highlighted, and the corresponding information is shown in each of the other three components; (i) The interactive breadcrumb trail shows the sequence of statuses on which the user is focused by hovering the mouse on the sunburst diagram. (ii) The line chart reflects the average amount of time it takes to transit from a status to another one in the specific sequence pattern. It also compares the trend to the average expiry time. (iii) Numeric values including the length and frequency ratio of the chosen sequence pattern are in the top-left of the visualization.

The data included in the visualization can be filtered based on the outcome of RBC units and the sequence pattern frequencies. Figure 9 shows a screenshot of the dashboard as the user has excluded all the sequence patterns with frequencies greater than 5% and having the last status as Transfused. Hovering on the patterns reveals the inefficiencies in the transfusion process for that particular pattern. For example, one

---

[1] A JavaScript library for building user interfaces (https://reactjs.org).

[2] Data-Driven Documents, a JavaScript library for manipulating documents based on data (https://d3js. org).

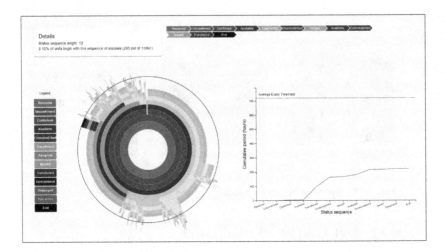

Fig. 8. The inventory dashboard showing a selected transition sequence

blood unit followed the selected pattern in the screenshot and was expired after several transitions prior to the Transferred status.

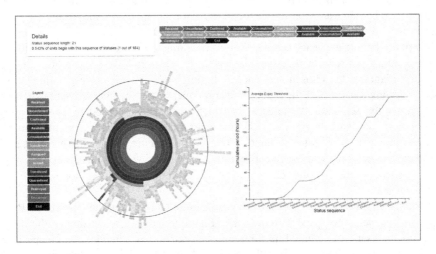

Fig. 9. Visualization showing the less frequent sequence patterns of discarded RBC units

8   Concluding Remarks

We investigated the lifecycle of RBCs and targeted wastage due to discard, which has barely been studied hitherto, to identify the underlying inefficiencies in the transfusion system to achieve appropriate management of RBC inventories. In this regard, we probed the lifecycle-independent attributes of RBC units to see if predicting the outcome of an RBC unit is possible from the very first step of its lifecycle when it is provided from a blood supplier. Our mathematical analysis showed that there is no correlation between those attributes and the target variable. Hence, we aimed to learn transition sequence patterns by building sequence learning models so that we could predict future transition steps in RBC units' lifecycle to identify discard patterns prospectively. Finally, we developed a web-based interactive dashboard that provides a high-level understanding of the type and frequency of RBC unit transitions.

The major limitations of this framework that should be the subject of further research are: (a) The investigated prediction models are not sufficiently competent to predict farther transition steps than the next immediate status since the accuracy drops to some extent each time we intend to predict one transition step farther. (b) In this study, RBC units' lifecycle is translated and simplified into sequences of statuses while there are other lifecycle-dependent attributes especially the time duration between two consecutive transition steps. Considering these attributes can lead us to achieving better sequence prediction accuracy. (c) Although the prediction models are generalizable to any BTS dataset, the value of the k parameter should be fine-tuned as a preliminary setting to ensure the highest accuracy.

Acknowledgements. This research is supported by the Blood Efficiency Accelerator Award by Canadian Blood Services. We thank the NSHA Central Zone Blood Transfusion Services for providing us the dataset and supporting the project.

References

1. Guan, L., et al.: Big data modeling to predict platelet usage and minimize wastage in a tertiary care system. Proc. Natl. Acad. Sci. U. S. A. **114**(43), 11368–11373 (2017). https://doi.org/10.1073/pnas.1714097114
2. Quinn, J., et al.: The successful implementation of an automated institution-wide assessment of hemoglobin and ABO typing to dynamically estimate red blood cell inventory requirements. Transfusion **59**(7), 2203–2206 (2019). https://doi.org/10.1111/trf.15272
3. van Dijk, N., Haijema, R., van Der Wal, J., Sibinga, C.S.: Blood platelet production: a novel approach for practical optimization. Transfusion **49**(3), 411–420 (2009)
4. Haijema, R., Van Dijk, N., Van Der Wal, J., Sibinga, C.S.: Blood platelet production with breaks: optimization by SDP and simulation. Int. J. Prod. Econ. **121**(2), 464–473 (2009). https://doi.org/10.1016/j.ijpe.2006.11.026
5. Stanger, S.H.W., Yates, N., Wilding, R., Cotton, S.: Blood inventory management: hospital best practice. Transfus. Med. Rev. **26**(2), 153–163 (2012)
6. Bertsimas, D., Kallus, N.: From predictive to prescriptive analytics. Manage. Sci. (2019). https://doi.org/10.1287/mnsc.2018.3253
7. Altman, N.S.: An introduction to kernel and nearest-neighbor nonparametric regression. Am. Stat. **46**(3), 175–185 (1992). https://doi.org/10.2307/2685209
8. Breiman, L.: Random forests. Mach. Learn. **45**(1), 5–32 (2001). https://doi.org/10.1023/A:101093340324
9. Cheng, C.K., Trethewey, D., Sadek, I.: Comprehensive survey of red blood cell unit life cycle at a large teaching institution in eastern Canada. Transfusion **50**(1), 160–165 (2010). https://doi.org/10.1111/j.1537-2995.2009.02375.x

10. Pedregosa, F., et al.: Scikit-learn: machine learning in {P}ython. J. Mach. Learn. Res. **12**, 2825–2830 (2011)
11. Bengio, S., Vinyals, O., Jaitly, N., Shazeer, N.: Scheduled sampling for sequence prediction with recurrent neural networks. In: Cortes, C., Lawrence, N.D., Lee, D.D., Sugiyama, M., Garnett, R. (eds.) Advances in Neural Information Processing Systems, vol. 28, pp. 1171–1179. Curran Associates, Inc. (2015)
12. Villegas, R., Yang, J. Hong, S. Lin, X., Lee, H.: Decomposing motion and content for natural video sequence prediction. In: 5th International Conference on Learning Representations, ICLR 2017 - Conference Track Proceedings (2019)
13. Qiao, Y., Si, Z., Zhang, Y., Abdesslem, F.B., Zhang, X., Yang, J.: A hybrid Markov-based model for human mobility prediction. Neurocomputing **278**, 99–109 (2018). https://doi.org/10.1016/j.neucom.2017.05.101
14. Pitkow, J., Pirolli, P.: Mining longest repeating subsequences to predict world wide web surfing. In: The 2nd USENIX Symposium on Internet Technologies & System (1999)
15. Deshpande, M., Karypis, G.: Selective Markov models for predicting web page accesses. ACM Trans. Internet Technol. **4**(2), 163–184 (2004)
16. Gueniche, T., Fournier-Viger, P., Tseng, V.S.: Compact prediction tree: a lossless model for accurate sequence prediction. In: Motoda, H., Wu, Z., Cao, L., Zaiane, O., Yao, M., Wang, W. (eds.) ADMA 2013. LNCS (LNAI), vol. 8347, pp. 177–188. Springer, Heidelberg (2013). https://doi.org/10.1007/978-3-642-53917-6_16
17. Gueniche, T., Fournier-Viger, P., Raman, R., Tseng, V.S.: CPT+: decreasing the time/space complexity of the compact prediction tree. In: Cao, T., Lim, E.P., Zhou, Z.H., Ho, T.B., Cheung, D., Motoda, H. (eds.) PAKDD 2015. LNCS (LNAI), vol. 9078, pp. 625–636. Springer, Cham (2015). https://doi.org/10.1007/978-3-319-18032-8_49

Towards Assigning Diagnosis Codes Using Medication History

Tomer Sagi[1](✉)(iD), Emil Riis Hansen[1](iD), Katja Hose[1](iD), Gregory Y. H. Lip[2,5](iD),
Torben Bjerregaard Larsen[2,4](iD), and Flemming Skjøth[2,3](iD)

[1] Department of Computer Science, Aalborg University, Aalborg, Denmark
{tsagi,emilrh,khose}@cs.aau.dk
[2] Aalborg Thrombosis Research Unit, Department of Clinical Medicine,
Aalborg University, Aalborg, Denmark
{tobl,fls}@rn.dk
[3] Thrombosis and Drug Research Unit, Department of Research and Innovation,
Aalborg University Hospital, Aalborg, Denmark
[4] Thrombosis and Drug Research Unit, Department of Cardiology,
Aalborg University Hospital, Aalborg, Denmark
[5] Liverpool Centre for Cardiovascular Sciences, University of Liverpool
and Liverpool Heart & Chest Hospital, Liverpool, UK
Gregory.Lip@liverpool.ac.uk

Abstract. Prior studies have manually assessed diagnosis codes and
found them to be erroneous/incomplete between 4–30% of the time. Previous methods to validate and suggest missing codes from medical notes
are limited in the absence of these, or when the notes are not written
in English. In this work, we propose using patients' medication data to
suggest and validate diagnosis codes. Previous attempts to assign codes
using medication data have focused on a single condition. We present
a proof-of-concept study using MIMIC-III prescription data to train a
machine-learning-based model to predict a large collection of diagnosis
codes assigned on four levels of aggregation of the ICD-9 hierarchy. The
model is able to correctly recall 58.2% of the ICD-9 categories and is
precise in 78.3% of the cases. We evaluate the model's performance on
more detailed ICD-9 levels and examine which codes and code groups can
be accurately assigned using medication data. We suggest a specialized
loss function designed to utilize ICD-9's natural hierarchical nature. It
performs consistently better than the non-hierarchical state-of-the-art.

1 Introduction

The practice of coding diagnoses of medical conditions using standardized coding
systems such as ICD-10 [25] has grown prevalent. However, while coding systems
are in wide-spread use, coding quality is uneven. Coding a medical diagnosis is
notoriously complex. There exist multiple hierarchies and choosing the appropriate code requires a deep understanding of their structure and the relationships.
For example, in a review of 1800 injury discharges from a New Zealand hospital,

© Springer Nature Switzerland AG 2020
M. Michalowski and R. Moskovitch (Eds.): AIME 2020, LNAI 12299, pp. 203–213, 2020.
https://doi.org/10.1007/978-3-030-59137-3_19

Davie et al. [6] found 2% to be uncoded, and 14% of PDx codes and 26% of external cause codes to be inaccurately coded. Wockenfuss et al. [26] determined that ICD-10 three and four level codes are too detailed to be reliable for general practitioners by measuring the Kappa inter-rater agreement scores. and found a sensitivity (recall) of 93.4 and positive predictive value (precision) of 88.9.

Some work exists on predicting diagnoses from laboratory results (e.g., [19]), but is limited to cases where such results are available and relevant. A large body of work exists on extracting diagnoses from clinical notes and reports (see review [23]). However, these systems' performance is reliant on techniques that tend to work much better in English, and must be retrained for every new language [17].

A patient's current medication can shed valuable light on their existing medical conditions. For example, observing that a patient has a chronic prescription for Metoprolol usually indicates that he/she is suffering from hypertension or ischaemic heart disease. Generalizing upon this observation, in this work we develop a machine-learning-based model able to predict the list of diagnoses assigned to a patient based upon his/her medications. Furthermore, in some countries (e.g., Denmark [20] and South Korea [15]) centralized medication repositories are comprehensive, while diagnosis codes are sporadic. Thus, such a model could provide emergency responders and critical care facilities with a rapid assessment of a patient's existing conditions in addition to the model's utility in diagnoses quality control. For example, an unconscious patient with a history of diabetes, will be first assessed for hyper/hypoglycemia, while one without a history of diabetes, but with a history of heart-disease, will be first assessed for acute heart conditions such as a heart-attack. We assess the viability of our approach using the publicly available MIMIC-III dataset [14]. The dataset contains rigorously anonymized and detailed medical records for over 50 K ICU patients.

2   Related Work

The need to perform quality control of diagnosis code assignment is justified by several studies. Cooke et al. [4] have shown that an ICD-9 code as a predictor of true COPD had a sensitivity of 76% and specificity of 67% using spirometry as their golden standard. A validity study of Danish national registry diagnoses [5] showed that only 75% of diabetic patients labeled with MI or stroke actually had such an event. Recent work attempted to predict ICD-9 assignment in MIMIC-III from discharge notes [12]. Their solution to the problem of multi-label, multi-level was to either limit the number of labels or aggregate predicted codes into categories, thereby solving two separate problems, namely to predict the top-10/50 codes or the top 10/50 categories. In this work, we aim to predict all codes, at different aggregation levels, in order to examine which codes and code groups can be predicted from medication data.

There have been a few attempts to use prescription data to predict a single or at most two conditions. Schmidt et. al. developed and validated an algorithm with 87% accuracy able to identify herpes zoster [22]. In another study, prescription data was used to classify whether or not patients had preexisting conditions of diabetes or hypertension [21]. In a recent review [8] of algorithms designed to extract cases for medical research from EMR data, some of the studies use medication data. However all studies extract cases for a single condition, often aggregating several diagnosis codes. In our scenario, we identify the probable diagnosis codes of multiple conditions at once and thus identify cases where improbable diagnosis codes have been used.

3   Methods and Data

3.1   Data

We use MIMIC-III [14] from PhysioNet [9], EHR data for 50 K patients from an American hospital's ICU departments over four years. MIMIC-III contains an extensive variety of data, including lab results, vital signs, medical notes, and most importantly for our needs, drugs administered and diagnoses ascertained. The prescriptions table (model input) contains 4 M rows of drugs prescribed during 50,216 admissions. There are 4,525 different drug names in the DRUG field, which are often the same drug, with different spelling or with an added comment, e.g., *Basiliximab* and **NF* Basiliximab*. To disambiguate and standardize the codes we use a mapping of MIMIC terms to the OMOP concepts [11] and group them by *Clinical Drug Form* to receive 1,602 RxNorm drug codes.

The diagnosis table (expected output) contains 651,048 diagnoses for 58,925 admissions using 6,841 different ICD-9 codes. ICD-9 is a hierarchical grouping of disease codes that consists of 5 levels starting from 0 (most general), to 4 (most specific). ICD-9 is built on the basis of grouping for similar disease. Upon review, we omit 5,994 codes for which less than 100 cases exist as it is typically not possible to generalize from such a low number. We further omit a number of codes focusing on diagnoses for chronic or persistent conditions. A complete and more detailed description of omissions can be found in the appendix. We use the patient data to add the *age* in years upon admission. MIMIC hides elderly (over 89) patient ages due to anonymization concerns and reports an average of 92.4 for the patients over 89. We use this age as the replacement age for these patients and further normalize the age by dividing it by 92.4, a practice that has been shown to be beneficial in machine learning techniques. After joining with the prescriptions table, the final table contains 52K admissions of 40K different patients using 567 unique codes, denoted labels in the following.

3.2   Task - Hierarchical Multi-label Classification (HMC)

Binary classification problems (e.g., will this person develop Sepsis) aim to correctly classify each task as either positive or negative. Single-label multi-class problems (e.g., is the following brain MRI normal, or does it contain a glioblastoma, a sarcoma, or a metastatic bronchogenic carcinoma?), extend the classification to allow more than one class for each task. These two types of ML tasks are, by far, the most commonly studied in the medical domain. Less common are multi-label classification problems which attempt to assign a set of labels to each example (e.g., which of the ICD-9 codes should be assigned following this medical report [1]), each of the labels is drawn from a possible set of classes. Since each person may have multiple co-morbidity, the task of assigning the correct set of diagnosis codes can be characterized as a multi-label classification problem [27]. The hierarchical nature of diagnoses both complicates the task and offers an opportunity to improve its applicability. If an algorithm predicts a patient suffering from non-specified chirosis (ICD-9 code 571.5) to be suffering from alcoholic chirosis (ICD-9 code 571.2) it should be more appreciated than if no chirosis related diagnosis are returned since both codes share a common ancestor. Further hierarchical constraints may dictate that a person cannot have more than one label from the same sub-tree of codes. Since ICD-9 is indeed hierarchical and imposes such constraints on some of its sub-trees, we can classify our task as an hierarchical multi-label classification (HMC) problem.

3.3   Machine Learning and Loss Functions

Many approaches to HMC include splitting the problem into multiple simple (single label) classification tasks, each of which is trained separately. Within these approaches, local and global approaches [7] differ by the amount of classifiers trained. In the local case, multiple classifiers are trained over a binary label pertaining to a single node in the hierarchy and the predictions of each level are subsequently propagated. In the global case, the labels are selected from a set of all possible labels. In this work we follow the observation of Cerri et al. [2] that by training a single global classifier based on a multi-level neural network representation, one can effectively reuse the high-level features learned to discriminate between high levels in the hierarchy and then refine these to more accurate code assignments using the subsequent levels of the neural network. Furthermore, deep neural networks (DNN) have repeatedly shown superiority over other techniques in the medical domain (e.g., [3,13]). We therefore employ a multi-layer perceptron, or fully connected neural network. The input layer for this network is comprised of one node for each RxNorm code in the data (and one for normalized age) and the output layer of one node for each ICD-9 code at the chosen roll-up level. The number of internal layers and the number of nodes in each layer are hyper-parameters over which we perform a classic grid-search.

Machine learning, in particular deep learning, uses a loss function during the training phase to quantify the error of the current iteration of the model with respect to the expected output. Choosing an appropriate loss function is crucial and in general must reflect the structure of the expected output. Thus, specific loss functions have been suggested for the multi-label case [16] as well as hierarchical multi-label functions [24]. However, these are tied directly to the structure of the global classifier, and none have been applied in the medical data setting using the inherent hierarchy of a medical ontology.

We therefore experiment with two types of loss functions. One suitable for the multi-label case, where each missed label is treated the same regardless of the extent of the mistake, and one designed for the HMC case. Our multi-label function is the multi-label soft margin loss function [28], defined as follows with C being the number of classes y being the class indicator and x the current value of the corresponding output node (i iterates over all classes).

$$loss(x, y) = -\frac{1}{C} \sum_i y[i] \cdot \log((1 + \exp(-x[i]))^{-1}) + (1 - y[i]) \cdot \log\left(\frac{\exp(-x[i])}{(1 + \exp(-x[i]))}\right) \tag{1}$$

We model our HMC loss function (hml, Eq. 2) after the one developed for HMCN-F [24], while adjusting it to account for the differences between a text-classification problem and our own task and minimize a function comprised of two components.

$$\mathcal{L}_{hml} = \mathcal{L}_L + \mathcal{L}_G \tag{2}$$

$\mathcal{L}_L$ is the local loss – calculation of Eq. 1 at the leaf level. $\mathcal{L}_G$ is calculated by rolling up the results one layer at a time until the ICD-9 chapter level (0). At each phase of the roll-up, the predictions for each inner node are set to the average of the predictions over its children. The loss of each level is calculated and summed to the other levels. Since our neural network does not directly predict the global scores, we do not suffer from hierarchical violations and do not require the third component that penalizes them in HMCN-F. We employ the Roll Up method to aggregate diagnoses given the ICD-9 hierarchy (see example in Fig. 1). A disease is only billable if it is a leaf-node of the ICD-9 hierarchy. However, not all leaves are on the same level. As an example, the code 322.2 is a billable level 4 code, which represents *Chronic meningitis,* whereas code 003.22 is a billable level 5 code for *Salmonella pneumonia.* Each patient initially starts with one or more billable disease from the ICD-9 hierarchy.

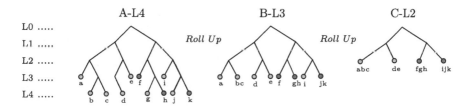

Fig. 1. Example of the roll up algorithm. An example level 4 code assignment is shown as tree A-L4. Disease codes {b, c, d, g, f, j, k} are level 4 billable codes, whereas codes {a, e, f, i} are billable codes on level 3. Red circles are the registered comorbidities of the patient. Green circles are diseases not recorded in the patient. (Color figure online)

3.4 Evaluation

A few previous ICD coding tasks have been evaluated by measures that consider its hierarchical nature as well [18]. To allow comparison of ICD code assignment using medication data to algorithms using medical notes, we use the more common micro-averaged precision and recall, and their harmonic mean F1. We perform the experiments on different prediction resolutions. With level 0 corresponding to the chapter level of ICD-9 (e.g., 520–579: diseases of the digestive system) and level 1 to the code group level (e.g., 401–405 Hypertensive Disease). Our last level corresponds to the most detailed available in the ICD-9 hierarchy (level 4) with 576 possible codes.

4 Results

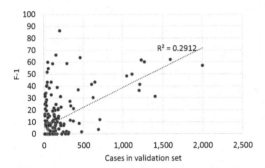

Fig. 2. F-1 by number of cases over level 2 codes.

Table 1 presents the best results (by F1) obtained by the ML-model following a standard hyper-parameter grid search. In each task, the code assignments were rolled up prior to both the training and the test phase and not only for the purpose of evaluation, such that the neural network encountered a different task for each level. For each ICD level we provide the number of codes in that level, the average branching factor, and the average number of eventual leaves a node in this level's sub-tree. In addition to precision, recall, and F1, we show the number of diagnosis codes for which F1 was equal to zero.

Since this is a relatively small dataset, the number of cases for many diagnoses is too low to expect reasonable performance. When examining the effect of the number of cases on the model's performance (Fig. 2) we find that at least some of the variance can be explained by the small number of cases (R^2 of 0.29 for a linear model). Top-5/top-10 results by code are available as an online appendix containing the full results [10].

4.1 Choice of Loss Function

To assess the effect of using *hml* versus a standard multi-label loss function (*ml*) we examine all experimental results where the F1 was at least 5.0 (Fig. 3). Models trained using *hml* consistently out-performed those trained using *ml* with an average F1 result between 3–8% better. This result holds when comparing the max values obtained in each levels as well with a 2–7% improvement for levels 2–4, although no significant improvement was seen for level 1. This last result is expected since the roll-up process for this level only rolls-up to level 0.

Table 1. MIMIC-III Diagnosis prediction results

Prediction task	Codes	Branching	Avg. leaves	Precision	Recall	F1	F1 = 0
ICD9 Top-10-groups (level 0)	10	NA	NA	82.6	52.4	64.1	0
ICD9 Top-10-codes (level 4)	10	NA	NA	61.3	50.5	55.4	0
ICD9-Rolled Up (Level 0)	15	5.7	565.1	78.3	58.2	66.8	2
ICD9-Rolled Up (Level 1)	65	8.4	108.3	60.9	43.0	50.4	15
ICD9-Rolled Up (Level 2)	236	6.6	14.0	56.6	31.5	40.5	122
ICD9-Rolled Up (Level 3)	461	1.6	1.6	52.3	19.9	28.8	297
ICD9-Raw (Level 4)	567	0	0	49.9	18.8	27.3	315

4.2   Discussion

The model's performance was able to recall 52.4% of the correct ICD-9 categories in a top-10 setting and assign the precise code in 82.6% of the cases. For this setting at the categorical level 0, and for the top-10 ICD codes (level 4), results are comparable to those published by Huang et al. [12] which predicted top-10 ICD-9 categories/codes by training deep neural networks over medical notes. In these

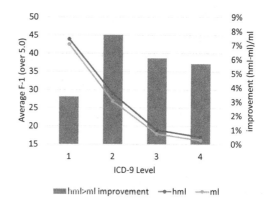

Fig. 3. F1 difference between models trained using hierarchical multi-label loss (hml) and multi-label loss (ml).

days of automated electronic health records, this approach offers a potential application to automatically assign a disease code on the basis of drugs prescribed. This may also provide opportunities to create quality control mechanisms for diagnosis code assignment.

F1 scores improve as the task is simplified with the worse performance obtained when the model is trying to assign the correct code from a set of 567 possible codes at level 4. The best performance is on level 0, when the model only has 15 possible labels. Consistently, in all experimental conditions, precision is higher than recall. This is partially explained by codes and groups that cannot be differentiated by their medication, and for which the model was unable to find any of the cases (F1 = 0). For example, at level 0, the model was unable to predict any assignment of chapters 780–799 (Symptoms, Signs, And Ill-Defined Conditions) and 710–739 (Diseases Of The Musculoskeletal System And Connective Tissue). These chapters may not be differentiable by medication, as the former is comprised of symptoms for many underlying conditions and the latter may be treated by orthopedic treatments and generic pain-relief medication. Further analysis shows that prediction of neoplasms mostly fails as well, as the treatment of cancer can be surgical or radiation-based. Furthermore, since MIMIC contains only ICU records, the patient may not be currently undergoing any medication-based cancer treatment. Further limitations include some drugs being prescribed for more than one diagnosis. For example, ACE inhibitors may be used for management of hypertension, heart failure, vascular disease and post-stroke. Also, doses of some drugs may vary depending on the disease indication, for example, rivaroxaban 2.5 mg BID is licensed for high risk patients with acute coronary syndrome, while rivaroxaban 20 mg OD is for stroke prevention in atrial fibrillation. Our analysis also does not consider changes in drugs over time, nor dose changes of the same drug. Also, some patients may swap their drug into another agent from the same class of drugs, causing a further dilution of the number of cases a model can learn from. Some drugs are also in combination therapies, for example, combining ACE inhibitors and a diuretic in a single *combo* pill for the treatment of hypertension.

5   Conclusion and Future Work

We presented a proof-of-concept study of the feasibility of using an ML-model to assign multiple diagnosis codes on multiple aggregation levels using a person's current medication. The model was able to correctly assign diagnosis codes on multiple levels and the detailed results allow to identify which codes and code-groups are predictable by medication data. The use of a hierarchical loss function has improved the model's performance by an average of 3–8%. The promising results support continued research into the ability to utilize larger medication datasets to create quality control mechanisms for diagnosis code assignment and to provide diagnostic information to caregivers in emergency situations that is language agnostic. We wish to pursue this expansion in future work, as well as experiment with additional hierarchical loss functions and methods to incorporate dosage and treatment regimen information in the model's input.

A   Appendix - Omitted codes and detailed results

Table 2 details the ommitted ocdes from the diagnosis table and the reasons for omission. We omit all codes with a low number of cases. We further omit 61 codes used to describe symptoms, as these are shared by multiple causes and will, most-probably, supplant a diagnosis code following medical investigation. Injuries and foreign bodies (30 codes) are omitted as well as their treatment is usually orthopedic or surgical, rather than medicinal. We omit the codes used in ICD-9 to classify birth-age and pre-term phase for infants (14 codes) as these are more descriptive than diagnostic. Finally, we omit the E and V series of codes that are used to provide additional details for statistical reasons and which do not cause differences in medicinal treatment. We remain with 567 codes and 54,423 cases (92.4%) that contain at least one of the remaining codes. Filtering out only admissions contained in both the diagnosis and prescription tables we remain with 50,211 admissions.

Table 2. List of Omitted ICD-9 Codes and Code Groups

Code(s)	Description	Reason
5994 different codes	A large collection of various codes	Low base rate (less than 100 cases)
765.X	Descriptive of gestation week or preterm weight	Will be accompanied by the specific results of pre-term birth if such exist
8XX and 9XX	Injury	Medical result would be Surgical or Orthopedic and impossible to accurately specify from medication
93.31,93.41	Foreign body	Undiscernable medicinally
99.X	Complications of medical care	Undiscernable medicinally
61 different codes	Collection of different symptoms such as pain, nausea, and nuances of mental state/ faculties	Should be accompanied by the symptom's cause which is the main diagnosis

Detailed results are available online [10].

References

1. Baumel, T., et al.: Multi-label classification of patient notes: case study on ICD code assignment. In: Workshops at the Thirty-Second AAAI Conference on Artificial Intelligence, June 2018. https://www.aaai.org/ocs/index.php/WS/AAAIW18/paper/viewPaper/16881
2. Cerri, R., Barros, R.C., De Carvalho, A.C.: Hierarchical multi-label classification using local neural networks. J. Comput. Syst. Sci. **80**(1), 39–56 (2014). https://doi.org/10.1016/j.jcss.2013.03.007
3. Cheng, X., Zhang, L., Zheng, Y.: Deep similarity learning for multimodal medical images. Comput. Meth. Biomech. Biomed. Eng. Imaging Vis. **6**(3), 248–252 (2018)
4. Cooke, C.R., et al.: The validity of using ICD-9 codes and pharmacy records to identify patients with chronic obstructive pulmonary disease. BMC Health Serv. Res. **11**(1), 37 (2011)
5. Dalsgaard, E.M., Witte, D.R., Charles, M., Jørgensen, M.E., Lauritzen, T., Sandbæk, A.: Validity of Danish register diagnoses of myocardial infarction and stroke against experts in people with screen-detected diabetes. BMC Public Health **19**(1), 228 (2019). https://doi.org/10.1186/s12889-019-6549-z
6. Davie, G., Langley, J., Samaranayaka, A., Wetherspoon, M.E.: Accuracy of injury coding under ICD-10-AM for New Zealand public hospital discharges. Inj. Prev. **14**(5), 319–323 (2008). https://doi.org/10.1136/ip.2007.017954

7. Fabris, F., Freitas, A.A., Tullet, J.M.: An extensive empirical comparison of probabilistic hierarchical classifiers in datasets of ageing-related genes. IEEE/ACM Trans. Comput. Biol. Bioinform. **13**(6), 1045–1058 (2016). https://doi.org/10.1109/TCBB.2015.2505288

8. Ford, E., Carroll, J.A., Smith, H.E., Scott, D., Cassell, J.A.: Extracting information from the text of electronic medical records to improve case detection: a systematic review. J. Am. Med. Inf. Assoc. **23**(5), 1007–1015 (2016). https://doi.org/10.1093/jamia/ocv180

9. Goldberger, A.L., et al.: Physiobank, physiotoolkit, and physionet: components of a new research resource for complex physiologic signals. Circulation **101**(23), e215–e220 (2000)

10. Hansen, E.R., Sagi, T., Hose, K., Lip, G.Y.H., Larsen, T.B., Skjøth, F.: MIMIC Prescriptions result files (2020). https://doi.org/10.7910/DVN/5VTBME

11. Hripcsak, G., et al.: Observational health data sciences and informatics (OHDSI): opportunities for observational researchers. Stud. Health Technol. Inf. **216**, 574–8 (2015)

12. Huang, J., Osorio, C., Sy, L.W.: An empirical evaluation of deep learning for ICD-9 code assignment using MIMIC-III clinical notes. Comput. Methods Programs Biomed. **177**, 141–153 (2019). https://doi.org/10.1016/j.cmpb.2019.05.024

13. Hung, C.Y., Chen, W.C., Lai, P.T., Lin, C.H., Lee, C.C.: Comparing deep neural network and other machine learning algorithms for stroke prediction in a large-scale population-based electronic medical claims database. In: Proceedings of the Annual International Conference of the IEEE Engineering in Medicine and Biology Society, EMBS, pp. 3110–3113. Institute of Electrical and Electronics Engineers Inc. September 2017. https://doi.org/10.1109/EMBC.2017.8037515

14. Johnson, A.E., et al.: MIMIC-III, a freely accessible critical care database. Sci. Data **3**(1), 1–9 (2016)

15. Kim, L., Kim, J.A., Kim, S.: A guide for the utilization of health insurance review and assessment service national patient samples. Epidemiol. Health **36**, e2014008 (2014). https://doi.org/10.4178/epih/e2014008

16. Martins, A.F.T., Astudillo, R.F.: From softmax to sparsemax: a sparse model of attention and multi-label classification. In: Balcan, M., Weinberger, K.Q. (eds.) Proceedings of the 33nd International Conference on Machine Learning, ICML 2016, New York City, USA, 19–24 June 2016. JMLR Workshop and Conference Proceedings, vol. 48, pp. 1614–1623. JMLR.org (2016). http://proceedings.mlr.press/v48/martins16.html

17. Névéol, A., Dalianis, H., Velupillai, S., Savova, G., Zweigenbaum, P.: Clinical natural language processing in languages other than English: opportunities and challenges. J. Biomed. Semant **9**(1), 12 (2018). https://doi.org/10.1186/s13326-018-0179-8

18. Perotte, A., et al.: Diagnosis code assignment: models and evaluation metrics. J. Am. Med. Inf. Assoc. **21**(2), 231–237 (2014). https://doi.org/10.1136/amiajnl-2013-002159

19. Razavian, N., Marcus, J., Sontag, D.A.: Multi-task prediction of disease onsets from longitudinal laboratory tests. In: Doshi-Velez, F., Fackler, J., Kale, D.C., Wallace, B.C., Wiens, J. (eds.) Proceedings of the 1st Machine Learning in Health Care, MLHC 2016, Los Angeles, CA, USA, 19–20 August 2016. JMLR Workshop and Conference Proceedings, vol. 56, pp. 73–100. JMLR.org (2016). http://proceedings.mlr.press/v56/Razavian16.html

20. Schmidt, M., et al.: The Danish health care system and epidemiological research: from health care contacts to database records. Clin. Epidemiol. **11**, 563–591 (2019). https://doi.org/10.2147/CLEP.S179083

21. Schmidt, M., Sørensen, H.T., Pedersen, L.: Diclofenac use and cardiovascular risks: series of nationwide cohort studies. BMJ **362**, k3426 (2018). https://doi.org/10.1136/bmj.k3426

22. Schmidt, S.A., Vestergaard, M., Baggesen, L.M., Pedersen, L., Schønheyder, H.C., Sørensen, H.T.: Prevaccination epidemiology of herpes zoster in Denmark: quantification of occurrence and risk factors. Vaccine **35**(42), 5589–5596 (2017). https://doi.org/10.1016/j.vaccine.2017.08.065

23. Wang, Y., et al.: Clinical information extraction applications: a literature review. J. Biomed. Inf. **77**, 34–49 (2018). https://doi.org/10.1016/j.jbi.2017.11.011

24. Wehrmann, J., Cerri, R., Barros, R.: Hierarchical multi-label classification networks. In: Dy, J., Krause, A. (eds.) Proceedings of the 35th International Conference on Machine Learning. Proceedings of Machine Learning Research, vol. 80, pp. 5075–5084. PMLR, Stockholmsmässan, Stockholm Sweden (2018). http://proceedings.mlr.press/v80/wehrmann18a.html

25. WHO: International Statistical Classification of Diseases and Related Health Problems, 10th Revision (ICD-10). Technical report., World Health Organization, Geneva, Switzerland (2004)

26. Wockenfuss, R., Frese, T., Herrmann, K., Claussnitzer, M., Sandholzer, H.: Three- and four-digit ICD-10 is not a reliable classification system in primary care. Scand. J. Prim. Health Care **27**(3), 131–136 (2009). https://doi.org/10.1080/02813430903072215

27. Xu, D., Shi, Y., Tsang, I.W., Ong, Y., Gong, C., Shen, X.: Survey on multi-output learning. IEEE Trans. Neural Netw. Learn. Syst. **(Early Access)**, 1–21 (2019)

28. Zhang, M.L., Zhou, Z.H.: A review on multi-label learning algorithms. IEEE Trans. Knowl. Data Eng. **26**(8), 1819–1837 (2014). https://doi.org/10.1109/TKDE.2013.39

Blockchain-Based Federated Learning in Medicine

Omar El Rifai[1,3]($\boxtimes$), Maelle Biotteau[2,3], Xavier de Boissezon[2,3], Imen Megdiche[1], Franck Ravat[1], and Olivier Teste[1]

[1] Institut de Recherche en Informatique de Toulouse SIG Team, Toulouse, France
{omar.el-rifai,imen.megdiche,franck.ravat,olivier.teste}@irit.fr
[2] ToNIC, Toulouse NeuroImaging Center, Université de Toulouse, Inserm, UPS, Toulouse, France
maelle.biotteau@inserm.fr
[3] Centre hospitalier universitaire de Toulouse, Toulouse, France
deboissezon.xavier@chu-toulouse.fr

Abstract. Worldwide epidemic events have confirmed the need for medical data processing tools while bringing issues of data privacy, transparency and usage consent to the front. Federated Learning and the blockchain are two technologies that tackle these challenges and have been shown to be beneficial in medical contexts where data are often distributed and coming from different sources. In this paper we propose to integrate these two technologies for the first time in a medical setting. In particular, we propose a implementation of a coordinating server for a federated learning algorithm to share information for improved predictions while ensuring data transparency and usage consent. We illustrate the approach with a prediction decision support tool applied to a diabetes data-set. The particular challenges of the medical contexts are detailed and a prototype implementation is presented to validate the solution.

1 Introduction

Researchers face ethical challenges when handling medical records. Indeed, medical records hold sensitive information about patients that can be prejudicial if leaked. A recent controversy involving unconsenting access to tens of millions of identifiable health records re-sparked an interest in the data ethics debate [27]. As a consequence, medical institutions are reticent in sharing medical records [21]. Researchers go through time consuming procedures to request and use medical data-sets, often at the expense of efficiently advancing research. This situation is exacerbated for data scientists who use large, and heterogeneous data-sets scattered across different sites.

At the forefront of the ethical challenges we find, *data privacy, transparency,* and *usage consent* [15,23,27]. Data privacy is often thought in terms of identity privacy or confidentiality [23,27]. Traditionally, anonymization techniques have been used for medical data processing [23]. Data transparency is about patients knowing and understanding how and by whom their data are used [27].

© Springer Nature Switzerland AG 2020
M. Michalowski and R. Moskovitch (Eds.): AIME 2020, LNAI 12299, pp. 214–224, 2020.
https://doi.org/10.1007/978-3-030-59137-3_20

And usage consent means that data subjects have the right to decide how and when their data are used. [23,27]. Unfortunately, these points are often glossed over when processing patients' data, and consent is sometimes implicitly assumed if data-sets are anonymized [15]. This situation is further complicated in the era of big-data where data-sets are scattered and processed by multiple often non-communicating parties [17].

Recently, two technologies emerged independently which address some of the issues highlighted above. Federated Learning (FL) emerged as a paradigm for training Machine Learning (ML) models across decentralized devices and minimizing the risk of exposing sensitive information [13,14,20]. While the blockchain emerged as a technology which offers unprecedented guarantees of reliability and usage transparency in decentralized settings [24].

This paper proposes a blockchain-based FL framework whereby the advantages of both technologies are put to use in the medical context. In particular, we propose a Smart Contract (SC) implementation of a coordinating server for a FL algorithm to ensure transparency and usage consent when sharing knowledge.

In Sect. 2, we discuss relevant related work for both FL and the blockchain in health care contexts. Then, in Sect. 3 we describe the problem in detail and explain how our solution can be applied for the medical setting. In Sect. 4, we show and discuss experiment results on a medical data-set for diabetes prediction. Finally, in Sect. 5, we present remaining challenges and open questions.

2   Health Care Analytics in Distributed Settings

Medical data offers a wealth of potential for improving the quality of care and reducing costs [21]. Nonetheless, medical data, as in other fields, is often "disorganized and distributed, coming from various sources and having different structures and forms" [21]. FL emerged as a response to these settings by providing a way to train ML models in heterogeneous and distributed settings while minimizing data transfers [13,14,20]. It has been used in health care settings because of the sensitive information of the data handled [32].

Likewise the blockchain has been studied in health care as it offers usage guarantees (transparency and immutability) not possible in traditional distributed data architectures [7,22]. In this section, we look at how these two technologies have been used in health care, and henceforth motivate the introduction of our blockchain-based FL medical decision support model.

2.1   Federated Learning Approaches

Although works on distributed computing have been around for decades [18,19], new contexts have brought up previously unaddressed challenges: instead of evenly, and moderately distributed data-sets, FL approaches deal with uneven, and massively distributed data-sets [13,14,20]. These contexts have been shown to be applicable in health care settings to "connect all the medical institutions and makes them share their experiences with privacy guarantee" [32]. The term

FL was coined by Konevcny et al. [14] and McMahan et al. [20] which proposed a variant to the Stochastic Variance Reduced Gradient Descent (SVRGD) algorithm to solve ML problems in a distributed setting.

Characteristically, participants' raw data never leaves the hosts' devices in FL settings. The only data that is shared with a coordinating server are model parameters of local ML models. These parameters are then aggregated efficiently by a central server and the result sent back to the participants for updating their own models.

A comparison between the FL framework and more traditional computing architectures [13] is shown in Fig. 1. Figure (1a) shows an architecture where all computations are performed on a centralized server. In such an architecture, end devices query the server to use the computation model. The centralized server needs to hold the entire data-sets at the moment of the learning phase for this architecture to work and is typical of a siloed health facility. Figure (1b) shows a distributed computing architecture where multiple servers (sometimes defined as a cloud) share data-sets and the workload. This architecture is common for data-sets that have been anonymized and need to be shared across institutions. Finally, in Figure (1c), the end devices become active participants in the computation and only upload partial information to the server. The new role of the server is then to aggregate the information of the different devices and broadcast back the aggregated information to the end devices.

(a) A centralized comput- (b) A distributed comput- (c) A federated computing architecture   ing architecture   ing architecture

Fig. 1. An illustration of the different computing architectures. A gearwheel indicates where the main computation tasks is executed and dashed lines indicate that minimal information is transferred.

Studies that have used FL in the health care settings have been recently reported in [32]. For instance in [3], the authors develop a binary classification problem to predict cardiac-related hospitalizations based on data from electronic health records. They find that their algorithm converges faster than a centralized one at the expense of increased communication costs. In [11], the authors use a FL-based approach to predict mortality and hospital stay time. They improve on the baseline FL models by first clustering the patients into communities and outperform the baseline FL approaches.

However, in the words of the authors in [20]: "Clearly, some trust of the server coordinating the training is still required [when using FL]". That is, server-side computations often suffer from a lack of transparency and it is difficult for users to verify the computations performed [26]. Specifically, cloud computing has been argued to decrease the sovereignty of users over their data and models [5].

2.2 Blockchain with Health Care Applications

The blockchain was invented to solve the consensus problem in a decentralized, trust-less network [24]. That is, given a distrusting network of peers, the blockchain provides a mathematically robust way of verifying that data stored on our device are identical with data held by other peers. Initially designed with a financial application in mind, the blockchain quickly evolved to different domains [12].

In particular, the blockchain has been heavily applied in the medical domain whereby the properties it possesses (decentralization, immutability, and transparency) are core issues [1,7,22]. One such application which has been studied extensively is Electronic Health Records (EHR) management [2,4,16]. EHR management is inherently decentralized as stakeholders are distributed between patients, medical institutions and government institutions in some cases. Blockchain-based solutions, allow the different stakeholders to manager EHR transparently while guaranteeing fairness and usage (records access) consent [16].

2.3 Discussion

Both the blockchain and FL algorithms address important ethical challenges and have been successfully used in many health care settings [9,32]. Indeed, by reducing the data that is shared when training models, FL algorithms reduce the risks of exposing sensitive patients data and hence address the privacy issue. Similarly, the blockchain addresses transparency and usage consent when dealing with medical records (c.f Appendix A.2).

Few papers have integrated ML models directly with the blockchain. Among those, Wang et al [29] for instance, have used the blockchain as a platform for hosting ML models, guaranteeing algorithmic correctness and usage traceability. Harris et al. [10] use the blockchain as a collaborative training platform for ML models. That is, the platform hosts ML models written in SCs and encourages data uploads from different users. Finally, the authors in [30] propose a domain agnostic setting where FL and blockchain technologies are successfully integrated. Indeed, a FL platform is used to train a deep learning network and instead of a centralized server, a SC is setup for federating the computations. The setup is said to provide "data confidentiality, computation auditability, and incentives for parties to participate in collaborative training" [30]. We bring these ideas for the first time to the medical context where privacy, transparency and usage consent are primordial. Indeed, medical records often fall under strict regulations such as the European General Data Protection Regulation (GDPR) and the suitability of such solutions needs to be investigated. Accordingly, we present in the next section, a medical decision support tool in a blockchain-based FL framework.

3   Problem Description

As discussed in previous sections, FL algorithms reduce the data communicated between participants and reduces the risks of exposing sensitive patients information compared to a cloud-based approach. Additionally, using SCs instead of a federating server, we can add transparency and usage guarantees to the setup. Specifically, at each federation round, the SC collects the values from the different participants and returns the aggregated parameters for participants to update their ML model in a completely automated way. Participants can also at any time verify the correct execution of this step in a completely transparent manner given the open and distributed execution of SCs. For this work, we choose to train an Artificial Neural Network (ANN) as previous results are available for comparison[1]. That said, other ML algorithms can easily be used in FL settings [14].

Let i be an index for the n different facilities that choose to participate in a collaborative ANN prediction model. The input to the model includes relevant patients' characteristics for the prediction model and the output is a binary variable. Each participant locally trains their ANN with weight parameters w and biases b [25] before sending the parameters to the federation SC. The loss function of such a ANN is of the following form.

$$f(w, b) \stackrel{def}{=} \frac{1}{2n} \sum_x \|y(x) - a\|^2 \tag{1}$$

Where x are the different data-set samples and $y(x)$ corresponds to the output of our model for the particular input x. In order to train the model, the cost function [25] needs to be minimized so that the difference between y(x) and the actual output a (indicating if the patient of sample x is diabetic or not) is minimized. Of course, more complex objective function are possible but this is not the focus of this work.

The overall architecture of our system is illustrated in Fig. 2. At each round τ, medical facilities send their trained parameters to a "Federating" SC uploaded on a blockchain network. The SC then aggregates the different parameters and sends back the result to the facilities for them to train the individual models again. As the aggregated parameters incorporate information from the different facilities, they have been shown to achieve near-optimal accuracy [20].

Let ω represent both the weights and the biases, we use the formulation of Mc Mahan et al [20] in a federated setting and define the following loss:

$$\min_{\omega \in \mathbb{R}^d} f(\omega) \text{ where } f(\omega) \stackrel{def}{=} \frac{1}{n} \sum_{i=1}^n f_i(\omega) \tag{2}$$

The objective in Eq. 2 is to minimize the overall loss function defined as the average of the participants' individual loss functions. The algorithm for minimizing the loss functions can be separated into "client" steps and "server" steps

---

[1] https://www.kaggle.com/ravichaubey1506/predictive-modelling-knn-ann-xgboost.

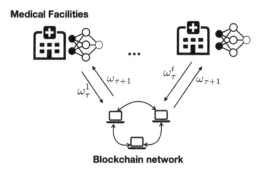

Fig. 2. An illustration of the proposed system. The blockchain is used instead of the coordinating server in the original FL architecture.

as defined in [14,20]. In our setting, the clients are the different facilities i and the "server" is the SC with access to all the local updates. The clients perform a normal gradient descent process and then send their parameters ω to the SC which proceeds with a weighted average of the different client parameters as in Eq. 3.

$$\omega_{\tau+1} = \sum_i \frac{n_i}{n} \omega_\tau \tag{3}$$

Next, we investigate in Sect. 4 some experiments to validate the solution explained in this section to a real-world medical problem.

4   Experiments

In this section we evaluate our solution using a diabetes data-set from the American National Institute of Diabetes and Digestive and Kidney Diseases available online[2]. First, we validate our ANN model's capacity to predict diabetes by using it in a centralized setting. Then, we test the FL setting by distributing the medical records between 15 participants and having the participants collaborate. Finally, we implement a small blockchain prototype based on Ethereum to validate feasibility. For all the tests, we use ANN consisting of 2 hidden layers with 32 and 16 neurons respectively and a binary output layer to indicate whether the patients is predicted diabetic or not. Also, we use 80% of the data for training the ANN and the remaining 20% for testing.

In the first experiment the centralized model is trained for 50 epochs and the accuracy of the trained model on the testing data-set is evaluated after each epoch. Figure 3 illustrates the results obtained after running the experiment 10 times with different data distribution scenarios between the test and validation set. For each scenario, we randomly select 80% of the entries for training and leave the remaining for testing. We note that the initial data distribution

---

[2] https://www.kaggle.com/uciml/pima-indians-diabetes-database.

markedly impacts the accuracy of the final model. This clearly indicates an influence of the training data-set and corroborate the idea of benefiting from a collaborative setting.

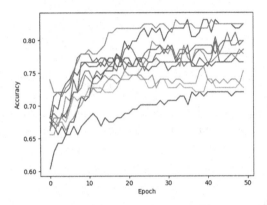

Fig. 3. The prediction accuracy of our ANN in a centralized setting for different training and testing scenarios.

Next, to test the FL model, we distribute the data-set among 15 participants to simulate a mid-sized collaboration scenario. After each 10 local epochs, the participants all share their model's parameters (weights and biases) with a centralized federation server. The data is distributed among the participants in a randomized way. Some example results are shown in Fig. 4. The blue line represents the accuracy with the federated parameters while the grey line is a baseline model without federation. The results show that individual models' accuracy are lower than in the centralized setting with an average accuracy of 73% (against 76% in the centralized setting) but all local models benefit from the data aggregates of the FL system with their accuracy improving between 1% to 5%. We choose to illustrate the results for 3 participants only as they had the most characteristic behavior, but other participants have similar trends. By having the ANN implemented in a FL setting, patients' raw medical records are protected from attackers as they never leave the medical facilities. However, some studies have shown that inference attacks are possible on aggregated data. That is, some unintended information can be leaked from model parameters only. Possible mitigation strategies are discussed in Sect. 5.

Finally, with the above setup, participants have no visibility on the computations performed on the federating server. To add transparency and facilitate usage consent, we implement the federating algorithm in a SC using the Ethereum blockchain [31] and several open source tools (c.f Appendix A.4). Our implementation is openly accessible on this link https://github.com/n-vcs/solidity-fl. By having the SC deployed instead of a federating server, participants can verify how the parameters submitted are used and track all the transactions that are happening. Of course, a full-fledged implementation is needed to solve setting-specific issues that are not addressed in this paper due to space limitation.

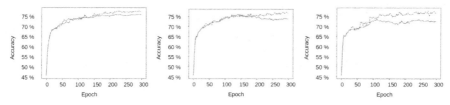

(a) Low improvement with federation

(b) Late improvement with federation

(c) Significant improvement with the federation

Fig. 4. Resulting accuracy for three participants in a FL setting. (Color figure online)

Clinical Relevance: Promptly testing for diabetes can help patients receive timely and accurate treatment and help prevent misdiagnosis. Furthermore, training the model in the setting explained here allows medical centers to transparently and securely share knowledge with other facilities. This knowledge sharing creates more robust models which are resilient to environment-specific biases.

5   Conclusion and Future Work

In this work, we presented a blockchain-based FL solution to a diabetic prediction model. Our solution capitalizes on the benefits of both the blockchain and FL algorithms and caters to the particularities of a medical context. We provided a prototype implementation and presented tests to validate the relevance of the solution in the medical context. However, although no raw training data ever leaves the participants' devices in FL settings, the updates sent to the coordinating server may still contain private information. Indeed, FL algorithms have been shown to be vulnerable to inference attacks [30]. These types of attacks can deduce information about the training population which was not intended by the model.

To deal with this problem, privacy preserving protocols have been developed to prevent leakage of sensitive data. For instance, differential privacy solutions, Differential Privacy [6]and Homomorphic Encryption [8] have been shown to be relatively useful against inferential attacks [33]. However, even without revealing any details of the model, it is quite hard to achieve perfect privacy [28]. Against this grim outlook, it is however essential to be aware of the risks of each particular settings and measure them against the potential benefits.

Acknowledgements. This work was supported by a grant from the Roche Institute 2018.

Appendix A   The Blockchain Technology: Key Concepts and Implementation

This appendix explains some key concepts related to the blockchain as well as technical implementation details relevant to the present work.

A.1   Main Characteristics

The blockchain is a distributed ledger technology managed by a network of peers. Data on the blockchain are visible and duplicated across participants. New data are added to the blockchain through a consensus protocol. As such, the blockchain is decentralized, immutable and transparent by design.

A.2   Smart Contracts (SCs)

SCs are set of instructions, specified in digital form and executed when predefined conditions are met. Contrary to regular software, SCs benefit from the main characteristics of the blockchain and can help attain transparency and usage consent.

As opposed to centralized algorithms, SCs allow data owners to verify, at any time, the implementation of the FedAvg algorithm and replicate the results locally. Additionally, usage consent can be facilitated through SCs by logging data ownership certification directly on the blockchain as described in [16].

A.3   Consensus and Incentive Mechanism

A consensus protocol is at the heart of the blockchain' mechanism. The Ethereum blockchain uses Proof-of-Work (PoW) by default. PoW relies on computational power to validate transactions or execute SCs. In our context, Medical institution are natural candidates for running the network as they will benefit from the resulting model. But to keep the network alive, nodes needs to be incentivized. As such, a possible set-up proposed in [30] is to have a reward mechanism for participants based on their data contribution.

A.4   Current Set-up

For the prototype[3] an Ethereum blockchain is set-up on the back-end using the Truffle suite (Ganache and the Truffle Development Environment). Also, the FedAvg [20] algorithm and a basic ANN are developed using Python's scientific computing library (Numpy). The front-end is developed in React. Data for each participant are stored locally in a MongoDB database.

---

[3] https://github.com/n-vcs/solidity-fl.

References

1. Agbo, C., Mahmoud, Q., Eklund, J.: Blockchain technology in healthcare: a systematic review. Healthcare **7**(2), 56 (2019)
2. Halamka, J.D., Andrew, M.D., Lippman A.E., Azaria, A.: A case study for blockchain in healthcare: "MedRec" prototype for electronic health records and medical research data. Technical report (2016)
3. Brisimi, T.S., Chen, R., Mela, T., Olshevsky, A., Paschalidis, I.C., Shi, W.: Federated learning of predictive models from federated electronic health records. Int. J. Med. Inf. **112**, 59–67 (2018)
4. Dagher, G.G., Mohler, J., Milojkovic, M., Babu, P.M.: Ancile: privacy-preserving framework for access control and interoperability of electronic health records using blockchain technology. Sustain. Cities Soc. **39**, 283–297 (2018)
5. De Filippi, S., McCarthy, S.: Cloud computing: Centralization and data sovereignty. *Euro. J. Law Technol.* **3**(2) (2012)
6. Dinur, I., Nissim, K.: Revealing information while preserving privacy. In: Proceedings of the Twenty-Second ACM SIGMOD-SIGACT-SIGART Symposium on Principles of Database Systems, pp. 202–210 (2003)
7. Drosatos, G., Kaldoudi, E.: Blockchain applications in the biomedical domain: a scoping review. Comput. Struct. Biotechnol. J. **17**, 229–240 (2019)
8. Gentry, C., Boneh, D.: A Fully Homomorphic Encryption Scheme, vol. 20. Stanford university Stanford, California (2009)
9. Gordon, W.J., Catalini, C.: Blockchain technology for healthcare: facilitating the transition to patient-driven interoperability. Comput. Struct. Biotechnol. J. **16**, 224–230 (2018)
10. Harris, J.D, Waggoner, B.: Decentralized and collaborative AI on blockchain. In: 2019 IEEE International Conference on Blockchain (Blockchain), pp. 368–375 (2020)
11. Huang, L., Shea, A.L., Qian, H., Masurkar, A., Deng, H., Liu, D.: Patient clustering improves efficiency of federated machine learning to predict mortality and hospital stay time using distributed electronic medical records. J. Biomed. Inf. **99**, 103291 (2019)
12. Janssen, M., Weerakkody, V., Ismagilova, E., Sivarajah, U., Irani, Z.: A framework for analysing blockchain technology adoption: integrating institutional, market and technical factors. Int. J. Inf. Manag. **50**, 302–309 (2020)
13. Kairouz, P., et al. Advances and open problems in federated learning. arXiv preprint arXiv:1912.04977 (2019)
14. Konečný, J., McMahan, B., Ramage, D.: Federated optimization: distributed optimization beyond the datacenter. arXiv preprint arXiv:1511.03575 (2015)
15. Larson, E.B.: Building trust in the power of "big data" research to serve the public good. Jama **309**(23), 2443–2444 (2013)
16. Leeming, G., Cunningham, J., Ainsworth, J.: A ledger of me: personalizing healthcare using blockchain technology. Front. Med. **6**(July), 1–10 (2019)
17. Leonelli, S.: Locating ethics in data science: responsibility and accountability in global and distributed knowledge production systems. Philos. Trans. R. Soc. Math. Phys. Eng. Sci. **374**(2083), 20160122 (2016)
18. Li, H., Ota, K., Dong, M.: Learning IoT in edge: deep learning for the internet of things with edge computing. IEEE Netw. **32**(1), 96–101 (2018)
19. Li, T., Sahu, A.K., Talwalkar, A., Smith, V.:. Federated learning: Challenges, methods, and future directions. arXiv preprint arXiv:1908.07873 (2019)

20. McMahan, B., Moore, E., Ramage, D., Hampson, S., et al.: Communication-efficient learning of deep networks from decentralized data. arXiv preprint arXiv:1602.05629 (2016)

21. Mehta, N., Pandit, A.: Concurrence of big data analytics and healthcare: a systematic review. Int. J. Med. Inf. **114**, 57–65 (2018)

22. Mettler, M.: Blockchain technology in healthcare: the revolution starts here. In: 2016 IEEE 18th International Conference on e-Health Networking, Applications and Services, Healthcom 2016, pp. 1–3 (2016)

23. Mittelstadt, B.D., Floridi, L.: The ethics of big data: current and foreseeable issues in biomedical contexts. Sci. Eng. Ethics **22**(2), 303–341 (2016)

24. Nakamoto, S.: Bitcoin: A peer-to-peer electronic cash system. Technical report, Manubot (2019)

25. Nielsen, M.A.: Neural Networks and Deep Learning, vol. 2018. Determination press San Francisco, CA (2015)

26. Santhosh, G., De Vita, F., Bruneo, D., Longo, F., Puliafito, A.: Towards trustless prediction-as-a-service. In: Proceedings - 2019 IEEE International Conference on Smart Computing, SMARTCOMP 2019, pp. 317–322 (2019)

27. Schneble, C.O., Elger, B.S., Shaw, D.M.: Google's project nightingale highlights the necessity of data science ethics review. EMBO Mol. Med. **12**(3), e12053 (2020)

28. Shokri, R., Stronati, M., Song, C., Shmatikov, V.: Membership inference attacks against machine learning models. In: 2017 IEEE Symposium on Security and Privacy (SP), pp. 3–18. IEEE (2017)

29. Wang, T.: A unified analytical framework for trustable machine learning and automation running with blockchain. In: Proceedings - 2018 IEEE International Conference on Big Data, Big Data 2018, pp. 4974–4983 (2018)

30. Weng, J., Weng, J., Zhang, J., Li, M., Zhang, Y., Luo, W.: DeepChain: auditable and privacy-preserving deep learning with blockchain-based incentive. IEEE Trans. Dependable Secure Comput. **8**, 1 (2018)

31. Wood, G., et al.: Ethereum: a secure decentralised generalised transaction ledger. Ethereum Proj. Yellow Pap. **151**(2014), 1–32 (2014)

32. Xu, J., Wang, F.: Federated learning for healthcare informatics. arXiv preprint arXiv:1911.06270 (2019)

33. Yang, Q., Liu, Y., Chen, T., Tong, Y.: Federated machine learning: concept and applications. ACM Trans. Intell. Syst. Technol. **10**(2), 1–19 (2019)

Image Processing

Forming Local Intersections of Projections for Classifying and Searching Histopathology Images

Aditya Sriram[1], Shivam Kalra[1], Morteza Babaie[1(✉)], Brady Kieffer[1],
W. Al Drobi[1], Shahryar Rahnamayan[2], Hany Kashani[1],
and Hamid R. Tizhoosh[1]

[1] KIMIA Lab, University of Waterloo, Waterloo, Canada
{mbabaie,tizhoosh}@uwaterloo.ca
[2] Electrical, Computer and Software Engineering, Ontario Tech University,
Oshawa, Canada

Abstract. In this paper, we propose a novel image descriptor called "Forming Local Intersections of Projections" (FLIP) and its multi-resolutional version (mFLIP) for representing histopathology images. The descriptor is based on the Radon transform wherein we apply parallel projections in small local neighborhoods of gray-level images. Using equidistant projection directions in each window, we extract unique and invariant characteristics of the neighborhood by taking the intersection of adjacent projections. Thereafter, we construct a histogram for each image, which we call the FLIP histogram. Various resolutions provide different FLIP histograms which are then concatenated to form the mFLIP descriptor. Our experiments included training common networks from scratch and fine-tuning pre-trained networks to benchmark our proposed descriptor. Experiments are conducted on the publicly available dataset KIMIA Path24 and KIMIA Path960. For both of these datasets, FLIP and mFLIP descriptors show promising results in all experiments. Using KIMIA Path24 data, FLIP outperformed non-fine-tuned Inception-v3 and fine-tuned VGG16 and mFLIP outperformed fine-tuned Inception-v3 in feature extracting.

Keywords: Radon projections histopathology · Image search · Feature extraction · Image descriptors

1 Introduction

Histopathology, is primarily concerned with the manifestations of a diseased human tissue [1]. A traditional diagnosis is based on examination of the tissue of concern mounted on a glass slide under various magnifications of a microscope [2]. More recently, digital pathology has connected the computer vision field to the diagnostic pathology by scanning the glass slide and creating a whole slide image (WSI). This allows easy storage, more flexibility in sharing information

© Springer Nature Switzerland AG 2020
M. Michalowski and R. Moskovitch (Eds.): AIME 2020, LNAI 12299, pp. 227–237, 2020.
https://doi.org/10.1007/978-3-030-59137-3_21

and eliminates the risk of losing specimens [3]. In digital pathology the images are extremely large and it takes an extremely long time to compute images. Hence, there is a need to develop a powerful image descriptor that can extract unique and invariant features from these large images to enable algorithms to retrieve and classify salient patterns and morphologies. In essence, the descriptor should suffice as an image representative such that one should be able to index the large scans with a limited number of descriptors.

In this work, we propose a novel descriptor called "Forming Local Intersections of Projections" (FLIP) which applies the Radon transform to small spatial windows (to extract the features of histopathology images). The FLIP descriptor enables fast image search while minimizing any extra storage requirement by storing the representation of an image as a compact histogram.

2   Related Works

Image descriptors quantify image characteristics such as shape, color, texture, edges, and corners. Local Binary Patterns (LBPs) [4] are a good example for image descriptor to classify texture with rotation invariance [5]. Designed as a particular case of texture spectrum model [6], LBPs are powerful image descriptors that have certainly set relatively high accuracy standards in the medical domain, including digital pathology scans [7].

Deep neural networks have been widely utilized to generate global image descriptors. These networks consist of functions in each layer to generate local features at different resolutions describing a particular image region. These local features may then be aggregated, to provide a global descriptor that is the entire image. Similar to LBP, deep descriptors have reported many promising results, specifically in the histopathology domain [7,8].

More recently, several approaches have been put forward to develop projection-based descriptors [9,10]. The Radon transform is a well-established approach [11]. A novel Radon barcode for medical image retrieval system was proposed in 2015 [12]. The Radon barcode is a binary vector generated from global projections with selected projection angles and projection binarization operation that can tag a medical image or its regions of interest. Using Radon barcodes, large image archives can be efficiently searched to find matches via Hamming distance, however, the performance of global projections is rather limited. More recently, local Radon projections and support vector machines (SVM) have been combined for medical image retrieval [9].

Tizhoosh et al. [13] have introduced Autoencoded Radon Barcode (ARBC) that used mini-batch stochastic gradient descent and binarizing the outputs from each hidden layer during training to produce a barcode per-layer. The ARBC was observed to achieve an Image Retrieval in Medical Application (IRMA) error of 392.09. More recently, Tizhoosh et al. [14] proposed MinMax Radon barcodes which were observed to retrieve images 15% faster compared to the "local thresholding" method.

A similar type of approach was proposed by Xiaoshuang et al. [15] which presented a cell-based framework for pathology images wherein they encode each cell into a set of binary codes using a hashing model [16]. The binary codes are then converted into a 1-dimensional histogram vector (which is the feature vector) used for learning using an SVM for image classification. In this paper, we attempt to design and test a "local" projection-based descriptor that should deliver good results for histopathology images.

3   Methods

The Radon transform provides scene/object projections (profiles) in different directions. The set of all projections can yield a reconstruction of the scene/objects when performing an inverse Radon transform (i.e., filtered back-projection).

Using the Dirac delta function $\delta(\cdot)$, the Radon transform of a two-dimensional image $f(x, y)$ can be defined as its line integral along a straight line inclined at an angle θ and at a distance ρ from the origin:

$$R(\rho, \theta) = \int_{-\infty}^{\infty} \int_{-\infty}^{\infty} f(x, y)\delta(x cos\theta + y sin\theta - \rho)dxdy \tag{1}$$

Here, $-\infty < \rho < \infty, 0 \leq \theta < \pi$, the Radon transform accentuate straight-line features from an image by integrating the image intensity over the straight lines to a single point [17]. Given the scan $\mathbf{S}$, we are interested in describing the (grayscale) image $\mathbf{I}(\subset \mathbf{S})$ via a short descriptor, or histogram, $\mathbf{h}$ using Radon projections $R(\rho, \theta)$ to transform the intensities $f(x, y)$ of $\mathbf{I}$. We can process all local neighbourhoods $\mathbf{W}_{ij} \subset \mathbf{I}$. For each neighbourhood $\mathbf{W}_{ij}$, we capture n_P projections with $0 < n_P \ll 180$: $\mathbf{p}_{ij}^1, \mathbf{p}_{ij}^2, \ldots, \mathbf{p}_{ij}^{n_P}$. One may find individual projections from different (and dissimilar) images to be quite similar. Hence, we take the "intersection of adjacent projections" to quantify the spatial correlations of a given neighbourhood pattern. The intersection of projections can be thought of as an approximation of the logical "AND" providing a unique characteristic of local patterns. Therefore, we receive n_P intersection vectors $\mathbf{V}_{k,m}$ as

$$\mathbf{V}_{k,m} = \min\left(\mathbf{p}_{ij}^k, \mathbf{p}_{ij}^{(k+1)\% n_P}\right), \tag{2}$$

with $k = 1, 2, \ldots, n_P$ and $m = 1, 2, \ldots, n_W$ where n_W is the total number of local windows of the image $\mathbf{I}$. Hence, we will have $n_P \times n_W$ intersections of local projections. The projections have different values which are also subject to intensity fluctuations. Hence we re-scale all projection values to be:

$$\bar{\mathbf{V}}_{k,m} = \left\lceil L \times \frac{\mathbf{V}_{k,m} - p_{\min}}{p_{\max} - p_{\min}} \right\rceil. \tag{3}$$

Now, we can count the values $\bar{\mathbf{V}} \in 1, 2, \ldots, L, \forall m = 1, 2, \ldots, n_P \times n_W$ to obtain $\mathbf{h}$, wherein L is the default histogram length of 128. Algorithm 1 provides the pseudo-code for calculating the FLIP descriptor. Figure 1 provides a simplified overview of extracting a FLIP histogram for sample images.

Algorithm 1. The FLIP algorithm

Input : An image **I** as part of a whole scan **S**: **I** $\subset$ **S**
Output: The FLIP histogram **h**
Set neighbourhood size and overlap;
$L \leftarrow 128$ (histogram default length);
$\mathbf{h} \leftarrow \emptyset$;
$\mathbf{F} \leftarrow \emptyset$;
$\mathbf{I}_g \leftarrow$ Convert image **I** to gray-scale;

foreach *window* $\mathbf{W}_i$ *in image* $\mathbf{I}_g$ **do**
   | $R_{(0,45,90,135)} \leftarrow$ RadonTransform($\mathbf{W}_i$);
   | $R_{min_1} \leftarrow \min(R_0, R_{45})$;
   | $R_{min_2} \leftarrow \min(R_{45}, R_{90})$;
   | $R_{min_3} \leftarrow \min(R_{90}, R_{135})$;
   | $R_{min_4} \leftarrow \min(R_{135}, R_0)$;
   | $R_{min} \leftarrow$ concatenate $(R_{min_1}, R_{min_2}, R_{min_3}, R_{min_4})$;
   | $\mathbf{F} \leftarrow$ AppendRow(R_{min});
end
$f_{min}, f_{max} \leftarrow$ FindMinMax($\mathbf{F}$);
$\mathbf{F} \leftarrow$ reScale($\mathbf{F}, f_{min}, f_{max}, L$);
$\mathbf{F} \leftarrow \mathbf{F}[1:128]$ (127 length histogram);
for $i = 1$ *to* $\mathbf{F}_{rows}$ **do**
   for $j = 1$ *to* $\mathbf{F}_{cols}$ **do**
     | $\mathbf{h}(\mathbf{F}(i,j)) \leftarrow \mathbf{h}(\mathbf{F}(i,j)) + 1$;
   end
end
Return **h**;

The intuition behind multi-resolutional representation is to capture structural changes that one observes in real-world objects [18]. Specifically, multiple scales for the same image capture the variation in visual appearances - providing a different representation of the same image for every scale. Furthermore, pathologists examine tissue samples at different magnifications to have a comprehensive perception of the specimen [19]. A multi-resolution FLIP is built using different image resolutions (inclusive of original and resized resolutions). These resolutions include: (i) original resolution, (ii) 0.75× the original resolution, (iii) 0.5× the original resolution, and (iv) 0.25× the original resolution. After obtaining a FLIP histogram for each of the resolutions, we concatenate them in descending order of resolution to form a final histogram, namely the mFLIP descriptor.

Indexing and Testing – For all training samples, the scan is first divided by a regular grid of proper size whereas each grid cell can be cropped into a new training image/patch. The mFLIP histogram of each image is calculated and saved in a database. Subsequently, all of the images are classified using an SVM algorithm. In the testing phase (which emulates the proposed system in action), every query scan can be processed in two possible ways: (i) a small number of locations within the query scan is selected to extract some patches, or (ii) one

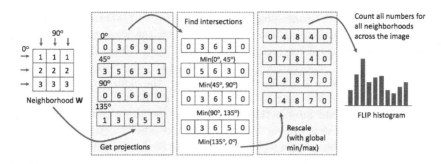

Fig. 1. A simplified overview of the FLIP histogram extraction. A small number of pairwise orthogonal projections, here $n = 4$, is computed for each neighbourhood, from which the intersection of all adjacent projections is computed. After re-scaling all intersections in the image based on global min/max projection values, the FLIP histogram can be assembled by counting the rescaled intersections.

patch within the query scan is selected either manually or automatically. In both cases the main task of the image search is to find the best match for a patch. The matching algorithm can be implemented in two ways: (i) using a proper distance measure, we quantify the (dis)similarity between the mFLIP descriptor of the query patch and the mFLIP descriptor of every image in the database, or (ii) we use the trained SVM to assign a class to the query patch. For the distance-based image search, several strategies were used including: χ^2, histogram intersection, Pearson coefficient, cosine similarity, and L_1 and L_2 metrics. As for classification, a generalized histogram intersection kernel SVM is adopted. In practice, pathologists prefer to inspect more than one retrieved case. Hence, we retrieve the top 3 images (patches) for a query scan (for visual inspection) and examine which one of the three images is the actual match for the query image to calculate the accuracy (only the first match is considered for accuracy calculation).

4   Experiments

We used two publicly available pathology datasets to validate the FLIP and mFLIP algorithms, namely: (i) KIMIA Path960, and (ii) KIMIA Path24. All images in these experiments are converted to grayscale.

KIMIA Path960 - Introduced by Kumar et al. [20]. KIMIA Path960 is a publicly available pathology dataset, comprised of 960 images of size 308×168 from 20 different classes (i.e. tissue types). Since this dataset is relatively small, we used leave-one-out approach to validate our proposed algorithms.

KIMIA Path24 - Introduced by Babaie et al. [7], is a digital pathology public dataset, published in 2017, that comprises of 24 scans depicting different tissue patterns and body parts. The scans have been converted to gray-scale using the Python library Scikit Learn. The dataset consists of 1,325 test patches of size 1000×1000 ($0.5\,\text{mm} \times 0.5\,\text{mm}$) for which the labels correspond to the scan number. The number of training patches can range from 27,000 to 50,000 patches, depending on the percentage of overlap selected by the algorithm designer. In our experiments, we have received 27,055 patches with %0 overlap.

For the proposed algorithms (FLIP and mFLIP), both image search and classification strategies are implemented to support the algorithms performance. For the image search, we compare the FLIP or mFLIP histogram using either chi-square or histogram intersection algorithm to retrieve a patch that is best-matched with the query image. We then use the label for the best-matched patch to determine the accuracy. The retrieved patch does not necessarily reside within the same WSI. Hence, in the KIMIA Path24 dataset, there are two accuracy measures – patch-to-scan accuracy (η_p) and whole-scan accuracy (η_W). The total accuracy (η_{total}) is a multiplication of both these accuracies. As for the classification, we train an SVM classifier for all the training patches. We down-sampled each image to 250×250 which resulted in 2–3% loss in accuracy, regardless of the histogram length, when compared to the accuracy results for the 1000×1000 images. Hence, we only report results for the gray-scaled 1000×1000 images as they yield better results.

Accuracy Measurement – For the KIMIA Path24, a total of $n_{tot} = 1,325$ test patches P_s^j are obtained which belong to either one of the 24 classes available $\Gamma_s = \{P_s^i | s \in S, i = 1, 2, \ldots, n_{\Gamma_s}\}$ with $s = 0, 1, 2, \ldots, 23$ [7]. In order to compare our method against other works, the accuracy calculation outlined in [7] is adopted. Hence, for a retrieved image R for any experiment, the patch-to-scan accuracy η_p and the whole-scan accuracy η_W can be given as:

$$\eta_p = \frac{\sum_{s \in S} |R \cap \Gamma_s|}{n_{tot}}, \eta_W = \frac{1}{24} \sum_{s \in S} \frac{|R \cap \Gamma_s|}{n_{\Gamma_s}} \tag{4}$$

The total accuracy η_{total} is obtained which is comprised of both patch-to-scan and whole-scan accuracies: $\eta_{total} = \eta_p \times \eta_W$.

As for the KIMIA Path960 the accuracy metrics were compliant with leave-one-out approach. Since this is a multi-class dataset, for each test image, we run it through the entire training set to obtain the image and its class with the highest probability. The test and the best-matched image classes are compared to determine if there is a match (i.e. 1) or a mismatch (i.e. 0). The overall accuracy is the percentage of all matched images with respect to the total number of test images (i.e. 960 in this case).

Experimentation on Deep Learning – For KIMIA Path24, we specifically computed four different deep learning structures to compare against the proposed mFLIP descriptor. These deep learning approaches are as follows: (i) VGG16: a pre-trained deep net as feature extractor, (ii) a fine-tuned VGG16 (transfer learning), (iii) Inception V3: a pre-trained deep net as feature extractor, and (iv) a fine-tuned Inception V3.

Pre-Trained CNN as a Feature Extractor. Specifically for the KIMIA Path24 dataset, the first set of experiments were developed using the Keras library in Python wherein we used pre-trained VGG16 and InceptionV3 for feature extraction without fine-tuning the parameters. In essence, the fully-connected layer (feature vector) for each of these pre-trained models were extracted and provided to an SVM for classification. For linear SVM classification, Python packages scikitlearn and LIBSVM were adopted [21,22]. Finally, Python libraries NumPy and SciPy were leveraged to manipulate and store the data [23,24].

Table 1. mFLIP and FLIP results for different retrieval strategies (χ^2, histogram intersection, and svm) for a histogram length of $L = 127$, generated using neighborhood size of with no-overlap ($\Delta = 3$). Best results are highlighted in bold.

	η_p	η_W	η_{total}
mFLIP$_{(508,3,3),\ \chi^2}$	77.28	77.55	59.93
mFLIP$_{(508,3,3),\ histInt}$	74.87	75.38	56.44
mFLIP$_{(508,3,3),\ svm}$	**84.68**	**85.52**	**72.42**
FLIP$_{(127,3,3),\ \chi^2}$	67.62	68.27	46.16
FLIP$_{(127,3,3),\ histInt}$	68.07	69.03	46.98
FLIP$_{(127,3,3),\ svm}$	**74.11**	**74.54**	**55.24**

Fine-Tuned CNN as a Classifier. For completion, we used the Keras library in Python to fine-tune the pre-trained networks VGG16 and Inception V3 as a classifier against the KIMIA Path24 dataset. For the VGG16 network, we first removed the fully-connected layers from the convolutional layers, after which, we fed the network with training images to extract bottleneck features through the convolutional layers. Thereafter, the new fully connected model is attached back onto the VGG16 convolutional layers and trained on each convolutional block, except the last block, in order to receive the adjusted classification weights.

Likewise for the InceptionV3 network, the originally fully connected layer is replaced with a single 1024 dense ReLU layer followed by a softmax classification layer. The new fully connected layers were trained on bottleneck features and then attached back onto the original convolutional layers for training the final two inception blocks.

Evaluation of mFLIP Descriptor – We performed multiple experiments with different mFLIP configurations in the form of "mFLIP$_{(L,w,\Delta),D}$" where $L = |\mathbf{h}|$ is the histogram length, w is the window size, Δ is the pixel stride (overlap), and D is the distance measure or classification scheme. Specifically, we experimented with $L = 127$ and 511 (after removing the first bin), $w = 3$ (3×3), and $\Delta = 3$ (no overlap).

Table 1 provides an overview of the performance of FLIP and mFLIP. When the FLIP is configured with utilizing the original dimensions, with a neighborhood size of 3×3 and $\Delta = 3$ pixel stride and a histogram length of $L = 127$, the best accuracy (η_{total}) of 55.24% is achieved using an SVM classifier (with generalized histogram intersection kernel) in the KIMIA Path24 dataset. On the other hand, we obtain a 46.98% accuracy when using histogram intersection distance metric for searching the best-matched image in the KIMIA Path24 dataset – determined by obtaining the lowest distance when comparing histograms. Currently, the benchmark score for the KIMIA Path24 is achieved by mFLIP - utilizing four image dimensions of 1000×1000, 750×750, 500×500, 250×250, with a neighborhood size of 3×3 and $\Delta = 3$ pixel stride and a histogram length of $L = 508$ (each dimension of which gets a FLIP descriptor of 127 concatenated together). The best total accuracy in the KIMIA Path24 dataset is an (η_{total}) of 72.42% which is achieved using an SVM classifier on mFLIP features and a 59.93% accuracy when using χ^2 distance metric (image search).

After numerous experiments, the best configuration for FLIP is to utilize the highest resolution of the dataset (namely 20x) which results in input images of 1000×1000 equivalent to 0.5×0.5 mm^2 that are processed in 3×3 neighbourhood windows with no overlap. Moreover, a 127 bin-size histogram was empirically selected as the size of the FLIP feature vector for each image. A window size of 3×3 is used, as it is the smallest window size that we can utilize for computing the histogram and appears to capture local changes of nuclei and other structures. Additionally, the window of 3×3 was observed to yield the best results when compared against 5×5, 8×8, 32×32, and 64×64 window sizes. Although one has the flexibility to change the window size for any application within the FLIP algorithm, for the purpose of our experimentation with KIMIA Path24, we chose a neighborhood of 3×3 as it yielded the best result.

Table 2 provides a comparison of FLIP and mFLIP against deep learning methods on the KIMIA Path24 based on gray-scale images. We also show the results of the ELP descriptor that also uses local projections. We explored the performance of a pre-trained deep features versus training from scratch. All the experiments were done on the same KIMIA Path24 dataset. We deduced that pre-trained networks are comparable to training a CNN from scratch. Also, fine-tuning VGG16 does not yield better results despite requiring more training time [25]. We also observed considerable improvement in image search and classification accuracy for the fine-tuned Inception structure. The fine-tuned InceptionV3 delivers $\eta_{total} = 56.98$ which is slightly higher than the FLIP accuracy, namely $\eta_{total} = 55.24$. However, all deep learning approaches are considerably lower when compared to the current benchmark, mFLIP$_{(508,3,3)}$ which achieves a $\eta_p = 85.53$, $\eta_p = 84.68$, and $\eta_{total} = 72.42$. The fact that a handcrafted algorithm can surpass deep learning methods, which are the result of substantial design and training efforts, is quite encouraging. However, the reason behind the relatively low performance of deep features might be due to the feeding of grey scale images to networks while deep networks tend to depend heavily on color. Also the reason for success of mFLIP may be due to the usage of projections in local windows across multiple magnifications.

For completion, Table 3 provides an overview of the top performing algorithms in the KIMIA Path960 dataset in comparison to the proposed mFLIP algorithm. Although mFLIP does not set the benchmark for the dataset, it certainly competes with the top methods with minimum computation time and resource.

Table 2. Results for a SVM classifier on FLIP and mFLIP against the literature.

Method	η_W	η_p	η_{total}
mFLIP$_{(508,3,3),\text{svm}}$	**85.52**	**84.68**	**72.42**
ELP$_{\text{svm}}$ [9]	82.70	79.90	66.01
Inception-v3 (Fine-Tuned) [25]	76.10	74.87	56.98
FLIP$_{(127,3,3),\text{svm}}$	**74.54**	**74.11**	**55.24**
Inception-v3 (Feature Extractor) [25]	71.24	70.94	50.54
VGG+RF [26]	67.12	64.66	43.40
VGG16 (Feature Extractor) [25]	64.96	65.21	42.36
VGG16 (Fine-Tuned) [25]	66.23	63.85	42.29
CNN (Trained from Scratch) [7]	64.75	64.98	41.80

Table 3. mFLIP accuracy against other methods for KimiaPath960.

Method	Accuracy
BoVW$_{(1200 \text{ codebooks}), \text{ IKSVM}}$ [20]	94.87
VGG16 $_{\text{L2}}$ [20]	94.72
AlexNet $_{\text{L1}}$ [20]	91.35
LBP $_{\text{L2}}$ [20]	90.62
mFLIP $_{\chi^2}$	88
mFLIP $_{\text{svm}}$	87

5   Conclusions

Here we introduced a new feature descriptor called *Forming Local Intersections of Projections* (FLIP) wherein we have shown that using element-wise intersections of local Radon projections, followed by re-scaling to create a histogram, can be used to construct a new image descriptor. In addition, a multi-resolution FLIP descriptor (mFLIP) is also introduced and validated against the publicly available, KIMIA Path24 and KIMIA Path960 datasets. Specifically, the mFLIP is observed to outperform deep solutions when tested on the KIMIA Path24 dataset. Furthermore, both FLIP and mFLIP provide a more compact image representation with 128 and 508 bins, respectively, compared to generally high-dimensionality of deep features (i.e., 4096 for CNN and VGG16) in the KIMIA Path24 dataset. It appears that mFLIP is particularly suitable for histopathology images as the proposed algorithm is observed to capture the texture of each image through the means of Radon projections and to quantify these projections onto a condensed histogram. In addition, the process of localizing and capturing the Radon transform for small neighborhood does not require learning or expensive training. The novel image descriptor (FLIP), and its multi resolutional version, the mFLIP descriptor have surpassed the current benchmark for the KIMIA Path24 dataset by achieving a total accuracy of ≈72% using an SVM classification with generalized histogram intersection kernel. We must mention that we have processed gray-scale images. Therefore, crucial information, such as staining that has chemical meaning in histopathology, may have been lost.

References

1. Lever, W.F.: Histopathology of the Skin. JB Lippincott Co. (1949)
2. Kayser, K., Hoffgen, H.: Pattern recognition in histopathology by orders of textures. Med. Inform. **9**(1), 55–59 (1984)
3. Farahani, N., Parwani, A.V., Pantanowitz, L.: Whole slide imaging in pathology: advantages, limitations, and emerging perspectives. Pathol. Lab Med. Int. **7**, 23–33 (2015)
4. Ojala, T., Pietikainen, M., Maenpaa, T.: Multiresolution gray-scale and rotation invariant texture classification with local binary patterns. IEEE Trans. Pattern Anal. Mach. Intell. **24**(7), 971–987 (2002)
5. Pietikäinen, M., Hadid, A., Zhao, G., Ahonen, T.: Local binary patterns for still images. In: Pietikäinen, M., Hadid, A., Zhao, G., Ahonen, T. (eds.) Computer Vision Using Local Binary Patterns. Computational Imaging and Vision, vol. 40. Springer, Heidelberg (2011). https://doi.org/10.1007/978-0-85729-748-8_2
6. Ojala, T., Pietikäinen, M., Harwood, D.: A comparative study of texture measures with classification based on featured distributions. Pattern Recogn. **29**(1), 51–59 (1996)
7. Babaie, M., et al.: Classification and retrieval of digital pathology scans: A new dataset, arXiv preprint arXiv:1705.07522
8. Xu, Y., et al.: Deep learning of feature representation with multiple instance learning for medical image analysis. In: 2014 IEEE International Conference on Acoustics, Speech and Signal Processing (ICASSP), pp. 1626–1630. IEEE (2014)
9. Tizhoosh, H., Babaie, M.: Representing medical images with encoded local projections. IEEE Trans. Biomed. Eng. 1 (2018). https://doi.org/10.1109/TBME.2018.2791567
10. Babaie, M., Tizhoosh, H.R., Khatami, A., Shiri, M.: Local radon descriptors for image search. In: 2017 Seventh International Conference on Image Processing Theory, Tools and Applications (IPTA), pp. 1–5. IEEE (2017)
11. Sanz, J.L., Hinkle, E.B., Jain, A.: Radon and Projection Transform-Based Computer Vision: Algorithms, a Pipeline Architecture, and Industrial Applications, vol. 16. Springer, Heidelberg (1988). https://doi.org/10.1007/978-3-642-73012-2
12. Tizhoosh, H.R.: Barcode annotations for medical image retrieval: a preliminary investigation. In: 2015 IEEE International Conference on Image Processing (ICIP), pp. 818–822. IEEE (2015)

13. Tizhoosh, H.R., Mitcheltree, C., Zhu, S., Dutta, S.: Barcodes for medical image retrieval using autoencoded Radon transform. In: 2016 23rd International Conference on Pattern Recognition (ICPR), pp. 3150–3155. IEEE (2016)
14. Tizhoosh, H.R., Zhu, S., Lo, H., Chaudhari, V., Mehdi, T.: MinMax Radon barcodes for medical image retrieval. In: Bebis, G., et al. (eds.) ISVC 2016. LNCS, vol. 10072, pp. 617–627. Springer, Cham (2016). https://doi.org/10.1007/978-3-319-50835-1_55
15. Shi, X., Xing, F., Xie, Y., Su, H., Yang, L.: Cell encoding for histopathology image classification. In: Descoteaux, M., Maier-Hein, L., Franz, A., Jannin, P., Collins, D.L., Duchesne, S. (eds.) MICCAI 2017. LNCS, vol. 10434, pp. 30–38. Springer, Cham (2017). https://doi.org/10.1007/978-3-319-66185-8_4
16. Shi, X., Xing, F., Xu, K., Xie, Y., Su, H., Yang, L.: Supervised graph hashing for histopathology image retrieval and classification. Med. Image Anal. **42**, 117–128 (2017)
17. Rey, M.T., Tunaley, J.K., Folinsbee, J., Jahans, P.A., Dixon, J., Vant, M.R.: Application of Radon transform techniques to wake detection in Seasat-A SAR images. IEEE Trans. Geosci. Remote Sens. **28**(4), 553–560 (1990)
18. Chan, C.-H., Kittler, J., Messer, K.: Multi-scale local binary pattern histograms for face recognition. In: Lee, S.-W., Li, S.Z. (eds.) ICB 2007. LNCS, vol. 4642, pp. 809–818. Springer, Heidelberg (2007). https://doi.org/10.1007/978-3-540-74549-5_85
19. Pantanowitz, L.: Digital Images and the Future of Digital Pathology, vol. 1. Wolters Kluwer-Medknow Publications (2010)
20. Kumar, M.D., Babaie, M., Zhu, S., Kalra, S., Tizhoosh, H.R.: A comparative study of CNN, BoVW and LBP for classification of histopathological images. In: IEEE Symposium Series on Computational Intelligence (SSCI), pp. 1–7. IEEE (2017)
21. Pedregosa, F., et al.: Scikit-learn: machine learning in python. J. Mach. Learn. Res. **12**(Oct), 2825–2830 (2011)
22. Chang, C.-C., Lin, C.-J.: LIBSVM: a library for support vector machines. ACM Trans. Intell. Syst. Technol. (TIST) **2**(3), 27 (2011)
23. van der Walt, S., Colbert, S.C., Varoquaux, G.: The NumPy array: a structure for efficient numerical computation. Comput. Sci. Eng. **13**(2), 22–30 (2011)
24. Jones, E., Oliphant, T., Peterson, P.: SciPy: open source scientific tools for Python
25. Kieffer, B., Babaie, M., Kalra, S., Tizhoosh, H.R.: Convolutional neural networks for histopathology image classification: training vs. using pre-trained networks. In: 2017 Seventh International Conference on Image Processing Theory, Tools and Applications (IPTA), pp. 1–6. IEEE (2017)
26. Bizzego, A., et al.: Evaluating reproducibility of ai algorithms in digital pathology with dapper. PLoS Comput. Biol. **15**(3), e1006269 (2019)

Difficulty Translation in Histopathology Images

Jerry Wei[1], Arief Suriawinata[2], Xiaoying Liu[2], Bing Ren[2],
Mustafa Nasir-Moin[1], Naofumi Tomita[1], Jason Wei[1],
and Saeed Hassanpour[1(✉)]

[1] Dartmouth College, Hanover, NH 03755, USA
saeed.hassanpour@dartmouth.edu
[2] Dartmouth-Hitchcock Medical Center, Lebanon, NH 03756, USA

Abstract. The unique nature of histopathology images opens the door to domain-specific formulations of image translation models. We propose a difficulty translation model that modifies colorectal histopathology images to be more challenging to classify. Our model comprises a scorer, which provides an output confidence to measure the difficulty of images, and an image translator, which learns to translate images from easy-to-classify to hard-to-classify using a training set defined by the scorer. We present three findings. First, generated images were indeed harder to classify for both human pathologists and machine learning classifiers than their corresponding source images. Second, image classifiers trained with generated images as augmented data performed better on both easy and hard images from an independent test set. Finally, human annotator agreement and our model's measure of difficulty correlated strongly, implying that for future work requiring human annotator agreement, the confidence score of a machine learning classifier could be used as a proxy.

Keywords: Deep learning · Histopathology images · Generative adversarial networks

1 Introduction

Automated histopathology image analysis has advanced quickly in recent years [1–4] due to substantial developments in the broader fields of deep learning and computer vision [5–7]. While histopathology imaging research typically applies these general computer vision models directly and without modification, there may be domain-specific models that might not generalize to broader computer vision tasks but can be useful for specifically analyzing histopathology images.

In this study, we formulate a difficulty translation model for histopathology images, i.e., given a histopathology image, we aim to modify it into a new image that is harder to classify. Our model is motivated by the observation that histopathology images exhibit a range of histological features that determines

© Springer Nature Switzerland AG 2020
M. Michalowski and R. Moskovitch (Eds.): AIME 2020, LNAI 12299, pp. 238–248, 2020.
https://doi.org/10.1007/978-3-030-59137-3_22

their histopathological label. For instance, both an image with small amounts of sessile serrated architectures and an image covered by sessile serrated architectures would be classified by a pathologist as a sessile serrated adenoma. We know that this range of features exists because normal tissue progressively develops precancerous or cancerous features over time, which differs from general domain datasets such as ImageNet [8], in which classes are distinct by definition (there is no range of cats and dogs, for instance). This continuous spectrum of features allows us to use the confidence of a machine learning classifier to determine the amount and intensity of cancerous features in an image. In other words, the confidence of the classifier can act as a proxy for the extent of histological features in an image). In this paper, we propose and evaluate a difficulty translation model that generates hard-to-classify images that are useful as augmented data, thereby demonstrating a new way to exploit the unique nature of histopathology images. We post our code publicly at https://github.com/BMIRDS/DifficultyTranslation.

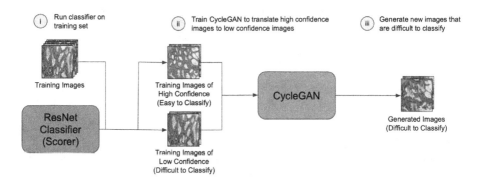

Fig. 1. Our proposed model for modifying colorectal histopathology images to be more challenging to classify.

2   Methods and Materials

Problem Set-Up and Model. Given some training image x_i of class X, we aim to generate $\tilde{x}_i$, which maintains the same histopathological class and general structure of x_i, but is more challenging to classify. We propose a model that comprises two networks: a scorer, which predicts the difficulty $c(x_i)$ of some image x_i [9,10], and an image translator, which translates images that are easy to classify into images that are harder to classify. In this study, we use ResNet-18 [5] as the scorer for colorectal histopathology images and train it to convergence on the downstream task of hyperplastic polyp/sessile serrated adenoma classification, and we assign $c(x_i)$ as the softmax output (confidence) of class X. For the image translator, we use a cycle-consistent generative adversarial network (CycleGAN), which learns the mapping $G : \hat{X} \rightarrow \tilde{X}$, where we assign $\hat{X}$ as the class for the set of images $\{\hat{x}\}$ such that $c(\hat{x}_i)$ is high for all $\hat{x}_i \in \{\hat{x}\}$ (easy-to-classify images) and $\tilde{X}$ as the class for the set of images $\{\tilde{x}\}$ where $c(\tilde{x}_j)$ is low for all $\tilde{x}_j \in \{\tilde{x}\}$ (hard-to-classify images). With this configuration, given some image $\hat{x}_i \in \{\hat{x}\}$, we can generate a similar example $\tilde{x}_i$ that maintains the same histopathological class but is harder to classify. An overview schematic of our model is shown in Fig. 1. Hereafter, we use the terms *easy images* and *hard images* to refer to images that are easy to classify and hard to classify based on a pre-trained classifier's confidence output.[1] When referring to easy and hard as perceived by annotators, we use the terms *high-agreement* images and *low-agreement images*, which represent 3/3 annotator agreement and 2/3 annotator agreement, respectively.

Table 1. Distribution of data in our training and test sets based on the level of annotation agreement among three pathologist annotators. HP: hyperplastic polyp, SSA: sessile serrated adenoma.

Level of agreement	Training set images			Test set images		
	HP	SSA	Total	HP	SSA	Total
2/3 annotators	670	173	843	316	89	405
3/3 annotators	860	348	1,208	492	204	696
Total	1,530	521	2,051	808	293	1,101

Dataset. For our experiments, we first collected and scanned 328 Formalin-Fixed Paraffin-Embedded (FFPE) whole-slide images of colorectal polyps, originally diagnosed as either hyperplastic polyps (HPs) or sessile serrated adenomas (SSAs), from patients at the Dartmouth-Hitchcock Medical Center, our tertiary medical institution. From these 328 whole-slide images, we then extracted 3,152 patches (portions of size 224×224 pixels from whole-slide images) representing diagnostically relevant regions of interest for HPs or SSAs. Three board-certified practicing gastrointestinal pathologists at the Dartmouth-Hitchcock Medical Center independently labeled each image as HP or SSA. The use of the dataset in this study was approved by our Institutional Review Board (IRB).

The gold standard label for each image was determined by the majority vote of the labels from three pathologists. Table 1 shows the distribution of high-agreement and low-agreement images for each class in the training and test set. Note that our dataset is imbalanced because SSAs naturally occur less frequently than HPs. Figure 2 shows examples of high-agreement and low-agreement images from each class. Images were split randomly by whole slide into the training set and test set, so images from the same whole slide either all went into the training

---

[1] In this metric for measuring the difficulty of images, the pre-trained classifier does not classify an image as easy or hard, but rather classifies an image as HP or SSA— the classifier's *confidence* on its HP or SSA prediction determines whether the image is considered to be easy or hard to classify.

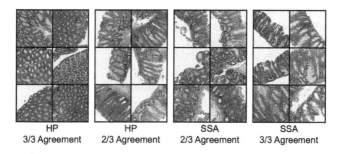

Fig. 2. Examples of high-agreement and low-agreement images for the hyperplastic (HP) and sessile serrated adenoma (SSA) classes.

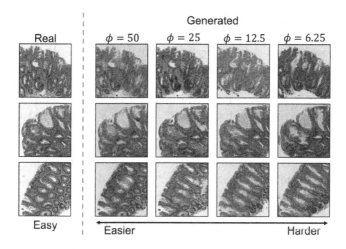

Fig. 3. Generated hyperplastic polyp (HP) images that are intended to be more difficult to classify than their real source images. ϕ is the selectivity parameter for target domain images used to train our image translator—images generated using a lower ϕ parameter are intended to be more difficult to classify.

set or all went into the test set. We chose the task of sessile serrated adenoma detection, which is challenging and clinically important for colonoscopy, one of the most common screening tests for colorectal cancer [11]. We used a training set of 2,051 images and a test set of 1,101 images, of which each image was labeled as either hyperplastic polyp (HP) or sessile serrated adenoma (SSA).

3   Experiments

3.1   Generating More Challenging Training Examples

In our image translation model, we define a selectivity parameter, ϕ, as the percent of training set images used as hard data in the target domain for training our image translator. For instance, at $\phi = 50$, the lower 50% of training set

Table 2. Pathologists agreed less on ground-truth labels for generated HP images compared with their real image counterparts. Most generated images maintained their HP label in that both source images and their corresponding generated images were classified as HP by the majority of annotators.

Image type	Pathologist annotator agreement (%)		Maintained Label (%)
	2/3 Agreement	3/3 Agreement	
Real–easy	13.3	86.7	100.0
Real–hard	26.7	64.0	90.7
Generated–hard	22.7	73.3	96.0

images by confidence, as determined by our pre-trained classifier, was used as target domain training data for the image translator. We train our image translation model for various ϕ to generate difficult HP images. We find that for a given image classifier, generated images—particularly those with lower ϕ—were indeed harder than their corresponding source images to classify (Fig. 6 in the Appendix). Figure 3 shows examples of images that were generated with varying ϕ. While images generated with lower ϕ are typically harder to classify, at very low ϕ, generated images are no longer representative of their target class (e.g., generated HP images begin to look like SSA images). We therefore recommend using the smallest ϕ possible such that the original class label is generally maintained and a sufficient number of images is provided to train the image translation model. For our dataset, we find that the model needs to be trained with at least 100 images ($\phi > 6.25$).

We also evaluated the difficulty of generated images by presenting them to three board-certified gastrointestinal pathologists for manual evaluation in a blinded test. Using labels where the top 50% and bottom 50% of images marked as HP by confidence of our pre-trained classifier were considered to be easy HP images and hard HP images, respectively, we randomly sampled 75 easy HP images, 75 hard HP images, 75 generated HP images that were translated from the selected easy HP images, and 75 SSA images from our training set. Each pathologist then independently classified each image as HP or SSA. As shown in Table 2, pathologists disagreed more (2/3 annotator agreement) on generated images (22.7%) than their real counterparts (13.3%), although not as much as they did for real hard images (26.7%). At the same time, generated images retained their original class label of HP 96% of the time based on annotator agreement.

To make Table 2 more readable, we omit the classification results for SSA images. The proportion of the 75 images in our blinded test with an original ground truth of SSA that were again marked as SSA by a majority of pathologists during the blinded test was 89.3%, indicating that our pathologist annotators were relatively consistent in their classifications. In Fig. 4, we show examples of translations that were successful and unsuccessful in making images more difficult to classify.

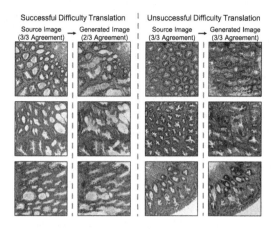

Fig. 4. Generated images, which were intended to be more difficult to classify, had lower inter-annotator agreement between pathologists. The left set of images shows translations that successfully lowered inter-annotator agreement, while the right set of images shows translations that did not lower the inter-annotator agreement. In all examples, generated images retained their original ground-truth labels.

3.2   Improving the Performance of Classifiers

We conduct further experiments to explore using the generated images as additional data to better train a classifier. Given a training set, we use the easy HP images as source images to generate harder HP images. Of these generated harder images, we use the images that maintain the HP class label, according to our pre-trained classifier, as additional data for training a new classifier.

For all image classifiers, we train ResNet-18 [5] for 50 epochs (far past convergence) using the Adam optimizer [12] with a L2 regularization factor of 10^{-4}. We use an initial learning rate of 10^{-3}, decaying by 0.91 every epoch. Every trained model used automatic data augmentation of online color jittering uniformly sampled from the range of ±0.5 for brightness, ±0.5 for contrast, ±0.2 for hue, and ±0.5 for saturation, as implemented in PyTorch.

Table 3 shows the performance of classifiers trained with our generated images as additional data compared with the baseline of standard training on the original dataset as well as a naïve data augmentation technique of directly combining parts from easy and hard images [13]. In order to account for variance in random weight initializations and performance fluctuations throughout training, we run each configuration for twenty random seeds, and for each seed we record the mean of the five highest AUC scores, which are calculated for every epoch. Notably, adding images generated at $\phi = 25$ consistently outperformed naïve data augmentation and no data augmentation for both low-agreement and high-agreement images. We posit that naïve data augmentation was unsuccessful in this case because the features in the augmented data were not reflected in the test set.

Table 3. Performance (% AUC ± standard error) of image classifiers trained with generated images as augmented data on test set images with high annotator agreement, low annotator agreement, and all test set images.

Training dataset	Test set performance		
	High-agreement	Low-agreement	All Images
Unmodified original dataset	91.3 ± 0.2	66.0 ± 0.6	83.1 ± 0.2
+ naïve data augmentation	90.9 ± 0.2	64.3 ± 0.5	82.1 ± 0.3
+ generated images, $\phi = 50$	91.9 ± 0.2	66.6 ± 0.4	83.8 ± 0.3
+ generated images, $\phi = 25$	**92.6 ± 0.2**	**68.1 ± 0.7**	**84.8 ± 0.4**
+ generated images, $\phi = 12.5$	91.7 ± 0.2	65.5 ± 0.4	83.4 ± 0.2

3.3 Comparing Machine and Human Difficulty Measures

While it was previously unconfirmed whether the confidence output of a machine learning model correlates with the human concept of difficulty, for our dataset, we find that the confidence of our pre-trained classifier indeed correlates strongly with human annotator agreement. As shown in Fig. 5, the predicted confidence distribution of images with high annotator agreement vastly differs from that of images with low annotator agreement. We compared these distributions using a Kolmogorov-Smirnov test for equality of two distributions [14] and computed a Kolmogorov-Smirnov statistic of 0.302 over all 1530 HP images in the training set with a statistically significant p-value of $p = 1.5 \times 10^{-30}$, indicating that the two distributions are not equal. The correlation between these two measures of difficulty implies that for tasks requiring human annotator agreement data, the confidence of a machine learning classifier, which is computed automatically, could be used as a reliable proxy.

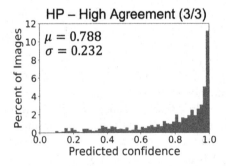 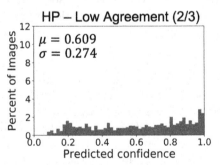

Fig. 5. Distributions of predicted confidences of a pre-trained classifier vastly differed for hyperplastic polyp (HP) images with low (2/3) and high (3/3) annotator agreement.

4   Related Work and Discussion

Generative adversarial networks (GANs) have been used in deep learning for medical imaging to generate synthetic data ranging from MRIs to CT scans [15–17]. For histopathology, several studies used GANs for both image generation and translation [18–23]. Along the same lines as our work, GAN-generated images have been used as augmented data for liver lesion [24], bone lesion [25], and rare skin condition classification [26]. While prior work generates augmented data as a means to improve general performance, the augmented data that we generate aims to help classifiers specifically on examples that are challenging. Moreover, our methodology, to our knowledge, substantially differs from previous work due to its focus on example difficulty.

This paper advances related work from our group, which is focused on colorectal polyp classification [27,28] and used image translation between different colorectal polyp types to address data imbalances [29]. In this study, we translate images within the same class to become more difficult to differentiate from other classes, arguing that the range of features in histopathology images can and should be utilized to train better-performing machine learning models.

Of possible limitations, measuring whether generated images maintained the same quality and realistic features as real images is challenging. Although generated images occasionally contained minor mosaic-like patterns, they remained readable and improved classifier training over baseline augmentation methods, suggesting that useful histologic features were retained. Also, another approach for difficulty translation could be to directly use human annotator agreement to translate images from high to low agreement. In our paper, however, we define difficulty according to the confidence of a pre-trained classifier, since this framework generalizes to cases where annotator agreement data is unavailable.

To conclude, this work shows how to generate difficult yet meaningful training data by exploiting the range of features in histopathology images. Future research could explore difficulty translation in the context of curriculum learning [30] or defending against adversarial attacks [31]. This study and its results encourage further research to make use of the range of features in histopathology images in more creative ways.

Acknowledgments. This research was supported in part by the National Institute of Health grants (R01LM012837, R01CA098286, and P20GM104416).

6   Appendix

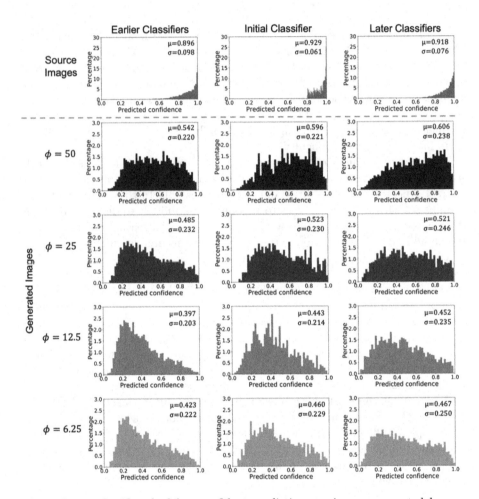

Fig. 6. Image classifiers had less confident predictions on images generated by our image translation model than on real source hyperplastic (HP) images in the training set. ϕ is the selectivity parameter for target domain images used to train our image translator—images generated using a lower ϕ parameter are intended to be more difficult to classify. We show the predicted distribution of machine learning classifiers with varying amounts of training, since classifiers with more training tend to have more confident predictions. The initial classifier is the original classifier used to define the training set for our image translation model, which was trained for 25 epochs. Earlier classifiers is the average of classifiers trained for 17, 19, 21, and 23 epochs, and later classifiers is the average of classifiers trained for 27, 29, 31, and 33 epochs.

References

1. Coudray, N., Moreira, A.L., Sakellaropoulos, T., Fenyö, D., Razavian, N., Tsirigos, A.: Classification and mutation prediction from non-small cell lung cancer histopathology images using deep learning. bioRxiv (2017)
2. Ehteshami Bejnordi, B., et al.: Diagnostic assessment of deep learning algorithms for detection of lymph node metastases in women with breast cancer. JAMA **318**(22), 2199–2210 (2017)
3. Tomita, N., Abdollahi, B., Wei, J., Ren, B., Suriawinata, A., Hassanpour, S.: Attention-based deep neural networks for detection of cancerous and precancerous esophagus tissue on histopathological slides. JAMA Netw. Open **2**(11), e1914645–e1914645 (2019)
4. Wei, J.W., Tafe, L.J., Linnik, Y.A., Vaickus, L.J., Tomita, N., Hassanpour, S.: Pathologist-level classification of histologic patterns on resected lung adenocarcinoma slides with deep neural networks. Sci. Rep. **9** (2019)
5. He, K., Zhang, X., Ren, S., Sun, J.: Deep residual learning for image recognition. CoRR (2015)
6. Krizhevsky, A., Sutskever, I., Hinton, G.E.: Imagenet classification with deep convolutional neural networks. In: Pereira, F., Burges, C.J.C., Bottou, L., Weinberger, K.Q. (eds.) Advances in Neural Information Processing Systems 25, pp. 1097–1105. Curran Associates, Inc. (2012)
7. Szegedy, C., et al.: Going deeper with convolutions. CoRR (2014)
8. Deng, J., Dong, W., Socher, R., Li, L.J., Li, K., Fei-Fei, L.: ImageNet: a large-scale hierarchical image database. In: CVPR 2009 (2009)
9. Hacohen, G., Weinshall, D.: On the power of curriculum learning in training deep networks. CoRR (2019)
10. Weinshall, D., Cohen, G.: Curriculum learning by transfer learning: Theory and experiments with deep networks. CoRR (2018)
11. Rex, D., et al.: Colorectal cancer screening: recommendations for physicians and patients from the U.S. multi-society task force on colorectal cancer. Am. J. Gastroenterol. **112**(7), 1016–1030 (2017)
12. Kingma, D., Ba, J.: Adam: a method for stochastic optimization. In: International Conference on Learning Representations (2014)
13. Summers, C., Dinneen, M.J.: Improved mixed-example data augmentation. In: 2019 IEEE Winter Conference on Applications of Computer Vision (WACV). IEEE (2019)
14. Massey, F.J.: The Kolmogorov-Smirnov test for goodness of fit. J. Am. Stat. Assoc. **46**, 68–78 (1951)
15. Dar, S.U., Yurt, M., Karacan, L., Erdem, A., Erdem, E., Cuker, T.: Image synthesis in multi-contrast MRI with conditional generative adversarial networks. IEEE Trans. Med. Imaging (2018)
16. Salehinejad, H., Valaee, S., Dowdell, T., Colak, E., Barfett, J.: Generalization of deep neural networks for chest pathology classification in X-Rays using generative adversarial networks. In: Proceedings of the IEEE International Conference on Acoustics, Speech and Signal Processing (2018)
17. Wang, J., Zhao, Y., Noble, J.H., Dawant, B.M.: Conditional generative adversarial networks for metal artifact reduction in CT images of the ear. In: Frangi, A.F., Schnabel, J.A., Davatzikos, C., Alberola-López, C., Fichtinger, G. (eds.) MICCAI 2018. LNCS, vol. 11070, pp. 3–11. Springer, Cham (2018). https://doi.org/10.1007/978-3-030-00928-1_1

18. Bayramoglu, N., Kaakinen, M., Eklund, L., Heikkila, J.: Towards virtual H & E staining of hyperspectral lung histology images using conditional generative adversarial networks. In: Proceedings of the IEEE International Conference on Computer Vision, pp. 64–71 (2018)

19. Bentaieb, A., Harmarneh, G.: Adversarial stain transfer for histopathology image analysis. Proc. IEEE Trans. Med. Imaging **37** (2017)

20. Burlingame, E.A., Margolin, A., Gray, J., Chang, Y.H.: SHIFT: speedy histopathological-to-immunofluorescent translation of whole slide images using conditional generative adversarial networks. In: Proceedings of IEEE Transactions on Medical Imaging, vol. 10581 (2018)

21. Cho, H., Lim, S., Choi, G., Min, H.: Neural stain-style transfer learning using GAN for histopathological images. J. Mach. Learn. Res.: Workshop Conf. Proc. (2017)

22. Jackson, C.R., Sriharan, A., Vaickus, L.J.: A machine learning algorithm for simulating immunohistochemistry: development of SOX10 virtual IHC and evaluation on primarily melanocytic neoplasms. Mod. Pathol. **33**, 1638–1648 (2020). https://doi.org/10.1038/s41379-020-0526-z

23. Ghazvinian Zanjani, F., Zinger, S., de With, P.H.N.: Deep convolutional Gaussian mixture model for stain-color normalization of histopathological images. In: Frangi, A.F., Schnabel, J.A., Davatzikos, C., Alberola-López, C., Fichtinger, G. (eds.) MICCAI 2018. LNCS, vol. 11071, pp. 274–282. Springer, Cham (2018). https://doi.org/10.1007/978-3-030-00934-2_31

24. Frid-Adar, M., Klang, E., Amitai, M., Goldberger, J., Greenspan, H.: GAN-based data augmentation for improved liver lesion classification. In: 2018 IEEE 15th International Symposium on Biomedical Imaging (ISBI 2018), pp. 289–293 (2018)

25. Gupta, A., Venkatesh, S., Chopra, S., Ledig, C.: Generative image translation for data augmentation of bone legion pathology. arXiv (2019)

26. Ghorbani, A., Natarajan, V., Coz, D., Liu, Y.: Dermgan: synthetic generation of clinical skin images with pathology (2019)

27. Korbar, B., et al.: Deep learning for classification of colorectal polyps on whole-slide images (2017)

28. Wei, J.W., et al.: Evaluation of a deep neural network for automated classification of colorectal polyps on histopathologic slides. JAMA Netw. Open **3**(4), e203398–e203398 (2020)

29. Wei, J., et al.: Generative image translation for data augmentation in colorectal histopathology images. In: Machine Learning for Health Workshop at the Thirty-third Conference on Neural Information Processing Systems (2019)

30. Bengio, Y., Louradour, J., Collobert, R., Weston, J.: Curriculum learning. In: Proceedings of the 26th Annual International Conference on Machine Learning, ICML 2009, pp. 41–48. Association for Computing Machinery, New York (2009)

31. Ma, X., et al.: Understanding adversarial attacks on deep learning based medical image analysis systems. CoRR (2019)

Weakly-Supervised Segmentation for Disease Localization in Chest X-Ray Images

Ostap Viniavskyi[1], Mariia Dobko[1,2(✉)], and Oles Dobosevych[1]

[1] The Machine Learning Lab at Ukrainian Catholic University, Lviv, Ukraine
{viniavskyi,dobko_m,dobosevych}@ucu.edu.ua
[2] SoftServe, Lviv, Ukraine
mdobk@softserveinc.com

Abstract. Deep Convolutional Neural Networks have proven effective in solving the task of semantic segmentation. However, their efficiency heavily relies on the pixel-level annotations that are expensive to get and often require domain expertise, especially in medical imaging. Weakly supervised semantic segmentation helps to overcome these issues and also provides explainable deep learning models.

In this paper, we propose a novel approach to the semantic segmentation of medical chest X-ray images with only image-level class labels as supervision. We improve the disease localization accuracy by combining three approaches as consecutive steps. First, we generate pseudo segmentation labels of abnormal regions in the training images through a supervised classification model enhanced with a regularization procedure. The obtained activation maps are then post-processed and propagated into a second classification model—Inter-pixel Relation Network, which improves the boundaries between different object classes. Finally, the resulting pseudo-labels are used to train a proposed fully supervised segmentation model.

We analyze the robustness of the presented method and test its performance on two distinct datasets: PASCAL VOC 2012 and SIIM-ACR Pneumothorax. We achieve significant results in the segmentation on both datasets using only image-level annotations. We show that this approach is applicable to chest X-rays for detecting an anomalous volume of air in the pleural space between the lung and the chest wall. Our code has been made publicly available (Implementation is available at https://github.com/ucuapps/WSMIS).

Keywords: Weakly-supervised learning · Segmentation · Deep learning · Chest X-rays · Disease localization · Explainable models

1 Introduction

Applications of Convolutional Neural Networks to medical images have recently produced efficient solutions for a vast variety of medical problems, such as

---

O. Viniavskyi and M. Dobko—Contributed equally to the work.

© Springer Nature Switzerland AG 2020
M. Michalowski and R. Moskovitch (Eds.): AIME 2020, LNAI 12299, pp. 249–259, 2020.
https://doi.org/10.1007/978-3-030-59137-3_23

segmentation of lung nodules in computed tomography (CT) scans [13], lesions detection in mammography images [1], segmentation of brain gliomas from MRI images [4] and others. One of the greatest challenges for using deep learning methods in medicine is the lack of large annotated datasets, especially with pixel-level labeled data. Creating such datasets is often very expensive and time consuming. For instance, Lin et al. [19] calculated that collecting bounding boxes for each class is about 15 times faster than producing a ground-truth pixel-wise segmentation mask; getting image-level labels is even easier. Moreover, domain expertise is required to label medical data, which poses another challenge as the doctor's time is costly and could more effectively be used for a patient's diagnosis and disease treatment. Working with image-level annotations also decreases the probability of disagreement between experts, since pixel-wise annotations tend to have more noise and vary among labelers.

To decrease the resources spent on labeling while preserving its quality, we propose a novel weakly-supervised approach to image segmentation that uses only image-level labels. Our method is domain-independent; we have tested it on several distant datasets, including popular PASCAL VOC 2012 [11], and a medical dataset, SIIM-ACR Pneumothorax [26]. We achieve 64.6 mean intersection-over union (mIoU) score on PASCAL VOC 2012 [11] validation set. Our method is capable of segmenting medical images with limited supervision achieving 76.77 mIoU score on the test set of SIIM-ACR Pneumothorax dataset [26]. The automatic approach to finding Pneumothorax can be used to triage chest radiographs requiring priority interpretation, to rapidly identify critical cases, and also provide a second-opinion for radiologists to make a more confident diagnosis.

2   Related Work

The objective of weakly-supervised segmentation is to create models capable of pixel-wise segmentation based on image-level labels. The existing approaches can be categorized by their methodologies into four groups: Expectation-Maximization, Multiple Instance Learning, Self-Supervised Learning, and Object Proposal Class Inference [6]. In this paper, we follow the self-supervised paradigm, which suggests training a fully supervised segmentation model on the created pseudo-pixel-level annotations, also known as Class Activation Maps (CAM) [30], which are extracted from the classification network. Judging from the quantitative performance on PASCAL VOC 2012 [11] validation set, the top five methods of weakly-supervised segmentation use the self-supervised learning approach [6].

Many methods of self-supervised learning for semantic segmentation have been recently suggested. Huang et al. [16] introduce Deep Seeded Region Growing (DSRG), which propagates class activations from high-confidence regions to adjacent regions with a similar visual appearance by applying a region-growing algorithm on the generated CAM. Lee et al. [18] present FickleNet, which trains a CNN at the image level with a regularization step represented as a center-fixed spatial dropout in the later convolutional layers, and then runs Grad-CAM [25] multiple times to generate a thresholded pseudo-labels for a segmentation step. Another approach, proposed by Ahn et al. [3], suggests using Inter-pixel Relations network (IRNet) [3], which takes the random walk from low-displacement field centroids in the CAM up until the class boundaries as the pseudo-ground-truths for training an FCN.

The weakly-supervised semantic segmentation on medical datasets has been explored in [2,5,10,21,23]. On the other hand, Ouyang et al. [22] combine the weakly-annotated data with well-annotated cases to segment Pneumothorax in chest X-rays. In our approach, we do not use any form of supervision besides image-level labels. We focus on developing a standardized method, which is efficient for various data types, especially for medical images.

3   Methodology

Our method can be split into three consecutive steps: Class Activation Maps generation, map enhancement with Inter-pixel Relation Network, and segmentation. After each step, we also apply one or more post-processing techniques such as Conditional Random Fields (CRF), thresholding, noise filtering (small regions with low confidence).

Step 1. CAM Generation. First, we train fully-supervised classification models on image-level labels. The two tested architectures for this step were ResNet50 [14] and VGG16 [27] with additional three convolutional layers followed by ReLU activation. We also replace stride with dilation [29] in the last convolutional layers to increase the size of the final feature map while decreasing the output stride from 32 to 8. We improve the classification performance by including a regularization term, inspired by FickleNet [18]. For this, we use DropBlock [12]—a dropout technique, which to our best knowledge has not been tried in previous works on weakly-supervised segmentation. The trained models are then used to retrieve activation maps by applying the Grad-CAM++ [7] method. The resulting maps serve as pseudo labels for segmentation task.

Step 2. IRNet. On the second step, IRNet [3] takes the generated CAM and trains two output branches that predict a displacement vector field and a class boundary map, correspondingly. They take feature maps from all five levels of the same shared ResNet50 [14] backbone. The main advantage of IRNet [3] is its ability to improve boundaries between different object classes. We train it on the generated maps, thus no extra supervision is required. This step allows us to obtain better pseudo-labels before segmentation. To our best knowledge, this approach has not been used in the medical imaging domain before.

Step 3. Segmentation. For the segmentation step, we train DeepLabv3+ [9] and U-Net [24] models with different backbones, which have proven to produce reliable results in fully supervised semantic segmentation on medical images [24].

The used backbones include ResNet50 [14] and SEResNeXt50 [15]. We modify the binary cross-entropy (BCE) loss during segmentation by adding weights to a positive class to prevent overfitting towards normal cases.

4   Experiments and Results

4.1   Datasets and Evaluation Metric

We conduct experiments on two datasets: PASCAL VOC 2012 [11] and SIIM-ACR Pneumothorax [26]. We evaluate the quality of our pseudo-ground-truth and the performance of the segmentation model trained on them using mIoU.

PASCAL VOC 2012 [11] is an image segmentation benchmark dataset containing 20 object classes, and a background class. As in other works on weakly-supervised segmentation, we train our models using augmented 10,582 training images with image-level labels. We report mIoU for 1,449 validation images.

SIIM-ACR Pneumothorax [26] is a competition that provides an open dataset of chest X-ray images with pixel-wise annotation for regions affected by Pneumothorax: a collapsed lung, where an abnormal volume of air is formed in the pleural space between the lung and the chest wall. This dataset was formed from a subset of ChestX-ray14 dataset [28], but relabeled by professional radiologists, and additionally annotated on a pixel level. The specified competition has two stages; ground truth labels are provided only for the first, the second is evaluated on the competition website. Thus, we divided images from the first stage into three sets: train, validation and test. Totally, 12,047 frontal-view chest X-ray cases are in the dataset. We use 2,379 positive and 8,296 negative images for training, 145 and 541 for validation, 145 and 541 for the test.

4.2   Data Challenges

SIIM-ACR Pneumothorax [26] dataset has a severe class imbalance problem. The number of normal cases exceeds approximately 4 times the number of positive ones. In order to prevent overfitting towards healthy patients we use various augmentation techniques such as scaling, rotation, blur, brightness adjustment, and horizontal flipping. We also add sampling in our data loader during training, which selects the constant ratio between negative and positive class. Another challenge in this dataset is the size of regions of interest. Pneumothorax usually affects a very small area of lungs resulting in a high disbalance among the image pixels. We solve this problem by adding weights for positive class to binary cross-entropy loss. In view of these data challenges we evaluate the performance of our method on SIIM-ACR Pneumothorax not only for all images in validation and test sets, but also separately for positive cases, see Table 2.

4.3   Experiments

Step 1. CAM Generation. For classification we implement ResNet50, and VGG16. As suggested in previous work [17], we added three convolutional layers

Table 1. Comparison of CAM generation techniques on PASCAL VOC 2012.

Classification model	CAM extraction method	mIoU train	mIoU val
VGG16	Cam-Grad	0.4137	0.3511
VGG16	Cam-Grad++	**0.4176**	**0.3941**

on the top of the fully-convolutional backbone, each of which is followed by a ReLU. The conducted experiments show that adding DropBlock regularization to our classification models improves their performance by 1.2% reaching 88.2% F1-score for multilabel classification. For both datasets, the best classification results were achieved using VGG16, which was, thus, selected as the final model for this task. We test two methods for generating pseudo-annotations: Grad-CAM [25], and Grad-CAM++ [7]. Our experiments show that Grad-CAM++, which utilizes a regularization that Grad-CAM is lacking, provides better object localization through visual explanations of model predictions; cf. Table 1.

Step 2. IRNet. For both datasets as post-processing of maps produced at Step 1, we use thresholding and then refine the pseudo-maps by dense CRF to better capture object shapes. The resulting annotations are used to train IRNet.

Step 3. Segmentation. The obtained maps after IRNet step are used as the pseudo-labels for segmentation. We implement three networks to complete this task: U-Net [24], DeepLabv3 [8], and DeepLabv3+ [9]. We report results on PAS-CAL VOC 2012 produced by DeepLabv3+, as it shows better performance than DeepLabv3 with the same ResNet50 backbone. For SIIM-ACR Pneumothorax, however, U-Net with SEResNeXt50 [15] backbone shows the best results.

4.4 Training and Optimization

For all the models, three optimization methods are examined: SGD, Adam and RAdam [20]. For Pneumothorax segmentation, SGD optimizer is applied with the learning rate initiated as 6e-5 and gradually decreasing each epoch, whereas momentum is set to 0.9, and weight decay to 1e-6. The size of network inputs is 512×512, the batch is 48, and it is balanced according to class distribution using augmentations to increase the sample of positive cases.

4.5 Results

The comparison of segmentation methods using the same chest X-ray datasets is not simple due to the challenge of finding public medical data. Moreover, this work is the first to present the results of weakly-supervised segmentation methods on SIIM-ACR Pneumothorax [26] data. However, the performance of our models is comparable to Ouyang et al. [22], who reported their scores on a collected closed dataset of Pneumothorax. These authors train their method with different combinations of well- and weakly-annotated data, whereas our method uses only image-level labels. In Table 2 we show how the results improve with each step of our approach; the result of Ouyang et al. [22] model trained on 400 weakly-annotated and 400 well-annotated cases is specified too.

We present method's explainability via disease localization regions; cf. Fig. 1. We provide qualitative results of segmentation on validation images from both datasets in Fig. 2 and Fig. 3. We demonstrate how the performance improves after each step. We achieve comparable results to state-of-the-art method on PASCAL VOC 2012; cf. Table 3.

We evaluate our method on the second stage test set on the competition server [26] to compare it against a fully-supervised upper-performance limit. We achieve 0.769 Dice score while the first place solution got 0.868 using pixel-level labels for training. Our method proves the capability of using only image-level

Table 2. Results on SIIM-ACR Pneumothorax validation and test sets after each step of our method. Calculated for only positive cases (pos.), and for the whole set, including the healthy patients (all). The [22], whose result is demonstrated, was trained on 400 weakly-annotated and 400 well-annotated cases.

Dataset	Method	mIoU val		mIoU test	
		pos	all	pos	all
SIIM-ACR Pneum. [26]	Step 1. CAM	0.117	0.7633	0.142	0.7590
SIIM-ACR Pneum. [26]	Step 2. IRNet	0.122	0.7645	0.154	0.7607
SIIM-ACR Pneum. [26]	Step 3. Segm	0.148	0.7649	0.162	0.7677
Custom [22]	Ouyang et al. [22]	-	-	-	0.669

(a) Image   (b) Step1.CAM   (c) Step2.IRNet   (d) Step3.Segm   (e) Mask

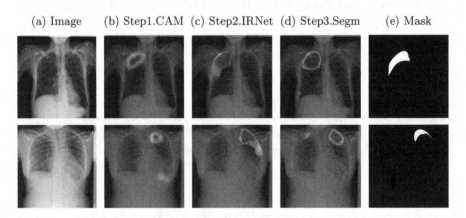

Fig. 1. Pneumothorax localization maps for (a) a random image from the test set at each consecutive step of our method: (b) map after CAM extraction, (c) improved map by IRNet trained on the outcomed of step 1, (d) prediction of U-Net trained on step 2 results, all compared to (e) ground truth mask.

(a) Image   (b) Step1.CAM   (c) Step2.IRNet   (d) Step3.Segm   (e) Mask

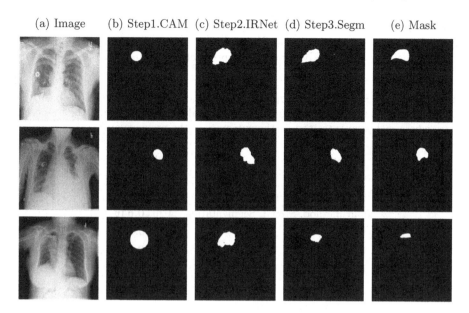

Fig. 2. Segmentation predictions for (a) a random image from test set of SIIM-ACR Pneumothorax produced at each step of our approach: (b) CAM extraction, (c) IRNet, (d) U-Net segmentation, compared to (e) ground truth mask.

(a) Image   (b) Step1.CAM   (c) Step2.IRNet   (d) Step3.Segm   (e) Mask

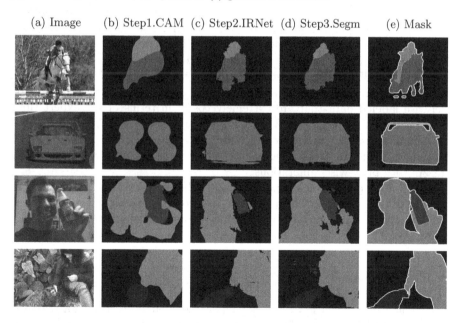

Fig. 3. Visualization of segmentation predictions on PASCAL VOC 2012 for (a) an image produced at each step of our approach: (b) CAM extraction, (c) IRNet, (d) DeepLabv3+ segmentation, compared to (e) ground truth mask.

annotations for semantic segmentation on chest X-rays, nevertheless, attaining as good or even better results than those produced by fully supervised networks is still a challenge for weakly-supervised approaches.

Table 3. Comparison of weakly-supervised semantic segmentation methods on PASCAL VOC 2012 validation set. Our approach is evaluated after each of the proposed steps, where each step is trained on the outcomes of the previous one.

Method	Year	mIoU
Our method. Step 1. CAM	2020	0.479
Our method. Step 2. IRNet	2020	0.631
Our method. Step 3. Segmentation	2020	**0.646**
IRNet [3]	2019	0.635
FickleNet [18]	2019	**0.649**
DSRG (ResNet101) [16]	2018	0.614

5   Clinical Relevance

During diagnosing procedure the final decision maker is a doctor, while AI-powered decision support systems can assist by detecting regions of interest and presenting the data in a convenient format that doctors can use. With an automatic image segmentation solution, the healthcare provider can reach higher efficiency by saving doctors' time spent on the primary analysis of images. At the same time, it will increase diagnosis accuracy by providing the second opinion. The major problem in building such a solution is the lack of large amounts of pixel-wise labeled data that are extremely costly in terms of the expert time required for their annotation. With our approach, which requires only image-level annotations, the costs can be reduced dramatically. In the long run, it also facilitates the research in the area by overcoming the problem of collecting datasets with pixel-wise annotations.

Our method can automate parts of the radiology workflow cutting operational costs for the hospitals. The proposed approach was designed to be general and applicable to other medical purposes; for example detection of various thoracic diseases.

6   Conclusions

We present a novel method of weakly-supervised semantic segmentation that demonstrated its efficiency for detecting anomalous regions on chest X-ray images. In particular, we propose a three-step approach to weakly-supervised semantic segmentation, which uses only image-level labels as supervision. Next, we customize and expand the previous works by including supplementary steps such as regularization, IRNet, and various post-processing techniques. Also, the method is general, domain independent and explainable via localization maps at each step. We evaluated it on two datasets of different nature; however, it can also be implemented in other medical problems.

Acknowledgements. This research was supported by SoftServe and Faculty of Applied Sciences at Ukrainian Catholic University (UCU), whose collaboration allowed to create SoftServe Research Group at UCU. The authors thank Rostyslav Hryniv for helpful and valuable feedback.

Appendix

For the evaluation we use $mIoU$ (mean intersection over union), Dice score and $F1$-score.

The mIoU metric was used to compare our result to already existing approaches. For all images and k classes is defined as follows:

$$mIoU = \frac{1}{k} \sum_{i=1}^{k} \frac{TP_{ii}}{\sum\limits_{j=1}^{k} FN_{ij} + \sum\limits_{j=1}^{k} FP_{ij} - TP_{ii}},$$

where TP, FP, FN are numbers of true positive, false positive and false negative pixels respectively.

The Dice score was used to compare our results of weakly-supervised model to the first place solution with fully-supervised approach on SIIM-ACR Pneumothorax [26]. The metric is defined as follows:

$$Dice = \frac{2 \times TP}{(TP + FP) + (TP + FN)}$$

where TP, FP, FN are numbers of true positive, false positive and false negative pixels respectively. If all pixels are true negative, prediction is considered correct and metric equals 1.

The $F1$-score was used to compare approaches on a first step of our pipeline— classification. The metric is defined as follows:

$$F1 = \frac{2 \times precision \times recall}{precision + recall}$$

where $precision = \frac{TP}{TP+FP}$, $recall = \frac{TP}{TP+FN}$, and TP, FP, FN are true positive, false positive and false negative rates respectively.

References

1. Abdelhafiz, D., Yang, C., Ammar, R., Nabavi, S.: Deep convolutional neural networks for mammography: advances, challenges and applications. BMC Bioinform. **20**(11), 281 (2019)
2. Agarwal, V., Tang, Y., Xiao, J., Summers, R.M.: Weakly supervised lesion cosegmentation on CT scans. arXiv preprint arXiv:2001.09174 (2020)
3. Ahn, J., Cho, S., Kwak, S.: Weakly supervised learning of instance segmentation with inter-pixel relations. In: Proceedings of the IEEE Conference on Computer Vision and Pattern Recognition, pp. 2209–2218 (2019)
4. Archa, S., Kumar, C.S.: Segmentation of brain tumor in MRI images using CNN with edge detection. In: 2018 International Conference on Emerging Trends and Innovations in Engineering and Technological Research (ICETIETR), pp. 1–4. IEEE (2018)
5. Cai, J., et al.: Accurate weakly-supervised deep lesion segmentation using large-scale clinical annotations: slice-propagated 3D mask generation from 2D RECIST. In: Frangi, A.F., Schnabel, J.A., Davatzikos, C., Alberola-López, C., Fichtinger, G. (eds.) MICCAI 2018. LNCS, vol. 11073, pp. 396–404. Springer, Cham (2018). https://doi.org/10.1007/978-3-030-00937-3_46
6. Chan, L., Hosseini, M.S., Plataniotis, K.N.: A comprehensive analysis of weakly-supervised semantic segmentation in different image domains. arXiv preprint arXiv:1912.11186 (2019)
7. Chattopadhay, A., Sarkar, A., Howlader, P., Balasubramanian, V.N.: Grad-CAM++: generalized gradient-based visual explanations for deep convolutional networks. In: 2018 IEEE Winter Conference on Applications of Computer Vision (WACV), pp. 839–847. IEEE (2018)
8. Chen, L.C., Papandreou, G., Schroff, F., Adam, H.: Rethinking atrous convolution for semantic image segmentation. arXiv preprint arXiv:1706.05587 (2017)
9. Chen, L.-C., Zhu, Y., Papandreou, G., Schroff, F., Adam, H.: Encoder-decoder with atrous separable convolution for semantic image segmentation. In: Ferrari, V., Hebert, M., Sminchisescu, C., Weiss, Y. (eds.) ECCV 2018. LNCS, vol. 11211, pp. 833–851. Springer, Cham (2018). https://doi.org/10.1007/978-3-030-01234-2_49
10. Demiray, B., Rackerseder, J., Bozhinoski, S., Navab, N.: Weakly-supervised white and grey matter segmentation in 3d brain ultrasound. arXiv preprint arXiv:1904.05191 (2019)
11. Everingham, M., Van Gool, L., Williams, C.K.I., Winn, J., Zisserman, A.: The PASCAL Visual Object Classes Challenge 2012 (VOC2012) Results. http://www.pascal-network.org/challenges/VOC/voc2012/workshop/index.html
12. Ghiasi, G., Lin, T.Y., Le, Q.V.: Dropblock: a regularization method for convolutional networks. In: Advances in Neural Information Processing Systems, pp. 10727–10737 (2018)
13. Gruetzemacher, R., Gupta, A., Paradice, D.: 3D deep learning for detecting pulmonary nodules in CT scans. J. Am. Med. Inform. Assoc. **25**(10), 1301–1310 (2018)
14. He, K., Zhang, X., Ren, S., Sun, J.: Deep residual learning for image recognition. In: Proceedings of the IEEE Conference on Computer Vision and Pattern Recognition, pp. 770–778 (2016)
15. Hu, J., Shen, L., Sun, G.: Squeeze-and-excitation networks. In: Proceedings of the IEEE Conference on Computer Vision and Pattern Recognition, pp. 7132–7141 (2018)

16. Huang, Z., Wang, X., Wang, J., Liu, W., Wang, J.: Weakly-supervised semantic segmentation network with deep seeded region growing. In: Proceedings of the IEEE Conference on Computer Vision and Pattern Recognition, pp. 7014–7023 (2018)
17. Jiang, P.T., Hou, Q., Cao, Y., Cheng, M.M., Wei, Y., Xiong, H.K.: Integral object mining via online attention accumulation. In: Proceedings of the IEEE International Conference on Computer Vision, pp. 2070–2079 (2019)
18. Lee, J., Kim, E., Lee, S., Lee, J., Yoon, S.: Ficklenet: weakly and semi-supervised semantic image segmentation using stochastic inference. In: Proceedings of the IEEE Conference on Computer Vision and Pattern Recognition, pp. 5267–5276 (2019)
19. Lin, T.-Y., et al.: Microsoft COCO: common objects in context. In: Fleet, D., Pajdla, T., Schiele, B., Tuytelaars, T. (eds.) ECCV 2014. LNCS, vol. 8693, pp. 740–755. Springer, Cham (2014). https://doi.org/10.1007/978-3-319-10602-1_48
20. Liu, L., et al.: On the variance of the adaptive learning rate and beyond. arXiv preprint arXiv:1908.03265 (2019)
21. Lu, Z., Chen, D.: Weakly supervised and semi-supervised semantic segmentation for optic disc of fundus image. Symmetry **12**(1), 145 (2020)
22. Ouyang, X., et al.: Weakly supervised segmentation framework with uncertainty: a study on pneumothorax segmentation in chest X-ray. In: Shen, D., et al. (eds.) MICCAI 2019. LNCS, vol. 11769, pp. 613–621. Springer, Cham (2019). https://doi.org/10.1007/978-3-030-32226-7_68
23. Qu, H., et al.: Weakly supervised deep nuclei segmentation using points annotation in histopathology images. In: International Conference on Medical Imaging with Deep Learning, pp. 390–400 (2019)
24. Ronneberger, O., Fischer, P., Brox, T.: U-net: convolutional networks for biomedical image segmentation. In: Navab, N., Hornegger, J., Wells, W.M., Frangi, A.F. (eds.) MICCAI 2015. LNCS, vol. 9351, pp. 234–241. Springer, Cham (2015). https://doi.org/10.1007/978-3-319-24574-4_28
25. Selvaraju, R.R., Cogswell, M., Das, A., Vedantam, R., Parikh, D., Batra, D.: Grad-CAM: visual explanations from deep networks via gradient-based localization. In: Proceedings of the IEEE International Conference on Computer Vision, pp. 618–626 (2017)
26. SIIM-ACR Pneumothorax Segmentation. https://www.kaggle.com/c/siim-acr-pneumothorax-segmentation
27. Simonyan, K., Zisserman, A.: Very deep convolutional networks for large scale image recognition. arXiv preprint arXiv:1409.1556 (2014)
28. Wang, X., Peng, Y., Lu, L., Lu, Z., Bagheri, M., Summers, R.M.: Chestx-ray8: hospital-scale chest x-ray database and benchmarks on weakly-supervised classification and localization of common thorax diseases. In: Proceedings of the IEEE Conference on Computer Vision and Pattern Recognition, pp. 2097–2106 (2017)
29. Yu, F., Koltun, V.: Multi-scale context aggregation by dilated convolutions. arXiv preprint arXiv:1511.07122 (2015)
30. Zhou, B., Khosla, A., Lapedriza, A., Oliva, A., Torralba, A.: Learning deep features for discriminative localization. In: Proceedings of the IEEE Conference on Computer Vision and Pattern Recognition, pp. 2921–2929 (2016)

A High-Throughput Tumor Location System with Deep Learning for Colorectal Cancer Histopathology Image

Jing Ke[1,2], Yiqing Shen[3,4], Yi Guo[5], Jason D. Wright[6],
Naifeng Jing[7,8], and Xiaoyao Liang[1,8(✉)]

[1] Department of Computer Science and Engineering,
Shanghai Jiao Tong University, Shanghai, China
kejing@sjtu.edu.cn, liang-xy@cs.sjtu.edu.cn
[2] School of Computer Science and Engineering,
University of New South Wales, Sydney, Australia
[3] School of Mathematical Sciences, Shanghai Jiao Tong University, Shanghai, China
shenyq@sjtu.edu.cn
[4] Zhiyuan College, Shanghai Jiao Tong University, Shanghai, China
[5] School of Computer, Data and Mathematical Sciences,
Western Sydney University, Sydney, Australia
y.guo@westernsydney.com.au
[6] Department of Obstetrics and Gynecology,
Columbia University, New York City, NY, USA
[7] Department of Micro-Nano Electronics,
Shanghai Jiao Tong University, Shanghai, China
[8] Biren Research, Shanghai, China

Abstract. Colorectal cancer is one of the major causes of morbidity and mortality worldwide, however, when discovered at an early stage, it is highly treatable. As the number of specimens increases every year, there has been a boost in the diagnostic workload on pathologists in recent years. In parallel to the development of digital pathology, deep learning has demonstrated its strong capability in feature extraction and interpretation in a variety of medical applications. In this paper, we propose a high-throughput whole-slide image (WSI) analysis system to localize tumor regions accurately with a patch-based convolutional neural network (CNN). We employ Monte Carlo adaptive sampling for a fast detection of tumors at slide level and a conditional random field (CRF) model to integrate spatial correlation for better classification accuracy. We use three datasets of colorectal cancer from The Cancer Genome Atlas (TCGA) for performance evaluation. Compared with the regular WSI analysis, the experimental benchmark shows an obvious decrease in processing time while a noticeable improvement in classification accuracy.

Keywords: Tumor region location · High-throughput diagnosis · Colorectal cancer · Whole-slide image · Convolutional neural network

© Springer Nature Switzerland AG 2020
M. Michalowski and R. Moskovitch (Eds.): AIME 2020, LNAI 12299, pp. 260–269, 2020.
https://doi.org/10.1007/978-3-030-59137-3_24

1   Introduction

Colorectal cancer ranked third among the most common cancer and also the third among the leading cause of cancer death in both men and women worldwide [16], yet the early detection and treatment have led to a drastic reduction of mortality rate. However, the number of pathologists can not follow the increase rate of histopathology specimens to be checked every year. Computational pathology utilizes mathematical models to generate diagnostic inferences and presents clinical knowledge to patients. It is not until the very recent years digital pathology [12,20] emerged when glass slides could be scanned into multi-gigapixel digital whole slide images (WSIs) with high resolution. Like microarray samples, computational pathology also has to inevitably face additional challenges due to its characteristic of huge size. Currently, a WSI has to be split into numerous patches in the most competent high performance computing (HPC) environments [3,6,9,11,21]. Tissue classifiers aggregate local information independently to make predictions at the WSI level, which is time-consuming and low efficient.

As a result, researches have noticed the challenges come with the processing of WSI, namely computational cost and spatial correlation [8,9,19]. Only a few researches have been conducted to address the former issue, such as a model is proposed to identify some of the diagnostically relevant regions of interest by following a parameterized policy [15] and a recurrent visual attention model is designed attending to the most discriminative regions in WSI [2]. For the latter, an end-to-end network is proposed to incorporate spatial correlations [11], yet with limited patterns of neighboring patches and higher computational cost. It is short of fast processing methodologies to locate tumor patches accurately by CNN, and in particular, when the slides contain few proportions of tumor tissues.

To address the two issues mentioned above, in this paper, we propose a novel system to quickly and accurately localize tumor tissues. The main contribution is summarized as follows: We design a Monte Carlo (MC) distribution approximation algorithm to efficiently locate the high-possibility tumor regions, by which only a small proportion of patches are required to be classified with deep learning approaches. The computational workload can be significantly reduced without a reduction in image resolution. The performance acceleration is even more obvious for WSIs which contain only a small proportion of tumor tissues. We integrate contextual information to further improve tumor tissue classification performance. To further improve the accuracy of patch-wise tumor identification, we employ a conditional random field (CRF) [11] model to incorporate the spatial correlation among the neighboring patches. Empirically tested on the three cohorts of colorectal cancer datasets from The Cancer Genome Atlas (TCGA), we obtain better tumor classification results with less computational time.

2   Methodology

We propose a high-throughput tumor region location system (HTRL) for WSI diagnosis. We use Monte Carlo (MC) adaptive sampling to select patches with the lowest fidelity in the approximated distribution and a CRF model to aggregate spatial correlation among neighboring patches. The overall architecture is mainly constructed with three functional modules, namely an MC distribution approximation, a tissue classifier, and a CRF model, depicted in Fig. 1.

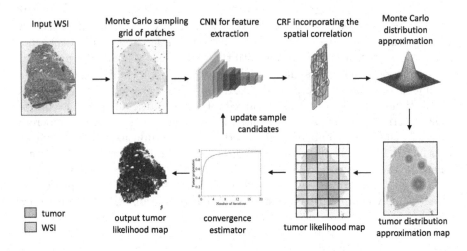

Fig. 1. An overall workflow of the proposed high-throughput tumor region location system HTRL.

2.1   A CRF Based Tissue Classifier

A patch-wise CNN based tissue classifier uses local information independently to make predictions for WSIs, where spatial correlation is absent. We attempt to use the prediction results of each sampled patch to represent a grid of patches around it by extracting the spatial correlation between the neighboring patches, i.e., to provide consistent patch-level predictions. To address these issues, we integrate a CRF [11] model in the tissue classification neural network. A CRF model takes the feature maps extracted by a CNN feature extractor as the input and then outputs the tumor possibility with respect to the central patch in a grid. We use DenseNet [7] in our system as the backbone because of its outstanding performance on histopathology among other prototypical CNN architectures, as well as its compact architecture and non-costly training time. The performance comparison will be elaborated in Sect. 3.1.

2.2   Monte Carlo Distribution Approximation

We denote the ground-truth tumor probability distribution in WSI as $P(L = l|\mathbf{x})$, where $l \in \mathcal{L}$ is a label from the set $\mathcal{L}$ and $\mathbf{x}$ is a patch in a slide. In the tumor detection task, we have $\mathcal{L} = \{tumor, nontumor\}$, and we only focus on the tumor distribution i.e., the case of $L = tumor$. With a tissue classier, we get tumor possibility distribution $p(\mathbf{x})$, and a Monte Carlo sampling method to efficiently sort out tumor areas of $P(L = tumor|\mathbf{x})$ by estimating patch-wise tumor possibility. With the prediction outcome from a neural network $\{p(\mathbf{x}_i)\}_{i=1}^N$ of N patches $\{\mathbf{x}_i\}_{i=1}^N$, which are randomly selected as initialization, we can estimate the target distribution $p(\mathbf{x})$ as follows [1]:

$$\widehat{p}_k(\mathbf{x}) = \frac{1}{Z(\mathbf{x})} \int_\Omega r(\mathbf{x}, \mathbf{t}) \cdot \mathbb{I}_{\{(\mathbf{x}_i, p(\mathbf{x}_i))\}}(\mathbf{t})d\mathbf{t}, \tag{1}$$

for all $\mathbf{x}$ that have not been sampled in the k-th iteration or previously. In (1), Ω denotes the entire region in a WSI, and $\mathbb{I}_{\{(\mathbf{x}_i, p(\mathbf{x}_i))\}}(\mathbf{t})$ is the indicator function, defined by:

$$\mathbb{I}_{\{(\mathbf{x}_i, p(\mathbf{x}_i))\}}(\mathbf{t}) = \begin{cases} p(\mathbf{x}_i), & \mathbf{t} \in \{\mathbf{x}_i | i = 1, 2...N\}, \\ 0, & \mathbf{t} \notin \{\mathbf{x}_i | i = 1, 2...N\}. \end{cases} \tag{2}$$

We use $r(\mathbf{x}, \mathbf{t})$ to denote the spatial distance kernel. We set $r(\mathbf{x}, \mathbf{t}) = \frac{1}{\|\mathbf{x}-\mathbf{t}\|}$ particular in this work. $Z(\mathbf{x}) = \int_\Omega r(\mathbf{x}, \mathbf{t}) \cdot \mathbb{I}_{\{(\mathbf{x}_i, 1)\}}(\mathbf{t})d\mathbf{t}$ is a normalization constant with respect to $\mathbf{x}$, which is used to make the approximation into a valid possibility map. To coincide with results predicted by CNN, we set $\widehat{p}_k(\mathbf{x}_i) = p(\mathbf{x}_i)$ for all the patches that have already been sampled in the previous iterations.

2.3 High-Throughput Tumor Region Location System

In the diagnostic task of a WSI by deep learning, we aim to sample a minimum proportion of patches that are later required to be processed by a CNN. First, a number of n_0 patches from a slide are randomly sampled as initialization, denoted as $\{\mathbf{x}_i^{(0)}\}_{i=1}^{n_0}$. Those patches, together with their neighboring patches are predicted by a fine-tuned CRF based neural network where the outcome $\{p(\mathbf{x}_i^{(0)})\}$ is tumor possibility of the sampled patches. Afterward, applying the Monte Carlo distribution approximation with Eq. (1), we reach an initialized approximated tumor distribution, denoted as $\widehat{p}_0(\mathbf{x})$. Denote the number of patches sample in the k-th iteration is n_k, and $\mathcal{S}_k = \cup_{j=0}^k \{\mathbf{x}_i^{(j)}\}_{i=1}^{n_j}$ represents all the sampled patches up to the k th iteration.

When k iterations have been completed, with tumor distribution approximation $\widehat{p}_k(\mathbf{x})$ in the $k + 1$ iteration, a total number of n_{k+1} patches $\{\mathbf{x}_i^{(k+1)}\}_{i=1}^{n_{k+1}}$ will be sampled progressively with the highest necessity to be sampled from the WSI based on following formulated query map:

$$q_{k+1}(\mathbf{x}) = (1 - \mathbb{I}_{S_k}(\mathbf{x})) \cdot \left[\int_\Omega H(\mathbf{t}) \cdot r(\mathbf{x}, \mathbf{t}) \cdot \mathbb{I}_{S_k}(\mathbf{t})d\mathbf{t} - \varepsilon \cdot \hat{p}_k(\mathbf{x}) \right]. \tag{3}$$

where ε is the penalty term, and $H(\mathbf{t})$ is the entropy to measure the fidelity of the approximation at $\mathbf{t} \in \mathcal{S}_k$, defined by

$$H(\mathbf{t}) = \sum_{l \in \mathcal{L}} \widehat{P}(L = l|\mathbf{t}) \ln[\widehat{P}(L = l|\mathbf{t})]. \tag{4}$$

Equation (3) is well-defined since we know the value of $\widehat{P}(L = l|\mathbf{t})$ at $\mathbf{t} \in \mathcal{S}_k$. We also have $q_{k+1}(\mathbf{x}) = 0$ for every $\mathbf{x} \in S_k$. A higher value of $q_k(\mathbf{x})$ is associated with a higher level of uncertainty of the approximation $\hat{p}_k(\mathbf{x})$.

Iteratively, the approximation distribution $\widehat{p}_{k+1}(\mathbf{x})$ will be updated via Eq. (1), based on the newest samples to coincide with the restriction that $\widehat{p}_{k+1}(\mathbf{x}_i^{(j)}) = p(\mathbf{x}_i^{(j)})$ for all the validate i and j, where the superscript denotes that a sampled patch is in the j-th iteration. As a result, the approximation will be more accurate with obvious reduction of samples than the routine method upon the convergence.

Algorithm 1 depicts the overall framework of the proposed HTRL, where the convergence criterion for the sampling procedure will be elaborated in the next section.

Algorithm 1. High-throughput Tumor Region Location System (HTRL)

Input: Ω whole slide image.
Output: $\hat{p}(\mathbf{x})$ approximated tumor distribution
$\{\mathbf{x}_i^{(0)}\}_{i=1}^{n_0}$ ←random sample n_0 patches from WSI
$\{p(\mathbf{x}_i^{(0)})\}_{i=1}^{n_0}$ ← attain predictions on sampled patches
construct the initialized MC approximation $\hat{p}_0(\mathbf{x})$
set $\mathcal{S}_0 = \{\mathbf{x}_i^{(0)}\}_{i=1}^{n_0}$ % $\mathcal{S}$ denotes all the sampled patches
set $k = 1$
while not converge **do**
    $q_k(\mathbf{x})$ ← updates the query map based on $\hat{p}_{k-1}(\mathbf{x})$
    $\{\mathbf{x}_i^{(k)}\}_{i=1}^{n_k}$ ← sample n_k patches from $\{\arg\max_{\mathbf{x}} q_{k-1}(\mathbf{x})\}_{i=1}^{n_k}$
    $\{p(\mathbf{x}_i^{(k)})\}_{i=1}^{n_k}$ ← attains predictions on sampled patches
    set $\mathcal{S}_k = \mathcal{S}_{k-1} \cup \{\mathbf{x}_i^{(k)}\}_{i=1}^{n_k}$
    $\hat{p}_k(\mathbf{x})$ ← updates the MC approximation map based on samples
    $k \leftarrow k + 1$ % counter for the iteration times
end while

2.4   Convergence Criterion for Sampling

As tumor proportion and distribution may vary a lot on different histopathology images, we propose a convergence criterion to effectively terminate the sampling processing. In the k-th iteration, we describe the high possibility tumor region $\mathfrak{D}_k$ from WSI as follows:

$$\mathfrak{D}_k = supp\{\mathbb{I}_{\{\widehat{p}_k(\mathbf{x}) \geq \mu_k\}}^{\{1\}}[\widehat{p}_k(\mathbf{x})]\}. \tag{5}$$

where $supp(f)$ denotes the support of f. μ_k functions as a threshold of "high possibility" tumor region. The arrival of the convergence in the estimation of tumor region will stop the iterative adaptive sampling. Then, the convergence criterion is given by:

$$\max\{|\#(\mathfrak{D}_{k-1}) - \#(\mathfrak{D}_{k-2})|, |\#(\mathfrak{D}_k) - \#(\mathfrak{D}_{k-1})|\} < \varepsilon(\mathfrak{D}_k), \tag{6}$$

where $\#(\cdot)$ denotes the number of clusters of aggregated tumor tissues i.e., the number of patches.

We employ $\varepsilon(\mathfrak{D})$ to determine the iterations to the convergence of the proposed sampling algorithm, which is tunable on the area of set $\mathfrak{D}$ up to the performance requirement. For instance, we value $\varepsilon(\mathfrak{D})$ with $C_1 \cdot \#(\mathfrak{D})^a$ for a fast convergence or conversely, $C_2 \cdot \exp(-b \cdot \#(\mathfrak{D}))$ for a gradual convergence. C_1 and C_2 are normalization constants, and a, b are the threshold ratios in the growth of sample areas to the area of the tumor region.

3   Results

We evaluate the performance of the proposed system by the classification accuracy and the processing time. Empirically, this method outperforms previous work in two aspects, namely the high-throughput location of tumor region in WSI with a significant reduction in auto-diagnostic time, and an improvement in the whole-slide tumor classification performance with an obvious reduction in the false predictions of patches. The experiment tests are performed on the state-of-the-art deep learning computing platform NVIDIA Tesla V100 GPU device

3.1   Dataset Pre-processing and Classifier Training

We evaluate the system on the large patient cohorts from The Cancer Genome Atlas (TCGA)[1], including three subsets: i) TCGA-STAD (n = 432 samples), ii) TCGA-COAD (n = 460 samples), iii) TCGA-READ (n = 171 samples). They are three-category classification of i) loose non-tumor tissue, ii) dense non-tumor tissue, iii) gastrointestinal cancer tissues, and their tumor proportions of ground truth are shown in Fig. 2.

The processing procedure of splitting WSIs into patches and color normalization is conducted with Matlab, while other computation tasks are written with Python. The hyper-parameters are set as follows: *learning rate* $= 1 \times 10^{-5}$ for the Adam optimizer, scheduler which reduces the *learning rate* by 10% each 10 epochs if the validation does not improve. The neural networks are pre-trained with a large number of images from the ImageNet dataset. We fine-tune the classifier on the last a couple of layers and train with 20 epochs at a learning rate of 1×10^{-5}, with the transfer learning. A proportion of 80% of the patches is used for training and the rest 10% for test and other 10%. In a previous tumor classification work [10], typical architectures e.g. ResNet-18 [5], and VGG-19 [17] have already achieved a high accuracy. To further improve the accuracy, we pre-process the images to 224×224 reserving the magnification rate of $20\times$. As shown in Table 1, DenseNet [7] outperforms the above mentioned neural networks. According to the performance of our fine-tuned network demonstrated in Table 1, we select DenseNet out of other neural networks for its comparatively short training time and comparable classification performance.

---

[1] The dataset is freely available at https://portal.gdc.cancer.gov.

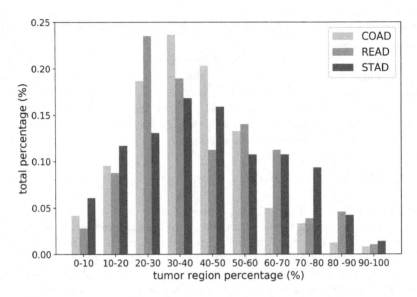

Fig. 2. The proportion of tumor tissue in WSI as ground truth in three datasets i.e., TCGA-COAD, TCGA-READ and TCGA-STAD.

Table 1. Accuracy comparison between the baseline tumor classifier [10] (patch size 512 × 512 resized to 224 × 224) and our fine-tuned network (pre-processing of patch size 224 × 224).

Network	Classifier in [10]	Fine-tuned classifier
ResNet-18 [5]	96.53%	97.04%
VGG-19 [17]	96.74%	97.02%
InceptionV3 [18]	87.02%	88.22%
DenseNet [7]	97.04%	**97.34%**

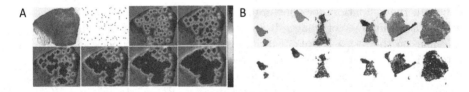

Fig. 3. A) An example of iterative adaptive sampling in WSI, from left to right are original WSI, random initialization, and distribution approximation maps. B) A couple of examples of the output predication and possibility maps of tumor tissue.

3.2   Performance Evaluation

The proposed HTRL model is implemented on Pytorch [14]. We employ DenseNet as the CNN feature extractor. The number of patches in the initial sampling and the iterative sampling is set to be 1% of valid patches in WSI (background excluded). Figure 3 shows the tumor likelihood map in the iterative sampling and the final predication results of a couple of WSI samples. Given the variance of tumor proportion and distribution, we profile the performance in Fig. 4, where our model also outperforms other approaches at their convergence. We demonstrate the performance improvement with the proposed system in Table 2.

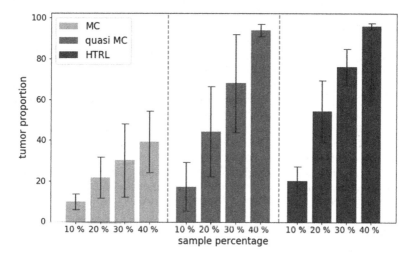

Fig. 4. Performance comparison between our HTRL and other methods, where MC stands for the Monte Carlo sampling [13] and quasi MC for the quasi Monte Carlo sampling [4]. Our model outperforms other baseline approaches with a smaller sample proportion while less error rate than the MC or quasi MC sampling based model.

Table 2. Average of accuracy improvement and performance speedup in the benchmark test.

Methods	Sampled patches proporation	Cost time	Performance speed-up	Sampled tumor proportion	WSI-level AUC
Regular[5]	100%	4 m 39 s	1.00×	**100%**	0.855
MC [13]	93.2%	3 m 52 s	1.20×	97.3%	
Quasi MC [4]	63.2%	2 m 51 s	1.63×	88.1%	
Our HTRL	**30.5%**	**1 m 50**	**2.53×**	98.3%	**0.873**

4 Conclusion

In this paper, we propose a high-throughput WSI analysis system to localize tumor regions accurately with a patch-based convolutional neural network and CRF model. The performance speedup is achieved by effectively sort out the tissues with the lowest fidelity. We also obtain a further accuracy improvement by incorporating spatial context with the predictions from the neighboring patches with a proposed CRF model. The performance acceleration will be even more obvious when only a small proportion of tumor tissues are contained in a WSI, hence the model is particularly efficient at identifying negative cases to save the waiting time.

References

1. Andrieu, C., de Freitas, N., Doucet, A., Jordan, M.I.: An introduction to MCMC for machine learning. Mach. Learn. **50**(1–2), 5–43 (2003)
2. BenTaieb, A., Hamarneh, G.: Predicting cancer with a recurrent visual attention model for histopathology images. In: Frangi, A.F., Schnabel, J.A., Davatzikos, C., Alberola-López, C., Fichtinger, G. (eds.) MICCAI 2018. LNCS, vol. 11071, pp. 129–137. Springer, Cham (2018). https://doi.org/10.1007/978-3-030-00934-2_15
3. Bychkov, D., et al.: Deep learning based tissue analysis predicts outcome in colorectal cancer. Sci. Rep. **8**(1), 3395 (2018)
4. Cruz-Roa, A., et al.: High-throughput adaptive sampling for whole-slide histopathology image analysis (HASHI) via convolutional neural networks: application to invasive breast cancer detection. PLoS ONE **13**(5), e0196828 (2018)
5. He, K., Zhang, X., Ren, S., Sun, J.: Deep residual learning for image recognition. In: Proceedings of the IEEE Conference on Computer Vision and Pattern Recognition, pp. 770–778 (2016)
6. Hou, L., Samaras, D., Kurc, T.M., Gao, Y., Davis, J.E., Saltz, J.H.: Patch-based convolutional neural network for whole slide tissue image classification. In: Proceedings of the IEEE Conference on Computer Vision and Pattern Recognition, pp. 2424–2433 (2016)
7. Iandola, F., Moskewicz, M., Karayev, S., Girshick, R., Darrell, T., Keutzer, K.: Densenet: implementing efficient convnet descriptor pyramids. arXiv preprint arXiv:1404.1869 (2014)
8. Janowczyk, A., Doyle, S., Gilmore, H., Madabhushi, A.: A resolution adaptive deep hierarchical (RADHical) learning scheme applied to nuclear segmentation of digital pathology images. Comput. Methods Biomech. Biomed. Eng.: Imaging Vis. **6**(3), 270–276 (2018)
9. Kather, J.N., et al.: Predicting survival from colorectal cancer histology slides using deep learning: a retrospective multicenter study. PLoS Med. **16**(1), e1002730 (2019)
10. Kather, J.N., et al.: Deep learning can predict microsatellite instability directly from histology in gastrointestinal cancer. Nat. Med. **25**, 1054–1056 (2019)
11. Li, Y., Ping, W.: Cancer metastasis detection with neural conditional random field. arXiv preprint arXiv:1806.07064 (2018)
12. Litjens, G., et al.: A survey on deep learning in medical image analysis. Med. Image Anal. **42**, 60–88 (2017)

13. Martino, L., Luengo, D., Míguez, J.: Independent Random Sampling Methods. SC. Springer, Cham (2018). https://doi.org/10.1007/978-3-319-72634-2
14. Paszke, A., et al.: Automatic differentiation in pytorch (2017)
15. Qaiser, T., Rajpoot, N.M.: Learning where to see: a novel attention model for automated immunohistochemical scoring. IEEE Trans. Med. Imaging **38**(11), 2620–2631 (2019)
16. Siegel, R., DeSantis, C., Jemal, A.: Colorectal cancer statistics. CA Cancer J. Clin. **64**(2), 104–117 (2014)
17. Simonyan, K., Zisserman, A.: Very deep convolutional networks for large-scale image recognition. arXiv preprint arXiv:1409.1556 (2014)
18. Szegedy, C., Vanhoucke, V., Ioffe, S., Shlens, J., Wojna, Z.: Rethinking the inception architecture for computer vision. In: Proceedings of the IEEE Conference on Computer Vision and Pattern Recognition, pp. 2818–2826 (2016)
19. Tokunaga, H., Teramoto, Y., Yoshizawa, A., Bise, R.: Adaptive weighting multi-field-of-view CNN for semantic segmentation in pathology. In: Proceedings of the IEEE Conference on Computer Vision and Pattern Recognition, pp. 12597–12606 (2019)
20. Wilbur, D.C.: Digital cytology: current state of the art and prospects for the future. Acta Cytol. **55**(3), 227–238 (2011)
21. Yan, J., Zhu, M., Liu, H., Liu, Y.: Visual saliency detection via sparsity pursuit. IEEE Signal Process. Lett. **17**(8), 739–742 (2010)

Unsupervised Learning

A Dual-Layer Architecture
for the Protection of Medical Devices
from Anomalous Instructions

Tom Mahler$^{(\boxtimes)}$ ⓘ, Erez Shalom ⓘ, Yuval Elovici ⓘ, and Yuval Shahar ⓘ

Department of Software and Information Systems Engineering (SISE),
Ben-Gurion University of the Negev, 8410501 Beersheba, Israel
mahlert@post.bgu.ac.il, {erezsh,elovici,yshahar}@bgu.ac.il
https://in.bgu.ac.il/en/

Abstract. Complex medical devices are controlled by instructions sent from a host PC. Anomalous instructions introduce many potentially harmful threats to patients (e.g., radiation overexposure), to physical components (e.g., manipulation of device motors) devices, or to functionality (e.g., manipulation of medical images). Threats can occur due to cyber-attacks, human errors (e.g., a technician's configuration mistake), or host PC software bugs. To protect medical devices, we propose to analyze the instructions sent from the host PC to the physical components using a new architecture for the detection of anomalous instructions. Our architecture includes two detection layers: (1) an unsupervised context-free (CF) layer that detects anomalies based solely on the instructions' content and inter-correlations; and (2) a supervised context-sensitive (CS) layer that detects anomalies with respect to the classifier's output, relative to the clinical objectives.

We evaluated the new architecture in the computed tomography (CT) domain, using 8,277 CT instructions that we recorded. We evaluated the CF layer using 14 different unsupervised anomaly detection algorithms. We evaluated the CS layer for four different types of clinical objective contexts, using five supervised classification algorithms for each context. Adding the second CS layer to the architecture improved the overall anomaly detection performance from an F1 score of 71.6% (using only the CF layer) to 82.3%–98.8% (depending on the clinical objective used). Furthermore, the CS layer enables the detection of CS anomalies, using the semantics of the device's procedure, which cannot be detected using only the purely syntactic CF layer.

Keywords: Anomaly detection · Medical devices · Medical imaging devices · CT scanner · Cyber-security

1 Introduction

Complex medical devices (e.g., medical imaging devices (MIDs)) often consist of an entire ecosystem of connected components (e.g., data processing servers,

© Springer Nature Switzerland AG 2020
M. Michalowski and R. Moskovitch (Eds.): AIME 2020, LNAI 12299, pp. 273–286, 2020.
https://doi.org/10.1007/978-3-030-59137-3_25

physical components, etc.), controlled by instruction sets[1] sent by a central host (controlling) PC. Anomalous instructions that the host PC might send, introduce many potentially harmful threats to the patients (e.g., radiation overexposure), to the physical components (e.g., manipulation of device motors), or to the device functionality (e.g., manipulation of medical images). Anomalous instructions might be sent due to either multiple types of cyber-attacks, or due to human errors (e.g., a technician's configuration mistake); or even due to host PC software or hardware bugs.

In a recent study, we have found the host PC of MIDs to be vulnerable to at least 21 different cyber-attacks [13]. Yaqoob *et al.* [22] surveyed numerous high-risk medical device vulnerabilities and published *Common Vulnerabilities and Exposures* (CVEs) or *Industrial Control Systems Medical Advisories* (ICSMAs) of complex medical devices (e.g., computed tomography (CT), magnetic resonance imaging (MRI), ultrasound), with *Common Vulnerability Scoring System* (CVSS) score of up to 10. For example, cyber-security vulnerabilities in Philips Brilliance CT scanners and General Electric Company (GE) medical imaging software were reported [5], allowing an attacker to execute privileged commands and to use hard-coded credentials that could impact the system integrity and availability. Recently, we have demonstrated how an adversarial attacker could tamper with medical images to insert or remove tumors [14]. Human errors (e.g., a technician's configuration mistake) and software bugs may also result in anomalous instructions. For example, incorrect settings on the CT host PC resulted in radiation overexposure of more than 200 patients for 18 months [21]. In another example, a critical software bug in the Therac-25 radiation therapy device for the treatment of cancer resulted in patients receiving massive amounts of direct radiation (sometimes a hundred times more than the usual dose) that even led to death [10].

Existing methods for mitigating the risk of anomalous instructions from cyber-attacks mainly focus on protecting the host PC from the *hospital networks*, and *not* on protecting the *inner physical components* from a potentially compromised *host PC*. Such methods are limited and are often breached (e.g., the WannaCry attack, exploitation of zero-day vulnerabilities, etc.), as they rely on constantly installing regular security updates, a challenging task in a clinical setting with numerous out-of-date devices. For example, Philips recommended users to "implement [a] multi-layer strategy" to protect systems from internal and external threats [5]. Furthermore, whether instructions are anomalous may also depend on the context of the *patient* or of the *clinical objective* (e.g., high radiation might be considered harmless in specific types of medical procedures that require it). Such a context is rarely used as part of protection methods.

We propose a *dual-layer architecture* for the protection of medical devices from anomalous instructions. The architecture focuses on detecting two types of anomalous instructions: (1) *Context-free (CF) anomalous instructions* (e.g., unlikely values or combinations of values, of instruction parameters [e.g., giving 100 times more radiation than usual]); and (2) *Context-sensitive (CS)*

---

[1] For simplicity, we will simply call them *instructions* throughout this paper.

anomalous instructions (e.g., normal values or combinations of values, of instruction parameters that are considered anomalous within a particular context [e.g., a wrong scan type, or mismatching patient age]).

To address the tasks of detecting the two anomaly types, our dual-layer architecture includes two algorithmic layers that are applied in serial fashion to analyze internal instructions sent from the host PC to the medical device, and that detect each type of anomaly using two very different layers: (1) An *unsupervised anomaly detection layer* (i.e., the *CF layer*); and (2) A *supervised anomaly detection layer* (i.e., the *CS layer*).

The second layer is applied only to instructions that were not already detected as anomalous by the first layer (i.e., CF anomalous instructions); thus, we do not need to evaluate, in the second layer, any instructions that were detected as anomalous by the CF layer.

As we show in the Methods section (Sect. 3), adding the second CS layer considerably enhances the sensitivity of the architecture to the detection of anomalous instructions, even when the original instruction was not detected as anomalous by the first CF layer, since the instruction might have made sense in some potentially plausible context.

2   Background

In this section, we start by providing the bare essentials necessary to understand anomaly detection and hybrid anomaly detection methods. Then, since in the current study, we are using CT host PC instructions for the evaluation, we explain several important details about these instructions that will facilitate the understanding of the computational methods.

2.1   Anomaly Detection Methods

Anomaly detection is used for various applications, such as fraud, intrusion detection, sensor networks, the Internet of Things (IoT), and in the medical domain, as in our case, for detecting anomalies in MIDs. Anomaly detection can use supervised methods or unsupervised methods. Using supervised anomaly detection methods requires data labeling (often by domain experts), an expensive and time-consuming task. Unsupervised anomaly detection methods can be used instead; however, the lack of labels makes CS anomaly detection harder for unsupervised methods. Hybrid anomaly detection combines unsupervised and supervised methods that can potentially be used for CS anomaly detection. For example, Denoising AutoEncoders are used for feature selection followed by a k-Nearest Neighbors (k-NN) ensemble (for anomaly detection in activity recognition) or a Convolutional Neural Network (CNN) (for malware detection in Android apps) [4]. Rawte *et al.* [17] used unsupervised clustering for disease detection followed by a supervised classification for medical fraud detection. Such methods are mostly used to improve CF anomaly detection and not for CS anomaly detection; as far as we know, we are the first to use hybrid anomaly detection for CS anomalous instruction detection for the protection of medical devices.

2.2   CT Host PC Instructions

In response to an instruction from the host PC, a CT scanner produces an imaging scan (also known as a *Series*) that consists of a sequence of 2D images (i.e., slices). There are three general *Scan Options* for each instruction: *Axial* (the slices are along the z-axes), *Helix* (the slices are moving like a screw), and *Surview* (an initial brief scan, with very low radiation, that allows the operator to configure the subsequent scans better, as well as apply various optimization techniques). The clinical procedure of a CT scan is called a *Study*. Usually, the CT operator does not configure the *Study* one scan at a time. Instead, specific sequences of CT scans are predefined as a set of *Protocols* from which the operator can choose from. A single *Study* can combine more than one *Protocol* (e.g., a *Chest/Abdomen Study* combines *Chest Protocol* and *Abdomen Protocol*). The *Study* usually depends on the *Body Part* being scanned. In this study, we consider the *Scan Options*, *Body Part*, *Study*, and *Protocol* as different abstractions of clinical objective contexts.

Our analysis of the collected instructions (see Sect. 3.1 next) also revealed that the clinical objective context abstractions uses a predefined finite set of classes (for each abstraction), and have the following, top-to-bottom, hierarchical relationship between them: *Scan Option* (3 classes), *Body Part* (11 classes), *Study* (34 classes), and *Protocol* (72 classes). A deeper hierarchy level provides more information about the clinical objective; however, limits the amount of available training data for each class.

3   Methods

In this section, we present the dual-layer architecture (Fig. 1), designed to detect anomalous instructions using two algorithmic layers (a CF one and a CS one), its implementation, and its evaluation. In this study, the implementation of each layer uses a set of specific classifiers, and the evaluation was done on CT host PC instructions (see Background Sect. 2.2), focusing on detection of clinical objective CS anomalous instructions (using the patient context is beyond the scope of the current study). Of course, it is possible to implement the architecture using other classifiers and evaluate it on different medical devices.

3.1   Data Collection

To collect CT instructions, we have partnered with a major CT manufacturer and developed a data collection tool that records instructions sent from the host PC to the gantry (i.e., the physical component of the CT). To collect real data, we partnered with a hospital and installed in its CT scanners the data collection tool. The collected data include the instruction parameters (233 features) and instruction metadata (77 features). The metadata is logged by the host PC but is not part of the instruction parameters and is not sent to the gantry. The metadata includes clinical objective context (e.g., *Scan Options, Body Part, Study*, and *Protocol*), and patient context (e.g., gender, age). In total, we collected 8,277 instructions that belong to 2,643 different *Studies* (which is roughly the number of patients) that we separated into a train set of 6,286 (75%) instructions and a test set of 1,991 (25%) instructions.

The test set includes 1,312 normal instructions and 679 anomalous instructions. Collecting labeled anomalous instructions (e.g., malicious instructions due to a cyber-attack) is very difficult since anomalous instructions are rare and unlabeled (i.e., the metadata does not include an *anomaly* label or whether the instruction [or even the entire *Study*] satisfied the clinical objective). Following, we explain how we collected the 679 CF and CS anomalous instructions.

CF Anomalous Instructions Collection. While analyzing the collected instructions, we noticed that 216 instructions looked suspicious (labeled as a *Physics Procedure* for the *Study* metadata). A technical discussion with the manufacturer verified that these instructions were part of a technical maintenance calibration procedure and should not be used on patients; thus, we considered them as CF anomalous instructions. In addition, we manually recorded 59 malicious anomalous instructions by asking an expert operator to, intentionally, execute malicious instructions (e.g., high radiation, high motor speed, long scan time, etc.) on a CT scanner (without a patient). These anomalous instructions are *CF*, as they should not be sent regardless of the patient being scanned or the clinical objective. In total, we collected 275 CF anomalous instructions.

CS Anomalous Instructions Collection. While analyzing the collected instructions, we discovered that 140 *Studies* (containing a total of 404 instructions), which make up 5% of all non-anomalous instructions, were repeated twice, one after the other, for the same patient, for no apparent reason; while there could be many reasons for repeating a *Study*, we assume that it was repeated since the first *Study* did not satisfy the clinical objective. Unlike the CF anomalous instructions, the repeated instructions are, in fact, normal instructions that are only considered anomalous given the clinical objective context; thus, we considered these 5% (i.e., 404) repeated instructions as CS anomalous instructions. In total, we collected 404 CS anomalous instructions.

Data Preprocessing. For each algorithm training, we cleaned the data (e.g., removed instructions that include parameters with NaN[2] value), encoded categorical features (one-hot encoding was used for neural networks), and applied standardization (i.e., Z-score normalization). Also, we used basic feature selection algorithms to drop features with a single value and features with a 100% correlation with other features. For the supervised CS layer, we dropped instructions of labels with less than 100 examples.

---

[2] Not a Number.

3.2   The Dual-Layer Architecture

The complete dual-layer architecture is presented in Fig. 1. The instructions are received from the host PC, and the clinical objective and patient-specific contexts from the operator and the Electronic Medical Record (EMR), respectively (for security reasons, the context must be sent from an isolated secure private channel and not directly from the host PC; else, a compromised host PC may send a malicious context matching the malicious instruction).

Implementation. We selected, for each layer, the algorithm with the highest F1 score on the test set without the CS anomalous instructions.

Evaluation. We compared the performance, with respect to overall anomalous instructions detection (both CF and CS), of (1) just the CF layer (representing the capabilities of current state-of-the-art unsupervised anomaly detection) to (2) the performance of the overall anomalous instructions detection when using, in addition to the first layer, also the CS layer.

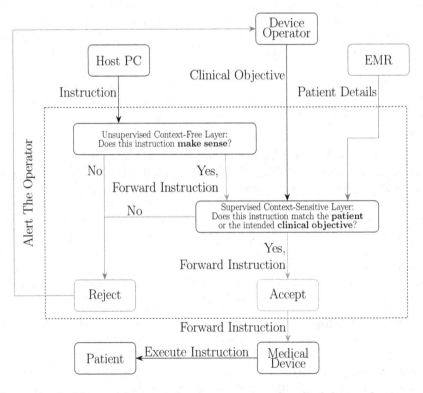

Fig. 1. The dual-layer architecture for the protection of medical devices from anomalous instructions, using both the context-free (CF) and the context-sensitive (CS) layers.

The Context-free (CF) Layer. The first layer receives the instructions (without context) as input and uses a pre-trained (using non-anomalous instructions) unsupervised anomaly detection algorithm to detect CF anomalous instructions and alert the operator. However, instructions that were not detected as anomalous might still be CS anomalous within a particular context; to detect these, we use the second layer.

Implementation. We have decided to use 11 state-of-the-art unsupervised anomaly detection algorithms (listed in Table 1), some of them implemented by the PyOD python toolbox [24]. Each algorithm calculates the anomaly score of an instruction, and if it is above the anomaly threshold, the instruction is detected as anomalous. The anomaly threshold is the $100 \cdot (1 - contamination)$ percentile of the training set anomaly scores, where *contamination* represents the expected proportion of anomalies in the data and is given during the initialization of the algorithms. We have also added three ensembles composed of the top (in terms of highest F1 score in our initial evaluation) four algorithms from the 11 algorithms that we evaluated (i.e., Angle-based Outlier Detector (ABOD), k-NN, One-Class Support Vector Machine (OCSVM), and Isolation Forest (IForest)): the Locally Selective Combination of Parallel Outlier Ensembles (LSCP) [23], and two that chooses either the maximal or the average anomaly score [2] of these four algorithms as the final anomaly score of the ensemble (which is compared to the ensemble's anomaly threshold, to determine its output, as is the case in the other algorithms).

Evaluation. Since we collected 275 CF anomalous instructions out of a total of 8,277 collected instructions, a *contamination* parameter of 0.01 (slightly lower than the actual portion of anomalies in the data) seemed to work well for most algorithms; thus, we decided to use it throughout the evaluation. We trained the algorithms using non-anomalous instructions and evaluated the performance using the CF and CS anomalous instructions. We included the evaluation of just the CF anomalous instructions to show the performance just on this type of anomalous instructions. We evaluated the performance using the confusion matrix, accuracy, recall, precision, and F1 score.

The Context-Sensitive (CS) Layer. The second layer receives the instructions (from the CF layer), and the intended instruction contexts (e.g., the clinical objective, provided by the technician operating the device, or the patient characteristics provided by the EMR) and uses pre-trained (using the context as its target labels) set of supervised classification algorithms to predict the contexts of the instructions. The predicted contexts are then compared to the intended contexts; if they do not match, the instruction is detected as anomalous within one or more contexts.

Implementation. We decided to use five state-of-the-art multi-class classification algorithms (listed in Table 1), implemented by the scikit-learn classification library [15], for the detection of CS anomalous instructions.

Evaluation. We evaluated, separately, each of the four scan type hierarchical abstraction levels of the clinical objective contexts (see Background Sect. 2.2) using the five supervised classification algorithms. We trained the algorithms using non-anomalous labeled instructions and evaluated the performance using the test set without the anomalous instructions (see Sect. 3.1). The comparison between the predictions of a CS instruction and the intended contexts (given as its input) should be *True* for non-anomalous instructions, and *False* for CS instructions (note that that the first layer already discarded CF instructions). Therefore, we evaluated the performance using the diagonal of the multi-class confusion matrix (representing the correctly classified instructions), accuracy, and weighted F1 score (due to class imbalance). We also included the evaluation of just the CS anomalous instructions (which are not part of the train or test sets) to show the performance of CS anomalous instructions detection. The comparison between the predictions of the correctly classified CS instructions will result in a set of contexts that will [correctly] not match the intended contexts.

4   Results

We applied the dual-layer protection algorithm on a CT scanner device to test it. We focused on the clinical objective context (the patient context is beyond the scope of the current study).

The CF Layer. In Table 1, we can see that the performance on the CF anomalous instructions is high for several algorithms, as expected, and highest for the Ensemble Average algorithm. However, the performance on the CF and CS anomalous instructions are much lower, since the algorithms fail to detect the CS anomalous instructions. Note that the number of instructions that were used is lower than the number of collected instructions due to the preprocessing (see Sect. 3.1); for example, *Surview* type instructions were removed, as these types of instructions are quick initial scans that have a lower potential of damage.

The CS Layer. For this layer, we can safely assume that the given instructions are not CF anomalous since the CF layer already detected them. In this study, we evaluated clinical objective CS anomalous instructions (patient context is beyond the scope of the current study). We represent the clinical objective using the four hierarchical abstractions of the scan type (see Sect. 2.2) and use it as the target labels of the supervised classification algorithms that we train. For each clinical context, we evaluated the performance on the test set (i.e., not including anomalous instructions) and on the test set with the CS anomalous instructions. Note that the preprocessing is slightly different since we have only used instructions of labels with at least 100 examples.

Scan Options Objective. In Table 3a, we can see that supervised classification seems to work extremely well for this level of abstraction, with most algorithms reaching an F1 score of 1. This implies that the topmost level of the scan type hierarchical separates instructions very well. Note that for this classification, we did not remove the *Surview* instructions during the preprocessing.

Table 1. The results of the unsupervised anomaly detection for the CF layer. The training set included 3,595 non-anomalous instructions, and the test set included 481 anomalous instructions (275 of which are CF) and 764 non-anomalous instructions.

Algorithm	TPR		FNR		FPR	TNR	Accuracy		F1	
	CF	CF+CS	CF	CF+CS			CF	CF+CS	CF	CF+CS
ABOD [9]	1	0.586	0	0.414	0.024	0.976	0.983	0.826	0.968	0.722
k-nn [16]	1	0.574	0	0.426	0.016	0.984	0.988	0.826	0.979	0.718
Ens. Avg. [2]	1	0.572	0	0.428	0.016	0.984	0.988	0.825	0.979	0.716
Ens. Max [2]	1	0.572	0	0.428	0.017	0.983	0.987	0.824	0.977	0.715
OCSVM [19]	1	0.572	0	0.428	0.017	0.983	0.987	0.824	0.977	0.715
IForest [11]	0.993	0.628	0.007	0.372	0.115	0.885	0.913	0.78	0.859	0.688
LOF [3]	1	0.64	0	0.36	0.132	0.868	0.903	0.78	0.845	0.692
LSCP [23]	0.898	0.486	0.102	0.514	0.016	0.984	0.961	0.792	0.923	0.644
SO-GAAl [12]	0.36	0.254	0.64	0.746	0.143	0.857	0.746	0.624	0.429	0.343
HBOS [6]	0.356	0.208	0.644	0.792	0.013	0.987	0.82	0.686	0.512	0.338
AE* [1]	0.313	0.179	0.687	0.821	0.012	0.988	0.809	0.676	0.465	0.299
PCA [20]	0.295	0.168	0.705	0.832	0.01	0.99	0.806	0.672	0.445	0.284
CBLOF [8]	0.291	0.166	0.709	0.834	0.008	0.992	0.807	0.673	0.443	0.282
MCD [7,18]	0.171	0.135	0.829	0.865	0.007	0.993	0.781	0.662	0.292	0.236

TPR = true positive rate, FNR = false negative rate, FPR = false positive rate, TNR = true negative rate, CF = context-free, CS = context-sensitive, RF = Random Forest, ABOD = Angle-based Outlier Detector, AE = AutoEncoder, CBLOF = Clustering-Based Local Outlier Factor, Ens = Ensemble, Avg = Average, HBOS = Histogram-based Outlier Score, IForest = Isolation Forest, k-nn = k-Nearest Neighbors, LOF = Local Outlier Factor, LSCP = Locally Selective Combination of Parallel Outlier Ensembles, MCD = Minimum Covariance Determinant, OCSVM = One-Class Support Vector Machine, PCA = Principal Component Analysis, SO-GAAl = Single-Objective Generative Adversarial Active Learning. *One Hot Encoded.

Body Part Objective. In Table 3b, we can see that supervised classification seems to work well for this level of abstraction, too, with some algorithms reaching an F1 score of 1. Note that the number of instructions is lower since not all instructions include the *Body Part* label, and we have removed instructions with *Surview Scan Option*.

Study Objective. In Table 3c, we can see that the performance of the supervised classification algorithms decreased compared with higher-level abstractions, with a maximal F1 score of 0.895 for Random Forest (RF). Note that classes with a relative high number of instructions available during training have a higher F1 score, implying that more data might improve the performance. Furthermore, we point out that by evaluating the confusion matrix we noticed that wrong classification was mostly between relatively similar *Study* types; for example, the Random Forest classifier was confused between *Abdomen*, *Chest*, and *Chest/Abdomen*, however, was not confused between *Abdomen* and *Head* or *Abdomen* and *CTA Cardiac*.

Protocol Objective. In Table 3d, we can see that the performance of the supervised classification algorithms also decreased compared with higher-level abstractions, with a maximal F1 score of 0.819 for RF. Similar to the *Study* class, classes with a relative high number of instructions available during training have a higher F1 score, and the wrong classifications were mostly between relatively similar *Protocols*.

4.1 Dual-Layer Architecture

In Table 2, we can see how adding the second CS layer improved the overall performance (F1 score and accuracy) for each clinical objective, relative to the performance of the CS layer. Note that while the detection of CS anomalous instructions was improved, miss-classification of the non-anomalous instructions (which are also analyzed by this layer) resulted in increased false positives. Note, for example, that using only the CF layer led to detecting only 57.2% of the anomalous instructions, while adding the CS layer and knowledge of the *Study* clinical objective led to a sensitivity of 94.7%.

Table 2. The results of the dual-layer architecture, showing the performance of the CF layer alone, and with the additional second CS layers. For the first CF layer, we used the Ensemble Average algorithms. For each second CS layer, we used the RF classifier (respectively).

CS Layer	TPR	FNR	FPR	TNR	Accuracy	F1
None	0.572	0.428	0.016	0.984	0.825	0.716
Scan Options	1	0	0.016	0.984	0.991	0.988
Body Part	1	0	0.022	0.978	0.986	0.978
Study	0.947	0.053	0.118	0.882	0.905	0.879
Protocol	0.939	0.061	0.19	0.81	0.856	0.823

TPR = true positive rate, FNR = false negative rate, FPR = false positive rate, TNR = true negative rate, CF = context-free, CS = context-sensitive, RF = Random Forest.

5   Discussion

In this study, we proposed a dual-layer architecture for the protection of medical devices from CF and CS anomalous instructions, and evaluated its performance using CT host PC instructions (that we collected from an operational CT at a hospital), for four, hierarchical, scan type abstractions of the clinical objective context. The CF layer detected all 275/275 CF anomalous instructions using unsupervised anomaly detection methods (e.g., ensemble average algorithm); however, it failed to detect the 206 CS anomalous instructions, resulting in an F1 score of 0.716. The CS layer detected 82%–100% of the CS anomalous instructions (depending on the clinical context used) by comparing supervised classification methods' predictions (e.g., RF) to the real context (received from a separated secure private channel); However, the low performance of some classifiers increased the false positive rate (FPR) due to wrong classifications of non-anomalous instructions. Overall, we can conclude that adding the second CS layer increased the overall F1 score from 71.6% to 82.3%–98.8%.

Our current study has three main limitations, mostly due to the difficulty in collecting sufficient data for evaluation. We intend to address these three limitations in our future work, as follows: (1) We shall validate our approach on additional medical devices (e.g., MRI, ultrasound); (2) We intend to strengthen the evaluation using a k-fold cross-validation; (3) We shall extend the CS layer and its evaluation so to also include the patient's context from the patient's full-fledged EMR.

From our results, we can conclude that for higher-level abstractions (i.e., *Scan Options* and *Body Part*) the CS layer performed very well with an F1 score of 99.4%–100%, while for lower-level abstractions (i.e., *Study* and *Protocol*) the performance was lower with an F1 score of 81.9%–89.5%. One reason for this is that lower-levels in the hierarchy limited the amount of available training data for each class; for example, for *Study* context RF classifier, the F1 score for *CTA Cardiac* class (trained using 854 instructions) was 0.96, compared with 0.842 for *CTA Head* class (trained using 269 instructions). Thus, we are confident that with more training data available, the performance of the CS layer for lower-level abstractions would increase. Furthermore, from the evaluation of the confusion matrices of classifiers of lower-levels in the hierarchy, we discovered that wrong classification was given mostly to relatively similar classes (e.g., between *Abdomen Routine (C+) /Abdomen* and *Abdomen Routine (C-) /Abdomen* classes of *Protocol*). While such classifications are considered wrong, there might not be a real significant difference between such classes. Therefore, by merging such classes, we could both increase the amount of available training data for the merged class and reduce the number of wrong classifications between similar classes.

Appendix A

Table 3. The results of the supervised classification of clinical objective contexts for the CS layer on the test set and the CS anomalous instructions, including the per-class F1 score, the total accuracy, and the total weighted F1 average. At the bottom of each table, we present, per-class, the number of instructions used during training and testing.

(a) Scan Options

Algorithm	Axial Test	Axial CS	Helix Test	Helix CS	Surview Test	Surview CS	Accuracy Test	Accuracy CS	Weighted F1 Test	Weighted F1 CS
RF	1	–	1	–	1	–	1	–	1	–
GB	1	–	1	–	1	–	1	–	1	–
k-NN	1	–	1	–	1	–	1	–	1	–
MLP	1	–	1	–	1	–	1	–	1	–
DT	0.998	1	0.999	1	1	1	0.999	1	0.999	1
Test Set	276	70	495	163	541	124			1,312	357
Train Set	1,006		2,590		2,520				6,116	

(b) Body Part

Algorithm	Abdomen Test	Abdomen CS	Brain Test	Brain CS	LSpine Test	LSpine CS	Accuracy Test	Accuracy CS	Weighted F1 Test	Weighted F1 CS
RF	0.994	1	1	1	0.990	1	0.994	1	0.994	1
GB	0.987	1	1	1	0.979	1	0.988	1	0.988	1
DT	0.987	0.978	1	1	0.979	0.903	0.988	0.903	0.988	0.970
MLP	0.977	0.978	1	0.994	0.970	0.903	0.979	0.970	0.979	0.970
k-NN	0.958	0.957	1	1	0.933	0.813	0.940	0.961	0.961	0.941
Test Set	153	70	86	15	98	15			337	100
Train Set	893		626		379				1,898	

(c) Study

Algorithm	Chest/Abdomen Test	CS	Chest Test	CS	Abdomen Test	CS	CTA Head Test	CS	CTA Cardiac Test	CS	Head Test	CS	Spine Lumbar Sacral Test	CS	Accuracy Test	CS	Weighted F1 Test	CS
RF	0.675	0.231	0.811	0.900	0.831	0.761	0.842	1	0.960	0.974	0.910	0.917	0.929	0.923	0.896	0.854	0.895	0.845
k-NN	0.632	0.333	0.808	0.800	0.812	0.754	0.800	0.625	0.961	0.967	0.935	0.909	0.883	0.857	0.889	0.835	0.887	0.830
GB	0.650	0.222	0.781	0.857	0.806	0.687	0.800	0.909	0.948	0.962	0.957	0.919	0.919	0.883	0.882	0.829	0.882	0.819
MLP	0.641	0.250	0.803	0.818	0.835	0.754	0.842	0.769	0.960	0.967	0.917	0.833	0.889	0.769	0.891	0.829	0.889	0.818
DT	0.452	0.345	0.683	0.613	0.672	0.613	0.640		0.932	0.901	0.710	0.804	0.571	0.741	0.793	0.741	0.799	0.747
Test Set	38	16	81	9	66	33	20	5	265	78	75	11	60	6	605	158	605	158
Train Set	246		318		391		269		854		554		211		2,843		2,843	

(d) Protocol

Algorithm	Abdomen Routine (C+)/Abdomen Test	CS	Abdomen Routine (C-)/Abdomen Test	CS	Brain Routine (C-)/Head Test	CS	Chest + Abdomen (C+)/Abdomen Test	CS	Chest (C-)/Thorax Test	CS	Coronary CTA HR<62/Cardiac Test	CS	Coronary Step & Shoot/Cardiac Test	CS	Lumbar Spine/Spine Test	CS	Lumbar Spine Trauma/Spine Test	CS	Stroke/Head Test	CS	TAVI CTA/Cardiac Test	CS	Accuracy Test	CS	Weighted F1 Test	CS
RF	0.611	0.483	0.291	0.682	0.636	0.898	0.494	0.636	1	1	0.98	0.977	0.958	0.952	0.679	0.6	0.677	0.571	0.833	1	0.927	0.857	0.823	0.826	0.819	0.827
GB	0.568	0.457	0.246	0.485	0.552	0.851	0.506	0.552	1	1	0.97	0.97	0.952	0.954	0.633	0.5	0.556	0.5	0.762	1	0.916	0.857	0.797	0.772	0.793	0.772
DT	0.56	0.692	0.262	0.294	0.432	0.86	0.517	0.432	0.986	0.818	0.96	0.96	0.915	0.934	0.444	0.633	0.449	0.571	0.714	0.788	0.896	0.647	0.779	0.698	0.776	0.692
k-NN	0.452	0.420	0.29	0.313	0.571	0.898	0.451	0.571	1	1	0.762	0.766	0.894	0.882	0.585	0.5	0.462	0.2	0.833	1	0.903	0.727	0.754	0.664	0.751	0.667
MLP	0.523	0.462	0.305	0.258	0.368	0.896	0.463	0.368	0.818	0.783	0.727	0.765	0.912	0.882	0.444	0.25	0.629	0.308	0.839	1	0.92	0.75	0.772	0.651	0.765	0.649
Test Set	37	13	39	23	48	8	31	11	36	13	52	21	128	31	34	7	26	2	26	5	91	15	553	149	553	149
Train Set	292		152		354		284		147		147		430		114		133		301		244		2,598		2,598	

CS=context-sensitive, DT=Decision Tree, GB=Gradient Boosting, k-NN=k-Nearest Neighbors, MLP=Multilayer Perceptron, RF=Random Forest.

References

1. Aggarwal, C.C.: Outlier analysis. Data Mining, pp. 237–263. Springer, Cham (2015). https://doi.org/10.1007/978-3-319-14142-8_8
2. Aggarwal, C.C., Sathe, S.: Theoretical foundations and algorithms for outlier ensembles. ACM SIGKDD Explor. Newsl. **17**(1), 24–47 (2015). https://doi.org/10.1145/2830544.2830549
3. Breunig, M.M., Kriegel, H.P., Ng, R.T., Sander, J.: LOF: identifying density-based local outliers. In: Proceedings of the 2000 ACM SIGMOD International Conference on Management of Data, SIGMOD, pp. 93–104. Association for Computing Machinery, New York (2000). https://doi.org/10.1145/342009.335388
4. Chalapathy, R., Chawla, S.: Deep learning for anomaly detection: a survey. arXiv preprint arXiv:1901.03407 (2019)
5. Donovan, F.: Philips CT scanner cybersecurity vulnerabilities pose PHI risk, April 2018. https://healthitsecurity.com/news/philips-ct-scanner-cybersecurity-vulnerabilities-pose-phi-risk
6. Goldstein, M., Dengel, A.: Histogram-based Outlier Score (HBOS): a fast unsupervised anomaly detection algorithm. KI-2012: Poster and Demo Track, pp. 59–63 (2012). https://www.dfki.de/fileadmin/user_upload/import/6431_HBOS-poster.pdf
7. Hardin, J., Rocke, D.M.: Outlier detection in the multiple cluster setting using the minimum covariance determinant estimator. Comput. Stat. Data Anal. **44**(4), 625–638 (2004)
8. He, Z., Xu, X., Deng, S.: Discovering cluster-based local outliers. Pattern Recogn. Lett. **24**(9), 1641–1650 (2003). https://doi.org/10.1016/S0167-8655(03)00003-5
9. Kriegel, H.P., Schubert, M., Zimek, A.: Angle-based outlier detection in high-dimensional data. In: Proceedings of the 14th ACM SIGKDD International Conference on Knowledge Discovery and Data Mining, KDD 2008, pp. 444–452. Association for Computing Machinery, New York (2008)
10. Leveson, N., Turner, C.: An investigation of the Therac-25 accidents. Computer **26**(7), 18–41 (1993). https://doi.org/10.1109/MC.1993.274940
11. Liu, F.T., Ting, K.M., Zhou, Z.: Isolation forest. In: 2008 Eighth IEEE International Conference on Data Mining, pp. 413–422 (2008). https://doi.org/10.1109/ICDM.2008.17
12. Liu, Y., et al.: Generative adversarial active learning for unsupervised outlier detection. IEEE Trans. Knowl. Data Eng. **32**, 1 (2019)
13. Mahler, T., Elovici, Y., Shahar, Y.: A new methodology for information security risk assessment for medical devices and its evaluation. arXiv preprint arXiv:2002.06938 (2020)
14. Mirsky, Y., Mahler, T., Shelef, I., Elovici, Y.: CT-GAN: malicious tampering of 3D medical imagery using deep learning. In: 28th USENIX Security Symposium (USENIX Security 19), pp. 461–478. USENIX Association, Santa Clara (2019). https://www.youtube.com/watch?v=_mkRAArj-x0
15. Pedregosa, F., et al.: Scikit-learn: machine learning in Python. J. Mach. Learn. Res. **12**, 2825–2830 (2011). https://scikit-learn.org/
16. Ramaswamy, S., Rastogi, R., Shim, K.: Efficient algorithms for mining outliers from large data sets. In: Proceedings of the 2000 ACM SIGMOD International Conference on Management of Data, SIGMOD 2000, pp. 427–438. Association for Computing Machinery, New York (2000)

17. Rawte, V., Anuradha, G.: Fraud detection in health insurance using data mining techniques. In: 2015 International Conference on Communication, Information & Computing Technology (ICCICT), pp. 1–5. IEEE (2015)
18. Rousseeuw, P.J., Driessen, K.V.: A fast algorithm for the minimum covariance determinant estimator. Technometrics **41**(3), 212–223 (1999). https://doi.org/10.1080/00401706.1999.10485670
19. Schölkopf, B., Platt, J.C., Shawe-Taylor, J., Smola, A.J., Williamson, R.C.: Estimating the support of a high-dimensional distribution. Neural Comput. **13**(7), 1443–1471 (2001)
20. Shyu, M.L., Chen, S.C., Sarinnapakorn, K., Chang, L.: A novel anomaly detection scheme based on principal component classifier. In: Proceedings of the International Conference on Data Mining (ICDM) (2003). https://apps.dtic.mil/docs/citations/ADA465712
21. Wintermark, M., Lev, M.: FDA investigates the safety of brain perfusion CT. Am. J. Neuroradiol. **31**(1), 2–3 (2010). https://doi.org/10.3174/ajnr.A1967
22. Yaqoob, T., Abbas, H., Atiquzzaman, M.: Security vulnerabilities, attacks, countermeasures, and regulations of networked medical devices-a review. IEEE Commun. Surv. Tutor. **21**(4), 3723–3768 (2019). https://doi.org/10.1109/COMST.2019.2914094
23. Zhao, Y., Nasrullah, Z., Hryniewicki, M.K., Li, Z.: LSCP: locally selective combination in parallel outlier ensembles, pp. 585–593 (2019). https://doi.org/10.1137/1.9781611975673.66
24. Zhao, Y., Nasrullah, Z., Li, Z.: Pyod: a python toolbox for scalable outlier detection. J. Mach. Learn. Res. **20**(96), 1–7 (2019). http://jmlr.org/papers/v20/19-011.html. https://pyod.readthedocs.io/

Multi-view Clustering with mvReliefF for Parkinson's Disease Patients Subgroup Detection

Anita Valmarska[1,2(✉)], Dragana Miljkovic[2], Nada Lavrač[2,3], and Marko Robnik–Šikonja[1]

[1] Faculty of Computer and Information Science,
University of Ljubljana, Ljubljana, Slovenia
{anita.valmarska,marko.robnik}@fri.uni-lj.si
[2] Jožef Stefan Institute, Jamova 39, Ljubljana, Slovenia
{dragana.miljkovic,nada.lavrac}@ijs.si
[3] University of Nova Gorica, Vipavska 13, Nova Gorica, Slovenia

Abstract. Parkinson's disease is a chronic neurodegenerative disease affecting people worldwide. Parkinson's disease patients experience motor symptoms and many other symptoms that affect the quality of their lives. Discovering groups of patients with similar symptoms from different symptom groups can improve the understanding of this incurable disease and advance the development of personalized treatment of Parkinson's disease patients. This paper proposes a multi-view clustering approach to discover groups of patients experiencing similar symptoms from different symptom groups (views). For that we modified ReliefF feature ranking algorithm to characterize subsets of most informative symptoms that maximize the similarity between the detected patient groups, described by symptoms from different views (i.e. different symptom groups). The adapted mvReliefF algorithm calculates the weight of features based on the values of their neighbors over multiple views. The current approach works for two views simultaneously, but can be extended to multiple views. The results of the experiments show that the proposed methodology, applied on a pair of data sets from the PPMI data collection, successfully identified lists of most important symptoms that divide patients into groups, ordered by the severity of patients' symptoms.

Keywords: Parkinson's disease · Multi-view clustering · Feature evaluation

1 Introduction

Parkinson's disease is a chronic neurodegenerative disease affecting people worldwide. The disease develops as a consequence of the death of nigral neurons and the decreased production of dopamine in the patient's brain. As a result, the most

© Springer Nature Switzerland AG 2020
M. Michalowski and R. Moskovitch (Eds.): AIME 2020, LNAI 12299, pp. 287–298, 2020.
https://doi.org/10.1007/978-3-030-59137-3_26

recognizable symptoms associated with Parkinson's disease are motor symptoms that include tremor, rigidity, slowness of movement, and in the later and more severe stages of the disease, postural instability.

The Parkinson's Progression Markers Initiative (PPMI) [7] is a landmark observational clinical study to comprehensively understand cohorts of significant interest including de nuovo Parkinson's disease patients. The data collected for over 400 Parkinson's disease patients for the period of up to 5 years. During their involvement, the patients pay regular visits to their assigned clinicians every 3–6 months where their symptoms are assessed, thus allowing the clinicians to monitor the disease progression over time. The clinical part of the data used in our work was gathered by PPMI using several standardized questionnaires, providing an opportunity to follow the status of the Parkinson's disease patients from several different points of view, which allows for multi-view data analysis that is the topic of this paper.

Parkinson's disease is still an incurable disease. The disease treatment consists of management of patients' symptoms, most significantly with the prescription of antiparkinson medications. Clinicians need to be aware of the overall patient's status, the severity of their symptoms, as well as the context of each patient's life, in order to prescribe a treatment offering the best trade-off between symptoms severity management and the corresponding side effects. The division of Parkinson's patients in groups is an important step towards more personalized treatment of Parkinson's disease patients. Patients associated with a particular group share a set of similar symptoms. The division of Parkinson's disease patients into groups is utilized mostly for the purpose of determining subtypes of Parkinson's disease. Further research into subtypes might provide insights about the mechanisms of neurodegeneration and assist clinical trial designs [12]. The current subtypes of Parkinson's disease patients, including the most popular division of patients into tremor-dominant, postural instability and gait dominant [4], are conducted on patients' data from a single time point, and do not investigate the possible commonalities between patients as the disease progresses.

In our previous work [14], we use the longitudinal clinical data from the PPMI study to divide Parkinson's disease patients into groups of patients with a similar overall status. The results of this study indicate that PPMI patients can be divided into clusters, ordered according to the severity of the patients' overall motor status (as represented by the sum of severity of symptoms from the standardized questionnaire MDS-UPDRS[1]). This approach proved to be successful for following the patients' disease progression by studying the changes of the patients' overall status between two consecutive visits to the clinician. However, the clustering of patients based on their overall status obscures the influence of the particular symptoms affecting the changes of the patient's overall status.

To study the effects of particular symptoms for finding groups of patients with similar symptoms across different views, this paper proposes a multi-view clustering approach applied to clinical data from the PPMI study, where

---

[1] MDS-UPDRS denotes Movement Disorder Society-sponsored revision of Unified Parkinson's Disease Rating Scale.

different views refer to different groups of symptoms, reflecting different aspects of patients' lives. In this paper, we focus on the following two views: motor symptoms (tremor, rigidity, slowness of movement, etc.), and motor experiences of daily living (speech, swallowing, dressing, eating, etc.).

In the proposed multi-view clustering approach, we aim to find clusters of patients that are best aligned, i.e. that are grouped into same patient groups based on the symptoms of the two views (motor symptoms and motor experiences of daily living). This is achieved by using an adapted multi-view ReliefF algorithm, named mvReliefF, to determine the quality of symptoms from the examined data views. The advantage of this novel method allows for detecting the most influential symptoms across symptom groups. While our research is still in its preliminary stage, we contribute to both the machine learning and the health-care community—we present a multi-view method for evaluation of feature importance over multiple views and we take a step forward towards defining groups of patients with similar symptoms over longer time periods.

The rest of this paper is structured as follows. In Sect. 2 we present the clinical data from the PPMI study used in this work. Section 3 presents the proposed methodology. Our results and findings are outlined in Sect. 4. We finish with concluding remarks and ideas for future work in Sect. 5. Appendix A includes the description of rules, representing the subgroups of data instances.

2   Data

In this paper, we use clinical data from the Parkinson's Progression Markers Initiative (PPMI) data collection [7]. The PPMI data collection records data for over 400 Parkinson's disease patients, who were involved in the study for up to 5 years. During their involvement, the patients pay regular visits to their assigned clinicians very 3–6 months, where their symptoms are assessed, thus allowing the clinicians to monitor the disease progression over time. The clinical data used in our work was gathered using several standardized questionnaires, briefly described below.

MDS-UPDRS (Movement Disorder Society-sponsored revision of Unified Parkinson's Disease Rating Scale) [2] is the most widely used, four-part questionnaire addressing 'non-motor experiences of daily living' (Part I, subpart 1 and subpart 2), 'motor experiences of daily living' (Part II), 'motor examination' (Part III), and 'motor complications' (Part IV). It consists of 65 questions, each addressing a particular symptom. Each question is anchored with five responses that are linked to commonly accepted clinical terms, ranging from $0 =$ normal (patient's condition is normal, the symptom is not present) to $4 =$ severe (symptom is present and severely affects the normal and independent functioning of the patient), where 1, 2, 3 denote intermediate symptom severity). Answers to the questions from each questionnaire form vectors of attribute values. All of the considered answers have ordered values, where increased values suggest higher symptom severity and decreased quality of life.

In this paper, we are interested whether there are similarities between patients with similar motor symptoms and their motor experiences of daily living, therefore we focus on two parts of MDS-UPDRS:

- MDS-UPDRS Part II questionnaire contains 13 questions, covering evaluation of patients' problems concerning swallowing, facial expression, dressing, freezing of gait, etc.
- MDS-UPDRS Part III questionnaire consists of 35 questions, covering the most characteristic symptoms of Parkinson's disease patients, i.e. the motor symptoms, including tremor, rigidity, slowness of movement, and in the later and more severe stages of the disease, postural instability.

3   Methodology

Our goal is to discover and describe groups of Parkinson's disease patients with similar symptoms pertained in different symptom views. In this section we propose a methodology that identifies subsets of symptoms, maximizing the similarity between groups of patients from different symptom views. Currently, the methodology works with two views, while the extension to multiple views is planned for further work. This section first presents the input data in Sect. 3.1, followed by the presentation of the five-step methodology in Sect. 3.2 and the proposed adaptation of the ReliefF feature ranking algorithm in Sect. 3.3.

3.1   Input Data

The input to the proposed methodology is the data described by a pair of views, each describing different aspects of life of a Parkinson's disease patient. The data set consists of patient data, where an individual instance (a row in a data table), marked as p_{ij}, corresponds to patient's p_i symptoms gathered at time j, denoting the patient's j-th consecutive visit to the clinician. The columns in table MDS-UPDRS Part II correspond to the symptoms gathered in the MDS-UPDRS Part II questionnaire, while the columns in table MDS-UPDRS Part III correspond to the symptoms gathered in the MDS-UPDRS Part III questionnaire. Note that for each patient p_i, each table contains several instances/rows, each corresponding to a separate consecutive visit to the clinician.

In this work we consider individual instances p_{ij} as being independent, regardless of the fact that instances p_{ij} refer to the same patient p_i, whose data has been gathered at visits to the clinician at consecutive time stamps $j = 1, 2, \ldots n$. In the two data tables, the instances are aligned across the two views, i.e. the rows with the same ID in the input data sets present the status of the same patient p_i from two different viewpoints in terms of their respective symptoms described in Part II and Part III symptoms data, respectively.

3.2 Methodology Description

The proposed methodology, illustrated in Fig. 1, consists of six steps, described below.

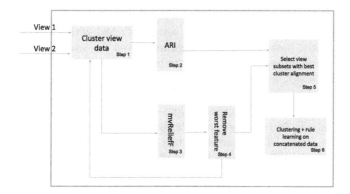

Fig. 1. Outline of the methodology for determining medications change scenarios in PPMI data using mvReliefF algorithm for multi-view feature selection, and rule learning based on the obtained aligned clusters from two views.

In the first step, performed separately for each of the two views, we perform clustering of data instances. In the experiments we used the *sklearn* implementation of the Birch hierarchical clustering algorithm [15] and set the number of clusters to 3 (this number of clusters was selected based on the results from our previous work [14]).[2]

In the second step, for the three clusters in each of the two separate views, we evaluate the quality of cluster alignment, calculated using Adjusted Rand Index (ARI) [3]. The original Rand Index [8] computes a similarity measure between two clusterings (vectors of cluster labels) by considering all pairs of samples and counting pairs that are assigned in the same or different clusters in the two cluster vectors. The Rand Index (RI) score is then "adjusted for chance" into the ARI score using the generalized hypergeometric distribution as a model of randomness. The ARI has expected value of 0 for random distribution of clusters, and value 1 for perfectly matching clusterings. ARI can also be negative.

In the third step, performed separately for each of the two views, we apply the developed mvReliefF algorithm for multi-view feature evaluation to determine the quality of features. In feature evaluation, class labels correspond to the three clusters, defined in the first step of the algorithm. Details of the proposed mvReliefF algorithm are provided in Sect. 3.3.

---

[2] Alternatively, the clustering algorithm and the number of clusters could be determined based on the value of the silhouette score [11]—a normalized measure for cluster quality; these experiments are left for further work.

In the fourth step, we remove the lowest weighted feature from the respective data set and repeat the steps 1–3 on the updated data sets. Note that in steps 1–3 we keep the record of all versions of data sets and of the quality of cluster alignments as calculated with the adjusted rand index (ARI). The loop is concluded when the number of features in each of the two views is reduced to two.

In the fifth step, we select the data sets, described with subsets of features that led to the best ARI cluster alignment, and concatenate the views—i.e. subsets of the input features with the best cluster alignment.

In the sixth step, we perform clustering on the new concatenated data set to get new clustering of patients, and perform classification rule learning to characterize these subgroups of patients. Here, the cluster labels are treated as class labels, and we use only the most important features. The choice of a classification learning algorithm is left to the user—in our experiments we used the Weka implementation of the Ripper rule learning algorithm [1], and the number of clusters was set to 3. These choices serve us to present the methodology and results. In future work we will present a systematic parameter tuning and comparison to relevant related methods.

3.3   The Proposed MvRefielfF Algorithm

The well-known Relief family of algorithms [10], Relief [5], ReliefF [6], and RReliefF [9], are highly competitive feature evaluation and ranking algorithms based on idea that good features distinguish between similar instances with different class labels.

The approaches from the Relief family are suitable for problems that involve feature interactions, as they do not make assumption about feature independence. The Relief algorithms randomly select an instance and find the nearest instances from the same class (*nearest hits*) and the nearest instances from different classes (*nearest misses*). When comparing attribute values of near instances, the algorithms rewards features that separate instances with different class labels and punishes features that separate instances with the same class label. The procedure is repeated for large enough sample of randomly chosen instances (or for all instances in case of small data sets). The algorithm returns weights between -1 and 1 for all attributes, where weights close to 0 or negative weights indicate uninformative attributes.

In the proposed mvReliefF algorithm, we adapted the ReliefF [10] algorithm to a multi-view scenario by defining the constraints for determining the closest neighbors of a chosen instance. For each view, the *nearest hit* of a given randomly chosen instance is defined as an instance that is close to the given instance and has the same class label as the given instance in both views. The instances that are close to the chosen instance, and are in both views labeled with different class labels than the chosen instance are designated as *nearest misses*.

In our implementation of mvReliefF, the selected similarity measure is the Euclidean distance, and the default number of nearest hits and nearest misses is set to 5.

Table 1. Statistics of the MDS-UPDRS Part II and MDS-UPDRS Part III data sets.

	MDS-UPDRS Part II	MDS-UPDRS Part III
Original number of features	13	35
Cluster quality using original features	0.2367	0.0522
Number of best features	12	15
Cluster quality using best features	0.2244	0.1429

4   Data Analysis Results

We tested the proposed methodology on the data presented in Sect. 2. We investigate the quality of clusters and the quality of cluster alignment for the selected pair of views (MDS-UPDRS Part II, MDS-UPDRS Part III)[3], each view describing different sets of symptoms affecting the Parkinson's disease patients.

The data sets consist of 1,345 instances, which our methodology divided into three clusters with sizes $(c_0, c_1, c_2) = (680, 386, 279)$. Table 1 presents the statistics of the original number of features, the quality of clusters obtained on the original data sets, the number of best features (according to mvReliefF, leading to clusters with best cluster alignments according to ARI—see Step 5 of the proposed methodology) describing the data set used for clustering in Step 6 of the proposed methodology, and the quality of clusters obtained on the data set described with the best features.

The quality of clusters on the newly obtained concatenated data set is improved over quality of clusters on the original concatenated data. We performed feature evaluation using the ReliefF algorithm on the original concatenated data set to select its 28 best features. The quality of clusters on this data set is lower than the quality of clusters of both the original and our new concatenated data set. The quality of clustering is calculated with the silhouette score.

Table 2 presents an ordered list of the most important features from MDS-UPDRS Part II (upper half of the table) and MDS-UPDRS Part III (bottom half of the table) according to the mvReliefF approach. In parenthesis, we present the weight of each feature normalized by the weight of the most important feature from each data set[4]. In their study, Hauser and McDermott showed that bradykinesia, rigidity, and bulbar symptoms (symptoms affecting the muscles of the throat, tongue, jaw and face) are associated with cluster membership[5].

---

[3] We chose the pair (MDS-UPDRS Part II, MDS-UPDRS Part III) for this analysis as it has the highest similarity with the cluster labels of groups of patients with similar overall status presented in [14].

[4] For reference, prior to normalization, feature NP2FREZ from MDS-UPDRS Part II has weight 42485.06, while feature NP3HMOVR from MDS-UPDRS Part III has weight of 32853.21.

[5] https://www.michaeljfox.org/grant/defining-pd-subtypes-based-patterns-long-term-outcome.

Table 2. Ordered list of the most important features and their descriptions from MDS-UPDRS Part II (upper half of the table) and MDS-UPDRS Part III (bottom half of the table) according to the mvReliefF. The most important features are on top of the table halves (NP2FREZ and NP3MOVR). The importance of each feature represents its weight normalized by the weight of the most important feature from each data set.

Feature name	Importance	Feature description
NP2FREZ	1.0000	Freezing
NP2HOBB	0.6641	Doing hobbies and other activities
NP2HWRT	0.4861	Handwriting
NP2WALK	0.4623	Walking and balance
NP2EAT	0.4420	Eating tasks
NP2SALV	0.3972	Saliva + drooling
NP2DRES	0.3928	Dressing
NP2TURN	0.3910	Turning in bed
NP2HYGN	0.3774	Hygiene
NP2SPCH	0.3417	Speech
NP2RISE	0.3265	Getting out of bed/car/or deep chair
NP2TRMR	0.2669	Tremor
NP3HMOVR	1.0000	Hand movements - right hand
NP3BRADY	0.7240	Global spontaneity of movement
NP3FACXP	0.7186	Facial expression
NP3SPCH	0.7164	Speech
NP3RIGN	0.6889	Rigidity - neck
NP3PRSPR	0.6766	Pronation-supination - right hand
NP3HMOVL	0.6062	Hand movements - left hand
NP3RTCON	0.6027	Constancy of rest
NP3POSTR	0.5935	Posture
NP3RIGLU	0.5852	Rigidity - lue
NP3FTAPL	0.5537	Finger tapping left hand
NP3TTAPR	0.5047	Toe tapping - right foot
NP3PRSPL	0.4344	Pronation-supination - left hand
NP3LGAGL	0.4343	Leg agility - left leg
NP3TTAPL	0.4303	Toe tapping - left foot

Nine of the reported most important symptoms[6] were also reported as the most influential symptoms in [13]. Classification rules describing clusters obtained on the data set concatenated from the MDS-UPDRS Part II and MDS-UPDRS Part III are presented in Table 4 in the Appendix.

Results from Table 4 show that patients assigned to cluster c_2 have problems with their motor symptoms, but do not experience problems with their motor activities—motor experiences of daily living. The rules for cluster c_1 suggest that these are patients who experience worsening of their motor symptoms and also problems with performing some of their daily motor activities. This is also evident from the average value of the sums of symptoms severity from MDS-UPDRS Part II and MDS-UPDRS Part III, presented in Table 3. Results suggest that the patients can be divided into groups that can be ordered according to the severity of the experienced motor symptoms and their affect on the patients' motor aspects of daily living. The worsening of symptoms NP2HOBB and NP2HWRT from MDS-UPDRS Part II can be seen as non-medical indicator of worsening of the patient's quality of live. Note that the study and results presented in this paper are in their preliminary state. More investigation is needed into the significance of the discovered groups and the quality of results produced by the presented method. Temporal analysis and analysis by medical specialists will be included in a later stage.

Table 3. Average sum of symptoms severity on cluster level and the corresponding standard deviation. The sum of symptoms severity from MDS-UPDRS Part II has a range of values from 0 to 52. The sum of symptoms severity from MDS-UPDRS Part III has range of values 0–138. Higher values indicate higher severity of symptoms.

	MDS-UPDRS Part II Sum	MDS-UPDRS Part III Sum
Cluster $c2$	5.3763 ± 3.4096	26.2158 ± 8.3822
Cluster $c1$	14.4663 ± 5.6495	36.2409 ± 10.6907
Cluster $c0$	6.4963 ± 3.6220	19.8265 ± 9.9557

5   Conclusions

This paper presents a multi-view clustering methodology that utilizes an adapted multi-view version of the ReliefF feature selection algorithm. We propose an adaptation of ReliefF, mvReliefF, which calculates the weight of features based on the values of their neighbors over multiple views. This enables the selection of high quality features favoring the alignment of clusters between the two views. The current approach works for two views simultaneously, but can be extended to multiple views.

---

[6] NP3BRADY, NP3TTAPL, NP3RTCON, NP3FACXP, NP3FTAPL, NP3PRSPL, NP3TTAPR, NP3PRSPR, and NP2HWRT.

We applied the proposed methodology to pairs of PPMI questionnaires (views), each describing different groups of symptoms of Parkinson's disease patients. For the given pair of views, we were able to identify lists of most influential symptoms—symptoms that produce the most similar divisions of patients from each view. The final clustering on both views is performed on the data set obtained by concatenation of the subsets of the most influential features.

Pair (MDS-UPDRS Part II, MDS-UPDRS Part III) of data sets resulted in the discovery of groups of patients that were the closest to the groups of patients detected in [14]. The proposed methodology detected 9 out of the 10 most influential symptoms from MDS-UPDRS Part II and MDS-UPDRS Part III reported in [13]. The description of the detected subgroups with classification rules revealed that the detected groups can be ordered based on the severity of symptoms as experienced by the involved patients, similar to the conclusions from [14]. In future work we will focus on adapting the proposed methodology to more than two views and altering the final clustering step to obtain better quality clusters.

Acknowledgments. We are grateful to Tadej Magajna who implemented the mvRe-liefF algorithm as part of his M.Sc thesis under supervision of Marko Robnik-Šikonja. The research was supported by the Slovenian Research Agency (research core funding programs P2-0209, P6-0411 and P2-0103, and project N2-0078) and the Slovenian Ministry of Education, Science and Sport (project R 2.1 - Public call for the promotion of researchers at the beginning of a career 2.1). Data used in the preparation of this article were obtained from the Parkinson's Progression Markers Initiative (PPMI) (www.ppmi-info.org/data).

A   Rules Describing Clusters of Patients on the Concatenated Data Set

Classification rules describing clusters obtained on the data set concatenated from the MDS-UPDRS Part II and MDS-UPDRS Part III are presented in Table 4. The rules are constructed using the Weka implementation of the Ripper [1] algorithm, with its default parameters.

Table 4. Rules describing clusters obtained by Birch clustering on the concatenated data set of most important attributes from MDS-UPDRS Part II and MDS-UPDRS Part II. Variables p and n denote the number of covered true positive and false positive examples respectively.

Rule	Cluster label	p	n
(NP2HWRT $\leq$ 0) & (NP3RIGLU $\geq$ 2) & (NP2RISE $\leq$ 0) & (NP3LGAGL $\geq$ 1)	$\Rightarrow$ cluster = c_2	62	3
(NP3HMOVL $\geq$ 1) & (NP3PRSPR $\leq$ 0) & (NP2WALK $\leq$ 0) & (NP3PRSPL $\geq$ 1) & (NP2HWRT $\leq$ 1) & (NP3RIGN $\geq$ 1) & (NP3FTAPL $\leq$ 2	$\Rightarrow$ cluster = c_2	32	3
(NP3FTAPL $\geq$ 1) & (NP2EAT $\leq$ 0) & (NP3PRSPL $\geq$ 2) & (NP3POSTR $\leq$ 0)	$\Rightarrow$ cluster = c_2	39	6
(NP3PRSPL $\geq$ 1) & (NP2HWRT $\leq$ 0) & (NP3HMOVR $\leq$ 0) & (NP3RIGLU $\geq$ 2) & (NP3POSTR $\leq$ 1) & (NP3TTAPR $\leq$ 0)	$\Rightarrow$ cluster = c_2	23	3
(NP3PRSPL $\geq$ 1) & (NP2HYGN $\leq$ 0) & (NP2TURN $\leq$ 0) & (NP3FTAPL $\geq$ 2) & (NP3HMOVR $\leq$ 1) & (NP2RISE $\leq$ 0) & (NP2HWRT $\leq$ 1)	$\Rightarrow$ cluster = c_2	26	4
(NP3FTAPL $\geq$ 1) & (NP2EAT $\leq$ 0) & (NP3RIGLU $\geq$ 1) & (NP3SPCH $\leq$ 0) & (NP3HMOVL $\geq$ 2) & (NP2TURN $\leq$ 1)	$\Rightarrow$ cluster = c_2	11	1
(NP3HMOVR $\leq$ 0) & (NP3LGAGL $\geq$ 1) & (NP2SALV $\leq$ 0) & (NP3RIGLU $\geq$ 1) & (NP3HMOVL $\leq$ 1) & (NP2SPCH $\leq$ 0) & (NP3RTCON $\geq$ 2)	$\Rightarrow$ cluster = c_2	13	2
(NP3FTAPL $\geq$ 1) & (NP2EAT $\leq$ 0) & (NP3RIGLU $\geq$ 2) & (NP3TTAPL $\leq$ 1) & (NP2HWRT $\leq$ 1)	$\Rightarrow$ cluster = c_2	11	3
(NP3TTAPL $\geq$ 2) & (NP3FACXP $\leq$ 1) & (NP2TURN $\leq$ 0) & (NP3HMOVR $\leq$ 0) & (NP3RIGN $\geq$ 1)	$\Rightarrow$ cluster = c_2	14	2
(NP3PRSPL $\geq$ 1) & (NP3HMOVR $\leq$ 0) & (NP2HWRT $\leq$ 1) & (NP2SPCH $\geq$ 1) & (NP2SPCH $\leq$ 1) & (NP3PRSPL $\leq$ 2) & (NP3TTAPL $\geq$ 2)	$\Rightarrow$ cluster = c_2	13	3
(NP3FTAPL $\geq$ 1) & (NP2EAT $\leq$ 0) & (NP3BRADY $\geq$ 2) & (NP2TURN $\leq$ 0) & (NP3HMOVR $\leq$ 1) & (NP2SPCH $\geq$ 2)	$\Rightarrow$ cluster = c_2	8	1
(NP3PRSPL $\geq$ 2) & (NP2WALK $\geq$ 1) & (NP3LGAGL $\geq$ 2)	$\Rightarrow$ cluster = c_1	111	2
(NP3HMOVL $\geq$ 1) & (NP2HYGN $\geq$ 1) & (NP2SPCH $\geq$ 2) & (NP3RIGLU $\geq$ 2)	$\Rightarrow$ cluster = c_1	420	
(NP3HMOVL $\geq$ 1) & (NP2HYGN $\geq$ 1) & (NP2SALV $\geq$ 2) & (NP2EAT $\geq$ 1) & (NP2RISE $\geq$ 1)	$\Rightarrow$ cluster = c_1	39	1
(NP3PRSPL $\geq$ 2) & (NP2SPCH $\geq$ 1)	$\Rightarrow$ cluster = c_1	56	11
(NP3BRADY $\geq$ 2) & (NP2FREZ $\geq$ 1)	$\Rightarrow$ cluster = c_1	37	5
(NP3LGAGL $\geq$ 1) & (NP2HOBB $\geq$ 1) & (NP2DRES $\geq$ 2)	$\Rightarrow$ cluster = c_1	24	6
(NP3SPCH $\geq$ 1) & (NP2DRES $\geq$ 1) & (NP3LGAGL $\geq$ 1) & (NP3FTAPL $\geq$ 2)	$\Rightarrow$ cluster = c_1	39	11
(NP2SPCH $\geq$ 2) & (NP2RISE $\geq$ 2)	$\Rightarrow$ cluster = c_1	14	2
(NP2HOBB $\geq$ 1) & (NP2SPCH $\geq$ 2) & (NP2SALV $\geq$ 3) & (NP2HWRT $\geq$ 2)	$\Rightarrow$ cluster = c_1	9	1
(NP3PRSPL $\geq$ 2) & (NP2TURN $\leq$ 0) & (NP3TTAPL $\geq$ 1)	$\Rightarrow$ cluster = c_1	13	4
(NP2FREZ $\geq$ 1) & (NP2SALV $\geq$ 3)	$\Rightarrow$ cluster = c_1	6	2
(NP3RIGN $\geq$ 2) & (NP3HMOVL $\geq$ 2) & (NP2EAT $\geq$ 2)	$\Rightarrow$ cluster = c_1	4	0
(NP2TURN $\geq$ 2) & (NP2SPCH $\geq$ 1)	$\Rightarrow$ cluster = c_1	6	1
(NP3FACXP $\geq$ 3) & (NP2SPCH $\geq$ 2)	$\Rightarrow$ cluster = c_1	2	0
(NP3BRADY $\geq$ 2) & (NP3RTCON $\geq$ 4) & (NP2HYGN $\geq$ 1)	$\Rightarrow$ cluster = c_1	3	0
	$\Rightarrow$ cluster = c_0	688	50

References

1. Cohen, W.W.: Fast effective rule induction. In: Proceedings of the 12th International Conference on Machine Learning, pp. 115–123 (1995)
2. Goetz, C., et al.: Movement disorder society-sponsored revision of the unified Parkinson's disease rating scale (MDS-UPDRS): scale presentation and clinimetric testing results. Mov. Disord. Soc. **23**(15), 2129–2170 (2008)
3. Hubert, L., Arabie, P.: Comparing partitions. J. Classif. **2**(1), 193–218 (1985)

4. Jankovic, J., et al.: Variable expression of Parkinson's disease - a base-line analysis of the DATATOP cohort. Neurology **40**(10), 1529 (1990)
5. Kira, K., Rendell, L.A.: The feature selection problem: traditional methods and a new algorithm. In: Proceedings of AAAI 1992, vol. 2, pp. 129–134 (1992)
6. Kononenko, I.: Estimating attributes: analysis and extensions of RELIEF. In: Bergadano, F., De Raedt, L. (eds.) ECML 1994. LNCS, vol. 784, pp. 171–182. Springer, Heidelberg (1994). https://doi.org/10.1007/3-540-57868-4_57
7. Marek, K., et al.: The Parkinson's Progression Markers Initiative (PPMI). Prog. Neurobiol. **95**(4), 629–635 (2011)
8. Rand, W.M.: Objective criteria for the evaluation of clustering methods. J. Am. Stat. Assoc. **66**(336), 846–850 (1971)
9. Robnik-Šikonja, M., Kononenko, I.: An adaptation of Relief for attribute estimation in regression. In: Machine Learning: Proceedings of the Fourteenth International Conference (ICML 1997), pp. 296–304 (1997)
10. Robnik-Šikonja, M., Kononenko, I.: Theoretical and empirical analysis of ReliefF and RReliefF. Mach. Learn. **53**(1–2), 23–69 (2003)
11. Rousseeuw, P.J.: Silhouettes: a graphical aid to the interpretation and validation of cluster analysis. J. Comput. Appl. Math. **20**, 53–65 (1987)
12. Thenganatt, M.A., Jankovic, J.: Parkinson disease subtypes. JAMA Neurol. **71**(4), 499–504 (2014)
13. Valmarska, A., Miljkovic, D., Konitsiotis, S., Gatsios, D., Lavrač, N., Robnik-Šikonja, M.: Symptoms and medications change patterns for Parkinson's disease patients stratification. Artif. Intell. Med. **91**, 82–95 (2018)
14. Valmarska, A., Miljkovic, D., Lavrač, N., Robnik-Šikonja, M.: Analysis of medications change in Parkinson's disease progression data. J. Intell. Inf. Syst. **51**(2), 301–337 (2018)
15. Zhang, T., Ramakrishnan, R., Livny, M.: Birch: an efficient data clustering method for very large databases. ACM Sigmod Record **25**(2), 103–114 (1996)

Unsupervised Grammar Induction for Revealing the Internal Structure of Protein Sequence Motifs

Olgierd Unold[1]([✉]) [ID], Mateusz Gabor[1] [ID], and Witold Dyrka[2] [ID]

[1] Department of Computer Engineering,
Wrocław University of Science and Technology, Wrocław, Poland
olgierd.unold@pwr.edu.pl, mateuszgabor95@gmail.com
[2] Department of Biomedical Engineering,
Wrocław University of Science and Technology, Wrocław, Poland
witold.dyrka@pwr.edu.pl

Abstract. Protein sequence motifs are conserved amino acid patterns of biological significance. They are vital for annotating structural and functional features of proteins. Yet, the computational methods commonly used for defining sequence motifs are typically simplified linear representations neglecting the higher-order structure of the motif. The purpose of the work is to create models of sequence motifs taking into account the internal structure of the modeled fragments. The ultimate goal is to provide the community with accurate and concise models of diverse collections of remotely related amino acid sequences that share structural features. The internal structure of amino acid sequences is modeled using a novel algorithm for unsupervised learning of weighted context-free grammar (WCFG). The proposed method learns WCFG both form positive and negative samples, whereas weights of rules are estimated using a novel Inside-Outside Contrastive Estimation algorithm. In comparison to existing approaches to learning CFG, the new method generates more concise descriptors and provides good control of the trade-off between grammar size and specificity. The method is applied to the nicotinamide adenine dinucleotide phosphate binding site motif.

Keywords: Grammar inference · Weighted context-free grammar · Unsupervised learning · Statistical methods · Protein sequence motifs · Amino acid patterns

1 Introduction

Protein sequence motifs are amino acid patterns conserved due to their biological significance [2]. Variability within a family of motifs is constrained by the requirement to maintain their functional or structural role. Even a single point mutation may disrupt the spatial fold of the motif or its propensity to a ligand, potentially causing a severe disorder [16]. Typical sequence motifs are linear stretches of ten to several dozens amino acids. Exemplary protein motifs of clinical importance are sequence patterns forming ligand binding sites [17].

© Springer Nature Switzerland AG 2020
M. Michalowski and R. Moskovitch (Eds.): AIME 2020, LNAI 12299, pp. 299–309, 2020.
https://doi.org/10.1007/978-3-030-59137-3_27

They may form a virtually whole structural feature, as in a case of some small ligand binding sites (e.g. EF hand or zinc finger), or just a part of it. In fact, searching for conserved motifs is one of the major steps in automated functional annotation of proteins [28]. Thus, the coverage and accuracy of annotations depend on highly sensitive and specific modelling methods of protein motifs.

Protein motifs are typically modelled as gapless or gapped linear stretches of residues (amino acids) without considering dependencies between particular positions [2]. The PROSITE patterns are the classic and still widely used example [23]. This representation is most efficient for highly conserved and longer patterns but cannot maintain high sensitivity and specificity for shorter and more diverse motifs. This also applies to probabilistic models, such as PROSITE profiles and profile Hidden Markov Models (HMM) [10]. The latter, while excelling in modeling longer domains, lack statistical power when dealing with short sequences due to insufficient information encoded the order of amino acids. A natural extension to profile HMMs are Random Markov Fields (RMF) and related models, which can capture information conveyed in the inter-position correlations [14]. However, RMFs cannot be used for searching gapped motifs due to combinatory explosion [21]. While there is ongoing research on heuristic approach to the problem, the solution is not available so far [27]. Moreover, both profile HMMs and RMFs require multiple sequence alignment for inferring their parameters. This effectively rules out modeling meta-families of motifs whose members do not share relevant homology yet still share the related principle.

An alternative line of research consists on using grammatical models which offer more flexibility needed for modeling nonalignable collections of motifs. A rare example of such an approach is Protomata, which can represent diverse families of motifs but as equivalent to the probabilistic regular grammar cannot directly take into account inter-position correlations [4]. This feature requires at least Context-Free Grammars (CFG), which are best known in bioinformatics for predicting RNA secondary structures [18]. In the realm of proteins, it has been shown recently that probabilistic CFGs are capable of representing a meta-family of evolutionarily unrelated Calcium binding sites [9]. However, CFGs, whether probabilistic or not, are notoriously difficult to learn from unstructured sequence data, which hampers their applications to bioinformatics. This works proposes a new unsupervised grammar induction approach for (weighted) CFG learning, with the goal of generating concise and accurate representation of proteins sequence motifs from positive and negative samples. As constructing the negative set is difficult, we show how the method benefits even from a very limited and crude sample.

The task of learning WCFGs and probabilistic context-free grammars (PCFGs) is divided into two subproblems: determining a discrete structure of the target grammar and estimating weights or probabilities of rules in the grammar. Unsupervised structure learning CFG is known to be a hard task [12], but is more practical than supervised learning due to the lack of annotated data (like a treebank or structured corpus). Unsupervised grammatical inference methods employ often the idea of substitutability, in which it applies replacing strings in the same contexts. From several methods available we mention here ABL [32], EMILE [1], ADIOS [25], or LS [33]. To establish stochastic parameters in the grammar, the Bayesian approach [15] or maximum likelihood estimation [20] are typically applied.

The novel algorithm for unsupervised learning of WCFG has been recently introduced [31]. Weighted Grammar-based Classifier system (WGCS) is one of the few grammatical inference approaches learning both structure and stochastic grammar parameters. Initially, the method was dedicated to learning crisp context-free grammar [29], and next was extended to a fuzzy version [30].

The main contribution of this paper is in defining and testing a new version of the WGCS approach, in which the substitutability concept has been employed. Moreover, the scope of direct negative evidence has been extended over the all available sets of negative samples. The paper also describes results from applying the proposed method on a set of protein sequence motif. The rest of the paper is organised as follows. Sections 2 gives some details about our approach. In Sect. 3, we present test environment, while the results are reported in Sect. 4. Section 5 concludes the paper.

2   WGCS

The Weighted Grammar-based Classifier System receives as the input labeled sentences, as the output the WCFG is returned. The core of the system is CKY (Cocke-Kasami-Younger) parsing algorithm that classifies whether a sentence belongs to grammar or not. CKY operates under the idea of dynamic programming, and its computational complexity is $O(n^3|G|)$, where n is the sentence length and $|G|$ is the grammar size. To discover new non-terminal rules in a grammar, a split algorithm is engaged.

The initial grammar is generated from two non-terminal symbols, start symbol S and one non-terminal symbol marked as A. Non-terminal rules are created by all possible combinations of the two initial non-terminal symbols A, S. Terminal rules are created by assigning each separate word (terminal symbol) from a dataset to a non-terminal symbol A in the form $A \rightarrow t$ where $t \in T$.

The WGCS, contrary to other approaches, makes use of direct negative evidence in learning WCFGs. Direct negative evidence is derived from language acquisition theory and depicts all ungrammatical sentences exposed to a language learner. In the case of WGCS, direct negative evidence covers negative sentences from the training dataset. Inspired by the research of Smith and Eisner [24], we extended the Inside-Outside algorithm by introducing negative sentences into the estimation mechanism, calling this method *Inside-Outside Contrastive Estimation* (IOCE) The main idea of IOCE is to move the weight mass from the direct negative evidence to positive one.

To prevent the grammar from growing too much and to improve its quality, grammar rules with weighs less than 0.001 are removed from the population (this value was determined experimentally). Note that the learned weights are not used in a further classification but only to prune the induced grammar.

2.1   Preliminaries

In this section, we introduce the concept of weighted context-free grammar. For this, we recall the notion of context-free grammar. A context-free grammar is a quadruple (N, T, S, R), where N - a finite set of non-terminals symbols disjoint from T, T - a finite set of terminals symbols, $S \in N$ is called the start symbol, R - a finite set of rules of the form $X \rightarrow \alpha$, where $X \in N$ and $\alpha \in (N \cup T)^*$. CFG is in Chomsky Normal Form (CNF) when each rule takes one of the two following forms: $X \rightarrow Y\ Z$ where $X, Y, Z \in N$ or $X \rightarrow t$ where $X \in N$ and $t \in T$.

A Weighted Context Free Grammar (WCFG) associates a positive number called the weight with each rule in R (assigning a weight of zero to a rule equates to excluding it from R). More formally the WCFG is a 5-tuple (N, T, S, R, W) where (N, T, S, R) is a CFG and W is a finite set of weights of each rule as the result of a function $\phi(X \rightarrow \alpha) \rightarrow w$ where $X \rightarrow \alpha \in R$ and $w > 0$ is a positive weight.

2.2   Split Algorithm

Here, we propose a new version of WGCS approach, in which the substitutability concept has been added. Instead of a genetic algorithm, model splitting has been proposed. Model splitting starts from a general language, and the learning algorithm searches for a more specific language, given the input sequences. As stated, splitting is the inverse of merging, in which we start with the most specific hypothesis, and by repeated merging two concepts into one, a more general hypothesis is obtained.

This method is strongly inspired by the works of [13,19]. In this approach, grammar is induced in an incremental way. We start with a small initial grammar with a small number of non-terminal symbols, adding a new non-terminal symbol (and relevant new grammar rules) in each iteration until the stop criterion is reached.

During each iteration of the algorithm, a new non-terminal symbol X_j from another non-terminal symbol X_i is created by *splitting* operation. The non-terminal symbol X_i selected for *splitting* operation is the symbol that is most often used (i.e., has the largest count) in parsing a dataset. This symbol is often called the split symbol. Then, for all terminal rules $X_i \rightarrow t$, the operation creates new terminal rules by replacing X_i with X_j. Next, new non-terminal rules are created in two ways:

1. Create all possible rules only two non-terminal symbols X_i and X_j in the form $X_a \rightarrow X_b, X_c$ where X_a, X_b or X_c is X_i or X_j. The number of such rules is always eight, because that is the number of all possible combinations of two non-terminal symbols for rule in CNF. For example for the split symbol A and the new non-terminal O, there are following rules are created: $A \rightarrow AA$, $A \rightarrow AO$, $A \rightarrow OA$, $A \rightarrow OO$, $O \rightarrow OO$, $A \rightarrow AO$, $O \rightarrow OA$, $O \rightarrow AA$.

2. For all non-terminal rules with X_i in the form $X_a \rightarrow X_b, X_c$, where X_a, X_b or X_c is X_i, create new non-terminal rules of the same form by replacing X_i with X_j. For multiple occurrences of X_i in rule, create all combinations. Consider two cases with the split symbol A and the new non-terminal symbol O. The first one is for a single occurrence of the split symbol in the rule, the second one for many.
   a) For a rule $A \rightarrow BC$ create new rule $O \rightarrow BC$.
   b) For rule $A \rightarrow AB$ create three new rules: $O \rightarrow OB$, $A \rightarrow OB$, and $O \rightarrow AB$.

2.3   Inside-Outside

Having established the topology of grammar, one can find the set of rules weights. The inside outside algorithm originally estimates the probabilities of rules, in our case we use it to estimate weights. The IO algorithm starts from some initial parameters setting, and iteratively updates them to increase the likelihood of the data (the training corpus). To estimate the rule probability, the algorithm counts the inside and outside probability. 4 The inside probability is the probability of deriving a particular substring from the given sentence $w_i \ldots w_j$ from a given left-side symbol $\alpha_{ij}(A) = P(A \longrightarrow w_i \ldots w_j)$, where A is any non-terminal symbol. The outside probability is the probability of deriving from the start symbol of the substring $w_i \ldots w_{i-1} A w_{j+1} \ldots w_n$ $\beta_{ij}(A) = P(S \longrightarrow w_1 \ldots w_{i-1} A w_{j+1} \ldots w_n)$.

Having the inside α and outside β probabilities for every sentence w_i in the training corpus, the occurrences of a given rule for a single sentence is calculated for non-terminal symbols: $c_\varphi(A \longrightarrow BC, W) = \frac{\varphi(A \longrightarrow BC)}{P(W)} \sum_{1 \le i \le j \le k \le n} \beta_{ik}(A) \alpha_{ij}(B) \alpha_{j+1,k}(C)$, and for terminal symbols: $c_\varphi(A \longrightarrow w, W) = \frac{\varphi(A \longrightarrow w)}{P(W)} \sum_{i \le 1} \beta_{ii}(A)$, where $P(W) = P(S \longrightarrow w_1 w_2 \ldots w_n)$ is the probability of deriving a sentence.

For each rule $A \longrightarrow \alpha$ the number $c_\varphi(A \longrightarrow \alpha, W_i)$ is added to the total count $count(A \longrightarrow \alpha) = \sum_{i=1}^{n} c_\varphi(A \longrightarrow \alpha, W_i)$ and then proceed to the next sentence.

After processing each sentence in this way the parameters are re-estimated to obtain new probability of the rule (maximization) $\varphi'(A \longrightarrow \alpha) = \frac{count(A \longrightarrow \alpha)}{\sum_\lambda count(A \longrightarrow \lambda)}$, where $count(A \longrightarrow \lambda)$ is as rule with the same left-hand symbol. One of the disadvantages of the algorithm is its computational complexity which is $O(n^3)$ both in terms of sentence length and the number of non-terminal symbols [20].

2.4   Inside-Outside Contrastive Estimation

We extended the Inside-Outside algorithm by introducing negative samples into the estimation mechanism, calling this method *Inside-Outside Contrastive Estimation* (IOCE). The main idea of IOCE is to move the weight mass from the direct negative evidence to positive one.

In IOCE method we introduce negative estimation factor:
$\psi(A \longrightarrow \alpha) = \frac{count(A \longrightarrow \alpha)}{count(A \longrightarrow \alpha) + count_{negative}(A \longrightarrow \alpha)}$, where: $count_{negative}(A \longrightarrow \alpha)-$
the estimated counts that a particular rule is used in negative samples in the
training dataset. The new weight of the rule is calculated as: $\varphi'(A \longrightarrow \alpha) = \frac{count(A \longrightarrow \alpha)}{\sum_{\beta} count(A \longrightarrow \beta)} \cdot \psi(A \longrightarrow \alpha)$.

2.5   WCFG Learning Algorithm

The first step in grammar induction with WGCS (see Algorithm 1) is to initialize
the grammar. During each induction step, new rules are added to grammar
through the operation of a split algorithm. Then the stochastic algorithm tunes
the weight values of all rules in grammar. The rule weight estimation algorithm
was stopped when the largest rule weight difference among all rules compared
to the previous iteration was less than 0.001. We end the main loop by cleansing
the grammar of the rules with low weights. The stop criterion in our experiment
is perform 30 iterations. Algorithm 1 will provide a description of the above
method in the form of a pseudo-code.

3   Test Environment

We evaluated the method through modeling a protein sequence motif based on
the PROSITE PS00063 pattern of the Nicotinamide Adenine dinucleotide Phosphate (NAP) binding site fragment from an aldo/keto reductase family [3]. The
positive sample (termed *NAP_pos*) was collected according to PS00063 true positive and false negative hits (four least consistent sequences were excluded) [9].
The set was complemented with a negative sample of 29 false positives matches
to PS00063 (*NAP_fp*), which can be seen as close neighborhood of the sample.
Within the sample, all sequences shared the same length of 16 amino acids, which
avoided sequence length effects on grammar scores (this could be resolved with
a null model). In addition, we used a large negative sample designed to roughly
approximate the entire space of protein sequences based on the negative set from
[7] cut into overlapping subsequences of the length of positive sequences [9]. The
resulting set *NAP_neg* consisted of 47,736 sequences. All samples were made
non-redundant at the level of sequence similarity around 70% to reduce sample
distribution bias in training and to avoid information leakage in cross-validation.

The performance of the two methods used to learn weights of WCFG: the
standard Inside–Outside method (IO), and the method that makes use of direct
negative evidence through the Contrastive Estimation (IOCE) was compared in
the 5-fold cross-validation procedure. Of note is that only a small part of the
NAP_neg training fold—equal to the number of positive samples—was used for
learning in the IOCE scheme.

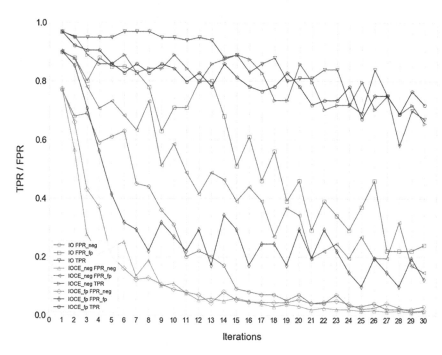

Fig. 1. Performance of unsupervised weighted CFG in the 5-fold cross-validation on protein sequence motif NAP. Average TPR and FPR values versus number of training steps are shown. IO stands for WGCS using standard Inside-Outside algorithm and only positive sample, whereas IOCE stands for WGCS using novel Inside-Outside Contrastive Estimation and direct negative evidence. The y-axis indicates the true positive rate (TPR) or false positive rate (FPR) value. (Color figure online)

Algorithm 1. Induction algorithm
_____
    Input: Training and validation set
    Output: Weighted context-free grammar
1: Initialize the grammar
2: **for** $i \leftarrow 1$ to *iterations* **do**   ▷ 30
3:    Run split algorithm
4:    **while** stop condition is not satisfied **do**
5:        Run the estimation algorithm on the training set   ▷ IO/IOCE
6:    Remove rules with low weights
7:    Evaluate grammar on the validation set
8: **return** $WCFG$
_____

4 Results

In all experiments, we used Intel Xeon CPU E5-2650 v2, 2.6 GHz, under Windows Server 2016 operating system with 62.5 GB RAM. The average execution time of induction Algorithm 1 for a single fold was 1 h and 50 min.

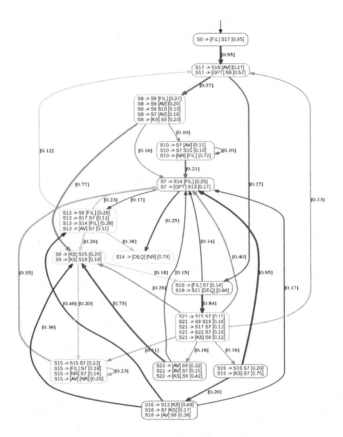

Fig. 2. Best performing grammar diagram, simplified. The grammar is pruned from rules with weight below 0.1 or rewriting rarely used non-terminals (sum of weights of rules generating the symbols below 0.1) to increase readability.

Figure 1 summarizes the performance of IO and IOCE methods during the learning process. The y-axis indicates the true positive rate (TPR, sensitivity) or false positive rate (FPR) value, depending on a curve. TPR is calculated according to the formula $\text{TPR} = tp/(tp+fn)$, whereas $\text{FPR} = fp/(fp+tn)$, and fp, tn, tp and fn are respectively the numbers of false positives, true negatives, true positives, and false negatives.

The learning process started with overly simplified grammar made with just 2 non-terminal symbols leading to FPR of around 0.80 for *NAP_neg* and 0.90 for harder *NAP_fp* set. With around ten non-terminal symbols, the IOCE scheme learned with *NAP_pos* and *NAP_neg* training sets (IOCE in Fig. 1, shades of green) achieved FPR below 0.10 and 0.50 for *NAP_neg* and *NAP_fp* test sets. Eventually, with around 30 non-terminals, the scheme approached FPR of 0.01 and 0.20, respectively. This is roughly 1.5 (*NAP_fp*) and 2 (*NAP_neg*) times lower FPR than achieved with grammars learned from the positive samples only (IO in Fig. 1, shades of red). Worth emphasizing is improvement of FPR against

NAP_fp test set achieved by simply using the unrelated *NAP_neg* set in the IOCE scheme. While using *NAP_fp* training set for learning grammars (IOCE_fp in Fig. 1, shades of blue) allowed for much lower FPR against *NAP_fp* test set in the beginning of learning, the advantage over IOCE diminished after 20th iteration. Sensitivity (TPR) of grammars obtained with IO, IOCE and IOCE_fp was in similar ranges and generally decreasing from around 0.90 for FPR against *NAP_neg* of 0.2 to below 0.70 towards the end of learning. The obtained results show that even small and crude negative dataset can improve the discriminatory power of the method (i.e., lower the FPR value for *NAP_neg* and *NAP_fp* datasets).

We analyzed the best-performing grammar obtained in the 30th iteration in one of the five cross-validation runs of the IOCE_fp scheme trained with *NAP_pos*, *NAP_fp* and *NAP_neg* (Fig. 2). This particular grammar consisted of 121 rules and achieved perfect TPR over the positive test set and perfect FPR over the *NAP_fp* negative test set, while FPR over *NAP_neg* test set was just below 0.01. The lexical rules of the grammar grouped the amino acids to six non-terminal symbols. Interestingly, there are two groups linking amino acids of high propensity to the NAP ligand binding: lysine and serine, and asparagine and arginine [5][7]. Another two non-terminal symbols group hydrophobic residues: alanine and valine, and phenyloalanine, isoleucine and leucine. Next, a non-terminal is dedicated to negatively charged aspartic and glutamic acids, and also hydrophilic glutamine. Finally, the last group consists of glycine, proline and threonine. The former two amino acids are often found in turns of the backbone chain, and indeed the PS00063 binding site involves a turn [5].

5   Conclusions

We have presented a novel approach for unsupervised learning of WCFG. The developed WGCS method with the split algorithm is based on the principles of incremental induction, starting from a small grammar with a small number of rules and non-terminal symbols, looking for a larger, more general grammar describing the training set in each subsequent step of induction. The introduced method was evaluated on a sample of protein sequence motif. We show how the method can benefit even from a very limited and crude negative sample.

Future work should examine the possibility of combining the split method with its opposite approach, i.e., the merge method [26], e.g., by alternately using them in the induction process. Another important issue would be to conduct experiments on more sophisticated amino acid patterns such as more complex binding sites and motifs forming amyloid-related oligomers and fibrils. The latter genre of motifs can be found in proteins controlling interaction with other organisms in archaea, bacteria and fungi, including species found in the human microbiome [6][8]. In the human proteome, several amyloid-forming proteins are associated with pathological conditions, such as neurodegenerative diseases and the type 2 diabetes [22]. Interestingly, the process of forming disease-related amyloids may be enhanced through cross-seeding with proteins secreted by the microbiome [11]. Searching pairs of structurally similar motifs that facilitate such interactions [22] is one of the aimed applications of the grammatical models.

Acknowledgements. The research was supported by the National Science Centre Poland (NCN), project registration no. 2016/21/B/ST6/02158.

References

1. Adriaans, P., Vervoort, M.: The EMILE 4.1 grammar induction toolbox. In: Adriaans, P., Fernau, H., van Zaanen, M. (eds.) ICGI 2002. LNCS (LNAI), vol. 2484, pp. 293–295. Springer, Heidelberg (2002). https://doi.org/10.1007/3-540-45790-9_24
2. Bailey, T.L., Elka, C.: Unsupervised learning of multiple motifs in biopolymers using expectation maximization. Mach. Learn. **21**, 51–80 (1995). https://doi.org/10.1007/BF00993379
3. Bohren, K.M., Bullock, B., Wermuth, B., Gabbay, K.H.: The aldo-keto reductase superfamily. cDNAs and deduced amino acid sequences of human aldehyde and aldose reductases. J. Biol. Chem. **264**(16), 9547–51 (1989)
4. Coste, F., Kerbellec, G.: A similar fragments merging approach to learn automata on proteins. In: Gama, J., Camacho, R., Brazdil, P.B., Jorge, A.M., Torgo, L. (eds.) ECML 2005. LNCS (LNAI), vol. 3720, pp. 522–529. Springer, Heidelberg (2005). https://doi.org/10.1007/11564096_50
5. Couture, J.F., Legrand, P., Cantin, L., Luu-The, V., Labrie, F., Breton, R.: Human 20α-hydroxysteroid dehydrogenase: crystallographic and site-directed mutagenesis studies lead to the identification of an alternative binding site for C21-steroids. J. Mol. Biol. **331**, 593–604 (2003)
6. Dyrka, W., et al.: Diversity and variability of NOD-like receptors in fungi. Genome Biol. Evol. **6**(12), 3137–3158 (2014)
7. Dyrka, W., Nebel, J.C.: A stochastic context free grammar based framework for analysis of protein sequences. BMC Bioinform. **10**, 323 (2009). https://doi.org/10.1186/1471-2105-10-323
8. Dyrka, W., et al.: Identification of NLR-associated amyloid signaling motifs in filamentous bacteria. bioRxiv p. 2020.01.06.895854, January 2020
9. Dyrka, W., Pyzik, M., Coste, F., Talibart, H.: Estimating probabilistic context-free grammars for proteins using contact map constraints. PeerJ **7**, e6559 (2019)
10. Eddy, S.R.: A probabilistic model of local sequence alignment that simplifies statistical significance estimation. PLoS Comput. Biol. **4**(5), e1000069 (2008)
11. Friedland, R.P., Chapman, M.R.: The role of microbial amyloid in neurodegeneration. PLoS Pathog. **13**, e1006654 (2017)
12. de la Higuera, C.: Grammatical Inference: Learning Automata and Grammars. Cambridge University Press, Cambridge (2010)
13. Hogenhout, W.R., Matsumoto, Y.: A fast method for statistical grammar induction. Nat. Lang. Eng. **4**(3), 191–209 (1998)
14. Hopf, T.A., Colwell, L.J., Sheridan, R., Rost, B., Sander, C., Marks, D.S.: Three-dimensional structures of membrane proteins from genomic sequencing. Cell **149**(7), 1607–21 (2012)
15. Johnson, M., Griffiths, T., Goldwater, S.: Bayesian inference for PCFGs via Markov chain Monte Carlo. In: Human Language Technologies 2007: The Conference of the North American Chapter of the Association for Computational Linguistics; Proceedings of the Main Conference, pp. 139–146 (2007)

16. Kim, P., Zhao, J., Lu, P., Zhao, Z.: mutLBSgeneDB: mutated ligand binding site gene DataBase. Nucleic Acids Res. **45**(D1), D256–D263 (2016)
17. Kinjo, A.R., Nakamura, H.: Comprehensive structural classification of ligand-binding motifs in proteins. Structure **17**(2), 234–246 (2009)
18. Knudsen, B., Hein, J.: RNA secondary structure prediction using stochastic context-free grammars and evolutionary history. Bioinformatics **15**, 446–54 (1999)
19. Kurihara, K., Sato, T.: Variational Bayesian grammar induction for natural language. In: Sakakibara, Y., Kobayashi, S., Sato, K., Nishino, T., Tomita, E. (eds.) ICGI 2006. LNCS (LNAI), vol. 4201, pp. 84–96. Springer, Heidelberg (2006). https://doi.org/10.1007/11872436_8
20. Lari, K., Young, S.J.: The estimation of stochastic context-free grammars using the inside-outside algorithm. Comput. Speech Lang. **4**(1), 35–56 (1990)
21. Lathrop, R.H.: The protein threading problem with sequence amino acid interaction preferences is NP-complete. Protein Eng. Des. Sel. **7**(9), 1059–1068 (1994)
22. Ren, B., et al.: Fundamentals of cross-seeding of amyloid proteins: an introduction. J. Mater. Chem. B **7**, 7267–7282 (2019)
23. Sigrist, C.J.A., et al.: New and continuing developments at PROSITE. Nucleic Acids Res. **41**(D1), D344–D347 (2013)
24. Smith, N.A., Eisner, J.: Guiding unsupervised grammar induction using contrastive estimation. In: Proceedings of IJCAI Workshop on Grammatical Inference Applications, pp. 73–82 (2005)
25. Solan, Z., Horn, D., Ruppin, E., Edelman, S.: Unsupervised learning of natural languages. Proc. Natl. Acad. Sci. **102**(33), 11629–11634 (2005)
26. Stolcke, A., Omohundro, S.: Inducing probabilistic grammars by Bayesian model merging. In: Carrasco, R.C., Oncina, J. (eds.) ICGI 1994. LNCS, vol. 862, pp. 106–118. Springer, Heidelberg (1994). https://doi.org/10.1007/3-540-58473-0_141
27. Talibart, H., Coste, F.: Using residues coevolution to search for protein homologs through alignment of Potts models. JOBIM (2019). https://hal.inria.fr/hal-02402687, poster
28. The UniProt Consortium: UniProt: the universal protein knowledgebase. Nucleic Acids Res. **45**(D1), D158–D169 (2017)
29. Unold, O.: Context-free grammar induction with grammar-based classifier system. Arch. Control Sci. **15**(4), 681–690 (2005)
30. Unold, O.: Fuzzy grammar-based prediction of amyloidogenic regions. In: International Conference on Grammatical Inference, pp. 210–219 (2012)
31. Unold., O., Gabor., M., Wieczorek., W.: Unsupervised statistical learning of context-free grammar. In: Proceedings of the 12th International Conference on Agents and Artificial Intelligence - Volume 1: NLPinAI, pp. 431–438. INSTICC, SciTePress (2020)
32. Van Zaanen, M.: ABL: alignment-based learning. In: Proceedings of the 18th Conference on Computational Linguistics, vol 2, pp. 961–967. Association for Computational Linguistics (2000)
33. Wieczorek, W.: A local search algorithm for grammatical inference. In: Sempere, J.M., García, P. (eds.) ICGI 2010. LNCS (LNAI), vol. 6339, pp. 217–229. Springer, Heidelberg (2010). https://doi.org/10.1007/978-3-642-15488-1_18

Temporal Data Analysis

Multi-scale Temporal Memory
for Clinical Event Time-Series Prediction

Jeong Min Lee[✉] and Milos Hauskrecht

University of Pittsburgh, Pittsburgh, PA 15260, USA
jlee@cs.pitt.edu, milos@pitt.edu

Abstract. The objective of this work is to develop and study dynamic patient-state models and patient-state representations that are predictive of a wide range of future events in the electronic health records (EHRs). One challenge to overcome when building predictive EHRs representations is the complexity of multivariate clinical event time-series and their short and long-term dependencies. We address this challenge by proposing a new neural memory module called Multi-scale Temporal Memory (MTM) linking events in a distant past with the current prediction time. Through a novel mechanism implemented in MTM, information about previous events on different time-scales is compiled and read on-the-fly for prediction through memory contents. We demonstrate the efficacy of MTM by combining it with different patient state summarization methods that cover different temporal aspects of patient states. We show that the combined approach is 4.6% more accurate than the best result among the baseline approaches and it is 16% more accurate than prediction solely through hidden states of LSTM.

Keywords: Electronic health records (EHRs) · Clinical event time-series prediction · Neural network · Sequence prediction

1 Introduction

Electronic health records (EHRs) are longitudinal collections of clinical information that cover many aspects of patient care in hospitals. It consists of patient demographics, records of the administration of medication, past procedures, lab test results, various physiological signals, and other significant events related to patient care. The EHRs and events recorded therein can be used for a variety of purposes, such as prediction of adverse events [25] and mortality risk scores [29], detection of deviations in care [8,9], automatic diagnosis [21,24], lab value estimation [18–20], or intelligent retrieval of similar patient cases from the database of past patients [28].

The objective of this work is to develop and study dynamic patient-state models and patient-state representations that are predictive of a wide range of future events in the electronic health records. Such representations can characterize well the patient state for many different problems mentioned above. Defining good

© Springer Nature Switzerland AG 2020
M. Michalowski and R. Moskovitch (Eds.): AIME 2020, LNAI 12299, pp. 313–324, 2020.
https://doi.org/10.1007/978-3-030-59137-3_28

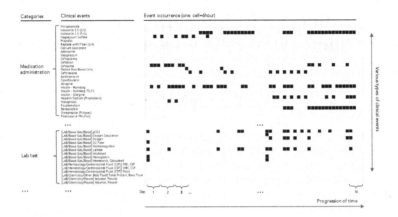

Fig. 1. A part of a patient's record in real-world EHRs (MIMIC-3 database) represented as a sequence of multi-hot vectors. Each vector indicates occurrence or non-occurrence of an event during a segmented time-window (e.g., 6 h).

predictive representation of EHRs is a challenging problem due to the inherent complexities of the EHR-based multivariate event time-series. In general, EHRs can consist of several thousands of clinical events corresponding to different types of medication, lab tests, medical procedures, physiological signals, etc. For example, MIMIC-3 [12], a widely used publicly available ICU Database, records more than 30,000 different types of clinical events. However, many clinical events are sparse and infrequent. Briefly, the average number of medication administration events per patient per admission is 10.1, lab test events 7.3, and procedures 1.5. To deal with the challenges of high-dimensionality and sparsity, deep learning based approaches have shown promising results in modeling EHRs-derived data and sequences. Two major deep learning approaches have been studied: latent-space embedding models [2,4,23] and neural temporal models based on RNNs and LSTMs [3,5,7,15,29].

One challenging issue related to predictive EHRs representations that have not been adequately addressed is how to properly model temporal dependencies among many different clinical events. More specifically, individual event-time-series in EHRs may have a different temporal dependency with respect to precursor events. Briefly, some events may strongly dependent on recently occurred events. For example, an administration of phenylephrine depends on the occurrence of hypotension (low blood pressure state) in connection with recent intubation. Lee and Hauskrecht [15] show that modeling such short-term dependency can improve the predictability of multivariate future events. However, other events may depend on more distant events. For example, valve replacement surgery in the distant past may impact the necessity of warfarin treatment. While neural temporal models (RNN or LSTM) can in principle model these long-range dependencies, the recurrent computations can easily dilute and attenuate such information in the hidden state [26]. In this work, we address the problem of modeling long-term dependencies in multivariate clinical event time-series by proposing a new type of information channel linking events in a distant past with the current prediction time. Through a novel mechanism called Multi-scale Temporal Memory (MTM), information about previous events on different time-scales is compiled and read on-the-fly for prediction through memory contents. The main benefit of this approach is that it is a modular and predictive signal from this module that can be combined with predictive signals from other patient state summarization modules.

We demonstrate the efficacy of MTM by combining it with different patient state summarization methods that cover different temporal aspects of patient states, including recent context module [15], recurrent temporal mechanism [16], and hidden states of LSTM [10]. We test the proposed approach on real-world clinical event time-series. We compare predictive performance (i.e., AUPRC) of the proposed combined approach with baseline models. We demonstrate that the combined approach is 4.6% more accurate than the best among the baselines and it is 16% more accurate than prediction solely through hidden states of LSTM.

2   Background

In this section, we introduce the EHR-based multivariate time-series and the prediction problem. Then, we review clinical event time-series based on neural temporal models.

2.1   Multivariate Clinical Event Time-Series

A patient's EHR is defined by a sequence of time-stamped clinical events $U = \{u_j\}_j$, where each event $u_j = (e_j, t_j)$ consists of a pair of type of the event $e_j \in E$ and timing of the event $t_j \in \mathbb{R}_{\geq 0}$. E is a set of all types of clinical events. As events in EHRs occur in continuous time, t_j is non-negative real value. One way to model the event time-series on real-valued continuous-time is by using point processes [27] such as a Poisson process or a Hawkes process [14]. However, point processes based approaches are hard to optimize directly, and existing works for clinical event time-series [17,22] explore multivariate event time-series with a relatively small number of events. Due to this limitation, multivariate event time-series are often converted to discrete-time event time-series. By sweeping the original time-series with a fixed-sized time window (e.g., 6 h), the time-series is segmented to a sequence of non-overlapping bins, where each bin represents events occur during the time-window. Then, events occurred or non-occurred during a time window are represented as a binary multi-hot vector $\mathbf{y} \in \{0,1\}^{|E|}$. With this discretization method, a patient's records in EHRs are represented as a sequence of the multi-hot vectors $\mathbf{y}_1, \cdots, \mathbf{y}_t$. Figure 1 shows an exemplar multi-hot vector representation of a patient's record.

The prediction problem we want to tackle can be then defined to predict the occurrence and non-occurrence of a wide range of EHR-related events in the future time step $\mathbf{y}_{t+1}$ given a sequence of patient history $\mathbf{y}_1, \cdots, \mathbf{y}_t$.

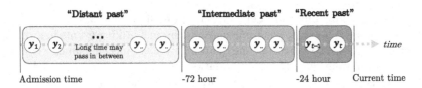

"Distant past"   "Intermediate past"   "Recent past"

Admission time   -72 hour   -24 hour   Current time

Fig. 2. MTM summarizes past history with multiple temporal scales

2.2 Neural Temporal Models for Clinical Event Time-Series

With the benefits of the flexible end-to-end training and combined feature representation learning capabilities, models based on neural architectures have been successfully adopted to various time-series modeling tasks. In the following, we summarize the approaches to clinical event time-series modeling and prediction.

Word-to-Vector Models (Word2Vec). The Word2Vec (e.g., CBOW, Skip-gram) [23] learns low-dimensional embeddings of words and documents in NLP. Continuous Bag-of-Word (CBOW) [23] predicts the probability distributions of a center (target) word given the word's neighborhood (context) words. Skip-gram [23] is similar to Word2Vec, but the context and the target are switched around. For clinical tasks, Word2Vec models have been adopted to process a sequence of clinical events instead of words. More specifically, for the CBOW-based approach, recent events in a fixed-size recent history window (e.g., 48 h) are set as the context and an event that occurs shortly after the history window is set as the target. Word2Vec models have been successfully applied to predict e.g. hospital visits [4]. One drawback of the Word2Vec models is that they cannot fully model the sequential information, as they treat the events in the past equally when pooling (summing or averaging) past event embeddings. Besides, the size of the neighborhood (context) window is limited to a certain number of events (e.g., 20 or 40). Hence, those events that occur outside of the window cannot be used for modeling.

RNN and LSTM Based Approaches. The sequential models based on RNN and LSTM [10] resolve the problems by summarizing the information from each past step via hidden states. The hidden states correspond to a real-valued (latent) representation of patient states. RNN and LSTM have been successfully applied to many clinical event predictions such as medication prescriptions [1,3], heart failure onset [6], and ICU mortality risk [29].

One advantage of RNNs is that it can model all events in the entire sequence without a length-span limit, unlike Word2Vec. However, RNN and LSTM models may encounter problems when modeling long sequences. Briefly, the loss (training objective function) is computed at the end of each sequence and the signal is passed to parameters at each time step via Back Propagation Through Time (BPTT). For RNN and LSTM, the length of the sequence is n. A long sequence (large n) can hinder the propagation of the loss signal to parameters, negatively

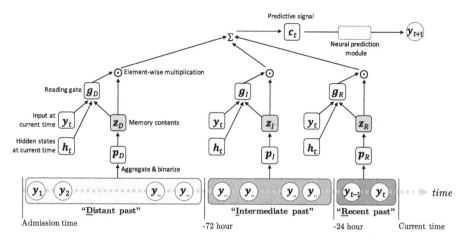

Fig. 3. Overview MTM's architecture: Given a sequence of multivariate patient state history $\mathbf{y}_1, \ldots, \mathbf{y}_t$, we **(1)** aggregate and binarize past history by each time-scale $\mathbf{p}_*$, $* \in \{D, I, R\}$, **(2)** compose memory contents $\mathbf{z}_*$, **(3)** compute reading gate $\mathbf{g}_*$, **(4)** read memory contents referring reading gate and merge contents of multi-scale temporal memory, and **(5)** make a predictive signal $\mathbf{c}_t$ for neural prediction module.

affecting their training [11]. Our proposed work tackles this challenge by creating a direct path of length 1 connecting the current time step with a predictive event that occurred in the distant past.

3   Methodology

In this work, we propose Multi-scale Temporal Memory (MTM), a new neural temporal based model that summarizes a clinical event history and generates a predictive signal for occurrence and non-occurrence of future multivariate clinical events (Fig. 3).

3.1   Multi-scale Temporal Memory

MTM summarizes patient history using multiple information channels where each channel covers the history in different temporal scales. We hypothesize that information on past event occurrences on different time range may have different importance for predicting future event occurrence. To process patient history on multiple time scales, MTM segments the patient history into three folds as shown in Fig. 2: distant past (e.g., from the beginning of admission to 72 h before current time), recent past (e.g., within 24 h from current time), and "intermediate past" (time range between boundaries of distant past and recent past). Contents of the memory are composed based on the types of events that occurred in each segmented window. Then, considering factors about current patient states, the model reads contents of the multi-scale memory and generates a predictive signal that will be combined with other neural temporal mechanisms that cover different aspects of clinical event time-series to generate a final prediction for next multivariate events. In the following, we describe MTM in detail and the neural framework for the next multivariate events prediction.

Composing Memory Contents. Given a segmented patient history (depicted in Fig. 2) on multiple time-scales, we compose memory contents for each time-scale with the following steps: (1) We aggregate patient states vectors $\{\mathbf{y}_i\}_i$ of each temporal segment $* \in \{D, I, R\}$ into a single multivariate vector $\mathbf{p}_*$ through binarization. $\{D, I, R\}$ denote distant, intermediate, and recent pasts respectively. (2) We compose contents of the memory $\mathbf{z}_* \in \mathbb{R}^{|E|}$ through linear projection followed by non-linear activation:

$$\mathbf{z}_* = \tanh\left(\mathbf{W}_*\mathbf{p}_* + \mathbf{b}_*\right) \tag{1}$$

where $\mathbf{W}_* \in \mathbb{R}^{|E|\times|E|}$ and $\mathbf{b}_* \in \mathbb{R}^{|E|}$ are trainable parameters for each time-scale. Through linear projection with $\mathbf{W}_*$, we extract information about the events that occurred in a specific temporal segment.

Reading Memory Contents. To comprehensively determine the amount of memory contents to be read for each prediction task (multivariate target events), MTM computes reading gates $\mathbf{g}_* \in \mathbb{R}^{|E|}$ considering three factors: (1) current patient state reflected on input $\mathbf{y}_t$, (2) recent dynamics of patient state reflected on hidden states $\mathbf{h}_t$ from LSTM, and the contents of the memory itself $\mathbf{z}_*$.

$$\mathbf{g}_* = \sigma(\mathbf{W}_h\mathbf{h}_t + \mathbf{W}_y\mathbf{y}_t + \tilde{\mathbf{W}}_*\mathbf{z}_*) \tag{2}$$

where σ denotes logistic sigmoid activation function and $\mathbf{W}_h \in \mathbb{R}^{|E|\times r}, \mathbf{W}_y \in \mathbb{R}^{|E|\times|E|}, \tilde{\mathbf{W}}_* \in \mathbb{R}^{|E|\times r}$ are parameters to learn and r is dimension of hidden state. The predictive signal $\mathbf{c}_t \in \mathbb{R}^{|E|}$ is computed as a linear combination of reading gates and memory contents for each temporal scale:

$$\mathbf{c}_t = \mathbf{g}_D \odot \mathbf{z}_D + \mathbf{g}_I \odot \mathbf{z}_I + \mathbf{g}_R \odot \mathbf{z}_R \tag{3}$$

where $\odot$ is element-wise multiplication.

3.2   Neural-Based Prediction Framework

We combine the predictive signal from MTM with additional patient history summarization methods that cover different temporal aspects of patient states. We use recent-context module [15], recurrent temporal mechanism [16] and hidden states of LSTM. Briefly the recent-context module projects current time-step input $\mathbf{y}_t$ to a target event space with a learnable parameters $\mathbf{W}_r \in \mathbb{R}^{|E|\times|E|}$ and $\mathbf{b}_r$ to get the "recent bias" term $\mathbf{b}_\kappa$:

$$\mathbf{b}_\kappa = \mathbf{W}_r\mathbf{y}_t + \mathbf{b}_r \tag{4}$$

The recurrent temporal mechanism captures information about periodic (repeated) events using a special recurrent mechanism based on probability distributions of inter-event gaps. It outputs two target event-specific periodicity-based predictive signals that use different sources of periodic information: $\boldsymbol{\alpha}^e \in \mathbb{R}$ signal is based on an interval of current patient's event time-series and $\boldsymbol{\beta}^e \in \mathbb{R}$ signal is compiled from a pool of past patient data in training set. Details of the signal generation processes can be found in [16]. We also use LSTM to derive dynamics of patient state through hidden state. To compute hidden state, we first project input $\mathbf{y}_t$ to low-dimensional space with embedding matrix: $\mathbf{W}_{emb} \in \mathbb{R}^{d \times |E|}$: $\mathbf{x}_t = \mathbf{W}_{emb}\mathbf{y}_t$. Based on previous time step's hidden state $\mathbf{h}_{t-1}$ and $\mathbf{x}_t$, we compute new hidden state $\mathbf{h}_t \in \mathbb{R}^r$:

$$\mathbf{h}_t = \text{LSTM}(\mathbf{h}_{t-1}, \mathbf{x}_t) \tag{5}$$

Given predictive signals $\{\boldsymbol{\alpha}^e, \boldsymbol{\beta}^e, \mathbf{h}_t, \mathbf{c}_t, \mathbf{b}_\kappa\}$, we first combine periodicity-based signals for each target event type with hidden state through concatenation:

$$\boldsymbol{\gamma}^e = [\mathbf{h}_t; \boldsymbol{\alpha}^e; \boldsymbol{\beta}^e] \tag{6}$$

Then, we project $\boldsymbol{\gamma}^e$ to a scalar $\lambda^e \in \mathbb{R}$ through $\mathbf{w}_e \in \mathbb{R}^{1 \times r+2}$ and $b_e \in \mathbb{R}$. We apply the same procedure to all events $e \in E$ and concatenate all λ^e:

$$\lambda^e = \mathbf{w}_e \boldsymbol{\gamma}^e + b_e \quad \boldsymbol{\lambda} = [\lambda^1; \dots; \lambda^{|E|}] \tag{7}$$

Final prediction for next multivariate event is computed as follows:

$$\hat{\mathbf{y}}_{t+1} = \sigma(\boldsymbol{\lambda} + \mathbf{b}_\kappa + \mathbf{c}_t) \tag{8}$$

We use the binary cross-entropy to compute loss $\mathcal{L}$ and parameters of the model are learned through a stochastic gradient descent optimization algorithm (Adam) [13].

$$\mathcal{L} = \sum_t -[\mathbf{y}_t \cdot \log \hat{\mathbf{y}}_t + (1 - \mathbf{y}_t) \cdot \log(1 - \hat{\mathbf{y}}_t)] \tag{9}$$

4 Experiments

In this section, we evaluate our approach on MIMIC-3, an ICU EHRs dataset.

4.1 Experiment Setup

Clinical Data. We extract 5137 EHRs of patients from MIMIC-3 database using the following criteria: (1) adult patient, (2) length of stay is between 48 and 480 h, (3) data are recorded in Meta Vision system, one of the two systems used to create MIMIC-3 database. We randomly split 5137 patients into train and test sets using 8:2 ratio. Then, multivariate event time-series are generated by

Table 1. Prediction results (AUPRC) for all events and each event category

Models	All-events	Medication	Lab test	Physio signal	Procedure
RC	18.26	35.52	4.11	45.23	34.67
HS	26.30	41.16	12.03	81.49	35.90
HS-RC	26.50	41.66	12.07	81.61	36.25
HS-RC-PP	26.76	42.82	12.04	81.70	36.29
HS-RC-PP-MTM	28.00	43.84	13.80	81.84	36.35

segmenting all sequences with a time-window ($W = 1$). As mentioned in Sect. 2.1, at each i-th window we obtain multi-hot vector $\mathbf{y}_i \in \{0,1\}^{|E|}$ by aggregating and making binary all events occur within the time range of the window. For the types of clinical events (E), we use events in the categories of medication administration, lab results, procedure, and physiological results. Among all types of events in the first three categories, we filter out those events observed in less than 500 different patients. For physiological events, we select 16 important event types with the help of a critical care physician. To this end, we get 63 medication events, 41 procedure events, 155 lab test events, and 16 physiological signal events ($|E| = 275$).

Baseline Methods. We compare our method (HS-RC-PP-MTM) with the following set of baseline models predicting a wide range of future events $\mathbf{y}_{t+1}$:

- **Logistic Regression with Recent Context (RC)** uses the current events. It amounts to use the recent bias term in Eq. (4): $\hat{\mathbf{y}}_{t+1} = \sigma(\mathbf{W}_p \mathbf{y}_t + \mathbf{b}_p)$
- **Hidden States from LSTM (HS)** uses hidden states of LSTM in Eq. (5) with linear projection and sigmoid activation: $\hat{\mathbf{y}}_{t+1} = \sigma(\mathbf{W}_q \mathbf{h}_t + \mathbf{b}_q)$
- **HS + Recent Context (HS-RC)** [15] uses hidden states of LSTM with the recent bias term $\mathbf{b}_\kappa$ in Eq. (4): $\hat{\mathbf{y}}_{t+1} = \sigma(\mathbf{W}_r \mathbf{y}_t + \mathbf{b}_r + \mathbf{b}_\kappa)$
- **HS + RC + Periodicity Predictor (HS-RC-PP)** [16] uses combination of hidden states of LSTM, the recent bias term $\mathbf{b}_\kappa$ and periodicity signal α, β from [16]. It computes prediction with λ in Eq. (7): $\hat{\mathbf{y}}_{t+1} = \sigma(\lambda + \mathbf{b}_\kappa)$

Evaluation Metrics. We evaluate the quality of predictions by calculating the area under the precision-recall curve (AUPRC). The reported AUPRC values (for the different methods) are averaged over all target events.

Implementation Details. For the experiments, we use embedding size $d = 64$, a fixed learning rate $= 0.005$ and minibatch size $= 128$. The size of the hidden state r is determined by the internal cross-validation from $(128, 256, 512)$. To prevent over-fitting, L_2 weight decay regularization is applied to all models and the weight is determined by the internal cross-validation.

4.2   Results

The second column of Table 1 shows the overall experiment results for predicting all types of events. The proposed model (HS-RC-PP-MTM) outperforms all baselines. Particularly, it outperforms HS-RC-PP by 4+%. With this, we can observe the benefit of multi-scale memory capturing dependencies that are not covered by other patient history summarization methods, including LSTM.

We further analyze the experiment results by dividing them into 4 event categories. As shown in Table 1 (column 3–6), we observe the performance gain of MTM is higher for medication and lab test events. Notably, lab tests are the hardest events to predict compared to other categories, 14+% performance gain from MTM for lab test prediction clearly shows its effectiveness.

We also experiment with two additional window sizes (W = 6,12). As Table 2 shows, larger segmentation window increases overall predictability. This is expected as larger window size results in the multivariate vector $\mathbf{y}_i$ with more event occurrences and it increases prior probability which directly affects the AUPRC score. Especially, we observe a pattern that the gap between HS-RC-PP-MTM and HS-RC-PP is decreasing as the window size is increased. An implication of this observation is that, for longer sequences (event time-series generated from $W = 1$ based window-segmentation), MTM brings more value than it does for shorter sequences (e.g., $W = 6,12$).

To validate learned weight matrices for multi-scale memory contents ($W_*, * \in \{D, I, R\}$ in Eq. (1)), we extract the top 3 events for an exemplar target event **extubation** in Table 3. We can see that MTM properly learns and gives higher weights to the intubation event, PEEP (setting of the mechanical ventilation), and fentanyl, analgesics used during mechanical ventilation. Additional examples are compiled in Table 4.

Table 2. Prediction results by varying time-series segmentation window settings

Models	$W = 1$	$W = 6$	$W = 12$
HS-RC-PP	26.76	36.68	40.07
HS-RC-PP-MTM	28.00 (4.6 +%)	37.28 (1.6 +%)	40.34 (0.6 +%)

Table 3. Top 3 past events predictive of **extubation**, based on the value from learned memory content parameter W_* for each temporal range in Eq. (1).

Distant past ($* = D$)	Intermediate past ($* = I$)	Recent past ($* = R$)
(Med) Potassium Chloride	(Proc) PEEP	(Proc) PEEP
(Med) KCL	(Med) Fentanyl	(Physio) Inspired O2 Fraction
(Proc) Intubation	(Proc) Intubation	(Med) Fentanyl

5   Conclusion

We proposed a novel mechanism called Multi-scale Temporal Memory (MTM) to model long-term dependencies in EHR-derived clinical event time-series. With MTM, information about past events on different time-scales is compiled and read on-the-fly for prediction through memory contents. We demonstrate the efficacy of MTM by combining it with different patient state summarization methods that cover different temporal aspects of patient states. We show that the combined approach is 4.6% more accurate than the baseline approaches and it is 16% more accurate than the prediction based on the popular LSTM summarization approach.

In the future we plan to study ways of relaxing hard segmentations of past history. That is, we plan to automatically identify the memory content and the timing information for past events that are important for predicting the next events. One possible direction is to design an attention mechanism capable of aggregating event history via a specialized kernel that considers both (a) the type of target and context events and (b) timing of events.

Acknowledgement. The work in this paper was supported by NIH grant R01GM088224. The content of the paper is solely the responsibility of the authors and does not necessarily represent the official views of NIH.

A   Examples of Top Past Events Predictive of Target Events

Table 4 shows top past events predictive of target events for the different temporal ranges (Distant, Intermediate, and Recent past) as identified by our methods. For example, the top predictive events for amiodarone (treats irregular heartbeat such as tachycardia) include metoprolol and diltiazem. Both of these are used to treat high blood pressure and heart issues. Similarly, past events predictive of diltiazem and labetalol (medications treating high blood pressure) include clinical events that are related to high blood pressure and heart function: digoxin, metoprolol, hydralazine, and nicardipine. Finally, the top past events predicting vasopressin (a medication treating a low blood pressure) include norepinephrine and phenylephrine that are also used to treat low blood pressure.

Table 4. Top 3 preceding events for example target events based on the value from learned memory content parameter W_* for each temporal range in Eq. (1).

Distant past ($* = $ D)	Intermediate past($* = $ I)	Recent past($* = $ R)
Target: (Med) Amiodarone		
(Med) Amiodarone	(Med) Amiodarone	(Med) Amiodarone
(Med) Diltiazem	(Med) Diltiazem	(Med) Metoprolol
(Lab) Urea Nitrogen, Urine	(Lab) Thyroid Stimulating Hormone	(Med) Diltiazem
Target: (Med) Diltiazem		
(Med) Diltiazem	(Med) Diltiazem	(Med) Diltiazem
(Lab) Digoxin	(Med) Metoprolol	(Med) Metoprolol
(Physio) Inspired O2 Fraction	(Med) Amiodarone	(Proc) EKG
Target: (Med) Labetalol		
(Med) Labetalol	(Med) Labetalol	(Med) Labetalol
(Med) Hydralazine	(Med) Hydralazine	(Med) Hydralazine
(Med) Nicardipine	(Med) Metoprolol	(Med) Haloperidol
Target: (Med) Vasopressin		
(Med) Vasopressin	(Med) Vasopressin	(Med) Vasopressin
(Proc) Ultrasound	(Med) Norepinephrine	(Med) Norepinephrine
(Med) Packed Red Blood Cells	(Med) Phenylephrine	(Med) Phenylephrine

References

1. Bajor, J.M., Lasko, T.A.: Predicting medications from diagnostic codes with recurrent neural networks. In: ICLR (2017)
2. Bengio, Y., et al.: A neural probabilistic language model. J. Mach. Learn. Res. **3**(Feb), 1137–1155 (2003)
3. Choi, E., et al.: Doctor AI: predicting clinical events via recurrent neural networks. In: Machine Learning for Healthcare Conference, pp. 301–318 (2016)
4. Choi, E., et al.: Multi-layer representation learning for medical concepts. In: The 22nd International Conference on Knowledge Discovery and Data Mining (2016)
5. Choi, E., et al.: Retain: an interpretable predictive model for healthcare using reverse time attention mechanism. In: Neural Information Processing Systems (2016)
6. Choi, E., et al.: Using recurrent neural network models for early detection of heart failure onset. J. Am. Med. Inform. Assoc. **24**, 361–370 (2017)
7. Esteban, C., et al.: Predicting clinical events by combining static and dynamic information using RNN. In: International Conference on Healthcare Informatics (ICHI) (2016)

8. Hauskrecht, M., et al.: Outlier detection for patient monitoring and alerting. J. Biomed. Inform. **46**(1), 47–55 (2013)
9. Hauskrecht, M., et al.: Outlier-based detection of unusual patient-management actions: an ICU study. J. Biomed. Inform. **64**, 211–221 (2016)
10. Hochreiter, S., Schmidhuber, J.: Long short-term memory. Neural Comput. **9**, 1735–1780 (1997)
11. Hochreiter, S., et al.: Gradient flow in recurrent nets: the difficulty of learning long-term dependencies (2001)
12. Johnson, A.E., et al.: MIMIC-III, a freely accessible critical care database. Sci. Data **3**, 160035 (2016)
13. Kingma, D.P., Ba, J.: Adam: a method for stochastic optimization. arXiv:1412.6980 (2014)
14. Laub, P.J., et al.: Hawkes processes. arXiv preprint arXiv:1507.02822 (2015)
15. Lee, J.M., Hauskrecht, M.: Recent-context-aware LSTM-based clinical time-series prediction. In: In Proceedings of AI in Medicine Europe (AIME) (2019)
16. Lee, J.M., Hauskrecht, M.: Clinical event time-series modeling with periodic events. In: The Thirty-Third International Flairs Conference. AAAI (2020)
17. Liu, S., Hauskrecht, M.: Nonparametric regressive point processes based on conditional Gaussian processes. In: Neural Information Processing Systems (2019)
18. Liu, Z., Hauskrecht, M.: Clinical time series prediction: toward a hierarchical dynamical system framework. Artif. Intell. Med. **65**(1), 5–18 (2015)
19. Liu, Z., Hauskrecht, M.: A regularized linear dynamical system framework for multivariate time series analysis. In: Twenty-Ninth AAAI Conference on Artificial Intelligence, pp. 1798–1804 (2015)
20. Liu, Z., Wu, L., Hauskrecht, M.: Modeling clinical time series using gaussian process sequences. In: Proceedings of the 2013 SIAM International Conference on Data Mining, pp. 623–631. SIAM (2013)
21. Malakouti, S., Hauskrecht, M.: Hierarchical adaptive multi-task learning framework for patient diagnoses and diagnostic category classification. In: International Conference on Bioinformatics and Biomedicine (BIBM) (2019)
22. Mei, H., Eisner, J.M.: The neural hawkes process: a neurally self-modulating multivariate point process. In: Neural Information Processing Systems (2017)
23. Mikolov, T., et al.: Distributed representations of words and phrases and their compositionality. In: Neural Information Processing Systems (2013)
24. Miotto, R., et al.: Deep patient: an unsupervised representation to predict the future of patients from the electronic health records. Sci. Rep. **6**, 1–10 (2016)
25. Nemati, S., et al.: An interpretable machine learning model for accurate prediction of sepsis in the ICU. Crit. Care Med. **46**(4), 547–553 (2018)
26. Pascanu, R., et al.: On the difficulty of training recurrent neural networks. In: International Conference on Machine Learning, pp. 1310–1318 (2013)
27. Rasmussen, J.G.: Lecture notes: temporal point processes and the conditional intensity function. arXiv preprint arXiv:1806.00221 (2018)
28. Wang, F., et al.: Composite distance metric integration by leveraging multiple experts' inputs and its application in patient similarity assessment. Stat. Anal. Data Min.: ASA Data Sci. J. **5**(1), 54–69 (2012)
29. Yu, K., et al.: Monitoring ICU mortality risk with a long short-term memory recurrent neural network. In: Pacific Symposium on Biocomputing. World Scientific (2020)

HYPE: Predicting Blood Pressure from Photoplethysmograms in a Hypertensive Population

Ariane Morassi Sasso[1]([✉]) [ID], Suparno Datta[1] [ID], Michael Jeitler[2,3] [ID],
Nico Steckhan[1,2,3] [ID], Christian S. Kessler[2,3] [ID], Andreas Michalsen[2,3],
Bert Arnrich[1], and Erwin Böttinger[1]

[1] Digital Health Center, Hasso Plattner Institute, University of Potsdam,
Potsdam, Germany
{ariane.morassi-sasso,suparno.datta}@hpi.de
[2] Institute for Social Medicine, Epidemiology and Health Economics,
Charité – Universitätsmedizin Berlin, Corporate Member of Freie Universität Berlin,
Humboldt-Universität zu Berlin, and Berlin Institute of Health, Berlin, Germany
[3] Department of Internal and Integrative Medicine,
Immanuel Hospital Berlin, Berlin, Germany

Abstract. The state of the art for monitoring hypertension relies on
measuring blood pressure (BP) using uncomfortable cuff-based devices.
Hence, for increased adherence in monitoring, a better way of measuring BP is needed. That could be achieved through comfortable wearables
that contain photoplethysmography (PPG) sensors. There have been several studies showing the possibility of statistically estimating systolic and
diastolic BP (SBP/DBP) from PPG signals. However, they are either
based on measurements of healthy subjects or on patients on (ICUs).
Thus, there is a lack of studies with patients out of the normal range of
BP and with daily life monitoring out of the ICUs. To address this, we
created a dataset (HYPE) composed of data from hypertensive subjects
that executed a stress test and had 24-h monitoring. We then trained and
compared machine learning (ML) models to predict BP. We evaluated
handcrafted feature extraction approaches vs image representation ones
and compared different ML algorithms for both. Moreover, in order to
evaluate the models in a different scenario, we used an openly available
set from a stress test with healthy subjects (EVAL). The best results for
our HYPE dataset were in the stress test and had a mean absolute error
(MAE) in mmHg of 8.79 (±3.17) for SBP and 6.37 (±2.62) for DBP; for
our EVAL dataset it was 14.74 (±4.06) and 7.12 (±2.32) respectively.
Although having tested a range of signal processing and ML techniques,
we were not able to reproduce the small error ranges claimed in the literature. The mixed results suggest a need for more comparative studies
with subjects out of the intensive care and across all ranges of blood

---

A. M. Sasso and S. Datta—The two authors contributed equally to this paper.

The original version of this chapter was revised: a new reference and a minor change
in conclusion section has been updated. The correction to this chapter is available at
https://doi.org/10.1007/978-3-030-59137-3_44

pressure. Until then, the clinical relevance of PPG-based predictions in daily life should remain an open question.

Keywords: Machine learning · Blood pressure · Photoplethysmogram

1   Introduction

According to the Global Disease Burden (GBD) study, high blood pressure (BP) (i. e. hypertension) is the risk factor that leads to more deaths worldwide [16]. The standard way of monitoring this condition is through the measurement of BP using an uncomfortable cuff-based device [25]. Fortunately, comfortable and common wearables can already detect changes in the flow of blood through a photoplethysmography (PPG) sensor [1]. The PPG signal (photoplethysmogram) obtained from it is already used with success to estimate heart rate [19] and, has the potential to go beyond that into accurate BP prediction [2,5].

Most of the work in this area focus on building predictive models for patients in intensive care units (ICUs) [12,20,24]. However, data collected from regular life contain motion artefacts that are not observed in intensive care. Additionally, models that work on healthy populations [17,18] should also be validated on hypertensive populations for guarantying their applicability in BP monitoring. Hence, in our work we focused on assembling a dataset containing data from subjects with hypertension (HYPE) during a stress test and 24-h monitoring.

We then evaluated machine learning (ML) models for predicting BP from PPG in the HYPE dataset and also in a dataset from healthy subjects during a stress test (EVAL). From the PPG signals, we extracted features from the time domain plus their image representations. Errors as low as the ones in the literature—for patients in the ICU or healthy subjects—could not be reproduced, even after processing the PPG signals with diverse time windows and filters.

This work is detailed as follows: Sect. 2 shows previous work in the field and Sect. 3 describes the datasets and methods we used to predict BP from the PPG signal. In Sect. 4 we convey our findings and results, followed by a discussion in Sect. 5 and Sect. 6 describing the implications of this work.

2   Related Work

Existing work focuses on predictive models using MIMIC [7], a dataset that contains physiological signals including PPG and ambulatory BP (ABP) from patients in ICUs. Kurylayak et al. [12] and Wong et al. [24] have both applied artificial neural networks (ANN) to predicted BP in this dataset and reported success. However, they used unknown or small sample sizes as can be seen in Table 1. Moreover, Kurylayak et al. only extracted time domain features from the PPG signal while Wong et al. also extracted frequency domain ones. Conversely, Slapničar et al. [20] tried a spectro-temporal ResNet with all features in a larger sample size but could not report the same success as his predecessors.

Others have tried to collect data from healthy subjects in daily life such as Lustrek et al. [17]. They have used the device empatica E4[1] and evaluated

---

[1] https://e4.empatica.com/e4-wristband.

Table 1. Blood Pressure Prediction from Photoplethysmograms

Work	Dataset	Features (PPG)	Method	MAE (mmHg)
[12]	MIMIC 15000 pulsations	Time Domain	ANN	SBP 3.80 (±3.46) DBP 2.21 (±2.09)
[24]	MIMIC 72 subjects	Time and Frequency Domain	ANN	SBP 4.02 (±2.79) DBP 2.27 (±1.82)
[20]	MIMIC 510 subjects	Time and Frequency Domain	Spectro-Temporal ResNet	SBP 9.43 (N/A) DBP 6.88 (N/A)
[17]	Healthy Subjects Daily Life 22 subjects	Time and Frequency Domain	Emsemble of Regression Trees	SBP 6.70 (N/A) DBP 4.42 (N/A)
[18]	Healthy Subjects Controlled 50 subjects	Time Domain	ANN	SBP 4.1 (N/A) DBP 1.7 (N/A)

a range of machine learning (ML) techniques, achieving the best results with an ensemble of regression trees and the leave-one-subject-out (LOSO) validation strategy. However, they had to use ground truth BP from each subject to personalize the algorithm. Lastly, there is the work of Manamperi et al. [18], in which they evaluated ANN in MIMIC and in a set with data from voluntary subjects (assumed as healthy). They claim to have done the second evaluation in a non-clinical scenario, but the subjects were mainly at rest in their experiment.

Therefore, the current state-of-the-art does not give yet conclusions about the use of PPG to predict BP in diverse populations and in daily life. There is a clear need for more comparative studies both with healthy and hypertensive subjects and in different scenarios, especially outside of controlled conditions.

3 Methods

3.1 Datasets

In our work we used two datasets: one created by us with data from a hypertensive population (HYPE) and one that is openly available containing data from a healthy population (EVAL). It should be noted that both datasets recorded patients during a stress test and HYPE also during 24-h monitoring. We describe the two datasets below.

HYPE. This dataset was created by us as part of the CardioVeg study (NCT03901183) approved by the Ethics Committee from Charité, Berlin (no. EA4/025/19). Data was collected from 12 subjects (6 female) in the age range of 31–75 (median 60) that had hypertension. The study collected data from a

(1) stress test and from (2) 24-h monitoring using the empatica E4 wristband as the PPG source and the Spacelabs (SL 90217) BP monitor.

(1) Stress Test. The subjects followed a protocol in which they watched a relaxing video for 5 min then had their BP taken by a physician five times with an interval of 1 min per measurement [25]. Then, the patients biked in an ergonomic bike from 5 to 10 min and relaxed again. During the second relaxation phase their BP was measured again 5 times with a 1 min interval. This dataset contains a total of 95 BP recordings. One subject could not bike due to extreme high BP and another one had a failure in the wearable device. Therefore, this experiment had 10 subjects (5 female).

(2) 24 H. In this phase, the same subjects from the stress test were monitored for 24-h during regular day activities. The Spacelabs monitoring device was configured to measure BP every 30 min during the day and every hour during the night. This dataset contains a total of 464 BP recordings and all 12 subjects were measured.

EVAL. This dataset was generated by Esmaili et al. [3]. The original paper tried to estimate BP based on pulse transit time (PTT) and pulse arrival time (PAT). Both variables are derived from the differences between the PPG and ECG signals. This data was collected from 26 healthy subjects in the age range of 21–50 years. The subjects were required to run for 3 min at the speed of 8 km/h to induce perturbations in their BP values. Directly after the exercise the subjects were made to sit upright and BP values were measured along with PPG and ECG. A force-sensing resistor (FSR) was used under the BP monitor cuff to measure the instantaneous cuff pressure. With the FSR it was possible to pin point the exact time when the SBP and DBP were measured. A total of 152 BP values were recorded in this dataset.

3.2 Handcrafted Feature Extraction Methods

Our first approach entailed extracting handcrafted features from the PPG signal. Time windows of 15, 30 and 45 s around the BP measurement were used for our experiments. To eliminate motion artefacts induced by wrist movements sections in which the Euclidean norm of x-, y- and z-acceleration lied outside of an interval of 25% of the standard deviation around the sample mean, for the current window, were removed from consideration. The motion removal was only done for the HYPE dataset as the EVAL dataset did not contain any motion signals corresponding to the PPG recording. We also experimented with signal normalization and filters such as Chebyshev II and Butterworth, since they were reported as the best filters for PPG signals [15]. For the processed signal, the PPG cycles were then identified with a standard peak detection function.

All detected cycles in the same window were combined into a custom PPG signal template (details in Sect. B), following a procedure described by Li and Clifford [13]. Individual cycles were then compared with the template using two signal quality indices (SQI): (1) direct linear correlation and (2) direct linear correlation between the cycle, re-sampled to match the template length, and the template itself. Only if both correlations lied above 0.8, the cycle was further processed to extract features. This resulted in some BP intervals not having any features extracted since no cycles matching the template were identified.

After the clean PPG cycles have been identified, time domain features were extracted and the detailed list can be found in the Appendix (Table 4). The first step was to identify the first peak in the cycle, which corresponded to the systolic peak. Then for various percentages of the peak amplitude, we extracted the time between systolic peak and end of the cycle (DW_n), start of the cycle and end of the cycle ($SW_n + DW_n$), and the ratio between the time in the cycle before and after the systolic peak (DW_n/SW_n). For every window, the mean and variance of each feature were computed and used as input for the models.

3.3   Image Representation Methods

An alternative approach to the manual feature extraction has recently gained much popularity involving convolutional neural networks (CNNs). The approach is to represent the waves as images and then use a transfer learning method based on pretrained CNNs to learn embedding from the images and use them to predict BP. The two different image-form representations of PPG signals that we tested were spectrograms and scalograms, described below.

Scalograms. A scalogram is usually plotted as a graph of time and frequency and it represents the absolute value of the Continuous Wavelet Transform (CWT) coefficients of a signal. The scalogram-CNN based approach was first discussed in Liang et al. [14]. However, it was only evaluated for hypertension stratification, not BP prediction. Before passing the signal to the CWT, we detrended it, i.e. subtracted the mean value from the input signal. CWT is a convolution of the input data sequence with a set of functions generated by the base wavelet. We used the complex Morlet wavelet function as the base wavelet, which is given by:

$$\Psi(t) = \frac{1}{\sqrt{\pi B}} exp^{-\frac{t^2}{B}} exp^{2\pi Ct} \tag{1}$$

The value of bandwidth frequency (B) and center frequency (C) was chosen to be 3 and 60 in the above equation, following the work of Liang et al. [14]. Compared to a spectrogram, a scalogram is usually better at identifying the low-frequency or fast-changing frequency component of the signal.

Spectrograms. A spectrogram displays changes in the frequencies in a signal over time. A third dimension indicating the amplitude of a particular frequency at a particular time is represented by the intensity or color of each point in the plot. The spectrogram-CNN approach to predict BP was first discussed in Slapničar et al. [20]. Similar to scalograms, we detrended our signal before generating the spectrogram plots. To generate a spectrogram, digitally sampled signals in the time domain are broken up into windows, which usually overlap, and they are Fourier transformed to calculate the magnitude of the frequency spectrum for each window [21].

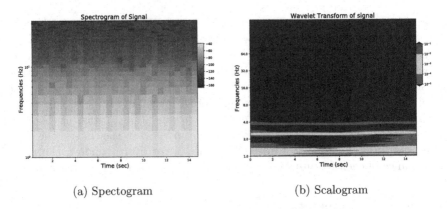

(a) Spectogram   (b) Scalogram

Fig. 1. (a) displays a sample spectrogram and (b) a scalogram with the complex Morlet wavelet function used as the base wavelet for the same PPG signal.

Figure 1 depicts a sample spectrogram and scalogram generated from a PPG snippet of 15 s. The image representations of the signal were then fed into a ResNet architecture to learn the image embeddings [9]. The Residual Network or ResNet design enables us to train very deep neural networks without running into the vanishing gradient problem. Since our datasize is very small, instead of training a network from scratch, we decided to take a network which was already trained on the ImageNet dataset [8]. In particular, we used the ResNet18 architecture and took the embeddings from the penultimate layer of the network. We also experimented with Alexnet, but Resnet18 always performed marginally better [11]. This might be due to the fact that the penultimate layer of the Resnet18 generates a 512 length embedding, whereas the AlexNet generates a embedding of length 4096. The larger size of the input vector, in spite of using feature selection and strong regularization techniques, might make it challenging for the models to learn from, due to the small data size.

3.4 Machine Learning Models

Previous works show that machine learning algorithms perform well in predicting BP from features derived from PPG and/or ECG. We have employed in our experiments three popular machine learning algorithms: (a) Generalised Linear Models (GLM) with Elastic Net regularisation [26], (b) Gradient Boosting Machines (GBM) [4], and (c) a recent more efficient implementation of GBM called LightGBM (LGBM) [10] to predict the systolic and diastolic BP.

For prediction from the image embeddings, we used a Recursive Feature Elimination (RFE) technique with a support vector machine (SVM) with linear kernel as the base estimator, before pushing the vectors into the models.

3.5 Experimental Settings

In order to train models that are robust and well generalizable, we used a leave-2-subjects-out cross validation for all models, i.e. at every iteration we use data from 2 subjects as the test set, trained our models on the remaining data and repeated this procedure till all subjects have been at some point used as the test set. All hyper parameters were optimized empirically. We evaluated the models based on the mean absolute error (MAE). The MAE was calculated at each iteration and we calculated the mean and standard deviation of these values.

4 Results

In this section we report our experimental results. In Table 2 and Table 3 we show the comparison of the MAEs for predicting systolic blood pressure (SBP) and diastolic blood pressure (DBP) respectively, between all models in the different datasets. The cells in these table contain the mean and standard deviation (in parenthesis) of the MAE of all cross validation folds. Noticeably, in the HYPE dataset feature extraction methods consistently outperformed the image based methods. In EVAL the spectrogram-representation method outperformed the other two approaches. In both datasets, the best results for the spectrogram based approach are usually marginally better than the best results of the scalogram based approach. For the image based methods the more advanced machine learning models such as LGBM and GBM clearly outperformed the GLM model. This is most probably due to the comparatively large dimension of the input image embeddings. For the feature extraction based methods this difference is not so prominent, and in some cases the GLM turns out to be the best performing model. In general, based on the MAE values, predicting SBP appears to be more difficult than DBP which is consistent with previous literature (see Table 1).

5 Discussion

5.1 Clinical Relevance

Cuff-less and continuous methods of measuring BP are particularly attractive as BP is one of the most important predictors of long term cardiovascular health [6]. Prediction models for BP based on PPG signals can be a very important stride in that direction. But for reliable continuous monitoring of BP, these models need to perform well during regular day-to-day activities and also for different patient populations. Apart from MIMIC there does exist a few other studies that try to collect PPG and corresponding BP signals following a strict protocol (similar to HYPE: stress-test). To the best of our knowledge, our 24-h dataset is the first attempt to collect this data from an uncontrolled environment where the subjects were free to do anything. Our evaluations do underline that it is indeed more challenging to accurately predict BP in such an uncontrolled environment.

Table 2. MAE of the different models for SBP prediction in different datasets

Dataset	Feature selection			Spectrograms-Resnet18			Scalograms-Resnet18		
	GLM	GBM	LGBM	GLM	GBM	LGBM	GLM	GBM	LGBM
EVAL Stress Test	16.71 (±4.32)	16.19 (±4.35)	16.00 (±4.63)	17.59 (±3.72)	15.24 (±4.01)	**14.74** (±4.06)	16.71 (±4.16)	15.68 (±4.34)	15.46 (±4.49)
HYPE Stress Test	10.26 (±1.18)	**8.79** (±3.17)	9.57 (±1.65)	15.58 (±1.21)	12.44 (±2.72)	12.15 (±2.72)	17.98 (±2.11)	12.91 (±2.57)	12.83 (±2.63)
HYPE 24-h	**14.44** (±2.96)	14.83 (±3.81)	14.74 (±4.07)	18.71 (±4.072)	16.92 (±5.22)	17.07 (±5.22)	18.97 (±5.15)	17.03 (±5.26)	17.30 (±5.21)

Table 3. MAE of the different models for DBP prediction in different datasets

Dataset	Feature selection			Spectrograms-Resnet18			Scalograms-Resnet18		
	GLM	GBM	LGBM	GLM	GBM	LGBM	GLM	GBM	LGBM
EVAL Stress Test	7.87 (±2.07)	7.86 (±2.27)	7.57 (±2.38)	13.18 (±13.01)	**7.12** (±2.32)	7.15 (±2.42)	8.67 (±3.08)	7.62 (±2.35)	7.53 (±2.48)
HYPE Stress Test	7.50 (±0.68)	**6.37** (±2.62)	7.22 (±2.69)	11.98 (±1.90)	9.55 (±2.74)	9.52 (±2.02)	12.18 (±1.98)	9.51 (±1.91)	9.35 (±1.93)
HYPE 24-h	11.52 (±3.05)	**11.48** (±3.57)	11.56 (±2.03)	13.93 (±3.20)	12.84 (±4.00)	12.79 (±4.10)	15.71 (±3.83)	12.94 (±4.00)	13.14 (±4.03)

5.2 Technical Relevance

In this work we impartially evaluated different models and approaches for BP prediction from PPG. Most of the methods have only been previously validated on MIMIC. Also, due to the large volume of existing work, very often the models were not compared against all available approaches. To the best of our knowledge, this is also the first work to compare the scalogram, spectrogram and feature extraction approaches. We employ strong cross validation methods to make sure our results are robust. Our models and code are available openly to make sure this results can be reproduced and also applied to similar datasets when needed.

5.3 Limitations and Future Work

The major limitation of our work is related to the small size of the datasets we used. For that reason, it was not possible to train a deep Long Short Term Memory (LSTM) network, which in a few recent papers have demonstrated very promising results [23]. In future work, we would like to extend our dataset with more diverse patient populations and also with a longer observation period per patient. We might consider data augmentation techniques as well. This will allow us to apply more data-demanding learning algorithms and, at the same time, to investigate how models trained in one population perform in a different one.

6   Conclusion

In conclusion, we presented a comprehensive comparison of different machine learning approaches to predict BP from PPG in two different datasets. We demonstrate that despite the plethora of work in this area, there exists a dearth of models that perform well in uncontrolled environments when the subjects indulge in various day-to-day activities. To achieve a MAE (≤ 5 mmHg), which is considered good by the Association for the Advancement of Medical Instrumentation® (AAMI) [22] we still have a long way to go. Moreover, we showed that for small to medium sized datasets feature extraction methods can produce better results than the recent image based approaches. We hope our work will inspire others to dig deeper into the generalizability and improve the accuracy of these models.

Acknowledgements. We would like to thank Manisha Manaswini, Felix Musmann, Juan Carlos Niño Rodriguez, and Carolin Müller for their help during data collection and, also Harry Freitas da Cruz and Attila Wohlbrandt for giving many valuable insights.

Appendix

A   Data and Code Availability

The code for the experiments is available at: https://github.com/arianesasso/aime-2020. Information on the HYPE dataset is also provided there. The EVAL dataset can be found at: https://www.kaggle.com/mkachuee/noninvasivebp.

B   Feature extraction

The features that were extracted from the PPG cycles are described in Table 4 and in Fig. 2 following the work of Kurylyak et al. [12].

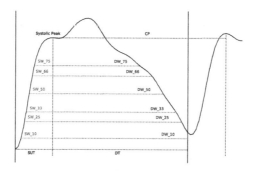

Fig. 2. Extracted features from PPG

Table 4. Features extracted from PPG [12]. Here $n \in \{10, 25, 33, 50, 66, 75\}$

Name	Description
SUT	Systolic Upstroke Time
DT	Diastolic Time
CP	Cardiac Period
DW_n	Diastolic Width at n% amplitude
$SW_n + DW_n$	Sum of Systolic Width and Diastolic Width at n% amplitude
DW_n/SW_n	Ratio between Systolic Width and Diastolic Width at n% amplitude

C   Experiments

More information and details on the methods and experiments can be found at: https://figshare.com/projects/AIME_2020/85166.

References

1. Challoner, A.V., Ramsay, C.A.: A photoelectric plethysmograph for the measurement of cutaneous blood flow. Phys. Med. Biol. **19**(3), 317–328 (1974). https://doi.org/10.1088/0031-9155/19/3/003
2. Elgendi, M., et al.: The use of photoplethysmography for assessing hypertension. NPJ Digit. Med. **2**(1), 60 (2019). https://doi.org/10.1038/s41746-019-0136-7
3. Esmaili, A., Kachuee, M., Shabany, M.: Nonlinear cuffless blood pressure estimation of healthy subjects using pulse transit time and arrival time. IEEE Trans. Instrum. Meas. **66**(12), 3299–3308 (2017)
4. Friedman, J.H.: Greedy function approximation: a gradient boosting machine. Ann. Stat. **29**, 1189–1232 (2001)
5. Ghamari, M.: A review on wearable photoplethysmography sensors and their potential future applications in health care. Int. J. Biosens. Bioelectron. **4**(4), 195–202 (2018)
6. Gholamhosseini, H., Meintjes, A., Baig, M.M., Lindén, M.: Smartphone-based continuous blood pressure measurement using pulse transit time. In: pHealth, pp. 84–89 (2016)
7. Goldberger, A.L., et al.: PhysioBank, physioToolkit, and physioNet: components of a new research resource for complex physiologic signals. Circulation **101**(23), e215–e220 (2000)
8. He, K., Zhang, X., Ren, S., Sun, J.: Delving deep into rectifiers: surpassing human-level performance on imagenet classification. In: Proceedings of the IEEE International Conference on Computer Vision, pp. 1026–1034 (2015)
9. He, K., Zhang, X., Ren, S., Sun, J.: Deep residual learning for image recognition. In: Proceedings of the IEEE Conference on Computer Vision and Pattern Recognition, pp. 770–778 (2016)
10. Ke, G., et al.: LightGBM: a highly efficient gradient boosting decision tree. In: Advances in Neural Information Processing Systems, pp. 3146–3154 (2017)

11. Krizhevsky, A., Sutskever, I., Hinton, G.E.: ImageNet classification with deep convolutional neural networks. In: Advances in Neural Information Processing Systems, pp. 1097–1105 (2012)
12. Kurylyak, Y., Lamonaca, F., Grimaldi, D.: A neural network-based method for continuous blood pressure estimation from a PPG signal. In: Conference Record - IEEE Instrumentation and Measurement Technology Conference, pp. 280–283. IEEE, May 2013. https://doi.org/10.1109/I2MTC.2013.6555424
13. Li, Q., Clifford, G.D.: Dynamic time warping and machine learning for signal quality assessment of pulsatile signals. Physiol. Meas. **33**(9), 1491 (2012)
14. Liang, Y., Chen, Z., Ward, R., Elgendi, M.: Photoplethysmography and deep learning: enhancing hypertension risk stratification. Biosensors **8**(4), 101 (2018)
15. Liang, Y., Elgendi, M., Chen, Z., Ward, R.: Analysis: an optimal filter for short photoplethysmogram signals. Sci. Data **5**, 1–12 (2018). https://doi.org/10.1038/sdata.2018.76
16. Lim, S.S., Vos, T., Flaxman, A.D., et al.: A comparative risk assessment of burden of disease and injury attributable to 67 risk factors and risk factor clusters in 21 regions, 1990–2010: a systematic analysis for the Global Burden of Disease Study 2010. Lancet **380**(9859), 2224–2260 (2012). https://doi.org/10.1016/S0140-6736(12)61766-8
17. Luštrek, M., Slapničar, G.: Blood pressure estimation with a wristband optical sensor. In: UbiComp, pp. 758–761 (2018). https://doi.org/10.1145/3267305.3267708
18. Manamperi, B., Chitraranjan, C.: A robust neural network-based method to estimate arterial blood pressure using photoplethysmography. In: 2019 IEEE 19th International Conference on Bioinformatics and Bioengineering (BIBE), pp. 681–685. IEEE, October 2019. https://doi.org/10.1109/BIBE.2019.00128
19. Shcherbina, A., et al.: Accuracy in wrist-worn, sensor-based measurements of heart rate and energy expenditure in a diverse cohort. J. Pers. Med. **7**(2), 3 (2017). https://doi.org/10.3390/jpm7020003
20. Slapničar, G., Mlakar, N., Luštrek, M.: Blood pressure estimation from photoplethysmogram using a spectro-temporal deep neural network. Sensors (Switz.) **19**(15) (2019). https://doi.org/10.3390/s19153420
21. Smith, J.O., III.: Mathematics of the Discrete Fourier Transform (DFT): With Audio Applications, 2nd edn. Booksurge, Charleston (2007)
22. Stergiou, G.S., et al.: A universal standard for the validation of blood pressure measuring devices. Hypertension **71**(3), 368–374 (2018). https://doi.org/10.1161/HYPERTENSIONAHA.117.10237
23. Su, P., Ding, X.R., Zhang, Y.T., Liu, J., Miao, F., Zhao, N.: Long-term blood pressure prediction with deep recurrent neural networks. In: 2018 IEEE EMBS International Conference on Biomedical & Health Informatics (BHI), pp. 323–328. IEEE (2018)
24. Wang, L., Zhou, W., Xing, Y., Zhou, X.: A novel neural network model for blood pressure estimation using photoplethesmography without electrocardiogram. J. Healthc. Eng. (2018). https://doi.org/10.1155/2018/7804243
25. Whelton, P.K., Carey, R.M., Aronow, W.S., et al.: 2017 ACC/AHA/AAPA/ABC/ACPM/AGS/APhA/ASH/ASPC/NMA/PCNA guideline for the prevention, detection, evaluation, and management of high blood pressure in adults. Hypertension **71**(6), e13–e115 (2018). https://doi.org/10.1161/HYP.0000000000000065
26. Zou, H., Hastie, T.: Regularization and variable selection via the elastic net. J. Roy. Stat. Soc.: Ser. B (Stat. Methodol.) **67**(2), 301–320 (2005)

Mortality Risk Score for Critically Ill Patients with Viral or Unspecified Pneumonia: Assisting Clinicians with COVID-19 ECMO Planning

Helen Zhou(iD), Cheng Cheng(iD), Zachary C. Lipton(iD), George H. Chen(iD), and Jeremy C. Weiss$^{(\boxtimes)}$(iD)

Carnegie Mellon University, Pittsburgh, PA 15213, USA
{hlzhou,ccheng2,zlipton,georgechen,jeremyweiss}@cmu.edu

Abstract. Respiratory complications due to coronavirus have claimed hundreds of thousands of lives in 2020. Extracorporeal membrane oxygenation (ECMO) is a life-sustaining oxygenation and ventilation therapy that may be used when mechanical ventilation is insufficient. While early planning and surgical cannulation for ECMO can increase survival, clinicians report the lack of a risk score hinders these efforts. We develop the PEER score to highlight critically ill patients with viral or unspecified pneumonia at high risk of mortality in a subpopulation eligible for ECMO. The score is validated across two critical care datasets, and predicts mortality at least as well as other existing risk scores.

Keywords: Mortality risk score · Pneumonia · COVID-19 · ARDS

1 Introduction

Coronavirus disease COVID-19 has infected millions globally. Many cases progress from Severe Acute Respiratory Syndrome (SARS-CoV-2) with viral pneumonia to acute respiratory distress syndrome (ARDS) to death. ECMO can temporarily sustain patients with severe ARDS when mechanical ventilation fails to facilitate with oxygenation via lungs. However, ECMO is costly and applicable only for patients healthy enough to recover and return to a high functional status.

While ECMO is more effective when planned in advance [7], applicable risk scores remain unavailable [2,17]. This paper introduces the Viral or Unspecified **P**neumonia **E**CMO-**E**ligible **R**isk (PEER) Score, using measurements from the time of would-be planning—early in the critical care stay. In contrast to existing pneumonia risk scores [6,8,18,19], the PEER score targets those with viral or unspecified pneumonia in the critical care setting, for a cohort potentially eligible

---

H. Zhou and C. Cheng—Equal Contribution.

M. Michalowski and R. Moskovitch (Eds.): AIME 2020, LNAI 12299, pp. 336–347, 2020.
https://doi.org/10.1007/978-3-030-59137-3_30

for ECMO. Unspecified pneumonia is included since the infectious etiology of pneumonia often cannot be determined, and it broadens the study population.

Though limited by geographic availability, ECMO usage has increased 4-fold in the last decade [22]. COVID-19 guidelines suggest ECMO as a late option in escalation of care for severe ARDS secondary to SARS-CoV-2 infection [1,17]. However, early epidemiological studies of coronavirus [27,30,31] have yet to establish ECMO's utility. A pooled analysis of four studies [13] showed mortality rates of 95% with ECMO vs. 70% without, but the number of ECMO recipients was small, and no studies described a protocol specifying indications for ECMO.

To better understand the role of ECMO as a rescue for ventilation non-responsive, SARS-CoV-2 ARDS, we study its broader use in ARDS. Treatment guidelines suggest ECMO use in severe ARDS alongside other advanced ventilation strategies [20,28], with the World Health Organization citing effectiveness for ARDS and reducing mortality of the Middle East Respiratory Syndrome (MERS). Despite these recommendations and allocated ECMO resources [22], risk scores tailored to ECMO consideration are lacking. Our study addresses this by drawing from viral and source unidentified cases of pneumonia that escalate to critical care admissions, guided by the intuition that ARDS from these pneumonia are expected to better resemble COVID-19 ARDS than all-comer ARDS.

Related Work. There are a number of pneumonia [6,8,11,18,26], COVID-19 [9,10,15], hospitalization mortality [32], and ECMO risk scores [23], but none center on the time of risk evaluation for ECMO candidacy. The pneumonia and COVID-19 risk scores are assessed on populations with lower acuity, while APACHE is not focused on respiratory illness. Our risk score is meant for use in ECMO planning rather than predicting outcomes among patients already receiving ECMO. Registry-based studies have also compared SARS-CoV-2 outcomes to that of other viral infections, including MERS, H1N1 flu, and seasonal flu. One MERS-related ARDS study of critically ill patients demonstrated higher mortality than those in studies on COVID-related ARDS, but may be attributed to sicker patients at enrollment [4]. A similar H1N1 study reported lower mortality (12–17%), albeit considering a younger population (average age 40) [3].

Physiologic concerns have also been raised about the use of ECMO for SARS-CoV-2. One argues that while ECMO is primarily beneficial for respiratory recovery, a spike in all-cause death but not ARDS-related death could indicate a limited role of ECMO [14]. Others point out that COVID-associated lymphopenia might be exacerbated by ECMO-induced lymphopenia which could mechanistically affect a healthy immune response to infection. Inflammatory cytokines and specifically interleukin 6 elevation is associated with COVID-19 mortality and rises with the use of ECMO [5,13]. These expert voices do not argue for the avoidance of ECMO, but rather call for additional study.

2   Data

The eICU Collaborative Research Database [21] contains 200,859 admissions to intensive care units (ICU) across multiple centers in the United States between 2014 and 2015. The MIMIC-III clinical database [16] consists of data from 46,476 patients who stayed in critical care units of the Beth Israel Deaconess Medical Center between 2001 and 2012. Model development and in-domain validation primarily use data from eICU, and out-of-domain validation uses MIMIC-III.

Cohort Selection. Inclusion criteria for the study cohort are delineated in Fig. 1. The population of interest is among patients with viral or otherwise unspecified non-bacterial, non-fungal, non-parasitic, and non-genetic pneumonia. While there are no absolute contraindications of ECMO, the therapy is reserved for patients likely to have functional recovery. Patients over 70 years old would not be good candidates for ECMO, and SARS-CoV-2 pneumonia progressing to hypoxic respiratory failure is exceedingly rare in patients under 18. Other relative contraindications to ECMO are also listed in Fig. 1. We select the first ICU stay within each patient's hospital stay, and exclude patients who died or were discharged within the first 48 h of being admitted. This is done to focus on the stage of critical care after initial entry when lower-risk oxygen supplementation strategies (*e.g.*, ventilation) are being performed, and, methodologically, to provide a richer set of features for prediction. Table 1 and Appendix Table 4 summarize characteristics of the cohorts.

Data Extraction. The study cohorts are extracted using string matching on diagnosis codes and subsequent clinician review. Features are merged through a process of visualization, query, and physician review. This includes harmonizing feature units, removing impossible values, and merging redundant data fields. Additional details are in Appendix B. All features are combined into a fixed-length vector, using the most recent value prior to 48 h after ICU admission. Before imputation, approximately half of the features had missingness below 5%, and 80% of the features had missingness below 30%, however multiple variables had high missingness (Appendix B). Missing values are imputed using MissForest [25], which we find PEER is insensitive to (Appendix B).

Features. Features are extracted from demographics, comorbidities, vitals, physical exams, and lab findings routinely collected in critical care settings. Numerical features are normalized, and categorical features are converted with dummy variables. All variables in Tables 1 and 4 are provided to the model.

Outcomes. Our primary outcome of interest is in-ICU mortality. Secondary outcomes indicating decompensation are vasopressor use and mechanical ventilation use. For each outcome, we define the time to event as the time to first outcome or censorship, where censorship corresponds to discharge from the ICU.

3   Methods

Lasso-Cox. To predict patient survival, we use the Cox proportional hazards model with L1 regularization, referred to as *Lasso-Cox* [24]. Lasso-Cox is chosen for its ease of interpretation and calculation, owing to its selection of sparse models.[1] For a patient with covariates $\mathbf{x} \in \mathbb{R}^d$, the predicted log hazard is $\beta^\top \mathbf{x}$, (higher hazard implies shorter survival time), where $\beta \in \mathbb{R}^d$ are coefficients that can be interpreted as log hazard ratios. L1 regularization $\lambda \sum_{j=1}^d |\beta_j|$ is used to encourage sparsity in β, where $\lambda > 0$ is a user-specified hyperparameter.

Table 1. Demographics and outcomes of patients with viral or unspecified pneumonia in eICU and MIMIC-III cohorts. Data are median (Q1–Q3) or count (% out of n).

	Variable	eICU (n = 3617)	MIMIC (n = 937)
Demographics	Age, years	58.0 (48.0–64.0)	54.5 (44.1–62.7)
	18–30	225 (6.2%)	83 (8.9%)
	30–39	277 (7.7%)	94 (10.0%)
	40–49	500 (13.8%)	159 (17.0%)
	50–59	1064 (29.4%)	281 (30.0%)
	60–70	1546 (42.7%)	320 (34.2%)
	Male	1949 (53.9%)	542 (57.8%)
	Female	1663 (46.0%)	395 (42.2%)
Out.	Deceased	270 (7.5%)	94 (10.0%)
	Vasopressors administered	589 (16.3%)	389 (41.5%)
	Ventilator used	1835 (50.7%)	758 (80.9%)

Evaluation Metrics. To evaluate model performance, we consider concordance and calibration. *Concordance* (c-index) is a common measure of goodness-of-fit in survival models [12], defined as the fraction of pairs of subjects whose survival times are correctly ordered by a prediction algorithm, among all pairs that can be ordered. Confidence intervals are computed using 1000 bootstrapped samples. We evaluate *calibration* by plotting the Kaplan-Meier observed survival probability versus the predicted survival probability. We construct our calibration plots (Fig. 3) [29] with 1000 bootstrap resamplings for internal calibration. Both internal and external calibrations use 5 groups for 7 days.[2]

Experimental Setup. The eICU cohort is divided into a training set (70% of the data, n = 2537) and test set (30%, n = 1080). The eICU training set is used for model development, whereas the eICU test set and entirety of the MIMIC cohort

---

[1] We also tried the Cox model with elastic-net regularization (combined L1 and L2 regularization) but found little to no gain in cross-validation concordance.

[2] We plot at day 7 instead of 30 because censorship level is too high beyond a week.

are used for model evaluation. Throughout our evaluation, we compare our risk score (PEER) to three pneumonia risk scores: CURB-65 [26], PSI/PORT [8], and SMART-COP [6]; and one COVID-19 risk score: GOQ [10].

Model Selection. We select λ via 10-fold cross validation and grid search on the eICU training set to maximize concordance subject to sufficient sparsity. We observe that $\lambda = 0.01$ gives the best trade-off between concordance (0.73) and number of features selected (18), as a 0.01 increase in concordance corresponds to 10 additional non-zero features. To check the stability of this hyperparameter choice, we impute our data using ten random seeds and run 10-fold cross validation on the resulting datasets. Across all runs, $\lambda = 0.01$ achieves concordance of approximately 0.73 and selects similar features and coefficients. Additional details about grid search, the concordance and sparsity tradeoff, and robust selection of coefficients can be found in Appendix B. Code for data extraction and all model results is available at https://github.com/hlzhou/peer-score.

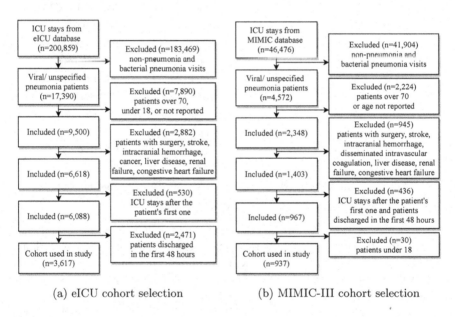

(a) eICU cohort selection   (b) MIMIC-III cohort selection

Fig. 1. Inclusion and exclusion criteria for cohorts extracted from eICU and MIMIC. Disseminated intravascular coagulation was highly missing from eICU.

4 Results

The hazard ratios from Lasso-Cox with $\lambda = 0.01$ are displayed in Table 2. For easy calculation of the PEER score, we also provide a nomogram (Fig. 2)[3].

The PEER score achieves concordance greater than or comparable to that of existing risk scores on all datasets (Table 3). On the eICU test set, PEER achieves the highest concordance among the risk scores, 0.77. On MIMIC, the maximum concordance degrades to 0.66, achieved by PEER and SMART-COP. The PEER calibration curves (Fig. 3) show one high risk group separate from low risk groups. While predicted survival of the high risk group is overestimated in the training set, it is within confidence intervals in both test sets.

We define low and high risk subpopulations by thresholding our model's predicted risks on the training set at the 90th percentile. Each group's Kaplan-Meier survival curves are plotted over a 30-day period (Fig. 4). For the first week, the low and high risk curves are clearly distinct (Fig. 4), with respective survival

Table 2. Hazard ratios (HR) for the Lasso-Cox model, i.e. the PEER score. HR and 95% confidence intervals (CI) are reported on normalized data. Means and standard deviations used for scaling are included for reference.

Feature	HR (95% CI)	Mean	Std. dev.
Age (years)	1.22 (1.04–1.43)	54.5	12.5
Heart rate (beats per minute)	1.13 (0.984–1.3)	89.4	17.8
Systolic blood pressure (mmHg)	0.928 (0.755–1.14)	122	22
Diastolic blood pressure (mmHg)	0.996 (0.745–1.33)	67.7	15.1
Mean arterial pressure (mmHg)	0.926 (0.673–1.27)	83.7	17.9
Glasgow Coma Scale	0.93 (0.803–1.08)	11.3	3.26
White blood cells (thousands/μL)	0.984 (0.871–1.11)	12.9	8.91
Platelets (thousands/μL)	0.924 (0.79–1.08)	208	108
Red blood cell dist. width (%)	1.24 (1.08–1.43)	15.8	2.47
Neutrophils (%)	0.972 (0.853–1.11)	79.1	13
Blood urea nitrogen (mg/dL)	1.07 (0.937–1.23)	25.1	19.5
Aspartate aminotransferase (units/L)	1.12 (1.06–1.18)	143	774
Direct bilirubin (mg/L)	1.03 (0.935–1.13)	0.385	0.816
Albumin (g/dL)	0.954 (0.82–1.11)	2.65	0.636
Troponin (ng/mL)	1.06 (0.985–1.14)	1.07	3.85
Prothrombin time (sec)	1.05 (0.909–1.2)	16.6	6.75
pH	0.856 (0.75–0.977)	7.38	0.0713
Arterial oxygen saturation (mmHg)	0.787 (0.723–0.856)	95.8	4.12

---

[3] To compute risk, look up a patient's values in the nomogram, match it to points listed across the top, add them up, and look up the total in the scale across the bottom.

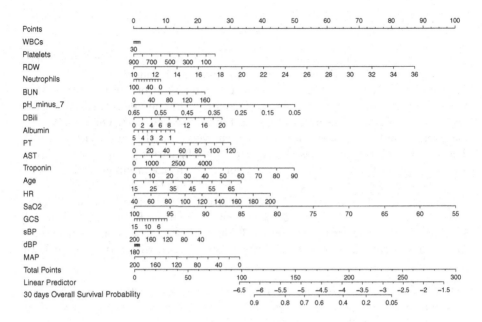

Fig. 2. Nomogram for manual calculation of the PEER score.

proportions 0.68 and 0.95 on eICU test, and 0.75 and 0.95 on MIMIC. Beyond the first week, censorship grows quickly and there is less data, resulting in increased uncertainty. Compared to low and high risk curves derived from related risk scores, those of the PEER score are the most separated (Appendix B). Secondary indicators of decompensation (i.e. ventilator and vasopressor use) are also more common in the high risk group than the low risk group (Fig. 5).

Table 3. Concordances (and 95% confidence intervals) of the PEER score, CURB-65, PSI/PORT, SMART-COP, and GOQ.

Score	Train eICU	Test eICU	MIMIC
PEER (ours)	**0.77 (0.72–0.81)**	**0.77 (0.69–0.83)**	**0.66 (0.57–0.74)**
CURB-65 [26]	0.66 (0.61–0.70)	0.62 (0.55–0.69)	0.59 (0.52–0.66)
PSI/PORT [8]	0.71 (0.66–0.76)	0.71 (0.63–0.78)	0.62 (0.55–0.69)
SMART-COP [6]	0.69 (0.64–0.73)	0.73 (0.67–0.80)	**0.66 (0.59–0.72)**
GOQ [10]	0.67 (0.63–0.71)	0.62 (0.54–0.70)	0.58 (0.50–0.66)

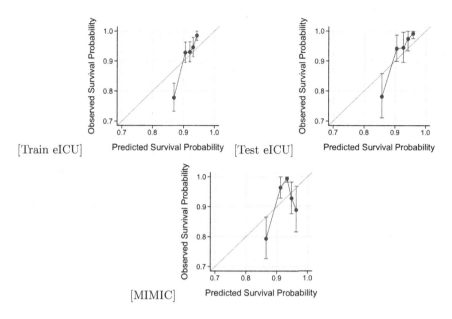

Fig. 3. Calibration plots with 95% confidence intervals.

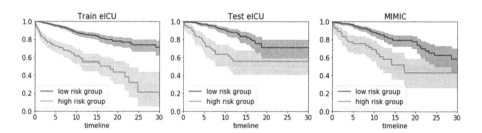

Fig. 4. Kaplan-Meier survival curves of high vs. low risk groups in train eICU, test eICU, and MIMIC. Shaded regions are the 95% confidence intervals.

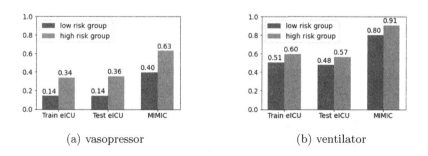

Fig. 5. Proportion of each subgroup that received vasopressors or ventilators.

Table 4. Summary characteristics per cohort, with median (Q1–Q3) or count (% of n).

		Variable	eICU (n = 3617)	MIMIC (n = 937)
Physical exam findings		Orientation		
		Oriented	1121 (31.0%)	411 (43.9%)
		Confused	1287 (35.6%)	76 (8.1%)
		Temperature (°C)	36.9 (36.6–37.3)	37.2 (36.6–37.7)
		Heart rate (beats per minute)	89.0 (77.0–101.0)	90.0 (78.0–104.0)
		Respiratory rate (breaths per minute)	20.0 (17.0–25.0)	20.0 (16.0–25.0)
		Systolic blood pressure (mmHg)	120.0 (106.0–136.0)	118.0 (104.0–134.0)
		Diastolic blood pressure (mmHg)	66.0 (57.0–76.0)	63.0 (54.0–72.0)
		Mean arterial pressure (mmHg)	81.0 (72.0–93.0)	79.0 (71.0–90.0)
		Glasgow Coma Scale	14.0 (10.0–15.0)	14.0 (9.0–15.0)
Laboratory findings (Abbrevations: Coagulation as Coag. and Blood Gas as B.G.)	Hemotology	Red blood cells (millions/µL)	3.5 (3.0–4.0)	3.4 (3.0–3.8)
		White blood cells (thousands/µL)	11.0 (7.9–15.6)	11.0 (8.0–15.1)
		Platelets (thousands/µL)	193.0 (136.0–261.0)	199.0 (128.8–276.0)
		Hematocrit (%)	31.1 (27.2–35.6)	30.2 (27.0–33.6)
		Red blood cell dist. width (%)	15.2 (14.0–16.8)	14.8 (13.8–16.4)
		Mean corpuscular volume (fL)	90.4 (86.0–95.0)	89.0 (85.0–93.0)
		Mean corpuscular hemoglobin/ MCH (pg)	29.7 (27.9–31.2)	30.2 (28.7–31.6)
		MCH concentration (g/dL)	32.7 (31.7–33.6)	33.8 (32.8–34.8)
		Neutrophils (%)	82.0 (73.3–89.0)	82.3 (73.8–88.5)
		Lymphocytes (%)	8.4 (5.0–14.0)	9.5 (5.8–15.7)
		Monocytes (%)	6.0 (3.7–8.6)	4.0 (2.7–5.9)
		Eosinophils (%)	0.1 (0.0–1.0)	0.4 (0.0–1.2)
		Basophils (%)	0.0 (0.0–0.3)	0.1 (0.0–0.3)
		Band cells (%)	8.0 (3.0–17.0)	0.0 (0.0–5.0)
	Chemistry	Sodium (mmol/L)	139.0 (136.0–142.0)	139.0 (136.0–142.0)
		Potassium (mmol/L)	3.9 (3.6–4.3)	3.9 (3.6–4.3)
		Chloride (mmol/L)	105.0 (101.0–109.0)	105.0 (101.0–109.0)
		Bicarbonate (mmol/L)	25.0 (22.0–28.0)	26.0 (23.0–29.0)
		Blood urea nitrogen (mg/dL)	19.0 (12.0–33.0)	17.0 (11.0–28.0)
		Creatinine (mg/dL)	0.8 (0.6–1.4)	0.8 (0.6–1.3)
		Glucose (mg/dL)	131.0 (105.0–165.0)	124.0 (104.5–151.5)
		Aspartate aminotransferase (units/L)	30.0 (19.0–57.0)	37.0 (22.0–70.0)
		Alanine aminotransferase (units/L)	27.0 (16.0–47.0)	28.0 (18.0–52.0)
		Alkaline phosphatase (units/L)	84.0 (62.0–117.0)	85.0 (62.0–121.0)
		Direct bilirubin (mg/L)	0.2 (0.1–0.5)	0.6 (0.2–2.2)
		Total bilirubin (mg/L)	0.5 (0.3–0.8)	0.6 (0.4–1.1)
		Total protein (g/dL)	6.0 (5.3–6.7)	6.1 (5.3–7.0)
		Calcium (mg/dL)	8.2 (7.7–8.6)	8.2 (7.8–8.6)
		Albumin (g/dL)	2.6 (2.2–3.1)	3.0 (2.6–3.5)
		Troponin (ng/mL)	0.1 (0.0–0.2)	0.0 (0.0–0.3)
	Coag.	Prothrombin time (sec)	14.5 (12.7–16.7)	13.9 (13.0–15.3)
		Partial thromboplastin time (sec)	33.0 (28.5–41.0)	30.2 (26.6–36.9)
	B.G.	pH	7.39 (7.33–7.43)	7.41 (7.36–7.45)
		Partial pressure of oxygen (mmHg)	83.0 (68.0–111.0)	97.0 (73.5–127.5)
		Arterial oxygen saturation (mmHg)	96.0 (94.0–99.0)	97.0 (95.0–98.0)

5 Discussion

The PEER score achieves greater or comparable concordance to baselines on the eICU (in-domain) and MIMIC (out-of-domain) test sets. Lasso-Cox selects 18 features, making for easy computation. Qualitatively, the score is consistent with clinical intuition. SaO2, associated with poorer oxygenation status, is predictive of decompensation. Old age is predictive of death. Red blood cell distribution width, associated with expanded release of immature red blood cells in response to insufficient oxygen delivery to tissues, is also a strong risk factor for death with COVID-19 [9]. However, the hazard ratios themselves should be interpreted with caution as three variables (pH, prothrombin time, and age) violate the proportional hazards assumption, and L1 regularization shrinks coefficients towards 0.

Stratifying each cohort into high and low risk subpopulations based the PEER score, we observe a clear separation in their survival curves (Fig. 4) across all three datasets. Additionally, secondary indicators of decompensation (e.g. vasopressor and ventilator use) are more prevalent in the high risk group (Fig. 5). Calibration plots for PEER also show a high risk group separated from the rest (Fig. 3). While the survival probability of the high risk group is overestimated on the eICU training set, it is within error bars on all test sets.

For ECMO allocation, practically, accurate *ranking* of risk, as measured by concordance, may be more important than the precise probabilities predicted. The PEER score outperforms other risk scores on the eICU test set, but there is a decline in performance on the MIMIC test set, and the performance of PEER becomes comparable to that of SMART-COP. One possible reason for this decline is that in MIMIC, an important feature for PEER, the arterial oxygen saturation (SaO2), has 72.6% missingness. In contrast, it has 1.5% missingness in eICU. This demonstrates the importance of thinking critically about how our risk score, which was trained on the eICU cohort and depends on 18 specific features, generalizes to the population to which the score is being applied.

Limitations and Future Work. Importantly our cohort is defined not by COVID-19 positive pneumonia patients but instead by viral or unspecified pneumonia patients who are ECMO-eligible. While our risk score demonstrates good discriminative ability and is interpretable, there are several additional decision-making considerations beyond the scope of this paper. Clinicians interested in applying the risk score to COVID-19 pneumonia should consider how representative this population is of their own. Because ECMO is a constrained resource, there are also ethical questions about who should get treatment. This risk score does not attempt to address these questions, but simply provides relevant information to those making such decisions. More broadly, we hope to provide this risk score as a potential resource for future SARS-like diseases that require ECMO consideration.

A   Summary Characteristics

B   Extended Version

Additional details can be found at https://arxiv.org/abs/2006.01898.

References

1. Alhazzani, W., Møller, M.H., et al.: Surviving sepsis campaign: guidelines on the management of critically ill adults with coronavirus disease 2019 (COVID-19). Intensive Care Med. **46**, 1–34 (2020). https://doi.org/10.1007/s00134-020-06022-5
2. American College of Cardiology: ACC's COVID-19 Hub (2020)
3. Aokage, T., Palmér, K., et al.: Extracorporeal membrane oxygenation for acute respiratory distress syndrome. J. Intensive Care **3**(1), 17 (2015). https://doi.org/10.1186/s40560-015-0082-7
4. Arabi, Y.M., Al-Omari, A., et al.: Critically ill patients with the Middle East respiratory syndrome: a multicenter retrospective cohort study. Crit. Care Med. **45**(10), 1683–1695 (2017)
5. Bizzarro, M.J., Conrad, S.A., et al.: Infections acquired during extracorporeal membrane oxygenation in neonates, children, and adults. Pediatr. Crit. Care Med. **12**(3), 277–281 (2011)
6. Charles, P., Wolfe, R., et al.: SMART-COP: a tool for predicting the need for intensive respiratory or vasopressor support in community-acquired pneumonia. Clin. Infect. Dis. **47**(3), 375–384 (2008)
7. Combes, A., Hajage, D., et al.: ECMO for severe acute respiratory distress syndrome. NEJM **378**(21), 1965–1975 (2018)
8. Fine, M.J., Auble, T.E., et al.: A prediction rule to identify low-risk patients with community-acquired pneumonia. NEJM **336**(4), 243–250 (1997)
9. Gong, J., Ou, J., et al.: Multicenter development and validation of a novel risk nomogram for early prediction of severe 2019-novel coronavirus pneumonia. Available at SSRN 3551365 (2020)
10. Gong, J., et al.: A tool to early predict severe 2019-novel coronavirus pneumonia (covid-19): a multicenter study using the risk nomogram in Wuhan and Guangdong, China. medRxiv (2020)
11. Guo, L., Wei, D., et al.: Clinical features predicting mortality risk in patients with viral pneumonia: the MuLBSTA score. Front. Microbiol. **10**(December), 1–10 (2019)
12. Harrell, F.E., et al.: Evaluating the yield of medical tests. JAMA **247**(18), 2543–2546 (1982)
13. Henry, B.M.: COVID-19, ECMO, and lymphopenia: a word of caution. Lancet Respir. Med. **8**, e24 (2020)
14. Henry, B.M., Lippi, G.: Poor survival with extracorporeal membrane oxygenation in acute respiratory distress syndrome due to coronavirus disease 2019: pooled analysis of early reports. J. Crit. Care (2020)
15. Jiang, X., Coffee, M., et al.: Towards an artificial intelligence framework for data-driven prediction of coronavirus clinical severity (2020)
16. Johnson, A.E.W., Pollard, T.J., et al.: MIMIC-III, a freely accessible critical care database. Sci. Data **3**(1), 160035 (2016)
17. Liang, T., et al.: Handbook of COVID-19 prevention and treatment. Zhejiang University School of Medicine. Compiled According to Clinical Experience, The First Affiliated Hospital (2020)
18. Lim, W.S., Van der Eerden, M., et al.: Defining community acquired pneumonia severity on presentation to hospital: an international derivation and validation study. Thorax **58**(5), 377–382 (2003)
19. Marti, C., Garin, N., et al.: Prediction of severe community-acquired pneumonia: a systematic review and meta-analysis. Crit. Care **16**(4), R141 (2012). https://doi.org/10.1186/cc11447

20. Matthay, M., Aldrich, J., Gotts, J.: Treatment for severe ARDS from COVID-19. The Lancet Respiratory Medicine (2020)
21. Pollard, T.J., Johnson, A.E., et al.: The eICU collaborative research database, a freely available multi-center database for critical care research. Sci. Data **5**, 1–13 (2018)
22. Ramanathan, K., Antognini, D., et al.: Planning and provision of ECMO services for severe ARDS during the COVID-19 pandemic and other outbreaks of emerging infectious diseases. Lancet Respir. Med. **8**, 518–526 (2020)
23. Schmidt, M., Burrell, A., et al.: Predicting survival after ECMO for refractory cardiogenic shock: the survival after veno-arterial-ECMO (SAVE)-score. Eur. Heart J. **36**(33), 2246–2256 (2015)
24. Simon, N., Friedman, J., et al.: Regularization paths for Cox's proportional hazards model via coordinate descent. J. Stat. Softw. **39**(5), 1–13 (2011)
25. Stekhoven, D.J., Bühlmann, P.: MissForest–non-parametric missing value imputation for mixed-type data. Bioinformatics **28**(1), 112–118 (2011)
26. Lim, W.S., et al.: Defining community acquired pneumonia severity on presentation to hospital: an international derivation and validation study. Thorax **58**(5), 377–382 (2003)
27. Wang, D., Hu, B., et al.: Clinical characteristics of 138 hospitalized patients with 2019 novel coronavirus-infected pneumonia in Wuhan, China. JAMA - J. Am. Med. Assoc. **323**(11), 1061–1069 (2020)
28. World Health Organization, et al.: Clinical management of severe acute respiratory infection (sari) when COVID-19 disease is suspected: interim guidance, 13 March 2020. Technical report (2020)
29. Xiao, N., Xu, Q.S., Li, M.Z.: hdnom: building nomograms for penalized Cox models with high-dimensional survival data. bioRxiv (2016)
30. Yang, X., Yu, Y., et al.: Clinical course and outcomes of critically ill patients with SARS-CoV-2 pneumonia in Wuhan, China: a single-centered, retrospective, observational study. Lancet Respir. Med. **8**, 475–481 (2020)
31. Zhou, F., Yu, T., et al.: Clinical course and risk factors for mortality of adult inpatients with COVID-19 in Wuhan, China: a retrospective cohort study. Lancet **395**, 1054–1062 (2020)
32. Zimmerman, J.E., Kramer, A.A., et al.: Acute physiology and chronic health evaluation (APACHE) IV: hospital mortality assessment for today's critically ill patients. Crit. Care Med. **34**(5), 1297–1310 (2006)

Diagnostic Prediction with Sequence-of-sets Representation Learning for Clinical Events

Tianran Zhang[1,2(✉)], Muhao Chen[3], and Alex A. T. Bui[1,2]

[1] Department of Bioengineering, UCLA, Los Angeles, USA
`tianranzhang@ucla.edu`
[2] UCLA Medical and Imaging Informatics (MII), Los Angeles, USA
[3] Department of Computer and Information Science, UPenn, Philadelphia, USA

Abstract. Electronic health records (EHRs) contain both ordered and unordered chronologies of clinical events that occur during a patient encounter. However, during data preprocessing steps, many predictive models impose a predefined order on unordered clinical events sets (e.g., alphabetical, natural order from the chart, etc.), which is potentially incompatible with the temporal nature of the sequence and predictive task. To address this issue, we propose DPSS, which seeks to capture each patient's clinical event records as sequences of event sets. For each clinical event set, we assume that the predictive model should be invariant to the order of concurrent events and thus employ a novel permutation sampling mechanism. This paper evaluates the use of this permuted sampling method given different data-driven models for predicting a heart failure (HF) diagnosis in subsequent patient visits. Experimental results using the MIMIC-III dataset show that the permutation sampling mechanism offers improved discriminative power based on the area under the receiver operating curve (AUROC) and precision-recall curve (pr-AUC) metrics as HF diagnosis prediction becomes more robust to different data ordering schemes.

Keywords: Clinical event sequences · Set learning · Diagnostic prediction

1 Introduction

Using the growing amounts of electronic health record (EHR) data, increasing attention has been paid to using data-driven machine learning (ML) methods for a range of classification and predictive tasks, including disease phenotyping and risk stratification [4,15]. Implicit to these ML-based approaches are a data representation that embodies the temporal nature of such data. One challenge of modeling clinical event data is to learn the representation that aligns with medical knowledge [6,8,19], where events (i.e., laboratory results, medications, diagnoses, etc.) can be extracted from time-stamped EHRs and other

M. Michalowski and R. Moskovitch (Eds.): AIME 2020, LNAI 12299, pp. 348–358, 2020.
https://doi.org/10.1007/978-3-030-59137-3_31

health-related information, such as claims data. However, many studies modeling such data fail to fully capture the nature of clinical events. For instance, studies modeling claim code sequences only consider temporality between visits, absent of within-visit dynamics [25] that contain essential contextual information. While other approaches utilizing time-stamped EHR events incorporate sequential order within-visit [12, 20], they model a patient's medical history as a fully ordered event sequence despite the fact that the sequence may contain unordered event sets when multiple events happen concurrently (i.e., sharing the same timestamp). An arbitrary ordering (e.g., random, alphabetical, etc.) is usually imposed on each event set during data preprocessing to establish a "structured" input (e.g., matrices, vectors or tensors) used in different ML models, including contemporary deep learning methods. Consequently, models trained on the corresponding data can be sensitive to the input sequence order as they assume elements from each input sequence to be strictly ordered [30].

The partially-unordered nature of event sequences in the EHR calls for permutation-invariant models: the prediction based on a patient's medical history should not be affected when the order of concurrent events is changed. In this study, we propose DPSS (Diagnostic Prediction with Sequence-of-Sets), an end-to-end deep learning architecture that incorporates set learning techniques [32] to model event sequences to support downstream diagnostic prediction. DPSS first introduces a permutation sampling technique on each set of concurrent clinical events. A self-attentive gated recurrent unit (GRU) model is then deployed on top of the permutation samples to characterize multiple sets of concurrent events in a patient visit history and correspondingly estimates the risk of specific diseases. To characterize the contextual features of a clinical event, DPSS also pre-trains an embedding model on a collection of unlabeled event sequences. The key contributions of DPSS are threefold: 1) an end-to-end framework modeling clinical temporal event sequences as *sequences of sets* (SoS) for next-visit disease code prediction, with the ability to capture the temporal patterns within each clinical visit; 2) a permutation-invariant prediction mechanism made possible by introducing a permutation sampling technique on SoS; and 3) a demonstration of the utility of a weighted loss function with additional regularization term enforcing permutation-invariant representation of SoS, which further improves the model predictive performance when using permuted sequences. In this way, DPSS is able to represent clinical event data as sequences of sets that are more consistent with the nature of clinical documentation processes.

We evaluate our proposed framework on a binary prediction task for next-visit diagnostic code prediction of heart failure (HF) using laboratory and diagnostic code data from the MIMIC-III dataset [16]. Our experimental results show that approaching clinical event sequence representation from a set learning perspective with permutation sampling more accurately characterizes the underlying disease dynamics and achieves better disease predictive performance. Techniques such as permutation sampling, sequence Laplacian regularization, and self-attention promote permutation invariance and contribute to robustness against different ordering schemes for concurrent events.

2   Related Work

Deep Learning on Clinical Event Sequences. Deep learning models, particularly variants of recurrent neural networks (RNN), have achieved some success in modeling sequential data for predictive tasks such as readmission and disease risk [1,6,7,12,31]. Early efforts in clinical event sequence representation learning focus on constructing low-dimensional representations of medical concepts through word embedding algorithms proposed for natural language processing (NLP) [10,31]. Key works improved concept embedding by incorporating EHR structures [5,6,8,9] and medical ontologies [29] to capture the inherent relations of medical concepts. More recent methods seek to utilize temporal information, instead of using the indexed ordering, to better characterize chronologies [2,20,26,28]. Still, these aforementioned models mostly assume a fixed temporal order among sequence elements as they serve as inputs, which can cause discrepancies when modeling inputs containing unordered elements.

Deep Set Learning. Characterizing heterogeneous feature sets was investigated for applications in point cloud analysis [21,27,32,33] and graph mining [13,23]. Essentially, a permutation-invariant function is needed for set learning to overcome the limitations of sequence models that are permutation-sensitive [24]. Some of these and other works [24,27,32] propose to compress sets of any size into a feature vector using a permutation-invariant pooling operation (e.g., sum/mean/max pooling), although such operations are prone to losing information contained in a feature set [33]. In contrast, permutation sampling-based methods [21,33] and attention-based methods [18] aim to resolve this issue. For example, Meng et al. [21] specifically use permutation sampling in a hierarchical architecture and concatenation to integrate set element embedding when modeling the structure as a set of sets.

Despite the partially-unordered nature of medical events, only a few studies [25] have been conducted to model clinical event sequences as sequence of sets using a permutation-invariant pooling method. There remains a lack of investigation in the use of permutation sampling strategies on corresponding tasks with EHR-based data, which is the focus of this paper.

3   Method

In this section, we first present the design of the proposed framework, DPSS, for next-visit diagnostic code prediction. Figure 1 illustrates the architecture of DPSS and its three components: 1) a pre-trained lab event embedding layer; 2) an event sequence handler with a permutation sampling mechanism for event sets; and 3) a self-attentive GRU predictor for diagnostic code classification.

3.1   Preliminary

We use E to denote the vocabulary of lab events, and P to denote the set of patient visit histories. A patient's visit history in the EHR is defined as a concatenation of lab event sets $S = [s_{t_1} \oplus s_{t_2} \oplus ... \oplus s_{t_n}] \in P$, where each set contains lab events with samples collected at the same time t_k, $s_{t_k} = \{e^1_{t_k}, e^2_{t_k}, ..., e^m_{t_k} \in E\}$. The goal of the diagnostic code prediction task is to provide a regression model to estimate the risk of developing a disease for a patient given the visit history S before the most recent visit. In this case, our goal is to predict codes related to HF.

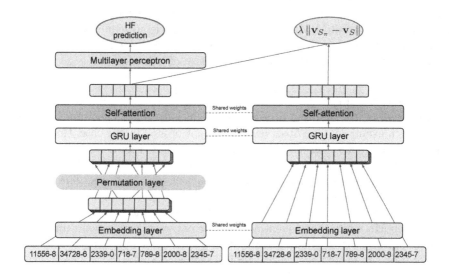

Fig. 1. Illustration of DPSS architecture

3.2   The DPSS Framework

Our DPSS framework sequentially incorporates three components to characterize and perform prediction on a given patient's visit history. We first pre-train a lab event embedding model on a large collection of unlabeled historical lab event sequences, which seeks to capture the contextual similarity of lab events. Next, with this pre-trained embedding representing the latent features of each lab event, the permutation sampling process then generates permutations for each event set in the visit history. Lastly, a downstream predictor is trained on the permutation-sampled data, learning to predict the risk for a specific disease while preserving the permutation invariance of concurrent events. Details of each model component is described as follows.

Pre-trained Lab Event Embeddings. To encode the non-numerical representations of lab events into numerical representations, we first conduct a pre-training process to obtain an embedding of LOINC codes. We trained a skip-gram language model [22] on a collection of unlabeled lab event sequences with the objective of representing the contextual similarity of lab events in a continuous vector space (obtained by minimizing log likelihood loss):

$$L_{SG} = -\frac{1}{|P|} \sum_{\text{seq}(S) \in P} \sum_{-C < j < C} \log p(\mathbf{e}_{p+j} | \mathbf{e}_p).$$

such that seq(S) is a temporally-ordered sequence of a visit history S, and where events in each concurrent set are arbitrarily ordered. Specifically, we extract lab event sequences (from MIMIC-III) as partially-unordered sequences to train the embedding model. $\mathbf{e}_p$ is the embedding vector of the t-th event $e_t \in$ seq(S), $\mathbf{e}_{p+j}$ is that of a neighboring event, and C is the size of half context.[1]

Permutation Sampling. Rather than training a decision making model on fixed sequences, the learning objective of DPSS is to make consistent decisions even if such events may be observed in different orders; in our case, this may be dependent on any number of factors as to how an EHR records the data. Inspired by the recent success of deep set learning on point clouds [21, 24, 27, 32], we introduce a permutation sampling strategy for patient visit histories. The principle of this process is to generate event sequences from a given patient's visit history such that events in a concurrent event set will be randomly ordered in each training epoch, while the sequential order across event sets remain unchanged. In detail, given a set of events s, we denote $\pi(s)$ as the set of its permutations. A permutation sample of a visit history S is a sequence $S_\pi \in \pi(S) = \{\bigoplus_{i=1}^{n} \pi(s_{t_i})\}$ that is obtained by sequentially concatenating a permutation of each concurrent event set in S. Specifically, $\pi(S)$ denotes the universal set of permutation samples for S. Based on this sampling strategy, the event sequence encoder introduced next follows an end-to-end learning process for predicting the target diseases, while remaining invariant to the order of concurrent events in a patient visit history.

Self-attentive GRU Encoder. We use $\mathbf{S}_\pi = [\mathbf{e}_1, \mathbf{e}_2, ..., \mathbf{e}_l]$ to denote an input vector sequence corresponding to an embedded lab event sequence after the permutation sampling process of the visit history, S. The self-attentive gated recurrent unit (GRU) encoder couples two techniques to represent the embedding representation of the permutation sampled visit history $\mathbf{v}_{S_\pi} = A(\mathbf{S}_\pi)$.

The GRU is an alternative to a long-short-term memory network (LSTM) [3], which consecutively characterizes sequential information without using separated memory cells [11]. Each unit consists of two types of gates to track the state of the sequence, a reset gate $\mathbf{r}_p$ and an update gate $\mathbf{z}_p$. Given the embedding vector $\mathbf{e}_p$ of an incoming event, the GRU updates the hidden state $\mathbf{h}_p^{(1)}$ of the sequence as a linear combination of the previous state, $\mathbf{h}_{p-1}^{(1)}$, and the candidate state, $\tilde{\mathbf{h}}_p^{(1)}$

---

[1] The *context* of a skip-gram refers to a subsequence of an ordered event sequence seq(S) such that the subsequence is of $2C + 1$ length.

of a new event e_p, calculated as follows:

$$\mathbf{h}_p^{(1)} = \text{GRU}(\mathbf{v}_p) = \mathbf{z}_p \odot \tilde{\mathbf{h}}_p^{(1)} + (1 - \mathbf{z}_p) \odot \mathbf{h}_{p-1}^{(1)}$$

$$\mathbf{z}_p = \sigma\left(\mathbf{M}_z \mathbf{v}_p + \mathbf{N}_z \mathbf{h}_{p-1}^{(1)} + \mathbf{b}_z\right)$$

$$\tilde{\mathbf{h}}_p^{(1)} = \tanh\left(\mathbf{M}_s \mathbf{v}_p + \mathbf{r}_p \odot (\mathbf{N}_s \mathbf{h}_{p-1}^{(1)}) + \mathbf{b}_s\right)$$

$$\mathbf{r}_p = \sigma\left(\mathbf{M}_r \mathbf{v}_p + \mathbf{N}_r \mathbf{h}_{p-1}^{(1)} + \mathbf{b}_r\right).$$

where $\odot$ denotes the element-wise multiplication. The update gate $\mathbf{z}_p$ balances the information of the previous sequence and the new item, where $\mathbf{M}_*$ and $\mathbf{N}_*$ denote different weight matrices, $\mathbf{b}_*$ are bias vectors, and σ is the sigmoid function. The candidate state $\tilde{\mathbf{h}}_p^{(1)}$ is calculated similarly to those in a traditional recurrent unit, and the reset gate $\mathbf{r}_p$ controls how much information of the past sequence contributes to $\tilde{\mathbf{h}}_p^{(1)}$.

Atop the GRU hidden states, the self-attention mechanism seeks to learn attention weights that highlight the clinical events that are important to the overall visit history. This mechanism is added to GRU as below:

$$\mathbf{u}_i = \tanh\left(\mathbf{M}_a \mathbf{h}_i^{(1)} + \mathbf{b}_a\right); \ a_i \quad = \frac{\exp\left(\mathbf{u}_i^\top \mathbf{u}_{S_\pi}\right)}{\sum_{x_i \in S_\pi} \exp\left(\mathbf{u}_i^\top \mathbf{u}_{S_\pi}\right)}; \ A(S_\pi) = \mathbf{v}_{S_\pi} = \sum_{i=1}^l a_i \mathbf{u}_i$$

where $\mathbf{u}_i$ is the intermediary representation of GRU output $\mathbf{h}_i^{(1)}$. $\mathbf{u}_X = \tanh(\mathbf{M}_a \mathbf{h}_X^{(1)} + \mathbf{b}_a)$ is the intermediary latent representation of the averaged GRU output $\mathbf{h}_X^{(1)}$ and can be interpreted as a high-level representation of the entire input sequence. By measuring the similarity of each $\mathbf{u}_i$ with $\mathbf{u}_X$, the normalized attention weight a_i for $\mathbf{h}_i^{(1)}$ is produced through a softmax function. The final embedding representation $\mathbf{v}_{S_\pi}$ of the visit history is then obtained as the weighted sum of the intermediary representation for each event in the sequence S_π.

Learning Objective. A multi-layer perceptron (MLP) with sigmoid activation is applied to the previous embedding representation of the visit history, whose output $\hat{c}^{S_\pi}$ is a scalar that indicates the risk of the target disease. The learning objective is to optimize the loss function defined below.

$$L = -\frac{1}{|P|} \sum_{S \in P} \frac{1}{|\pi(S)|} \sum_{S_\pi \in \pi(S)} x^S \log \sigma(\hat{c}^{S_\pi}) + (1 - x^S) \log\left(1 - \sigma(\hat{c}^{S_\pi})\right) + \lambda \|\mathbf{v}_{S_\pi} - \mathbf{v}_S\|$$

The main loss function uses binary cross-entropy, where $x^S \in \{0, 1\}$ is the training label indicating if the disease code exists in the disease code list from the next patient visit $s_{t_{n+1}}$. Optimizing for the main loss enforces predictions to be invariant to the input within-set order. The last term of the loss function corresponds to a graph Laplacian regularization term, where λ is a small positive coefficient. Notably, this regularization term teaches the self-attentive GRU encoder to generate similar representations for different permutation samples of the same visit history record, and helps differentiate such representations from those of unrelated records in the embedding space. We show below that this regularization mechanism is able to improve the prediction accuracy of the target disease in various experiments.

4   Experiments

We hereby evaluate DPSS on the next-visit HF diagnosis prediction task.

4.1   Dataset

We evaluated DPSS using data from MIMIC-III [16], a publicly available clinical dataset associated with patients admitted to critical care units of Beth Israel Deaconess Medical Center between 2001 and 2012. MIMIC-III contains records from different sources including demographics, lab results, medications, CPT (Current Procedural Terminology) procedures, and ICD-9 (International Classification of Diseases) diagnostic codes. The within-visit temporal information for diagnostic and procedure codes is not available in MIMIC-III as they are only specific to a patient visit; and while medications are tagged with timestamps, they are recorded with a duration (i.e., start and end times), which poses further challenges on determining the relative ordering between medication and lab events. To simplify our task, we choose to model only lab event sequences as they are less vague with respect to temporal ordering when defined as sequence of sets. Specifically, the timestamp recorded for lab events in MIMIC-III indicates sample acquisition time so a set of lab events with shared timestamps inform patient status at a given time point.

To perform next-visit HF diagnosis prediction, we extracted 7,235 sequences of abnormal lab events for adult ($age \geq 18$) patients with at least two hospital admissions from the MIMIC-III dataset by concatenating all abnormal lab events from each visit history. These sequences, each representing a unique patient, are divided into training (75%, 5,426 patients), validation (12.5%, 904 patients) and test (12.5%, 905 patients) datasets. Based on the existence of the level 3 ICD-9 code representing HF, 428, in the diagnostic codes of the most recent visit, we identified a total of 2,495 HF cases.We used LOINC codes as the lab event ontology, with 187 unique codes present in our data. During data preprocessing, all eligible event codes for a patient are extracted by patient ID and admission ID matching, sorted by chart time. Concurrent events during the same patient admission are usually imposed with an arbitrary order (e.g., random or alphabetically ordered event codes) when inputted as part of the sequence.

4.2   Experimental Configuration

We set the pre-trained skip-gram embedding model on LOINC codes with a context size of 5 and dimensionality of 256. For all reported models, we use the Adam optimizer [17] with a learning rate of 0.001. For each model variant or baseline, we select hyperparameters that lead to the lowest validation loss during training for testing, with the maximum number of epochs set to 100. Training may also be terminated before 100 epochs based on early stopping with a patience of 10 epochs on the validation area under the receiver operator characteristic curve (AUROC) metric. The best combination of GRU layer dimension (candidate values from {64, 128, 256, 512}) and sequence length (candidate values: {128, 256, 512}) is selected based on the AUROC score on the validation set.

We compared the proposed method with the following baseline methods: 1) *GRU*, a single-layer GRU, as defined in Sect. 3.2; 2) *self-attentive GRU*, a GRU model incorporating the self-attention mechanism; and 3) *Pooling GRU*, following previous work [25,32], we apply a sum-pooling based or a max-pooling based set function on the set element embedding to acquire a permutation invariant feature aggregation. To show the effects of different model components of DPSS, we also evaluate different variants of DPSS, where we remove the sequence Laplacian regularization or self-attention.

4.3   Results

Experiments for baseline models and DPSS are each evaluated on the same hold-out test set. We repeated the evaluations 10 times to calculate 95% confidence intervals (CIs) for test AUROC and pr-AUC. Table 1 summarizes test performance of the baseline models and DPSS.

Table 1. Model comparison on next-visit HF risk prediction using MIMIC-III data

Method	AUROC (95% CI)	pr-AUC (95% CI)
GRU	0.7421(±0.00331)	0.6133(±0.00564)
Self-attentive GRU	0.7405(±0.0034)	0.6386(±0.0074)
Sum-pooling GRU	0.7070(±0.00101)	0.5839(±0.00173)
Max-pooling GRU	0.6954(±0.00116)	0.5730(±0.00361)
DPSS w/o self-attention& Sequence Laplacian	0.7741(±0.00277)	0.6659(±0.00615)
DPSS w/o Sequence Laplacian	0.7748(±0.00176)	0.6752(±0.00309)
DPSS	**0.7766(±0.00185)**	**0.6801(±0.00453)**

DPSS significantly outperforms the other models in terms of AUROC and pr-AUC metrics. By comparing all of our permutation sampling based model variants with the baseline, we show that the effectiveness of addressing the partially-unordered nature through a permutation sampling mechanism. Specifically, being able to model within-set element interactions, DPSS is shown to be more suitable for modeling lab events as a sequence of sets compared to other permutation-invariant aggregation methods like sum- and max-pooling, with improvements of 9.8% and 11.7% in AUROC, 16.5% and 18.7% in pr-AUC, respectively and relatively. Comparing DPSS variants, we also see that sequentially adding the self-attention mechanism and the sequence Laplacian for permutation-invariant regularization boosted the model's discriminative power, with greater improvement observed in pr-AUC, which is a metric that considers the model's ability to cope with imbalanced data [8]. As for the impact of the self-attention mechanism, when added to a basic GRU and DPSS without self-attention and Laplacian loss, the pr-AUC of both models has increased by 4.1% and 1.4%, respectively, while the AUROC metric remained comparable.

We observe that in the raw data of MIMIC-III, concurrent events are ordered randomly in the extracted event sequence. In other data processing scenarios, the event set elements are ordered by the primary key (when applicable) or alphabetically ordered by code strings. The imposed order could lead to bias toward certain data storage methods or a specific coding scheme, which is ultimately irrelevant to the underlying disease. Such inconsistencies may also impair a model's generalizability when the ordering scheme adopted in training differs from that used during inference. We hypothesized that our set learning framework is able to alleviate the aforementioned bias, as the sequence representation is not restricted to any event set ordering scheme. To test this hypothesis, as our previous experiments are trained and tested on data with random within-set order, we further compared DPSS and the best baseline model against a different event set ordering scheme using test sequences with alphabetically-ordered event sets. These evaluation results are presented in Table 2.

Table 2. Comparison against the best baseline method on the test data with a different ordering scheme (alphabetical) for concurrent events.

Method	AUROC (95% CI)	pr-AUC (95% CI)
Self-attentive GRU	0.7364(±0.00953)	0.6214(±0.00878)
DPSS	0.7755(±0.00305)	0.6721(±0.00379)

The best baseline model, self-attentive GRU, is trained on set sequences with an imposed arbitrary random order. When tested on alphabetically-ordered set sequences, it suffers from 0.6% decrease in AUROC and 2.7% decrease in pr-AUC. In contrast, DPSS's performance experienced a smaller decline: 0.1% in AUROC and 1.2% in pr-AUC. The results suggest that DPSS benefited from its permutation sampling mechanism and is more robust against different set ordering schemes.

In summary, the experimental results show that DPSS achieved better performance than the non-permutation sampling-based baseline models on the HF prediction task. The proposed techniques are shown to better capture the clinical events in the visit history according to their partially-unordered nature, hence better supports the downstream decision making.

5   Conclusion

We introduce DPSS, a permutation-sampling-based RNN architecture that supports diagnostic prediction with sequence-of-set learning on clinical events. Our proposed method uses a permutation-sampling technique, sequence Laplacian regularization, and self-attention to learn a permutation invariant representation that allows for more accurate prediction for a binary disease prediction task. We also demonstrated the robustness of DPSS against arbitrary set orderings by comparing performance on a test set with an altered set order. For future work, we plan to extend DPSS to jointly model lab event sequences with medication and demographic information. We also seek to better support multi-disease prediction by incorporating structured label representations [14] and leveraging pre-training [34] to improve domain adaptation of DPSS.

References

1. Barbieri, S., Kemp, J., Perez-Concha, O., et al.: Benchmarking deep learning architectures for predicting readmission to the ICU and describing patients-at-risk. Sci. Rep. **10**(1), 1–10 (2020)
2. Cai, X., Gao, J., Ngiam, K.Y., Ooi, B.C., Zhang, Y., Yuan, X.: Medical concept embedding with time-aware attention. In: IJCAI (2018)
3. Cho, K., et al.: Learning phrase representations using RNN encoder-decoder for statistical machine translation. In: EMNLP (2014)
4. Choi, E., Bahadori, M.T., Schuetz, A., Stewart, W.F., Sun, J.: Doctor AI: Predicting clinical events via recurrent neural networks. In: MLHC, pp. 301–318 (2016)
5. Choi, E., et al.: Multi-layer representation learning for medical concepts. In: KDD (2016)
6. Choi, E., Bahadori, M.T., Song, L., Stewart, W.F., Sun, J.: Gram: graph-based attention model for healthcare representation learning. In: KDD 2017 (2017)
7. Choi, E., Schuetz, A., Stewart, W.F., Sun, J.: Using recurrent neural network models for early detection of heart failure onset. J. Am. Med. Inform. Assoc. **24**(2), 361–370 (2016)
8. Choi, E., Xiao, C., Stewart, W.F., Sun, J.: Mime: multilevel medical embedding of electronic health records for predictive healthcare. In: NIPS (2018)
9. Choi, E., Xu, Z., Li, Y., et al.: Learning the graphical structure of electronic health records with graph convolutional transformer. In: AAAI (2020)
10. Choi, Y., Chiu, C.Y.I., Sontag, D.A.: Learning low-dimensional representations of medical concepts. AMIA Summits Transl. Sci. Proc. **2016**, 41–50 (2016)
11. Dhingra, B., Liu, H., et al.: Gated-attention readers for text comprehension. In: ACL (2016)
12. Farhan, W., Wang, Z., Huang, Y., et al.: A predictive model for medical events based on contextual embedding of temporal sequences. JMIR Med. Inform. **4**, e39 (2016)
13. Hamilton, W.L., Ying, Z., Leskovec, J.: Inductive representation learning on large graphs. In: NIPS (2017)
14. Hao, J., Chen, M., Yu, W., et al.: Universal representation learning of knowledge bases by jointly embedding ontological concepts and instances. In: KDD (2019)
15. Harutyunyan, H., Khachatrian, H., Kale, D.C., Galstyan, A.: Multitask learning and benchmarking with clinical time series data. Sci. Data **6**, 1–18 (2019)
16. Johnson, A.E.W., Pollard, T.J., et al.: MIMIC-III, a freely accessible critical care database. Sci. Data **3**, 1–9 (2016)
17. Kingma, D.P., Ba, J.: Adam: a method for stochastic optimization. In: ICLR (2014)

18. Lee, J., Lee, Y., Kim, J., Kosiorek, A.R., Choi, S., Teh, Y.W.: Set transformer: a framework for attention-based permutation-invariant neural networks. In: ICML (2018)
19. Ma, F., You, Q., Xiao, H., Chitta, R., Zhou, J., Gao, J.: Kame: knowledge-based attention model for diagnosis prediction in healthcare. In: CIKM (2018)
20. Ma, T., Xiao, C., Wang, F.: Health-atm: a deep architecture for multifaceted patient health record representation and risk prediction. In: SDM (2018)
21. Meng, C., Yang, J., Ribeiro, B., Neville, J.: Hats: a hierarchical sequence-attention framework for inductive set-of-sets embeddings. In: KDD, pp. 783–792 (2019)
22. Mikolov, T., Sutskever, I., Chen, K., Corrado, G.S., Dean, J.: Distributed representations of words and phrases and their compositionality. In: NIPS, pp. 3111–3119 (2013)
23. Moore, J., Neville, J.: Deep collective inference. In: AAAI (2017)
24. Murphy, R.L., Srinivasan, B., Rao, V.A., Ribeiro, B.: Janossy pooling: learning deep permutation-invariant functions for variable-size inputs. In: ICLR (2019)
25. Nguyen, P., Tran, T., Venkatesh, S.: Resset: a recurrent model for sequence of sets with applications to electronic medical records. In: IJCNN, pp. 1–9 (2018)
26. Peng, X., Long, G., Shen, T., Wang, S., Jiang, J., Blumenstein, M.: Temporal self-attention network for medical concept embedding. In: ICDM, pp. 498–507 (2019)
27. Qi, C.R., Su, H., Mo, K., Guibas, L.J.: Pointnet: deep learning on point sets for 3d classification and segmentation. In: CVPR, pp. 652–660 (2017)
28. Rajkomar, A., Oren, E., Chen, K., et al.: Scalable and accurate deep learning with electronic health records. NPJ Digit. Med. **1** (2018)
29. Song, L., Cheong, C.W., Yin, K., Cheung, W.K.W., Fung, B.C.M., Poon, J.: Medical concept embedding with multiple ontological representations. In: IJCAI (2019)
30. Vinyals, O., Bengio, S., Kudlur, M.: Order matters: sequence to sequence for sets. In: ICLR (2015)
31. Xiao, C., Ma, T., Dieng, A.B., Blei, D.M., Wang, F.: Readmission prediction via deep contextual embedding of clinical concepts. PLoS ONE **13**, e0195024 (2018)
32. Zaheer, M., Kottur, S., Ravanbakhsh, S., Poczos, B., Salakhutdinov, R.R., Smola, A.J.: Deep sets. In: NIPS, pp. 3391–3401 (2017)
33. Zhang, Y., Hare, J.S., Prügel-Bennett, A.: Fspool: learning set representations with featurewise sort pooling. In: ICLR (2020)
34. Zhou, G., Chen, M., Ju, C., et al.: Mutation effect estimation on protein-protein interactions using deep contextualized representation learning. NAR Genom. Bioinform. (2020)

Sepsis Deterioration Prediction Using Channelled Long Short-Term Memory Networks

Peter Svenson[1]([✉]) [iD], Giannis Haralabopoulos[2] [iD],
and Mercedes Torres Torres[2] [iD]

[1] University Hospitals of Derby and Burton, Derby, UK
peter.svenson@nhs.net
[2] University of Nottingham, Nottingham NG7 2RD, UK
{giannis.haralabopoulos,mercedes.torrestorres}@nottingham.ac.uk

Abstract. Sepsis is a severe medical condition that results in millions of deaths globally each year. In this paper, we propose a Channelled Long-Short Term Memory Network model tasked with predicting 48-hour mortality in sepsis against the Sequential Organ Failure Assessment (SOFA) score. We use the MIMIC-III critical care database. Our research demonstrates the viability of deep learning in predicting patient outcomes in sepsis. When compared with published literature for similar tasks, our channelled LSTM models demonstrated a comparable AUROC with superior precision score. The results showed that deep learning models outperformed the SOFA score in predicting 48-hour mortality in sepsis in AUROC (0.846–0.896 vs 0.696) and average precision score (0.299–0.485 vs 0.110). Finally, our Fully-Channelled LSTM outperforms a baseline LSTM by 5.4% in AUROC and 59.9% in average precision score.

Keywords: Sepsis · Machine learning · Deep learning · Recurrent neural networks · Long short-term memory networks

1 Introduction

Sepsis is a serious medical condition with an estimated 48.9 million cases worldwide in 2017, resulting in 11.0 million sepsis-related deaths [27]. Since a validated diagnostic test for sepsis does not exist, multiple risk assessment scores are used for diagnosis. These share an underlying shortcoming: all use a single snapshot of a patient's clinical state. Scoring can be performed repeatedly, but there is no direct tracking of the clinical measurements over time. This can be important, as a patient's deterioration can often be viewed as a trend of changing vital signs.

The relatively recent adoption of Electronics Health Records (EHR) means that medical data is more readily available for analysis. These records, combined with advances in artificial intelligence (AI) and machine learning (ML), lay the foundations of a modern medical practice that is both data-oriented and

© Springer Nature Switzerland AG 2020
M. Michalowski and R. Moskovitch (Eds.): AIME 2020, LNAI 12299, pp. 359–370, 2020.
https://doi.org/10.1007/978-3-030-59137-3_32

computer-assisted, in which automated tools can be used perform tasks such as diagnosis, patient monitoring, or outcome prediction [30].

Deep Neural Networks (DNN) are an evolution of Artificial Neural Networks (ANN) inspired by biological systems [15]. They consists of multiple hidden layers between the input and the output that learn meaningful representations from the data and use it for prediction [15]. DNNs have been successfully used in Computer Vision [24] and Natural Language Processing [5,36]. They have also showed potential in health informatics [23,32]. Recurrent Neural Networks (RNN) are a type of deep network where each neuron can process variable temporal information. The temporal nature of health-informatics, such as the data in EHRs, pose a great fit for these temporal networks, which have the ability to learn patterns in sequential data [33]. Scientists have also implemented an improved RNN with the ability to learn long-term dependencies: Long Short-Term Memory (LSTM) networks, which demonstrate state-of-the-art performance with time-series across a wide variety of machine learning problems [14].

In this paper, we develop a novel LSTM-based network, Channelled LSTM, to predict mortality from sepsis within 48 h. By predicting these short-term outcomes, we make sure that patients at risk of deterioration are promptly identified. At the same time, an accurate model can direct treatment and interventions to those in need. We use MIMIC-III, a critical care database with data from more than forty thousand patients who stayed in Intensive Care Units (ICU) [11]. It includes a diverse range of data, such as vital signs, demographics, medication, laboratory testing results and mortality. Results show that our Channelled LSTM outperforms the SOFA (Sequential Organ Failure Assessment) score and two baseline models, a RNN and a LSTM in terms of AUROC (Area Under the Receiver Operating Characteristics) and average precision score.

2   Related Work

Sepsis: Sepsis is the primary cause of death from infection and requires early diagnosis and treatment. Until recently, sepsis was defined as a host's Systemic Inflammatory Response Syndrome (SIRS) due to an underlying infection, based on patient temperature, heart rate, respiratory rate and white cell count [1]. However, Singer et al. [31] recognised sepsis to have a much broader impact on physiological pathways [28], and proposed using the Sequential Organ Failure Assessment (SOFA). SOFA has been shown to positively correlate with mortality [34]. The authors also recommended the use of the quick SOFA (qSOFA) score, which consists of only three components (respiratory rate, mental status, and systolic blood pressure), as a bedside screening tool [19].

Another sepsis scoring system that has been developed is the PIRO (Predisposition Insult/Infection Response and Organ dysfunction) score [26]. The PIRO score has proven mortality prediction capabilities [8], it can outperform SOFA [18], but has not been adopted for widespread use. PIRO includes a mix of temporal and non-temporal variables (e.g. comorbidities, age) that render it inefficient for time-series learning.

Machine Learning and Electronic Health Records: Machine Learning has been used in healthcare from the early 90s [3]. From medical diagnosis [22], to epidemiology [35] and health monitoring [10], machine learning has proven to be a powerful tool, particularly when dealing with complex data, be it images [32] or text [22]. In recent years, deep learning methods have consistently obtained state-of-the-art results in many healthcare problems [4]. Deep Learning Networks are powerful mathematical models able to learn generalisable information from large quantities of data [15].

Deep learning methods have been used with EHR in wide variety of problems, including information extraction, representation learning, de-identification and, such as the subject of this paper, outcome or deterioration prediction [30]. For outcome prediction, two major approaches can be considered. First, a static approach, which can be beneficial when predicting particular outcomes such as breast cancer [30]. For these approaches, Convolutional Neural Networks or Multi-layer perceptrons obtain highly accurate results. Second, temporal approaches can applied to more complex outcomes, such as readmission prediction [22]. In these approaches, temporal networks, such as Recurrent Neural Networks (RNNs) and Long Short-Term Memory Networks (LSTMs), consistently obtain state-of-the-art results [30]. Examples of successful applications of temporal networks include [23], where EHRs were transformed to Fast Healthcare Interoperability Resources (FHIR) standards to predict mortality, readmission and prolonged hospital stays. Unfortunately, the data used for training is not publicly available and the computational overhead make reproducing such a study unfeasible. [17] used LSTM to predict over 128 diagnoses using target replication and auxiliary targets for less-common diagnostic labels.

Another study developed multiple machine learning models using the MIMIC-III database, tasked with predicting in-hospital mortality, deterioration, phenotyping, and length of stay [6]. The deterioration task aimed to predict 24-hour mortality from all causes, not just sepsis. Their best performing model for deterioration was their channel-wise LSTM with deep supervision, with an AUROC of 0.911. However, it only achieved an AUC-PR (Area Under the Precision-Recall Curve) of 0.344, and other metrics such as recall, precision and F1 score are not included. This limits the study, as in clinical practice the recall (known as sensitivity) is important in determining how a test or tool is applied. In this paper, we approach sepsis deterioration prediction as a temporal prediction task, and build on the channel-wise LSTM method.

Machine Learning and Sepsis: Authors in [9] examine machine learning for early prediction of sepsis. They look at studies using Support Vector Machines, Logistic Regression, APeX, and InSight, a system developed using MIMIC-II which works by mapping vital sign measurements to finite discrete hyperdimensional space [2]. Performance is varied: the worse performing model, Logistic Regression, obtained an AUROC of 0.78, while the best performing model, Deep Neural Networks, obtained an AUROC of 0.92. One of the InSight studies achieved an AUROC of 0.92, but the other two InSight studies had lower scores: 0.88 and 0.83. The Deep Learning study demonstrated improved performance of both a deep feed-forward network and an LSTM model compared to InSight [13]. There have also been efforts in developing LSTM models for septic shock prediction. In [16], authors combined CNNs and LSTMs to predict septic shock using data from the Christiana Care Health System. However, to address class imbalance, significant under-sampling of the 'shock negative visits' was used for both training and testing. This means the evaluation was not representative of performance in the real-world, where significant class imbalance exists. Furthermore, they display results for 'visit level early diagnosis' of septic shock, which used the first 12 h of patient sequence data, but more than 50% of the Septic Shock population had a Shock Onset time well within this 12 h. As such, for the majority of cases, these models are only identifying that shock has occurred, and are not diagnosing it early, as the presentation of the results would suggest. This explains why the models perform significantly worse on their 'event level early prediction', in which they use varying 'hold-off' window sizes, prior to the onset of septic shock, to align their data sequences.

On the other hand, [29] aimed to predict sepsis incidence as defined by Angus criteria, an International Classification of Diseases (ICD) coding system. They used MIMIC-III and compared traditional machine learning against LSTM. However, none of these techniques performed particularly well: Random Forest was the best model with an AUROC of 0.699, outperforming their LSTM models. This may in part be due to their data pre-processing, which re-sampled the time-series data into 6-hour bins. This could have lacked the granularity to predict the changes in clinical condition that indicate sepsis.

3   Methodology

3.1   Data Pre-processing: MIMIC-III

In our study, we focus on a subset of the Medical Information Mart for Intensive Care III v1.4 (MIMIC-III) database [11]. We use a cohort of sepsis cases defined by the Sepsis-3 criteria [31], as identified by [12]. This subset dates from 2008 to 2012, and consists of 5,784 critical care admissions, with an in-hospital mortality rate of 14.5% (836 cases). We extract variables for assessment tools currently used in clinical practice: SOFA, qSOFA, NEWS2 and SIRS. We also include Lactate and Base Excess as they have proven valuable in assessing acute illness and infection [20,25]. We remove clinically impossible outliers and standarised each variable.

These variables form a complex, irregular pattern of time-series data. The time interval between data points of a single variable can fluctuate, and for different variables these intervals can range from less than an hour (e.g. Heart Rate), to over 24 h (e.g. Bilirubin). Furthermore, each ICU admission is of a different length and varies from several hours to days or weeks. This complexity leads to significant challenges in both efficient training of RNNs, and their ability to learn patterns, which is still an area of active research [21]. Instead, to achieve efficient training of our models, we resample variables, resulting in regular and synchronous data [6]. This presents its own challenge, as frequent re-sampling results in a significant rate of imputed data, whereas infrequent re-sampling loses granular detail that may be useful to the model. We took two approaches to this, resulting in Set A and Set B, described below. For both sets, cases with less than 24 h of data were excluded.

Set A: It includes 10 variables of our subset: Heart Rate, Systolic BP, Mean Arterial Pressure, Respiratory Rate, FiO_2, Oxygen Saturation, the three components of the GCS, and Temperature. All samples are re-sampled into constant one-hour timeslots. Multiple measurements within an hour are averaged. This loses some granular detail, though so would alternatives (e.g. choosing the most recent value) and the short timeslots minimises this risk. Missing measurements, before or after the initial sampling, are imputed with backward and forward filling respectively. In summary, Set A includes 4,975 cases with an inpatient mortality rate of 11.4% (565 cases), and 38.7% of the processed data is imputed.

Set B: Set B is split into three groups, re-sampling the data at three different frequencies. This allows it to include all 17 of the variables of our subset whilst also reducing the imputed data rate. Group one consists of the most frequently measured vital signs: Heart Rate, Systolic BP, Mean Arterial Pressure, Respiratory Rate, FiO_2, and Oxygen Saturation; re-sampled into one-hour timeslots. The second group includes the less frequently measured vital signs: the three components of the GCS, and Temperature; re-sampled into four-hour timeslots. While the final group includes the 'blood tests' variables: White Blood Cells, Bilirubin, Platelets, Creatinine, PaO_2, Lactate and Base Excess; re-sampled into 24-hour timeslots. In summary, Set B includes 5,019 cases with and inpatient mortality rate of 11.3% (566 cases), and 17.3% of the processed data is imputed.

For both sets, each case is split into multiple 24-hour snapshots of the same variable length. This is done by moving a 24-hour window across each case, using 12-hour time-steps. Each 24-hour snapshot is treated as input for our classification. The output is a binary value representing patient mortality within 48 h. Since we expand the dataset by treating multiple daily windows of the same patient as separate cases, the 48-hour mortality rate falls to 4.84% for Set A, and 4.83% for Set B. Due to class imbalance and to improve training efficiency we apply oversampling in both training subsets. When a patient is within the 48-hour window of mortality, the case is split by moving the window using 1-hour timesteps. This increased the mortality rate of Set A to 31.23% and Set B to 31.22%. This oversampling is not performed on the testing subsets, so that the evaluation of the models will be indicative of real-world performance on significantly class-imbalanced data. Additionally, we calculate the SOFA scores for each testing instance in Set B, using the values in the first timeslot. SOFA score represents the current best clinical practice and is used as baseline. A process chart summarising our data processing steps is available in Appendix A.

3.2 Deep Learning Models

Recurrent Neural Networks (RNN): RNNs are ANNs in which nodes are unidirectionaly connected through time. RNN nodes have an internal memory module which allows them to process inputs of variable sequence. It updates weight matrices via backpropagation through time (BPTT). We use a baseline RNN model with two layers using the default hyperbolic tangent function.

Long Short-Term Memory Networks (LSTM): LSTMs are refined RNNs able of learning long term dependencies. At their core, LSTMs have a module with four single layered NNs and a cell state. The cell state receives the input from a previous LSTM module and is controlled by a set of three gates. These gates interact directly with the cell state and can: forget past unneeded information, store newly learnt information, and save the cell state that is passed to the next LSTM module. Our LSTM model [7] includes two such stacked LSTM layers.

Channelled LSTM: We create 'Channelled LSTM' [6] models. Rather than feed the input data into a single RNN or LSTM layer, it is instead split into multiple channels where each input is fed into its dedicated LSTM layer. The first model of the modified LSTM models group is a Fully Channelled LSTM (FC-L) with 10 channels, each of which includes two LSTM stacked layers. The FC-L accommodates Set A by having a separate channel for each of the 10 clinical variables. The second model is a Reduced Channelled LSTM (RC-L). Its structure is identical to FC-L, but the number of channels reduced from 10 to 4. Variables are grouped together according to their physiological system. The RC-L aims to reduce training time and attempts to maintain a similar performance to FC-L. Both FC-L and RC-L are capable of handling the multiple data streams present in Set B, but need some modifications to handle the 'blood tests' stream. We apply two modifications per model that result in four modified LSTM models for Set B. The first method feeds the 'blood tests' input stream at the same layer as the other inputs, only replacing the RNN layer with a dense layer, named FC-S and RC-S. While, the second method feeds the 'blood tests' input stream at a deeper level, separating further the input streams, noted as FC-D and RC-D.

4   Experiments

Our task is a binary classification task, aiming to predict 48-hour mortality. All models are evaluated using 10-fold cross validation. Oversampling is applied in each of the training sets for each fold. All DNN models use Adam optimization with Learning Rate = 0.001, $\beta_1 = 0.9$, Binary Cross-entropy loss function. Each fold is trained for 30 epochs with a batch size of 32. In addition, class weights are used to counteract class imbalance in the datasets. The binary 48-hour mortality (class 1) has a weight of 2, whilst 48-hour survival (class 0) has a weight of 1. This results in a class weight ratio of 2:1, versus the approximate 30:70 split class imbalance. Models are developed in Keras[1].

Metrics: To obtain a detailed overview of the performance of our models, we calculate the following combination of metrics: AUROC, average precision score,

---

[1] https://keras.io/.

recall, precision, negative predictive value, F1 score, and accuracy. We present the results in Receiver Operating Characteristic (ROC) curve for each of the models, which portray the positive rate (recall) against the false positive rate (FPR) at various thresholds. A greater weighting is given to the AUROC and average precision score of the models. AUROC is a useful summary of the ROC curve, informing us how well the model is able to distinguish between classes. The average precision score summarises the precision-recall curve by calculating the average precision of the model at different thresholds weighted against the recall. This too measures how well the model can distinguish between classes. Finally, additional weighting is given to recall (sensitivity) due to how such a tool would be used in clinical practice. A False Negative (a patient who is at high risk of dying, that the model identifies as low risk) would be more likely to result in poor outcomes than a False Positive (a patient who is low risk that the model identifies as high risk). A high recall ensures the model does not miss patients at risk of deterioration.

4.1   Results

Table 1 shows our results when comparing the performance of all LSTM-based models, the SOFA score, and two baseline methods (RNN and LSTM).

Within Set A, the Fully Channelled LSTM demonstrated the best performance in the three key metrics AUROC, average precision score, and recall, as well as negative predictive value. It shows a significant improvement when compared to the baseline LSTM model, with an increase of 5.4% in AUROC, 59.9% in average precision score, and 59.3% in recall.

Set B underwent a modified processing scheme, which results in significant drop of the imputed data rate from 38.7% to 17.3%. It also includes additional data in the 'blood tests' input stream that is not present in Set A. The FC-S model demonstrates minor improvements in AUROC (0.896 vs 0.892), average precision score (0.485 vs 0.478), recall (0.794 vs 0.779) and negative predictive value (0.988 vs 0.987) compared to its Set A counterpart, while FC-D achieves the overall highest recall score (0.810). The Reduced Channelled models improved training times, but with a respective performance cost. This was of little benefit due to the overall low training times, which did not impede model development. All the deep learning models outperform the baseline SOFA score in AUROC (0.846-0.896 vs 0.696) and average precision score (0.299–0.485 vs 0.110). Figure 1 summarises the ROC curves of the SOFA score, baseline LSTM, and FC-S LSTM models. ROC curves for Set A and B models separately are shown in Appendix A.

5   Discussion

The prior studies discussed have aimed to predict the early onset of sepsis [9], but not short term outcomes. Other studies aim to predict all-cause inpatient mortality [23] or deterioration [6] but have not been focused on sepsis. Our

Table 1. Model evaluation for Recurrent Neural Network (RNN), Long Short-Term Memory (LSTM), Fully-Channelled (FC-L) and Reduced-Channelled (RC-L) LSTM. -S/-D: modifications of the FC-L and RC-L models.

Metric	Set A				Set B				
	RNN	LSTM	FC-L	RC-L	SOFA	FC-S	FC-D	RC-S	RC-D
AUROC	0.869	0.846	**0.892**	0.862	0.696	**0.896**	0.892	0.865	0.864
Avg. Precision	0.378	0.299	**0.478**	0.425	0.110	**0.485**	0.457	0.435	0.416
Recall	0.758	0.489	**0.779**	0.714	–	0.794	**0.810**	0.708	0.755
Precision	0.171	**0.256**	0.197	0.188	–	0.195	0.182	**0.200**	0.186
Neg. Pred. Value	0.985	0.973	**0.987**	0.983	–	**0.988**	**0.988**	0.983	0.985
F1 Score	0.277	**0.333**	0.313	0.296	–	**0.311**	0.295	0.310	0.296
Accuracy	0.806	**0.905**	0.834	0.836	–	0.829	0.811	**0.847**	0.824
Epoch Train	68s	**14s**	83s	37s	–	76s	78s	**35s**	36s

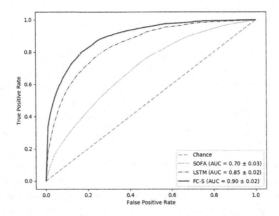

Fig. 1. ROC curves of SOFA, baseline and Fully-Channelled LSTM (FC-S)

LSTM-based approach outperforms the SOFA score by at least 24.1%, and it significantly outperforms the baseline models, previously shown to be effective in deterioration tasks [6]. FC-S, our best performing model, when compared to [6], achieves a AUROC of 0.896 with a much higher average precision of 0.485. Whilst the precision scores of our models were low, this is explained by our focus on maximising recall, and the significant class imbalance of the data. This does not impact the potential use of the models, as we discuss next. The value of our models can be illustrated by indicating the pre- and post-test probabilities of deterioration using our FC-S model. The pre-test probability of 48-hour mortality is 4.8%. If the FC-S model identifies a patient as high-risk, the post-test probability is increased to 19.5% - this is a four-fold increase in risk. Whereas if the FC-S model identifies a patient as low-risk, the post-test probability is reduced to 1.2% (the FC-S model achieved a negative predictive value of 0.988). We can therefore argue that the FC-S model has achieved its aim, and is able to identify those patients at a greatest risk of short term deterioration.

Clinical Relevance: The clinical relevance of a retrospective study such as this is limited. Prospective studies are required if benefit to patients is to be demonstrated. Our work acts as a proof of concept, to hopefully allow further studies to be done. Furthermore, the value of predicting deterioration in a critical care setting, in which patients are already undergoing significant intervention, has its own limits. The goal of future work is to evaluate such deep learning models front of house, in the Emergency Department or on a general medical ward, to better identify patients that require escalation of care.

Limitations: There are two main limitations. First, the SOFA score had to be approximated using the data available. Despite this, SOFA achieved an AUROC of 0.696 on Set B, not substantially lower than the SOFA AUROC of 0.74 in the Sepsis-3 review [31]. Secondly, MIMIC-III is collected from a single site. External validation of models, at different hospitals, are required, to assess the generalisation capabilities of our approach.

6   Conclusions

We present Channelled LSTMs, a novel method for sepsis mortality prediction within 48 h using MIMIC-III. Our models outperform the baseline SOFA score, as well as baseline RNN and LSTM models. When compared with current literature on similar deterioration tasks, our channelled LSTM models demonstrate a comparable AUROC, with an improvement in precision score. Our models either match or exceed the current state-of-the-art for predicting patient deterioration. However, our work comes with certain limitations: further work is needed with prospective clinical trials, multiple sites and wards, to assess the real-world performance of our Deep Learning models in clinical practice.

Appendix A

See Figs. 2 and 3.

Fig. 2. Process chart summarising the data processing steps.

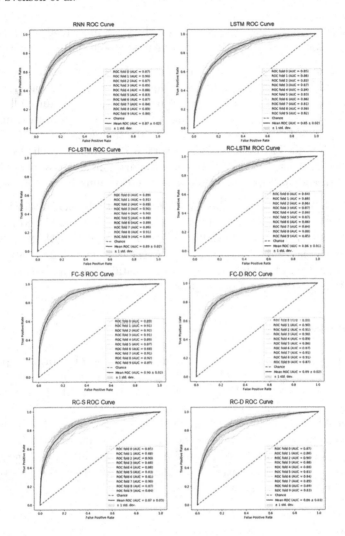

Fig. 3. ROC curves for each fold during cross-validation for the individual models.

References

1. Bone, R.C.: Modulators of coagulation: a critical appraisal of their role in sepsis. Arch. Intern. Med. **152**(7), 1381–1389 (1992)
2. Calvert, J., et al.: Cost and mortality impact of an algorithm-driven sepsis prediction system. J. Med. Econ. **20**(6), 646–651 (2017)
3. Cestnik, B., et al.: Estimating probabilities: a crucial task in machine learning. In: ECAI, vol. 90, pp. 147–149 (1990)
4. Esteva, A., et al.: A guide to deep learning in healthcare. Nat. Med. **25**(1), 24–29 (2019)
5. Haralabopoulos, G., et al.: Ensemble deep learning for multilabel binary classification of user-generated content. Algorithms **13**(4), 83 (2020)

6. Harutyunyan, H., et al.: Multitask learning and benchmarking with clinical time series data. Sci. Data **6**(1), 1–18 (2019)
7. Hochreiter, S., Schmidhuber, J.: Long short-term memory. Neural Comput. **9**(8), 1735–1780 (1997)
8. Howell, M.D., et al.: Proof of principle: the predisposition, infection, response, organ failure sepsis staging system. Crit. Care Med. **39**(2), 322–327 (2011)
9. Islam, M.M., et al.: Prediction of sepsis patients using machine learning approach. Comput. Methods Programs Biomed. **170**, 1–9 (2019)
10. Jagannatha, A.N., Yu, H.: Bidirectional RNN for medical event detection in electronic health records. In: ACL, p. 473 (2016)
11. Johnson, A., et al.: MIMIC-III, a freely accessible critical care database. Sci. Data **3**, 160035 (2016)
12. Johnson, A., et al.: A comparative analysis of sepsis identification methods in an electronic database. Crit. Care Med. **46**(4), 494 (2018)
13. Kam, H.J., Kim, H.Y.: Learning representations for the early detection of sepsis with deep neural networks. Comput. Biol. Med. **89**, 248–255 (2017)
14. Karim, F., et al.: LSTM fully convolutional networks for time series classification. IEEE access **6**, 1662–1669 (2017)
15. LeCun, Y., et al.: Deep learning. Nature **521**(7553), 436–444 (2015)
16. Lin, C., et al.: Early diagnosis and prediction of sepsis shock by combining static and dynamic information using convolutional-LSTM. In: IEEE ICHI, pp. 219–228 (2018)
17. Lipton, Z.C., et al.: Learning to diagnose with LSTM recurrent neural networks. arXiv preprint arXiv:1511.03677 (2015)
18. Macdonald, S.P., et al.: Comparison of PIRO, SOFA, and MEDS scores for predicting mortality in emergency department patients with severe sepsis and septic shock. Acad. Emer. Med. **21**(11), 1257–1263 (2014)
19. Marik, P.E., Taeb, A.M.: SIRS, qSOFA and new sepsis definition. J. Thorac. Dis. **9**(4), 943 (2017)
20. Mikkelsen, M.E., et al.: Serum lactate is associated with mortality in severe sepsis independent of organ failure and shock. Crit. Care Med. **37**(5), 1670–1677 (2009)
21. Neil, D., Pfeiffer, M., Liu, S.C.: Phased LSTM: accelerating recurrent network training for long or event-based sequences (2016)
22. Nguyen, P., et al.: Deepr: a convolutional net for medical records. IEEE J. Biomed. Health Inform. **21**(1), 22–30 (2016)
23. Rajkomar, A., et al.: Scalable and accurate deep learning with electronic health records. NPJ Digi. Med. **1**(1), 18 (2018)
24. Rawat, W., Wang, Z.: Deep convolutional neural networks for image classification: a comprehensive review. Neural Comput. **29**(9), 2352–2449 (2017)
25. Rocktaeschel, J., et al.: Acid-base status of critically ill patients with acute renal failure: analysis based on s-f methodology. Crit. Care **7**(4), R60 (2003). https://doi.org/10.1186/cc2333
26. Rubulotta, F., et al.: Predisposition, insult, infection, response, organ dysfunction: a new model for staging severe sepsis. Crit. Care Med. **37**(4), 1329–1335 (2009)
27. Rudd, K.E., et al.: Global, regional, and national sepsis incidence and mortality 1990–2017: analysis for the Global Burden of disease study. Lancet **395**(10219), 200–211 (2020)
28. Sagy, M., et al.: Definitions and pathophysiology of sepsis. Curr. Prob. Pediatr. adolesc. Health Care **43**(10), 260–263 (2013)
29. Saqib, M., et al.: Early prediction of sepsis in EMR records using traditional ml techniques and deep learning LSTM networks. In: EMBC, pp. 4038–4041 (2018)

30. Shickel, B., et al.: Deep EHR: a survey of recent advances in deep learning techniques for electronic health record (EHR) analysis. IEEE J-BHI **22**(5), 1589–1604 (2017)
31. Singer, M., et al.: The third intl consensus definitions for sepsis and septic shock (sepsis-3). JAMA **315**(8), 801–810 (2016)
32. Torres Torres, M., et al.: Postnatal gestational age estimation of newborns using small sample deep learning. Image Vis. Comput. **83**, 87–99 (2019)
33. Vasilakos, A.V., et al.: Neural networks for computer-aided diagnosis in medicine: a review. Neurocomputing **216**, 700–708 (2016)
34. Vincent, J.L., et al.: Use of the sofa score to assess the incidence of organ dysfunction/failure in intensive care units: results of a multicenter, prospective study. Crit. Care Med. **26**(11), 1793–1800 (1998)
35. Wiens, J., Shenoy, E.S.: Machine learning for healthcare: on the verge of a major shift in healthcare epidemiology. Clin. Infect. Dis. **66**(1), 149–153 (2018)
36. Young, T., et al.: Recent trends in deep learning based natural language processing. IEEE Comput. Intell. Mag. **13**(3), 55–75 (2018)

Neural Topic Models with Survival Supervision: Jointly Predicting Time-to-Event Outcomes and Learning How Clinical Features Relate

Linhong Li[ID], Ren Zuo[ID], Amanda Coston[ID], Jeremy C. Weiss[ID], and George H. Chen[✉][ID]

Carnegie Mellon University, Pittsburgh, PA 15213, USA
{linhongl,acoston,jweiss2,georgech}@andrew.cmu.edu,
renzuo.wren@gmail.com

Abstract. In time-to-event prediction problems, a standard approach to estimating an interpretable model is to use Cox proportional hazards, where features are selected based on lasso regularization or stepwise regression. However, these Cox-based models do not learn how different features relate. As an alternative, we present an interpretable neural network approach to jointly learn a survival model to predict time-to-event outcomes while simultaneously learning how features relate in terms of a topic model. In particular, we model each subject as a distribution over "topics", which are learned from clinical features as to help predict a time-to-event outcome. From a technical standpoint, we extend existing neural topic modeling approaches to also minimize a survival analysis loss function. We study the effectiveness of this approach on seven healthcare datasets on predicting time until death as well as hospital ICU length of stay, where we find that neural survival-supervised topic models achieves competitive accuracy with existing approaches while yielding interpretable clinical "topics" that explain feature relationships.

Keywords: Survival analysis · Topic modeling · Interpretability

1 Introduction

Predicting the amount of time until a critical event occurs—such as death, disease relapse, or hospital discharge—is a central focus in the field of survival analysis. Especially with the increasing availability of electronic health records, survival analysis data in healthcare often have both a large number of subjects and a large number of features measured per subject. In coming up with an interpretable survival analysis model to predict time-to-event outcomes, a standard approach is to use Cox proportional hazards [6], with features selected using lasso regularization [25] or stepwise regression [12]. However, these Cox-based models do not inherently learn how features relate.

© Springer Nature Switzerland AG 2020
M. Michalowski and R. Moskovitch (Eds.): AIME 2020, LNAI 12299, pp. 371–381, 2020.
https://doi.org/10.1007/978-3-030-59137-3_33

To simultaneously address the two objectives of learning a survival model for time-to-event prediction and learning how features relate specifically through a topic model, Dawson and Kendziorski [8] combine latent Dirichlet allocation (LDA) [3] with Cox proportional hazards to obtain a method they call SURVLDA. The idea is to represent each subject as a distribution over topics, and each topic as a distribution over which feature values appear. The Cox model is given the subjects' distributions over topics as input rather than the subjects' raw feature vectors. Importantly, the topic and survival models are jointly learned.

In this paper, we propose a general framework for deriving neural survival-supervised topic models that is substantially more flexible than SURVLDA. Specifically, SURVLDA estimates model parameters via variational inference update equations derived specifically for LDA combined with Cox proportional hazards; to use another other sort of combination would require re-deriving the inference algorithm. In contrast, our approach combines any topic model and any survival model that can be cast in a neural net framework; combining LDA with Cox proportional hazards is only one special case. Importantly, our framework yields survival-supervised topic models that are interpretable so long as the underlying topic and survival models are interpretable. As a byproduct of taking a neural net approach, we can readily leverage many deep learning advances. For example, we can avoid deriving a special inference algorithm and instead use any neural net optimizer such as Adam [17] to learn the joint model in mini-batches, which scales to large datasets unlike SURVLDA's variational inference algorithm.

As numerous combinations of topic/survival models are possible, for ease of exposition, we demonstrate how to combine LDA with Cox proportional hazards in a neural net framework, yielding a neural variant of SURVLDA. We refer to our neural variant as SURVSCHOLAR since we build on SCHOLAR [5], a neural net approach to learning LDA and various other topic models. We benchmark SURVSCHOLAR on seven datasets, finding that it can yield performance competitive with various baselines while also yielding interpretable topics that reveal feature relationships. For example, on a cancer dataset, SURVSCHOLAR learns two topics that are associated with longer survival time, and one topic associated with lower survival time. The first two pro-survival topics provide different explanations for patients attributes correlated with surviving longer: one topic is associated with normal vital signs and laboratory measurements, while the other includes vital sign and laboratory derangements of sodium and creatinine. SURVSCHOLAR can help discover such feature relationships that clinicians could then verify. Meanwhile, when SURVSCHOLAR's prediction is inaccurate, examining the topics learned could help with model debugging.

2 Background

We begin with some background and notation on topic modeling and survival analysis. For ease of exposition, we phrase notation in terms of predicting time until death; other critical events are possible aside from death.

We assume that we have access to a training dataset of n subjects. For each subject, we know how many times each of d "words" appears, where the dictionary of words is pre-specified (continuous clinical feature values are discretized into bins). As an example, one word might correspond to "white blood count reading in the bottom quintile"; for a given subject, we can count how many such readings the subject has had recorded in the past. We denote $X_{i,u}$ to be the number of times word $u \in \{1, \ldots, d\}$ appears for subject $i \in \{1, \ldots, n\}$. Viewing X as an n-by-d matrix, the i-th row of X (denoted by X_i) can be thought of as the feature vector for the i-th subject.

As for the training label for the i-th subject, we have two recordings: event indicator $\delta_i \in \{0, 1\}$ specifies whether death occurred for the i-th subject, and observed time $Y_i \in \mathbb{R}_+$ is the i-th subject's "survival time" (time until death) if $\delta_i = 1$ or the "censoring time" if $\delta_i = 0$. The idea is that when we stop collecting training data, some subjects are still alive. The i-th subject still being alive corresponds to $\delta_i = 0$ with a true survival time that is unknown ("censored"); instead, we know that the subject's survival time is at least the censoring time.

Topic Modeling. A topic model transforms the i-th subject's feature vector X_i into a topic weight vector $W_i \in \mathbb{R}^k$, where $W_{i,g}$ is the fraction that the i-th subject belongs to topic $g = 1, 2, \ldots, k$. The $W_{i,g}$ terms are nonnegative and $\sum_{g=1}^k W_{i,g} = 1$. For example, LDA models topic weight vectors W_i's to be generated i.i.d. from a user-specified k-dimensional Dirichlet distribution. Next, to relate feature vector X_i with its topic weight vector W_i, let $\overline{X}_{i,u}$ denote the fraction of times a word appears for a specific subject, meaning that $\overline{X}_{i,u} = X_{i,u}/\left(\sum_{v=1}^d X_{i,v}\right)$. Then LDA assumes the factorization

$$\overline{X}_{i,u} = \sum_{g=1}^k W_{i,g} A_{g,u} \tag{2.1}$$

for a topic-word matrix $A \in \mathbb{R}^{k \times d}$. Each row of A is a distribution over the d vocabulary words and is assumed to be sampled i.i.d. from a user-specified d-dimensional Dirichlet distribution. In matrix notation, $\overline{X} = WA$. Standard LDA is unsupervised and, given matrix $\overline{X}$, estimates the matrices W and A.

Survival Analysis. Standard topic modeling approaches like LDA do not solve a prediction task. To predict time-to-event outcomes, we turn toward survival analysis models. Suppose we take the i-th subject's feature vector to be $W_i \in \mathbb{R}^k$ instead of X_i. As this notation suggests, when we combine topic and survival models, W_i corresponds to the i-th subject's topic weight vector; this strategy for combining topic and survival models was first done by Dawson and Kendziorski [8], who worked off of the original supervised LDA formulation by McAuliffe and Blei [23] (which is not stated for survival analysis). We treat the training data as $(W_1, Y_1, \delta_1), \ldots, (W_n, Y_n, \delta_n)$, disregarding the "raw" feature vectors X_i's.

The standard survival analysis prediction task can be stated as using the training data $(W_1, Y_1, \delta_1), \ldots, (W_n, Y_n, \delta_n)$ to estimate, for any test subject with feature vector $w \in \mathbb{R}^k$, the subject-specific survival function

$$S(t|w) = \mathbb{P}(\text{subject survives beyond time } t \mid \text{subject's feature vector is } w).$$

Importantly, unlike standard regression where, for any test feature vector w, we predict a single real number, here we predict a whole function $S(\cdot|w)$.

Our neural survival-supervised topic modeling framework crucially requires that the we can construct a predictor $\widehat{S}(\cdot|w)$ for the subject-specific survival function $S(\cdot|w)$ by minimizing a differentiable loss. Numerous survival models satisfy this criterion. For example, consider the classical Cox proportional hazards model [6]. We learn a parameter vector $\beta \in \mathbb{R}^k$ that weights the features, i.e., prediction for an arbitrary feature vector $w \in \mathbb{R}^k$ is based on the inner product $\beta^\top w$. The differentiable loss function for the Cox model is

$$L_{\mathsf{Cox}}(\beta) = \sum_{i=1}^n \delta_i \left[-\beta^\top W_i + \log \sum_{j=1 \text{ s.t. } Y_j \geq Y_i}^n \exp(\beta^\top W_j) \right]. \qquad (2.2)$$

After computing parameter estimate $\widehat{\beta}$ by minimizing $L_{\mathsf{Cox}}(\beta)$, we can estimate survival functions $S(\cdot|w)$ via the following approach by Breslow [4]. Denote the unique times of death in the training data by $t_1, t_2, \ldots, t_m$. Let d_i be the number of deaths at time t_i. We first compute the so-called hazard function $\widehat{h}_i := d_i/(\sum_{j=1 \text{ s.t. } Y_j \geq Y_i}^n e^{\widehat{\beta}^\top W_j})$ at each time index $i = 1, 2, \ldots, m$. Next, we form the "baseline" survival function $\widehat{S}_0(t) := \exp(-\sum_{i=1 \text{ s.t. } t_i \leq t}^m \widehat{h}_i)$. Finally, subject-specific survival functions are estimated to be powers of the baseline survival function: $\widehat{S}(t|w) := [\widehat{S}_0(t)]^{\exp(\widehat{\beta}^\top w)}$.

3   Neural Survival-Supervised Topic Models

We now present our proposed neural survival-supervised topic modeling framework. Our framework can use any topic model that has a neural net formulation (e.g., neural versions of LDA [3], SAGE [10], and correlated topic models [20] are provided by Card et al. [5]; recent topic models like the Embedded Topic Model [9] can also be used). Moreover, our framework can use any survival model learnable by minimizing a differentiable loss (e.g., Cox proportional hazards [6] and its lasso/elastic-net-regularized variants [25], the Weibull accelerated failure time (AFT) model [15], and all neural survival models we are aware of). For ease of exposition, we focus on combining LDA with the Cox proportional hazards model, similar to what is done by Dawson and Kendziorski [8] except we do this combination in a neural net framework.

We first need a neural net formulation of LDA. We can use the SCHOLAR framework by Card et al. [5]. Card et al. do not explicitly consider survival analysis in their setup although they mention that predicting different kinds of real-valued outputs can be incorporated by using different label networks. We use their same setup and have the final label network perform survival analysis. We give an overview of SCHOLAR before explaining our choice of label network.

The SCHOLAR framework specifies a generative model for the data, including how each individual word in each subject is generated. In particular, recall that $X_{i,u}$ denotes the number of times the word $u \in \{1, 2, \ldots, d\}$ appears for the i-th subject. Let v_i denote the number of words for the i-th subject, i.e., $v_i = \sum_{u=1}^{d} X_{i,u}$. We now define the random variable $\psi_{i,\ell} \in \{1, 2, \ldots, d\}$ to be what the ℓ-th word for the i-th subject is (for $i = 1, 2, \ldots, n$ and $\ell = 1, 2, \ldots, v_i$). Then the generative process for SCHOLAR with k topics is as follows, stated for the i-th subject:

1. Generate the i-th subject's topic distribution:
   (a) Sample $\widetilde{W}_i$ from a logistic normal distribution with mean vector $\boldsymbol{\mu} \in \mathbb{R}^k$ and covariance matrix $\boldsymbol{\Sigma} \in \mathbb{R}^{k \times k}$.
   (b) Set the topic weights vector for the i-th subject to be $W_i = \text{softmax}(\widetilde{W}_i)$.
2. Generate the i-th subject's words:
   (a) Set word parameter $\phi_i = f_{\text{word}}(W_i)$, where f_{word} is a generator network.
   (b) For word $\ell = 1, 2, \ldots, v_i$: Sample $\psi_{i,\ell} \sim \text{Multinomial}(\text{softmax}(\phi_i))$.
3. Generate the i-th subject's output label:
   Sample Y_i from a distribution parameterized by label network $f_{\text{label}}(W_i)$.

Different choices for the parameters $\boldsymbol{\mu}$, $\boldsymbol{\Sigma}$, f_{word}, and f_{label} lead to different topic models. The parameters are learned via amortized variational inference [18,24]. To approximate LDA where topic distributions are sampled from a symmetric Dirichlet distribution with parameter $\alpha > 0$, we set $\boldsymbol{\mu}$ to be the all zeros vector, $\boldsymbol{\Sigma} = \text{diag}((r-1)/(\alpha r))$, and $f_{\text{word}}(w) = w^\top H$ where $H \in \mathbb{R}^{k \times d}$ has a Dirichlet prior per row. We describe how to set f_{label} to obtain survival supervision next.

Survival Supervision. To incorporate the Cox survival loss, we change step 3 of the generative process above to be deterministic and output the variable $\Xi_i = f_{\text{label}}(W_i) := \beta^\top W_i$ for parameter vector $\beta \in \mathbb{R}^k$. In particular, we do not model how observed times Y_i's are generated; modeling Ξ_i's is sufficient. Then we can minimize the Cox proportional hazards loss from Eq. (2.2), rewritten to use the variables Ξ_i's that are parameterized by β:

$$L_{\text{Cox}}(\beta) = \sum_{i=1}^{n} \delta_i \Big[-\Xi_i + \log \sum_{j=1 \text{ s.t. } Y_j \geq Y_i}^{n} \exp(\Xi_i) \Big], \text{ where } \Xi_i = \beta^\top W_i. \quad (3.1)$$

For a hyperparameter $\eta > 0$ that weights the importance of the survival loss, the final overall loss that gets minimized is the sum of $\eta L_{\text{Cox}}(\beta)$ and SCHOLAR's topic model loss (given by the negation of Eq. (3.1) in the SCHOLAR paper [5]). We refer to the resulting model as SURVSCHOLAR.

We remark that rewriting the Cox loss to use Ξ_i variables (for which we can replace the inner product $\Xi_i = \beta^\top W_i$ with a neural net $\Xi_i = g(W_i)$) is by Katzman et al. [16] and also works for the Weibull AFT model.

Model Interpretation. For the g-th topic learned, we can look at its distribution over words $A_g \in \mathbb{R}^d$ (given in Eq. (2.1)) and, for instance, rank words by their probability of appearing for topic g (our experiments later rank words using a notion of comparing to background word frequencies). The g-th topic is also associated with Cox regression coefficient β_g, where $\beta = (\beta_1, \beta_2, \ldots, \beta_k) \in \mathbb{R}^k$ is the parameter from Eq. (3.1). Under the Cox model, β_g being larger means that the g-th topic is associated with *shorter* survival times.

Table 1. Basic characteristics of the survival datasets used.

Dataset	Description	# subjects	# features	% censored
SUPPORT-1	Acute resp. failure/multiple organ sys. failure	4194	14	35.6%
SUPPORT-2	COPD/congestive heart failure/cirrhosis	2804	14	38.8%
SUPPORT-3	Cancer	1340	13	11.3%
SUPPORT-4	Coma	591	14	18.6%
UNOS	Heart transplant	62644	49	50.2%
METABRIC	Breast cancer	1981	24	55.2%
MIMIC(ICH)	Intracerebral hemorrhage	1010	1157	0%

4   Experimental Results

Data. We conduct experiments on seven datasets: data on severely ill hospitalized patients from the Study to Understand Prognoses Preferences Outcomes and Risks of Treatment (SUPPORT) [19], which—as suggested by Harrell [11]—we split into four datasets corresponding to different disease groups (acute respiratory failure/multiple organ system failure, cancer, coma, COPD/congestive heart failure/cirrhosis); data from patients who received heart transplants in the United Network for Organ Sharing (UNOS);[1] data from breast cancer patients (METABRIC) [7]; and lastly patients with intracerebral hemorrhage (ICH) from the MIMIC-III electronic heath records dataset [14]. For all except the last dataset, we predict time until death; for the ICH patients, we predict time until discharge from a hospital ICU. Basic characteristics of these datasets are reported in Table 1. We randomly divide each dataset into a 80%/20% train/test split. Our code is available and includes data preprocessing details.[2]

Experimental Setup. We benchmark SURVSCHOLAR against a total of 7 baselines: 4 classical methods (Cox proportional hazards [6], lasso-regularized Cox [25], k-nearest neighbor Kaplan-Meier [2,22], and random survival forests (RSF) [13],

---

[1] We use the UNOS Standard Transplant and Analysis Research data from the Organ Procurement and Transplantation Network as of September 2019, requested at: https://www.unos.org/data/.

[2] https://github.com/lilinhonglexie/NPSurvival2020.

2 deep learning methods (DeepSurv [16] and DeepHit [21], and a naive two-stage decoupled LDA/Cox model (fit unsupervised LDA first and then fit a Cox model). For all methods, 5-fold cross-validation on training data is used to select hyperparameters (if there are any) prior to training on the complete training data. Hyperparameter search grids are included in our code. For both cross-validation and evaluating test set accuracy, we use the time-dependent concordance C^{td} index [1], which roughly speaking is the fraction of pairs of subjects in a validation or test set who are correctly ordered, accounting for temporal and censoring aspects of survival data. Similar to area under the ROC curve for classification, a C^{td} index of 0.5 corresponds to random guessing and 1 is a perfect score. For every test set C^{td} index reported, we also compute its 95% confidence interval, which we obtain by taking bootstrap samples of the test set with replacement, recomputing the C^{td} index per bootstrap sample, and taking the 2.5 and 97.5 percentile values.

For SURVSCHOLAR, we also include a variant SURVSCHOLAR-FEW that instead of picking whichever hyperparameters (number of topics k and the survival loss importance weight η) achieve the highest training cross-validation C^{td} index, we instead favor choosing a hyperparameter setting with the fewest number of topics that achieves a cross-validation C^{td} index within 0.005 of the best score. We empirically found that often a much fewer number of topics achieves a training cross-validation score that is nearly as good as the max found. For ease of model interpretation, using a fewer number of topics is preferable.

Results. Test set C^{td} indices are reported in Table 2 with 95% confidence intervals. The main takeaways are that: (a) the two SURVSCHOLAR variants are the best or nearly the best performers on SUPPORT-1, SUPPORT-3, and METABRIC; (b) even when the SURVSCHOLAR variants are not among the best performers, they still do as well as some established baselines; (c) the two SURVSCHOLAR variants have very similar performance (so for interpretation, we use SURVSCHOLAR-FEW), and (d) no single method is the best across all datasets.

Table 2. Test set C^{td} indices with 95% bootstrap confidence intervals.

Model	Dataset						
	SUPPORT-1	SUPPORT-2	SUPPORT-3	SUPPORT-4	UNOS	METABRIC	MIMIC(ICH)
COX	0.630 (0.606, 0.655)	0.571 (0.538, 0.604)	**0.569** (0.531, 0.607)	0.592 (0.537, 0.649)	0.583 (0.575, 0.592)	0.664 (0.622, 0.706)	0.610 (0.564, 0.652)
LASSO-COX	0.627 (0.604, 0.652)	0.567 (0.535, 0.600)	0.556 (0.517, 0.594)	0.603 (0.538, 0.666)	0.557 (0.548, 0.565)	0.664 (0.623, 0.708)	**0.667** (0.621, 0.712)
K-NN	0.601 (0.577, 0.628)	0.581 (0.545, 0.614)	0.557 (0.517, 0.592)	0.501 (0.432, 0.576)	0.584 (0.576, 0.592)	0.669 (0.627, 0.708)	0.563 (0.518, 0.612)
RSF	0.602 (0.575, 0.628)	**0.604** (0.570, 0.636)	0.568 (0.530, 0.601)	0.492 (0.414, 0.575)	0.587 (0.579, 0.595)	**0.697** (0.659, 0.736)	0.651 (0.602, 0.697)
DEEPSURV	0.636 (0.611, 0.660)	0.555 (0.521, 0.589)	0.555 (0.517, 0.591)	0.602 (0.548, 0.659)	0.580 (0.572, 0.589)	0.686 (0.644, 0.725)	0.616 (0.571, 0.661)
DEEPHIT	0.633 (0.607, 0.660)	0.579 (0.548, 0.609)	0.547 (0.511, 0.585)	0.590 (0.518, 0.657)	**0.598** (0.590, 0.606)	0.683 (0.644, 0.721)	0.598 (0.553, 0.649)
NAIVE LDA/COX	0.586 (0.559, 0.611)	0.565 (0.533, 0.595)	0.525 (0.486, 0.563)	**0.607** (0.541, 0.672)	0.537 (0.528, 0.545)	0.661 (0.622, 0.698)	0.599 (0.549, 0.646)
SURVSCHOLAR	0.630 (0.604, 0.655)	0.587 (0.553, 0.618)	0.568 (0.528, 0.605)	0.567 (0.509, 0.625)	0.588 (0.580, 0.595)	0.690 (0.649, 0.731)	0.619 (0.572, 0.661)
SURVSCHOLAR-FEW	**0.637** (0.612, 0.662)	0.580 (0.547, 0.610)	0.568 (0.528, 0.605)	0.586 (0.532, 0.640)	0.588 (0.581, 0.596)	0.695 (0.656, 0.735)	0.590 (0.547, 0.632)

Next, we interpret the learned topic models. We plot the topics learned by SURVSCHOLAR-FEW for the SUPPORT-3 dataset on cancer patients in Fig. 1: each topic is a column in the plot, where above each topic, we denote its Cox β regression coefficient (higher means shorter survival time); rows correspond to features. Deeper red colors indicate features that occur more for a topic; color intensity values are multiplicative ratios compared to background word frequencies and are explained in more detail in the appendix. The three topics in this SUPPORT-3 cancer dataset indicate one anti-survival and two pro-survival topics. There is a primary anti-survival topic described by old age, multicomorbidity, hyponatremia, and hyperventilation. The first pro-survival topic describes vital sign and laboratory derangements including hypernatremia, elevated creatinine, hypertension, and hypotension. The second pro-survival topic with slightly stronger pro-survival association suggests otherwise-healthy patients with normal vital signs and laboratory measurements.

We summarize our findings for the other datasets. For SUPPORT-1, SUPPORT-2, SUPPORT-4, UNOS, and METABRIC, only two topics (corresponding to healthy and unhealthy) are identified per dataset by SURVSCHOLAR-FEW. For the MIMIC(ICH) dataset, SURVSCHOLAR-FEW has similar prediction performance as deep learning baseline DeepHit (c.f., Table 2) but neither method performs as well as lasso-regularized Cox. By inspecting the 5 topics learned by SURVSCHOLAR-FEW, we find the topics difficult to interpret as too many features are surfaced as highly probable. In this high-dimensional setting where the number of features is larger than the number of subjects, we suspect that regularizing the model (e.g., by replacing LDA with SAGE [10] is essential to obtaining interpretable topics. Our interpretations of learned topic models for all datasets along with additional visualizations are available in our code repository.

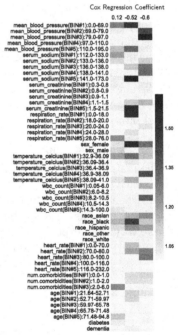

Fig. 1. Topics learned for SUPPORT-3. Rows index features, columns index topics.

5   Discussion

Despite many methodological advances in survival analysis with the help of deep learning, these advances have mostly not focused on interpretability. Model interpretation can be especially challenging when there are many features and how they relate is unknown. In this paper, we show that neural survival-supervised topic models provide a promising avenue for learning structure over features in terms of "topics" that help predict time-to-event outcomes. These topics can be used by practitioners to check if learned topics agree with domain knowledge and, if not, to help with model debugging. Rigorous evaluations of other neural survival-supervised topic models aside from fusing LDA with Cox are needed to better understand which combinations of topic and survival models yield both highly accurate time-to-event predictions and clinically interpretable topics.

Acknowledgments. This work was supported in part by Health Resources and Services Administration contract 234-2005-370011C. The content is the responsibility of the authors alone and does not necessarily reflect the views or policies of the Department of Health and Human Services, nor does mention of trade names, commerical products, or organizations imply endorsement by the U.S. Government.

A   Interpreting Topic Heatmaps

In this appendix, we explain how to interpret our topic heatmaps (Fig. 1 and additional plots in our code repository). For many topic models including LDA, a topic is represented as a distribution over d vocabulary words. SCHOLAR [5] (and also our survival-supervised version SURVSCHOLAR) reparameterizes these topic distributions; borrowing from SAGE [10], SCHOLAR represents a topic as a deviation from a background log-frequency vector. This vector accommodates common words that have similar frequencies across data points. When we visualize a topic, we take this modeling approach into account and only choose to highlight features that have positive log-deviations from the background. Given a topic, having positive log-deviation is analogous to having higher conditional probabilities in the classic topic modeling case but explicitly is relative to background word frequencies (rather than being raw topic word probabilities).

To fill in the details, in step 2(a) of SURVSCHOLAR's generative process (stated in Sect. 3), each word is drawn from the conditional distribution softmax($\gamma + w^T B$), where $\gamma \in \mathbb{R}^d$ is the background log-frequency vector, $w \in \mathbb{R}^k$ contains a sample's topic membership weights, and $B \in \mathbb{R}^{k \times d}$ encodes (per topic) every vocabulary word's log-deviation from the word's background. This is a reparameterization of how LDA is encoded, which has each word drawn from the conditional distribution softmax($w^T H$) for $H \in \mathbb{R}^{k \times d}$. In particular, note that $H_g = \gamma + B_g$ for every topic $g \in \{1, 2, \dots, k\}$. The background log-frequency vector γ is learned during neural net training. Note that SAGE [10] further encourages sparsity in B by adding ℓ_1 regularization on B.

We found ranking words within a topic by their raw probabilities (A_g in Eq. (2.1)) to be less interpretable than ranking words based on their deviations from their background frequencies (B_g) precisely because commonly occurring background words make interpretation difficult. In fact, when Dawson and Kendziorski [8] introduced SURVLDA, they used an ad hoc pre-processing step to identify background words to exclude from analysis altogether. We avoid this pre-processing and use log-deviations from background frequencies instead.

In heatmaps such as the one in Fig. 1, each column corresponds to a topic. For the g-th topic, instead of plotting its raw log-deviations (encoded in $B_g \in \mathbb{R}^d$), which are harder to interpret, we exponentiated each word's log-deviation to get the word's multiplicative ratio from its background frequency (i.e., we compute $\exp(B_g)$); the color bar intensity values are precisely these multiplicative ratios of how often a word appears relative to the word's background frequency.

To highlight features that distinguish topics from one another, we also sort rows in the heatmap by descending differences between the largest and smallest values in a row. Thus, features whose deviations vary greatly across topics tend to show up on the top. A technical detail is that we sorted with respect to the original features, rather than the one-hot encoded or binned features. Therefore, as an example, all bins under mean blood pressure stay together. For features associated with multiple rows in the heatmap, we computed the difference between the largest and smallest values for each row, and used the largest difference (across rows) for sorting.

References

1. Antolini, L., Boracchi, P., Biganzoli, E.: A time-dependent discrimination index for survival data. Stat. Med. **24**(24), 3927–3944 (2005)
2. Beran, R.: Nonparametric regression with randomly censored survival data. Technical report, University of California, Berkeley (1981)
3. Blei, D.M., Ng, A.Y., Jordan, M.I.: Latent Dirichlet allocation. J. Mach. Learn. Res. **3**, 993–1022 (2013)
4. Breslow, N.: Discussion of the paper by D. R. Cox cited below. J. R. Stat. Soc. Ser. B **34**(2), 216–217 (1972)
5. Card, D., Tan, C., Smith, N.A.: Neural models for documents with metadata. In: Proceedings of Association for Computational Linguistics (2018)
6. Cox, D.R.: Regression models and life-tables. J. Roy. Stat. Soc. B **34**(2), 187–202 (1972)
7. Curtis, C., et al.: The genomic and transcriptomic architecture of 2,000 breast tumours reveals novel subgroups. Nature **486**(7403), 346 (2012)
8. Dawson, J.A., Kendziorski, C.: Survival-supervised latent Dirichlet allocation models for genomic analysis of time-to-event outcomes. arXiv preprint arXiv:1202.5999 (2012)
9. Dieng, A.B., Ruiz, F.J., Blei, D.M.: Topic modeling in embedding spaces. arXiv preprint arXiv:1907.04907 (2019)
10. Eisenstein, J., Ahmed, A., Xing, E.P.: Sparse additive generative models of text. In: International Conference on Machine Learning, pp, 1041–1048 (2011)
11. Harrell, F.E.: Regression Modeling Strategies: With Applications to Linear Models, Logistic and Ordinal Regression, and Survival Analysis. Springer, Cham (2015)
12. Harrell, F.E., Lee, K.L., Califf, R.M., Pryor, D.B., Rosati, R.A.: Regression modelling strategies for improved prognostic prediction. Stat. Med. **3**(2), 143–152 (1984)
13. Ishwaran, H., Kogalur, U.B., Blackstone, E.H., Lauer, M.S.: Random survival forests. Ann. Appl. Stat. **2**(3), 841–860 (2008)
14. Johnson, A.E., et al.: MIMIC-III, a freely accessible critical care database. Sci. Data **3**(1), 1–9 (2016)
15. Kalbfleisch, J.D., Prentice, R.L.: The Statistical Analysis of Failure Time Data, 2nd edn. Wiley, Hoboken (2002)

16. Katzman, J.L., Shaham, U., Cloninger, A., Bates, J., Jiang, T., Kluger, Y.: Deep-Surv: personalized treatment recommender system using a Cox proportional hazards deep neural network. BMC Med. Res. Methodol. **18**(1), 24 (2018)

17. Kingma, D.P., Ba, J.: Adam: a method for stochastic optimization. arXiv preprint arXiv:1412.6980 (2014)

18. Kingma, D.P., Welling, M.: Auto-encoding variational Bayes. In: International Conference on Learning Representations (2014)

19. Knaus, W.A., et al.: The SUPPORT prognostic model: objective estimates of survival for seriously ill hospitalized adults. Ann. Intern. Med. **122**(3), 191–203 (1995)

20. Lafferty, J.D., Blei, D.M.: Correlated topic models. In: Advances in Neural Information Processing Systems, pp. 147–154 (2006)

21. Lee, C., Zame, W.R., Yoon, J., van der Schaar, M.: DeepHit: a deep learning approach to survival analysis with competing risks. In: AAAI Conference on Artificial Intelligence (2018)

22. Lowsky, D.J., et al.: A K-nearest neighbors survival probability prediction method. Stat. Med. **32**(12), 2062–2069 (2013)

23. McAuliffe, J.D., Blei, D.M.: Supervised topic models. In: Advances in Neural Information Processing Systems, pp. 121–128 (2008)

24. Rezende, D.J., Mohamed, S., Wierstra, D.: Stochastic backpropagation and approximate inference in deep generative models. In: International Conference on Machine Learning (2014)

25. Simon, N., Friedman, J., Hastie, T., Tibshirani, R.: Regularization paths for Cox's proportional hazards model via coordinate descent. J. Stat. Softw. **39**(5), 1 (2011)

Medical Time-Series Data Generation Using Generative Adversarial Networks

Saloni Dash[1](✉), Andrew Yale[2], Isabelle Guyon[3], and Kristin P. Bennett[2]

[1] Department of CSIS, BITS Pilani, Goa Campus, Goa, India
salonidash77@gmail.com
[2] Rensselaer Polytechnic Institute, Troy, NY, USA
[3] UPSud/INRIA U. Paris-Saclay, Paris, France

Abstract. Medical data is rarely made publicly available due to high de-identification costs and risks. Access to such data is highly regulated due to it's sensitive nature. These factors impede the development of data-driven advancements in the healthcare domain. Synthetic medical data which can maintain the utility of the real data while simultaneously preserving privacy can be an ideal substitute for advancing research. Medical data is longitudinal in nature, with a single patient having multiple temporal events, influenced by static covariates like age, gender, comorbidities, etc. Extending existing time-series generative models to generate medical data can be challenging due to this influence of patient covariates. We propose a workflow wherein we leverage existing generative models to generate such data. We demonstrate this approach by generating synthetic versions of several time-series datasets where static covariates influence the temporal values. We use a state-of-the-art benchmark as a comparative baseline. Our methodology for empirically evaluating synthetic time-series data shows that the synthetic data generated with our workflow has higher resemblance and utility. We also demonstrate how stratification by covariates is required to gain a deeper understanding of synthetic data quality and underscore the importance of including this analysis in evaluation of synthetic medical data quality.

Keywords: Synthetic data · Generative Adversarial Networks · Time-series

1   Introduction

Medical data in the form of Electronic Medical Records (EMR) has been widely used by hospitals in the United States for aiding hospital processes like quality improvement, monitoring patient safety etc. [14]. Furthermore, EMR data has also been used for advancing healthcare research for decades [13]. However, high de-identification costs and risks severely limit public access to such data. This imposes huge restrictions on data-driven clinical research and makes studies that use this data difficult to reproduce.

© Springer Nature Switzerland AG 2020
M. Michalowski and R. Moskovitch (Eds.): AIME 2020, LNAI 12299, pp. 382–391, 2020.
https://doi.org/10.1007/978-3-030-59137-3_34

A generative model that samples from the distribution of the health data, while simultaneously preserving it's privacy is an ideal solution to the problem. Generative models like Generative Adversarial Networks (GANs) [4,15] (Health-GAN, medGAN) explicitly generate snapshots of EMR type data. However, real EMR data is longitudinal in nature and falls in the domain of time-series generative modelling. A key aspect of medical data is static covariates that heavily influence temporal variables. For instance, a patient record not only contains details of hospital visits over a period of time but also static demographic details like gender, ethnicity, comorbidities etc. We characterize a good medical time-series generative model as one that jointly models the distribution of the static as well as the temporal variables.

We address this problem in the paper and provide a simple baseline that can be used for comparison against future medical time-series generative models. The primary contributions of this paper[1] are:

1. Illustration of an efficient, flexible workflow to facilitate joint modelling and synthesis of static and temporal variables.
2. Explicit qualitative evaluation of influence of static covariates on time-series variables.
3. Reproducing clinical time-series benchmarks on synthetic versions of a publicly available and widely used medical dataset.

2   Related Work

An open source synthetic patient generator called Synthea [5] uses hand-crafted modules aided by health care practitioners and statistics derived from real data to generate patients from their birth day to the present day. It does not violate any privacy restrictions because it does not use real patients to generate the data. It also claims to maintain utility as the generator uses underlying rules manually derived from the real data. However, the time-series generated for a record are not necessarily representative of real patient trajectories. Additionally, the custom designed rules severely limit the type of data that can be generated and are not easily extendable to other distributions of data.

Recurrent (Conditional) GAN (RCGAN) [8] uses recurrent neural networks (RNNs) in the GAN framework, to generate real valued time-series medical data like respiratory rate, heart rate etc. In the conditional setting, both the generator and discriminator are conditioned on labels sampled from the real data during training, and generated from independent distributions during the generation process. The labels guide the generative process but are not modelled jointly with the time-series variables.

Time-Series GAN [16] explicitly models time-series distributions as a joint distribution of static and temporal variables. It produces realistic time-series by jointly optimizing adversarial and supervised losses. We found TimeGAN to be the only time-series generative model that addresses the problem of jointly

---

[1] This paper is an extension of [6] submitted to the ML4H workshop at NeurIPS 2019.

modelling static and temporal variables, and use it as a comparative baseline for our methodology.

3   Method

We illustrate our approach by generating time-series datasets for three time-series datasets where the covariates have a strong influence on the time-series variables. The datasets are (1) PJM Hourly Energy Consumption Dataset (Kaggle) [12],(2) Sleeping Patterns from American Time Use Survey (ATUS) [1], (3) Medical Information Mart for Intensive Care (MIMIC-III) [10].

The workflow is as follows:

1. Identify appropriate summary statistic(s) for time-series variables (e.g. mean, median, skew, count etc.)
2. Compute summary statistic(s) for fixed time-intervals over the whole time period.
3. Append summary statistic(s) to static variables. This maps the time-series data frame to a cross-sectional data frame.
4. Use a generative model of your choice to generate this transformed data.

For (1), we choose summary statistics inspired by downstream applications of the synthetic data. For (4), we use HealthGAN, a Wasserstein GAN [15], to generate the transformed data. The HealthGAN includes encoding mappings for categorical, numerical and ordinal variables of which the ordinal mappings particularly boosted our results for the ATUS dataset. We evaluate resemblance of the synthetic data to real data by assessing summary statistics conditioned on covariates and the utility by reproducing published research results.

This workflow is best suited for data where the time-series can be dissected into meaningful intervals to compute summary statistics relevant to the downstream application of synthetic data. An appropriate transformation of computing mean, variance, skew, kurtosis etc. for the whole time-series is always feasible.

We use TimeGAN[2] as a comparison for this workflow for two of the above datasets. It should be noted that we use the default parameters for TimeGAN and do not fine-tune the model while generating the data.

4   Results

4.1   PJM Hourly Energy Consumption

Our first dataset is not a medical dataset, but it provide an illustration of the challenge of synthesizing time series datasets with covariates. PJM Interconnection LLC (PJM) is a transmission organization (RTO) which is part of the

---

[2] We use the source code available at https://bitbucket.org/mvdschaar/mlforhealthlab pub/src/master/alg/timegan/.

Eastern Interconnection grid operating an electric transmission system serving specific regions of the United States. The dataset[3] primarily comprises of a datetime stamp and the average energy consumed in Mega Watts (MW) in that hour.

A natural summary statistic for the dataset is the average energy consumed per hour. We set the time period to be one day. We hence get twenty-four time-series statistics, one for each hour of the day which we append to the static variables of day of week and month derived from the datetime stamp. The transformed data now has twenty-six variables which are generated by HealthGAN. We also separately generate the original data using TimeGAN.

We then qualitatively evaluate the synthetic datasets by comparing trends in the real data. Figure 1 shows close resemblance of the summary statistic of average hourly energy consumption across twenty-four hours for the real and synthetic datasets (derived from HealthGAN and TimeGAN). A Welsch t-test of the samples binned by hour shows that the hourly means from HealthGAN (p-value = 0.51) as well as the hourly means from TimeGAN (p-value = 0.18) do not significantly differ from the hourly means of the real data. Both generations methods are seemingly performing well.

Figure 2 analyses the influence of the static covariates of day of week and month on the average energy consumption in the real data, which reports highest energy consumption during the weekdays in the evening hours and the summer months. These trends are mimicked in the synthetic data generated by HealthGAN. In the synthetic data generated by TimeGAN, the hourly and weekly trends are captured reasonably well but when examining by the covariate months, the peak at months 7 and 8 is missed.

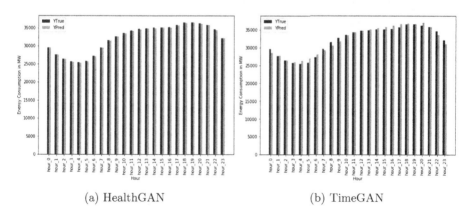

(a) HealthGAN   (b) TimeGAN

Fig. 1. Hourly Energy Consumption - YTrue is the hourly energy consumption in the real data and YPred is the hourly energy consumption in the generated data. The values in both the real and synthetic datasets match closely.

---

[3] https://www.kaggle.com/robikscube/hourly-energy-consumption.

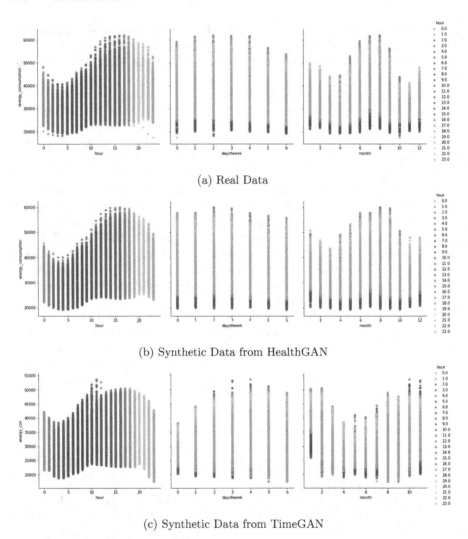

(a) Real Data

(b) Synthetic Data from HealthGAN

(c) Synthetic Data from TimeGAN

Fig. 2. Average Energy Consumed vs Hour, Day of Week and Month for real and synthetic data sets shows HealthGAN synthetic data highly resembles real data even when covariates are considered.

4.2 American Time Use Survey (ATUS)

We generate sleeping patterns from American Time Use Survey (ATUS), a federally administered, annual survey on time use in the United States [1]. The survey records how Americans divide their time among life's activities in a nationally representative sample. There are many different types of events per person. However, we choose to restrict our data to only the sleep activities of the people over a period of thirty hours. (Please refer to [6] for more details). We divide the thirty hours into thirty events of one hour each, and compute average sleep in that hour. We then append them to the static variables of age, sex, day of the week and month of the year. The data is now in matrix form, consisting of 34 (including covariates) features per patient, ready for the HealthGAN to generate this data. Sleep data is synthesized from TimeGAN as well to use as a comparative baseline.

Figure 3 shows the average sleep trends in the real data and synthetic datasets. The average sleep per hour in the HealthGAN synthetic data closely resembles the real data. Most people are awake by 10:00 am and asleep by 12:00 am. The synthetic data from TimeGAN does not follow this distribution. A Welsch t-test of the sleep times binned by hour shows that the mean sleep time per hour in the data from HealthGAN (p-value = 0.58) is not statistically different from that in the real data. However, for the data from TimeGAN (p-value = 0.012), the means appear to be significantly different.

To analyse the relationship between the static covariates and time-series variables, we reproduce a sleep study [2] which analyses the average sleeping time stratified by age and day of week. Figure 4 shows the variations in sleep depending on age group and day of the week. In the real data, the average sleeping time on weekends is significantly different from that on weekdays. Overall, teenagers and young adults sleep (age group 15–24) significantly more than other age groups, whereas adults between 35–54 years in general require less sleep than other age groups. These trends are captured well in the synthetic data from HealthGAN. The variations in average sleep times binned by group and day of the week are so high in the real data that although the synthetic plot appears to be shifted by an hour, it still falls within the 95% confidence intervals of the means in the real data. For the synthetic data from TimeGAN, the distributions for days Tuesday through Saturday are missing completely, suggesting mode collapse, and the trends captured are significantly different from those in the real data.

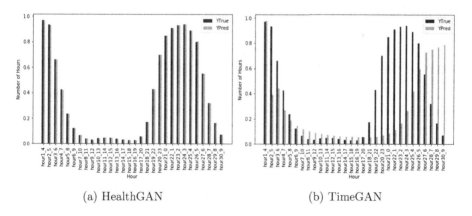

(a) HealthGAN   (b) TimeGAN

Fig. 3. Average Hourly Sleep Trends of real (blue) and synthetic data (yellow) generated by HealthGann and TimeGAN (Color figure online)

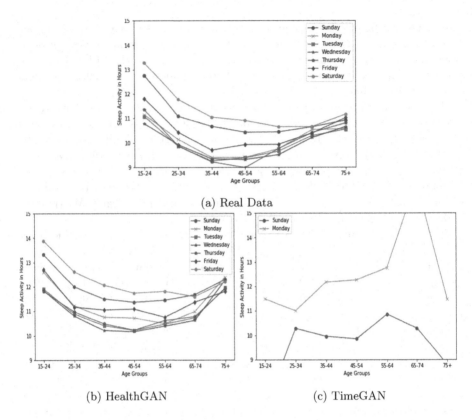

(a) Real Data

(b) HealthGAN   (c) TimeGAN

Fig. 4. Average hours of sleep grouped by age and day of the week for real and synthetic HealthGAN and TimeGAN data

4.3   MIMIC - III

The MIMIC - III dataset [10] is a publicly available critical care database which is widely used in research studies [3,7,11]. Specifically we use three clinical prediction time-series benchmarks derived from MIMIC - III [9]. These tasks consist of:

1. **In-Hospital Mortality Prediction** - Predicting In-Hospital mortality based on 48 hours of ICU data.
2. **Decompensation Prediction** - Predicting whether a patient's health will worsen over the next 24 hours.
3. **Phenotype Classification** - Predicting which of the 25 acute care conditions are present in a patient record

For each of the above tasks, the logistic regression baseline specifies which summary statistics to extract from the time-series. We attempt to reproduce these baseline results in the generated data as well. The paper identifies 17 clinical variables as primary temporal variables. For each variable, six different sample statistic features (mean, std dev, skew etc.) are computed on seven different subsequences of a given time series (full, first 10%, first 25% etc.). Please refer to [9] for more details. In total there are 714 temporal variables. In (1) the static covariates are age, gender and the mortality label. This results in a total of 717 variables for each patient. The task is a binary classification task with the primary metric being AUC-ROC. In (2) the static variables are age, gender and decompensation label, resulting in 717 variables. This is also a binary classification task with the primary metric being AUC-ROC. The results for (1) and (2) are summarized in Fig. 5. In (3) the static variables are age, gender and the 25 acute care conditions, resulting in 741 variables. This is a multi-label classification problem to predict phenotype with the primary metric being AUC - ROC for each variable treated independently. The results for the 25 prediction tasks are illustrated in Fig. 6.

In Figs. 5 and 6, RR refers to train on real test on real, RS to train on real test on synthetic, etc. RS scores indicate whether the synthetic data can be substituted for the real data for a downstream application, while SR score indicates whether the synthetic data has realistic features. Overall, across all 27 MIMIC-III tasks, we report RS and SR scores to be reasonably close to RR scores. Note, however, that the SS scores tend to usually overshoot the RR scores indicating that the generated distribution is more regular than the real data.

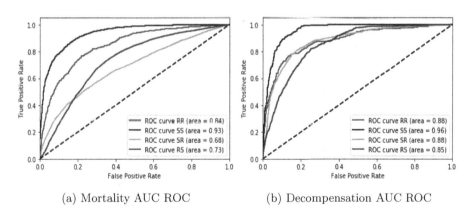

(a) Mortality AUC ROC   (b) Decompensation AUC ROC

Fig. 5. AUC ROC for MIMIC-III Mortality and Decompensation tasks. First letter indicates training set (Real or Synthetic). Second letter indicates testing set

PHENOTYPE	AUC-ROC RR	AUC-ROC SS	AUC-ROC SR	AUC-ROC RS
Hypertension with complications	0.964	0.994	0.721	0.912
Diabetes mellitus with complications	0.929	0.976	0.749	0.765
Chronic kidney disease	0.927	0.992	0.725	0.960
Shock	0.917	0.992	0.689	0.794
Respiratory failure	0.870	0.977	0.769	0.832
Acute cerebrovascular disease	0.870	0.956	0.754	0.690
Coronary atherosclerosis and related	0.866	0.977	0.749	0.896
Septicemia (except in labor)	0.848	0.988	0.720	0.915
Acute and unspecified renal failure	0.845	0.962	0.716	0.870
Diabetes mellitus without complication	0.840	0.978	0.616	0.727
Acute myocardial infarction	0.828	0.977	0.577	0.622
Essential hypertension	0.822	0.969	0.655	0.719
Congestive heart failure	0.816	0.981	0.649	0.441
Pneumonia	0.815	0.942	0.578	0.757
Disorders of lipid metabolism	0.789	0.976	0.682	0.895
Cardiac dysrhythmias	0.780	0.975	0.622	0.631
Conduction disorders	0.765	0.966	0.591	0.644
Fluid and electrolyte disorders	0.764	0.963	0.656	0.881
Other liver diseases	0.756	0.993	0.582	0.412
Gastrointestinal hemorrhage	0.748	0.988	0.565	0.782
Chronic obstructive pulmonary disease	0.701	0.969	0.541	0.672
Complications of surgical/medical care	0.690	0.982	0.597	0.492
Other upper respiratory disease	0.690	0.956	0.521	0.551
Pleurisy; pneumothorax; pulmonary collapse	0.665	0.968	0.542	0.625
Other lower respiratory disease	0.628	0.976	0.526	0.398

Fig. 6. Phenotype Classification Heatmap - In general, RR, RS, SR scores fall in the same range. SS scores tend to overshoot RR scores significantly, suggesting increased regularity in synthetic data as compared to the real data.

5   Conclusion and Future Work

Medical time-series data sets are characterized by both static as well as temporal variables which must be modelled jointly to generate realistic medical time-series data. We provide a simple, flexible and effective workflow to generate this kind of data. We test our methodology by synthesizing three different time-series datasets, two of which are health datasets. We empirically show that the data generated by HealthGan not only shows close univariate resemblance with the real data but also captures trends influenced by static covariates. We use a state-of-the-art benchmark[4] as a comparative baseline. We highlight the importance of evaluating synthetic medical data with respect to critical covariates and the importance of including such analysis in time-series generative models for medical data. We plan to use super-resolution multivariate GANs trained on varying length interval summaries to capture more complex EMR data in the future.

References

1. American time use survey. https://www.bls.gov/tus/home.htm. Accessed 10 Sept 2019
2. Basner, M., et al.: American time use survey: sleep time and its relationship to waking activities. Sleep **30**(9), 1085–1095 (2007)
3. Bose, S., Johnson, A.E.W., Moskowitz, A., Celi, L.A., Raffa, J.D.: Impact of intensive care unit discharge delays on patient outcomes: a retrospective cohort study. J. Intensive Care Med. **34**(11–12), 924–929 (2019)
4. Choi, E., Biswal, S., Malin, B., Duke, J., Stewart, W.F., Sun, J.: Generating multi-label discrete electronic health records using generative adversarial networks. CoRR, abs/1703.06490 (2017)
5. MITRE Corporation. Synthetic patient generation. https://synthetichealth.github.io/synthea/. Accessed 16 May 2019
6. Dash, S., Dutta, R., Guyon, I., Pavao, A., Yale, A., Bennett, K.P.: Synthetic event time series health data generation. arXiv preprint arXiv:1911.06411 (2019)
7. Deliberato, R.O., et al.: Severity of illness scores may misclassify critically ill obese patients. Crit. care Med. **46**(3), 394–400 (2018)
8. Esteban, C., Hyland, S.L., Rätsch, G.: Real-valued (medical) time series generation with recurrent conditional GANs. arXiv preprint arXiv:1706.02633 (2017)
9. Harutyunyan, H., Khachatrian, H., Kale, D.C., Steeg, G.V., Galstyan, A.: Multitask learning and benchmarking with clinical time series data. Sci. Data **6**(1) (2019)
10. Johnson, A.E.W., et al.: MIMIC-III, a freely accessible critical care database. Sci. Data **3**, 160035 (2016)
11. Lokhandwala, S., et al.: One-year mortality after recovery from critical illness: a retrospective cohort study. PloS One **13**(5) (2018)
12. Mulla, R.: https://www.kaggle.com/robikscube/hourly-energy-consumption
13. Nordo, A.H., et al.: Use of EHRs data for clinical research: Historical progress and current applications. Learn. Health Syst. **3**(1), e10076 (2019). e10076 LRH2-2018-04-0019.R3
14. Parasrampuria, S., Henry, J.: Hospitals' use of electronic health records data, 2015–2017 (2019)
15. Yale, A., Dash, S., Dutta, R., Guyon, I., Pavao, A., Bennett, K.P.: Privacy preserving synthetic health data. In: Proceedings of the 27th European Symposium on Artificial Neural Networks ESANN, pp. 465–470 (2019)
16. Yoon, J., Jarrett, D., van der Schaar, M.: Time-series generative adversarial networks. In: Advances in Neural Information Processing Systems, pp. 5509–5519 (2019)

---

[4] We would like to thank the authors of TimeGAN for their help with implementation and evaluation details.

Acute Hypertensive Episodes Prediction

Nevo Itzhak[1]([✉]), Aditya Nagori[2], Edo Lior[1], Maya Schvetz[1], Rakesh Lodha[3], Tavpritesh Sethi[3,4], and Robert Moskovitch[1]

[1] Software and Information Systems Engineering,
Ben Gurion University of the Negev, Beer-Sheva, Israel
{nevoit,edoli,schvetz}@post.bgu.ac.il, robertmo@bgu.ac.il
[2] Institute of Genomics and Integrative Biology, New Delhi, India
aditya.kumar@igib.in
[3] Pediatric Medicine, All India Institute of Medical Sciences, New Delhi, India
rlodha1661@gmail.com
[4] Indraprastha Institute of Information Technology Delhi, Delhi, India
tavpriteshsethi@iiitd.ac.in

Abstract. Predicting outbursts of hazardous medical conditions and its importance has arisen significantly in recent years, particularly in patients hospitalized in the Intensive Care Unit (ICU). In hospitals worldwide, patients are developing life-threatening complications, which might lead to organ dysfunctions and, if not treated properly, to death. In this study, we use patients' longitudinal vital signs data from the ICUs, focusing on predicting Acute Hypertensive Episodes (AHE). In this study, two approaches were used for prediction: predicting continuously whether a patient will experience an AHE in a pre-defined time period ahead using an observation sliding window, or predicting whether it will generally occur during the ICU admission, given a fixed time period from the admission. Temporal abstraction was employed to transform the heterogeneous multivariate temporal data into a uniform representation of symbolic time intervals, and frequent Time Intervals Related Patterns (TIRPs), which are used as features for classification. For comparison, Convolutional Neural Network (CNN) and Recurrent Neural Network (RNN) are used. Our results show that using frequent temporal patterns leads to a better AHE prediction.

Keywords: Acute Hypertensive Episodes · Intensive care units · Symbolic Time Intervals · Temporal patterns · Outcome prediction

1 Introduction

Outcomes prediction in Electronic Health Records (EHR) is an essential topic of interest in recent years with the growing availability and access to patients' data [3]. Unlike most clinical data domains, in which data are relatively sparse, in critical care, a meaningful part of the data is regularly sampled, since patients are closely monitored. One of the known severe medical complications in ICU is

© Springer Nature Switzerland AG 2020
M. Michalowski and R. Moskovitch (Eds.): AIME 2020, LNAI 12299, pp. 392–402, 2020.
https://doi.org/10.1007/978-3-030-59137-3_35

Acute Hypertensive Episodes (AHE) - a long-term medical condition in which severe elevations in blood pressure are likely to cause damage to one or more organ systems. Early prediction of AHE is of high clinical importance in ICU admissions and may enable taking precautions to prevent interventions that may be too late potentially. The longitudinal intrinsic nature of critical care data may benefit from the use of temporal data analytics [4,8,15].

In a supervised learning task, to construct a binary classifier, positive or negative examples should be specified. We use *sliding windows*, in which the extracted windows contain multivariate longitudinal values series that were measured in patients. Our goal is to evaluate the patient's risk of experiencing an outcome by dividing the data streams into overlapping time windows and examining each window individually. Several studies reported using different study designs for outcomes prediction using ICU data [11,13], or generally in EHR data [5]. However, not all the studies used the proper study design, as we elaborate in the background, in which we explain how to perform a case-crossover-control design that reflects real-life application conditions in continuous prediction. The second option is generally predicting whether an AHE will occur based on a fixed observation time period relative to the admission time, for which case-control relative to an earlier event is appropriate. In this study, we experiment with both approaches and discuss their pros and cons.

In recent years, the use of temporal abstraction was proposed to transform heterogeneous multivariate temporal data, such as in ICUs [8], into a uniform representation of Symbolic Time Intervals (STIs), which facilitates reasoning, pattern discovery and analytics. Once a database of STIs is created, whether raw or abstracted, time intervals analytics can be performed, such as the discovery of frequent patterns [8] and match similar sequences [4] or queries. In this paper, we apply the use of frequent time intervals patterns as features for AHE prediction in ICU data. We compare this model to the use of a Convolutional Neural Network (CNN) and a Recurrent Neural Network with Long Short-Term Memory (RNN-LSTM). These models are neural network architectures that were designed for the classification of multivariate temporal data.

The main contributions of this paper are the following:

1. A continuous prediction model, based on temporal abstraction and frequent time intervals patterns for AHE prediction.
2. Comparison of predicting whether or not a patient will experience AHE in a pre-defined prediction time versus based on a fixed observation time period relative to the admission.

2   Background

2.1   Acute Hypertensive Episodes

Acute Hypertensive Episode (AHE) is a long-term medical condition in which severe elevations in blood pressure are likely to cause damage to one or more organ systems. Its severity levels are divided into three groups: the pre-hypertensive stage, stage 1 and stage 2 [10]. In this study, we predict the pre-hypertensive stage. The reasons for AHE occurring in hospitalized patients are uncertain to this day, and there is a great deal of demand for medical teams to predict when the onset will abrupt [2]. Treating the event is usually performed in an invasive approach to downgrade the blood pressure, and therefore using a prediction model will allow the medical team to plan ahead and use non-invasive treatments.

2.2   Outcomes Prediction in ICU

Several studies have attempted to predict the onsets of various ICU complications. Not many studies have been conducted on acute hypertensive episodes. Notice however, how previous studies have examined acute hypotensive episodes, which in contrast to hypertensive episodes, are characterized by abnormal low blood pressure. Many of these studies were based on the PhysioNet challenge, in which the goal was to predict which patients will experience an acute hypotensive episode's beginning within a 60 min window, based on a data of at least 600 min per patient. Some studies used logistic regression [5] or neural networks [11,15]. Studies that reviewed Sepsis, a life-threatening condition caused by the body's response to an infection, used overlapping sliding windows [13]. A part of these studies applied sequential patterns based models [4] or neural networks [3].

2.3   Temporal Abstraction and Time Intervals Analytics

Temporal Abstraction refers to transforming a series of raw continuous time point values into a series of symbolic time intervals. *Symbolic Time Intervals* (STIs) can be raw, or the result of temporal abstraction, defined by a triplet of a start-time, end-time and a symbol-id. Temporal abstraction is often performed in two ways: (1) *state abstraction*, in which values are classified into states, are given cutoffs and are later concatenated into STIs; (2) *gradient abstraction*, determines the slope of the changing values. More about the process of temporal abstraction can be found in [8].

Analyzing STIs data has been increasingly investigated over the past decade [8]. Allen had defined seven temporal relations and their inverse [1], between a pair of STIs, namely *before* ($<$), *meets* (m), *overlaps*(o), *starts*(s), *finished-by*(fi), *equals*($=$), and *contains*(c). Once the data are transformed into a series of STIs, patterns can be discovered. *Time Intervals Related Patterns (TIRPs)* discovery methods were developed in the past two decades [8,9], in which most of them use Allen's definition to represent the relation between a pair of STIs, which are used to define a TIRP. A TIRP is called frequent if its vertical support exceeds a minimum threshold. *Vertical support* is denoted by the cardinality of the distinct entities within which TIRP holds at least once, divided by the total number of entities in the database. Moskovitch et al. [8,9] introduced the KarmaLego algorithm that exploits the transitivity of temporal relations to generate candidates efficiently and completely. More details about the evolution of time intervals mining can be found in [9].

3 Methods

Before explaining the TIRPs based prediction framework, it is important to explain the problem setup. We had two approaches to address the prediction of AHE, which influence the way the data is created for the classifier. The first intention was to predict whether or not the AHE will occur using a sliding observation time window that predicts the occurrence of the outcome within a prediction time period ahead, for which the *case-crossover-control* study design was used. Alternatively, we wanted to predict whether AHE will generally occur during the ICU stay based on a fixed observation time period relative to the admission, for which the *case-control relative to an earlier event* was used.

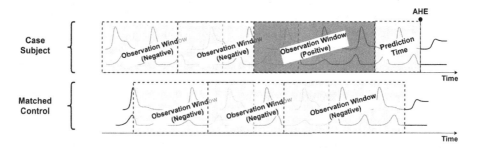

Fig. 1. Case-crossover-control design. The positive examples are the observation time window is the last extracted prior to the prediction time period, which ends with the outcome, while the negative examples are taken from the earlier observation time windows in the cases, or from the matched controls.

However, in study such as ours, first the group of patients having the outcome in their records which are called *cases* are defined, as well as a group of matched patients, called *controls*, who do not have the outcome. Then, observation time windows are extracted, which have to be labeled as positive or negative, to train a classifier. In case-crossover-control, as illustrated in Fig. 1, positive examples are defined as the observation time window that is last prior to the outcome of interest. Negative examples are defined as observation time windows that appear earlier in the cases, or any observation time windows in the controls. *Prediction time* is defined before the outcome to give enough time to intervene and hopefully prevent the outcome, but from an experimental perspective, it verifies that no data is used that can be related directly or implicitly to the outcome, such as a relevant test. Another option is to predict generally whether the AHE will happen in the ICU stay, in which a case-control relative to an earlier event study design can be used, and in this case, one observation time window relative to the ICU admission is extracted from the cases, which is positive, and from the controls, which is negative.

Figure 2 describes the entire flow of the AHE prediction framework, starting with the cases and controls data, and proceeding with the observation time periods. The model learns from historical data composed of time windows-based measurements and is used to predict AHE onset occurrence. We extended the Maitreya framework [7,9] to use time point series variables in addition to raw STIs. Thus, as shown in Fig. 2, after the time point series are abstracted into STIs (Sect. 2.3), frequent TIRPs are discovered from the cases' observation time windows in the training set (Fig. 2.1), using the KarmaLego algorithm [8,9]. The result is a Bag-Of-TIRPs that will be used as features for the classifier. Once the TIRPs' instances are also detected [8] in the control patients, a matrix of TIRP-features is created (Fig. 2.2). The feature-matrix rows stand for patients' observation time windows, and the columns are the TIRP features along with the class (whether the observation time is positive or negative). Each matrix entry has a value representing the TIRP for a given patient observation time window. For each TIRP, several metrics are extracted and used to represent in the classification: *Binary*, which represents whether the TIRP was detected at the observation time period; *Horizontal Support*, which represents the number of the TIRP instances that were detected; and *Mean Duration*, which refers to their average duration from the first start time till the last end time, as defined more formally in [8]. In addition, ten more metrics of the instances' duration of a TIRP were used, including the median, standard deviation and maximal duration. Thus, the number of features will be the number of discovered TIRPs multiplied by thirteen (the number of metrics), as shown in Fig. 2, for example, column $T_1 R_1$, represents the first metric (R_1) of the first TIRP T_1. Afterward, the classifier is trained on the training feature matrix (Fig. 2.3). The process of detecting TIRPs is done on the testing set to build the test feature matrix (Fig. 2.4, 2.5), based on Bag-Of-TIRPs that were discovered in the training set.

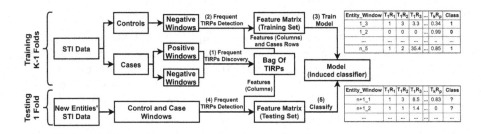

Fig. 2. The overall AHE prediction framework using frequent TIRPs.

4   Evaluation

We had defined three research questions and two corresponding experiments to answer them. The main *research questions* for this study were:

(A) Which prediction model, the TIRPs based (Sect. 3), 1D-CNN or RNN-LSTM will perform best in for the AHE prediction?

(B) What will be the performance of predicting whether a patient will experience AHE in a pre-defined prediction time?

(C) What will be the performance of generally predicting whether AHE will occur during the ICU stay given an observation time period from admission?

To evaluate the performance of the outcome prediction models, we created a Receiver Operating Characteristics (ROC) curve and computed the Area Under the Curve (AUC). However, the number of negative windows was bigger than the positive windows (i.e., imbalance); thus, we also use a Precision-Recall (PR) curve and compute the Area Under the PR Curve (AUPRC).

4.1 Dataset

The data collected for this study was extracted from the MIMIC-III database [6], which contains an ICU from a single-center in Boston, United States. However, the core mechanisms of physiological regulation of blood pressure are similar across the populations across the globe. Therefore, blood pressure interventions and protocols are not expected to be divergent across multiple ICUs in the pre-emptive phase. We used vital signs data, which is a time series-based sampled each 1-minute and is comprised of: systolic arterial blood pressure (ABP), heart rate (HR), arterial oxygen saturation (SpO2), and respiratory rate (RESP). The patient's files are filtered to at least 12 h of data for each patient, leaving the data to a total of 2,688 patients; 1,344 case-subjects experienced AHE. A patient is matched to each case-subject to fit a real-world ratio standard. In our study designs, we have maintained the originality of the ratio between case and control groups, 1:1 ratio, as similar to the actual ratio that was found in [12]. We calculate the similarity score between the patients by their demographic parameters (see Appendix A). Patients are matched based on the closest distance of the similarity score out of the entire dataset and not in a greedy comparison approach. For this study, we extracted the data until the first occurrence of AHE. We define the AHE target onset for overlapping epoch intervals of 30 min that includes more than half of data points with Systolic ABP greater than 130 mmHg.

4.2 Experimental Setup

To answer the research questions, two experiments were designed. We ran the experiments with ten-fold cross-validation, using group folding, which means that the same case or control observation time windows appeared in the training set or in the test set (but not both).

Baselines. We trained a 3D structure, known as a tensor representation, from the raw data of patients for 1D-CNN and RNN-LSTM. The architecture introduced in [14] was used (see Appendix B). RNN is designed to learn dependencies of data in a sequence, while LSTM is explicitly designed to avoid the long-term dependency problem. CNN is particularly suited for learning local patterns in raw features from inputs such as images (2D, width and height). CNN can also be used for applying and sliding a filter over temporal data, which as opposed to images, the filters will exhibit only one dimension (time), known as 1D-CNN.

TIRPs Based. In both experiments, the time-based data for the TIRPs discovery processes went through a temporal abstraction phase using cut-offs defined by domain experts (knowledge-based discretization) with three bins (low, normal and high values). STIs concatenated into a larger STI in cases where the gap was shorter than ten minutes and there were not any STIs between them from the same variable. The normal level values were: HR 60–100 bpm, RESP 7–15 breath/min, SpO2 88–92% and ABP 110–140 mmHg. We ignored the patients' demographic data for the prediction and used only the time-based data. We discovered TIRPs using KarmaLego [8] with minimum vertical support of 70% and we used Random Forest as the classifier (see Appendix B).

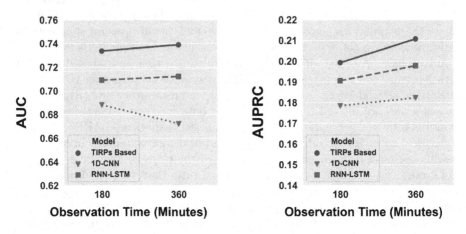

Fig. 3. The TIRPs based prediction outperformed and using an observation time of 360 min performed better than using 180 min.

5   Experiments and Results

Experiment 1. Continuous AHE Prediction Time Period Ahead. The goal was to evaluate how we can predict, continuously, whether AHE will occur within a pre-defined prediction time period ahead. The case-crossover-control study design was used to extract the observation time windows and label them, as described earlier (research question B). For this purpose, we applied the TIRPs based model, 1D-CNN and RNN-LSTM for comparison (research question A).

The models were evaluated with a prediction time of 60 or 180 min, observation time of 180 or 360 min, and a non-overlapping time of 120 min. One positive window was extracted from each case-patient, while five negative windows were

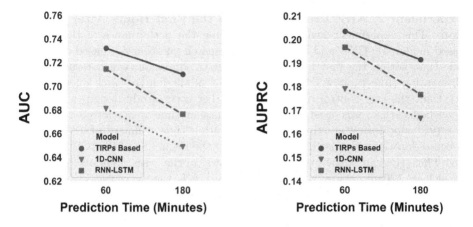

Fig. 4. The TIRPs based prediction performed better, and predicting 60 min ahead was slightly better than predicting 180 min ahead.

extracted from each case and control. The control's first-time window was taken randomly and its following windows overlapped one another.

Figures 3 and 4 show the results when using the various parameters on the TIRPs based prediction, 1D-CNN and RNN-LSTM models. The TIRPs based model led to a slightly better performance than 1D-CNN and RNN-LSTM. In Fig. 3 an observation window of 360 min performed better than using 180 min, while as shown in Fig. 4, predicting 60 min ahead performed better than predicting 180 min ahead.

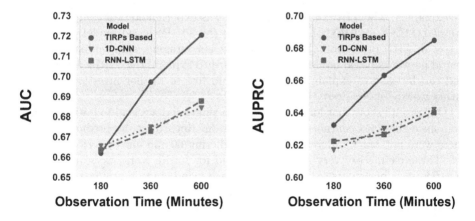

Fig. 5. Observation times versus models using case-control relative to an earlier event. Predicting based on longer observation time performed better.

Experiment 2. AHE Prediction Given the First Hours After Admission. This experiment focused on evaluating the performance of the TIRPs based model, 1D-CNN and RNN-LSTM (research question A), based on a specific time within the first hours since the ICU admission, which results in an observation time window relative to the ICU admission (research question C). For that, the case-control relative to an earlier event study design, which was described earlier, was used. The ICU admission time was referred to as the observed monitoring starting time. Observation time windows of 180, 360 or 600 min duration were used, which started at the beginning of the ICU monitoring. Thus, the end of the observation windows, in the cases, were always earlier by at least 120 min before the AHE onset. In Fig. 5, the use of the TIRPs based performed best, especially when the observation time increased.

6   Discussion and Conclusions

In this paper, we experimented with AHE prediction, which was relatively under-researched in the literature. The use of temporal abstraction, using knowledge-based state abstraction, and time intervals mining for the prediction of AHE in ICU data based on frequent TIRPs was introduced. Since we predict the first onset of AHE, before any clinical interventions were undertaken, we do not anticipate much differences in protocols across sites as our operating window is still in the pre-emptive stage rather than post-detection.

The option of AHE prediction was tested in two setups, the first predicting whether or not a patient will experience the complication in a pre-defined prediction time using a sliding observation time window, and the second in predicting whether generally it will occur within the ICU stay, based on several hours since the admission. Each experiment requires a different study design, as we explained. The TIRPs based prediction was experimented in comparing the use of 1D-CNN and RNN-LSTM, in which the TIRPs based model performed better in AHE prediction. These results are encouraging, since unlike artificial neural networks based models, which are hard to explain, the TIRPs are explicit and can be easy to explain, which we would like to further investigate in our future work.

Our results show that longer observation time windows lead to better prediction since the data contains more information. Regarding the prediction time, there were not much differences, while predicting 60 min ahead performed better. These results are very encouraging, and for future work, we would like to expand our experiments to predict AHE stage 1 and stage 2, and use additional temporal abstraction methods, a different number of states, more sizes of the observation time windows and prediction periods.

Acknowledgments. This research was supported by a collaboration grant of the Israeli Ministry of Science and Technology grant #8760441 and a donation from the Prof. Avram and Anat Bar-Cohen Project Desert Nova. Nevo Itzhak was funded by Kreitman School of Advanced Graduate Studies and the Israeli Ministry of Science and Technology Jabotinsky scholarship grant #3-16643.

A   Appendix 1 - Demographic and Physiological Characteristics

The demographic information of the patients (see Table 1).

Table 1. Demographics and physiological characteristics of the entire population, AHE cases, and the matched controls

Variables	All stays (2,688 patients)	Cases (1,344 patients)	Controls (1,344 patients)
Female, # (%)	1,144 (42.56)	569 (42.34)	575 (42.79)
Age (years), median (IQR)	66 (56–78)	67 (55–78)	66 (54–77)
ICU length of stay (days), median (IQR)	3.89 (2.15–7.83)	4.56 (2.43–10.17)	3.24 (2.02–5.79)
Initial HR (bpm), median (IQR)	88.00 (77.00–102.00)	88.30 (77.50–102.23)	87.00 (76.25–101.00)
Initial RESP (breath/min), median (IQR)	17.72 (13.42–22.22)	17.00 (13.00–22.13)	18.00 (14.00–22.45)
Initial SpO2 (%), median (IQR)	97.00 (93.00–99.50)	97.00 (93.00–99.90)	97.00 (93.00–99.00)
Initial ABP (mmHg), median (IQR)	115.62 (107.19–125.18)	115.00 (106.06–123.75)	116.18 (108.31–126.87)

B   Appendix 2 - Model Parameters

The parameters of each model are selected after testing the performance of each combination (not in a greedy comparison approach), and here we describe the parameters that performed best. For the TIRPs based model, we used Random Forest as the classifier with 100 trees in the forest and a maximum depth of 5 for the tree. Bootstrap is used when building trees and out-of-bag samples to estimate the generalization accuracy. Random Forest was implemented with Python 3.6 Scikit-Learn (https://scikit-learn.org/stable/) version 0.22.1. For the parameters we did not specify, we used the default.

For the CNN, we used one conventional layer with 48 filters and a kernel size of 7, stride length of one, same padding (i.e., the output size is the same as the input size), ReLU as the activation function and max-polling size of 3. Then, we defined two more fully connected hidden layers with 128 and 8 neurons. Activation function LeakyReLU with a dropout rate of 0.4 and 0.3, respectively, and a batch normalization in each layer. The batch size was 128, the initial learning rate was set to 0.0005 and the used optimizer was Nadam. Softmax activation was used in the output layer.

For all RNN, we used a 1 hidden dense layer LSTM with 64 hidden units per layer and with a recurrent dropout of 0.2. Activation function LeakyReLU with a dropout rate of 0.3 and a batch normalization in each layer. Then, we defined fully connected hidden layers with 128 neurons. We trained all models using a maximum epoch of 80, a batch size of 128 and a learning rate of 0.001. We used early stopping for both RNN and CNN, with a minimum change of 0.001, lower than this value, it is not considered as an improvement for the loss and is tolerated for 5 epochs. Both RNN and CNN models were implemented with Keras (https://keras.io/) version 2.2.5. For the parameters we did not specify, we used the default.

References

1. Allen, J.: ªmaintaining knowledge about temporal intervals, ° comm (1983)
2. Benken, S.T.: Hypertensive emergencies. Medical Issues in the ICU (2018)
3. Futoma, J., et al.: An improved multi-output gaussian process RNN with real-time validation for early sepsis detection. arXiv preprint arXiv:1708.05894 (2017)
4. Ghosh, S., Li, J., Cao, L., Ramamohanarao, K.: Septic shock prediction for ICU patients via coupled hmm walking on sequential contrast patterns. J. Biomed. Inform. **66**, 19–31 (2017)
5. Jin, K., Stockbridge, N.: Predicting acute hypotensive episodes from ambulatory blood pressure telemetry. Stat. Interface **5**(4), 425–429 (2012)
6. Johnson, A.E., et al.: MIMIC-III, a freely accessible critical care database. Sci. Data **3**, 160035 (2016)
7. Moskovitch, R., Choi, H., Hripcsak, G., Tatonetti, N.P.: Prognosis of clinical outcomes with temporal patterns and experiences with one class feature selection. IEEE/ACM Trans. Comput. Biol. Bioinform. **14**(3), 555–563 (2016)
8. Moskovitch, R., Shahar, Y.: Classification-driven temporal discretization of multivariate time series. Data Min. Knowl. Discov. **29**(4), 871–913 (2015)
9. Moskovitch, R., Walsh, C., Wang, F., Hripcsak, G., Tatonetti, N.: Outcomes prediction via time intervals related patterns. In: 2015 IEEE International Conference on Data Mining, pp. 919–924. IEEE (2015)
10. Pak, K.J., Hu, T., Fee, C., Wang, R., Smith, M., Bazzano, L.A.: Acute hypertension: a systematic review and appraisal of guidelines. Ochsner J. **14**(4), 655–663 (2014)
11. Rocha, T., Paredes, S., De Carvalho, P., Henriques, J.: Prediction of acute hypotensive episodes by means of neural network multi-models. Comput. Biol. Med. **41**(10), 881–890 (2011)
12. Vuylsteke, A., et al.: Characteristics, practice patterns, and outcomes in patients with acute hypertension: European registry for Studying the Treatment of Acute hyperTension (Euro-STAT). Crit. Care **15**(6), R271 (2011)
13. van Wyk, F., Khojandi, A., Davis, R.L., Kamaleswaran, R.: Physiomarkers in real-time physiological data streams predict adult sepsis onset earlier than clinical practice. bioRxiv, p. 322305 (2018)
14. Zhao, J., et al.: Learning from longitudinal data in electronic health record and genetic data to improve cardiovascular event prediction. Sci. Rep. **9**(1), 1–10 (2019)
15. Zhou, Y., Zhu, Q., Huang, H.: Prediction of acute hypotensive episode in ICU using Chebyshev neural network. JSW **8**(8), 1923–1931 (2013)

Falls Prediction in Care Homes
Using Mobile App Data Collection

Ofir Dvir[1(✉)], Paul Wolfson[2], Laurence Lovat[2], and Robert Moskovitch[1]

[1] Software and Information Systems Engineering, Ben Gurion University,
Beersheba, Israel
ofirdvi@post.bgu.ac.il, robertmo@bgu.ac.il
[2] Interventional and Surgical Sciences, University College London, London, UK
{p.wolfson,l.lovat}@ucl.ac.uk

Abstract. Falls are one of the leading causes of unintentional injury related deaths in older adults. Although, falls among elderly is a well documented phenomena; falls of care homes' residents was under-researched, mainly due to the lack of documented data. In this study, we use data from over 1,769 care homes and 68,200 residents across the UK, which is based on carers who routinely documented the residents' activities, using the Mobile Care Monitoring mobile app over three years. This study focuses on predicting the first fall of elderly living in care homes a week ahead. We intend to predict continuously based on a time window of the last weeks. Due to the intrinsic longitudinal nature of the data and its heterogeneity, we employ the use of Temporal Abstraction and Time Intervals Related Patterns discovery, which are used as features for classification. We had designed an experiment that reflects real-life conditions to evaluate the framework. Using four weeks of observation time window performed best.

Keywords: Temporal data mining · Outcomes prediction · Falls prediction

1 Introduction

The ageing population portion in the society has grown rapidly and becomes a major challenge in healthcare worldwide. It is estimated that, by 2025, this group will number approximately 1.2 billion and expand to 2 billion by 2050[1]. More than a third of the population above 65 years old fall each year. Approximately 1 in 10 falls results in a serious injury, such as hip fracture, major soft-tissue injury and head injury. The mortality rate following a fall, increases dramatically with the age, exceeding to 70% of accidental deaths in adults above 75 years old. Therefore, predicting fall risk will allow ideally prevention or interventions, which potentially reduce fall occurrences and fall-related costs [1]. Most studies,

---

[1] World Health Organization and others:Active ageing, A policy framework. World Health Organization. Geneva, (2002).

© Springer Nature Switzerland AG 2020
M. Michalowski and R. Moskovitch (Eds.): AIME 2020, LNAI 12299, pp. 403–413, 2020.
https://doi.org/10.1007/978-3-030-59137-3_36

if not all, till now in falls prediction focused on hospital or nursing environments, consisting on demographic descriptives and data from electronic health records, and not based on continuously documented data in care homes – as we do in this paper. Thus, typically existing studies focused on the characterization of risk profiles, based on broad demographics and the adults conditions, rather than their daily data. Numerous parameters associated with the risk of falling have already been identified [2,3], such as physical characteristics or their medical history. Although, previous studies had reported models that predict falls in elderly [4] they typically used heterogeneous samples that included prior fall events. While falls history is the strongest single indicator and the most frequently used factor for fall prediction, it cannot be used in the identification of individuals at risk of falling for the first time – which we investigate in this study. Therefore, there is still a need for assessment tools that can predict the risk of a first fall onset. Nevertheless, most of these studies were limited by their use of summary metrics, and relying solely on data collected infrequently, rather than considering longitudinal data, such as risk factors that change over time [5]. This happened, since there was no data typically available of the adults daily routine, which makes our database exceptional in the opportunities it provides. In this study we explore for the first time the ability to predict falls in care homes, based on careres documentation through a mobile app. We use data from over 1,769 care homes and 68,200 residents across the UK, routinely collected by the Mobile Care Monitoring (MCM) over three years. The goal is to predict a fall a week ahead, based on a sliding observation time window continuously. Due to the nature of the longitudinal data, we extend in this study the use the Maitreya framework [16] to perform prediction using a case-crossover-control evaluation design to reflect real life conditions. The contributions of the paper are the following:

- A rigorous evaluation of the falls prediction on a novel real life large database
- Investigating for the first time First Fall Prediction, based on secondary use of care homes daily documentation based on a mobile app
- A comprehensive framework for falls prediction based on TIRPs extracted from a sliding window

2   Background

2.1   Falls Risk Assessment

According to the World Health Organization, approximately one third of the population over the age of 70 will fall and the likelihood rises with the age and frailty. Falls account for more than half of injury-related hospital admissions and 40% of injury-related deaths in the elderly[2]. In addition to the human cost of

---

[2] World Health Organization and World Health Organization. Ageing and Life Course Unit: WHO global report on falls prevention in older age. World Health Organization. (2008).

falling (distress, pain, fractures, loss of confidence and loss of independence), fall pose a substantial financial burden on healthcare systems and estimated to cost the NHS (National Health Care) more than £2.3 billion per year. Therefore falling has an impact on quality of life, health and healthcare costs [7]. Identifying care home residents risk for falls can facilitate targeted prevention [8], potentially reducing incidence and associated costs.

2.2   Temporal Abstraction and TIRPs Mining

In order to analyze heterogeneous multivariate temporal data, it was proposed to use Temporal Abstraction in order to transform the various variables into a uniform representation of symbolic time intervals, which enables later to perform time intervals analytics [6]. In this study we used only state temporal abstraction, in which values are being discretized into states based on given cutoffs, which are later being concatenated, when adjacent and having the same symbol, into symbolic time intervals. There are several relevant discretization methods including, Equal Width Discretization (EWD), in which the cutoffs are determined by dividing the values range into equal bins; Symbolic Aggregate approXimation (SAX) [9], which consists on the gaussian distribution of the data, and the cutoffs are derived from its mean and standard deviation; Temporal Discretization for Classification (TD4C), which is a supervised temporal discretization method that searches for cutoffs that result in the states having the most different distribution between the classes (TD4C) [10] and more. Previous studies had shown the advantages of TD4C in comparison to EWD and SAX. Once the data is in a uniform format of symbolic time intervals, TIRPs can be discovered and used as features for classification. To discover TIRPs several time intervals mining methods were developed in the past two decades often using a subset of Allen's temporal relations, which often used to represent the temporal relations between pairs of time intervals in symbolic time intervals mining. There are seven temporal relations: before, meet, overlap, contain, starts, finishes, and equals. Therefore, a TIRP P is defined as $P = IS, R$, where $IS = I_1, I_2, .., I_k$ is a set of k symbolic time intervals ordered lexicographically and R defines all the temporal relations among each of the $(k^2 - k)/2$ pairs of symbolic time intervals in I. Several TIRPs mining methods were developed along the past two decades, in which the TIRP representation was improved, as well as the mining structures [6,12].

2.3   TIRPs Based Classification

Outcomes prediction modelling while employing the analysis of the longitudinal data is one of the most important and challenging research fields in medical data. The use of frequent patterns as features for classification and prediction of data and specifically in multivariate temporal data is increasingly reported in the past decades [11]. It is based on the idea that richer predictive information is in the temporal order of the data, rather than independent features. Interestingly, several studies proposed using TIRPs as features for classifying multivariate time series quite simultaneously [12]. A recent study introduced the Maitreya framework for the classification of multivariate time series via TIRPs [16], which is used in this paper. The Maitreya framework provides novel TIRP metrics for classification, representing the TIRPs' number of instances, and their average duration, in addition to binary, as well as an efficient TIRPs detection method called SingleKarmaLego [6].

3   Methods

We introduce here the TIRPs based falls prediction, starting with the data creation, and proceed with the falls prediction framework.

3.1   Mobile App Data Collection

In many countries, such as the United Kingdom, documenting care home's residents is mandatory, but despite technological advances the data collection process today, commonly, is still written manually on paperwork. Alternatively, our study consists on data that was collected from elderly using the Mobile Care Monitoring (MCM) mobile app. This mobile app enables a care home carers to continuously record residents' activities and statuses. Documentation may include from fluid consumption and medication administration, to exercise activities and mental stimulation, recorded continuously using a simple and intuitive interface (Fig. 1). The system is icon driven with limited need in typing to insert information, designed to be suitable for non-native English speakers. The data is automatically uploaded to a central database, integrated to provide a comprehensive, detailed, up-to-date data available for the careres, residents' families, which allows later secondary analytics, which is demonstrated in this paper. Using the Care App, data about the residents are recorded in real-time rather than at the end of the shift, or a day. Every day, 1,769 care homes around the UK use the MCM app to document more than 2.4 million care notes over 68,200 residents.

Fig. 1. Data collection using the icon driven mobile app.

3.2   Problem Definition and Study Evaluation Design

Given a database of documented data from the mobile app, the goal was to learn predictive models, based on a recent observation time window (of several weeks) that continuously provide a risk assessment a prediction time ahead (a week in this study). For that we create a cohort of cases of residents who had a fall in their records, and a group of matched controls, who are residents without a fall in their records. The controls are matched based on their age, gender and Care Quality Commission (CQC) score set by the independent regulator of health and social care in England. However, in order to create the prediction model which is an induced classifier, the observation time windows have to be labeled, which is illustrated in Fig. 2.

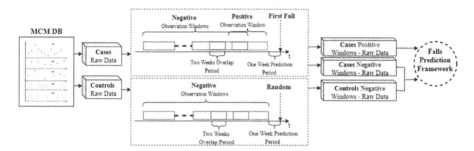

Fig. 2. Labeled observation time windows creation. From the MCM longitudinal database, the cases (residents who fell) are extracted, and their matched controls. Given the first fall in the cases' records, an observation time window is sled, in which the time window a week before the fall in each case is labeled as positive, and the earlier time windows are labeled as negative. In addition, random observation time windows from each matched control are labeled as negative, which enables a case-crossover-control design, reflecting real application conditions.

Thus, in each case the last observation time window prior to the prediction time period, which is just before the outcome event (which leaves potentially enough time for prevention) is labeled as positive. The negative time windows are taken from both the cases' earlier observation time windows, and from the controls' random observation time windows (since they do not have an outcome in their records). This corresponds to a case-crossover-control study design, which reflects real life conditions that are evaluated both on residents having a fall in their data, or not.

3.3   FallPry

We introduce FallPry for falls prediction, based on the MCM longitudinal database, that includes the time windows creation and their labeling, as was shown in Fig. 2, and a TIRPs based prediction framework that extends the use of the Maitreya framework [16]. The Maitreya framework includes four main components: Temporal Abstraction, Time Intervals Mining, TIRP-based Feature Representation, and Classifier induction. We explain here each of the framework components.

Temporal Abstraction. The input raw data includes multiple variables that are described by time-point values series, or event (i.e., measurement, actions or risk assessment). Therefore, after dividing the raw data into positive and negative time windows the raw data being abstracted into a uniform format of symbolic time intervals, based on the use of state abstraction [10]. In this study we employed Equal Width Discretization (EWD); Symbolic Aggregate Approximation (SAX) [9] which creates states based on the gaussian distribution of the values; and Temporal Discretization for Classification (TD4C) [10], which is a supervised approach that determines the cutoffs for the states that increase the divergence of the classes values distribution.

TIRPs Discovery. After the variables went through temporal abstraction and are represented by symbolic time intervals, TIRPs can be discovered, which is done by the KarmaLego algorithm [12]. As mentioned in the background, we use non ambiguous TIRPs that use Allen's temporal relations to represent the relations between each pairwise symbolic time intervals. The discovered frequent TIRPs, which are referred as a Bag-of-TIRPs, are used as features for the classifier.

TIRPs Based Classification. For that a features-matrix is constructed, in which the rows are the observation time windows, and the columns are the TIRPs (each time window is represented by a vector of values for each TIRP-feature, and the class). The TIRP-features' values are calculated according to the chosen TIRP representation metric: Boolean (whether the TIRP was detected in the time window), Horizontal Support (how many instances of the TIRP were detected), and Mean Duration (what was their average duration) [12]. Eventually, a classifier is induced from the training feature matrix. After a classifier was induced, given a new time window, the TIRPs are detected using the SingleKarmaLego algorithm [6] which results in a corresponding features vector which is given to the classifier to perform classification.

4   Evaluation and Results

We define here the research questions for the study, and its corresponding experimental plan and results. The experiment focused on examining the ability to predict first falls in care homes, based on the data from the mobile app entered by the careres.

4.1   Research Questions

1. How accurate can first fall be predicted a week ahead, and what is the best observation time window size?
2. What are the best FallPry settings for falls prediction? (i.e., state abstraction method, bins number, TIRP metric and classifier)

4.2   Data

The data for this study were routinely collected from over, 1769 care homes across the UK over three years, from 2017 to 2019 by the icon-driven Mobile Care Monitoring (MCM) system from Person Centered Software, Guildford, UK (Subsect. 4.1). The total population includes 68,200 residents, of whom 13,412 are above 75 years old, not bed bound and had at least one full year of documentation. We define two groups of residents: the fallers and the non-fallers, which will be referred as the cases and controls group respectively. The non-fallers (control) were residents who have not had a documented fall during the entire documentation period while the fallers (case) are residents who had documentation of a fall with an injury of some severity level (in this study no separation was made between serious and moderate injury). Moreover, since this study focused on predicting residents' first fall, the fallers group was defined so throughout the entire period prior to the fall (at least six months) no falls were documented (Table 1).

Table 1. Cases and controls statistics

	All residents	Case	Control
N	13,412	4894	8518
Age (std)	87.3 (±6.5)	88 (±6.1)	86.9 (±6.7)
Female (%)	71.2%	73.9%	69.7%
Female age, mean (std)	87.8 (±6.4)	88.4 (±6)	87.8 (±6.4)
Male age, mean (std)	85.9 (±6.5)	86.8 (±5.9)	85.0 (±6.5)
BMI under 18.5 (%)	11.9%	13%	10.8%

4.3   Experimental Setup

We used the Maitreya framework (Subsect. 3.2) for all the experiments, having the following settings: a 50% minimal vertical support threshold, maximal gap of 7 days and the full set of Allen's seven temporal relations. In the experiments, we tested four discretization methods: EWD, SAX, TD4C with Kullback Leibler (TD4C-KL), and TD4C with cosine as the distance measure function (TD4C-Cosine). The number of states determines the level of granularity of the abstraction method, which we experimented with two, three, and four states. Three different TIRP metrics were used: binary (BIN), horizontal support (HS) and mean duration (MeanD) and four types of classifiers: K-Nearest Neighbor (KNN), XGBoost (XGB), Logistic Regression (LR) and Neural Net (NN) using the default settings of sklearn. We ran the experiments using 10-fold cross-validation, and using AUC as the metric to evaluate the prediction performance. To evaluate our models in conditions that reflect the real application of a potential time-dependent falls prediction model, we train and test our model by positive and negative time windows taken from the case and control groups defined above (Subsect. 3.2). The positive and negative time windows were selected in ration of 1:6.

4.4   Falls Prediction Framework Baseline - Features Based Approach

As an alternative to the TIRPs based prediction, we implemented a feature based approach, in which features were extracted from each time window. Data aggregation has been widely applied as effective techniques to reduce data redundancy and simplify complex data, especially in longitudinal data [13,14]. Due to the heterogeneity of the various variables, and sparseness we decided to perform Piecewise Aggregate Approximation [15], based on ten time periods, for which the mean is calculated, as done in PAA, which will become the features. Although in PAA only the mean is calculated, we used count for specific variables that contain events, such as the exercises frequency, nutritional risk assessment frequency and more per observation time period. Eventually these ten time periods representative values become features. Performing the aggregation on each time window individually, rather than preforming it on the whole time window's data which enables to capture the temporal relations to yield meaningful results (Fig. 3).

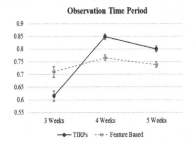

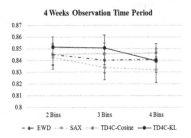

Fig. 3. Using FallPry preform bather then the feature based approach when using both 4 and 5 weeks observation time window

Fig. 4. The use of TD4C KL with two bins (states) performed best using 4 weeks observation time period.

4.5   Results

Experiment – Elderly First Injury Fall Prediction. In this experiment the performance in predicting the first fall with a prediction time of one week ahead was evaluated (research question A). For the observation time periods, three, four or five weeks were used. Additionally, we compared the combinations of the state abstraction methods, bins number, and TIRP representation metrics, with the four different classifiers (research question B). As a baseline, the feature based approach (Subsect. 4.5) was used that were extracted from the same time windows. Figure 4 shows the mean results, including 95% confidence intervals, for each observation time period of 3, 4, or 5 weeks, while using FallPry with TIRPs or Features for comparison. The use of the TIRPs based was better with 4 or 5 weeks, while the Features based was better with 3 weeks. However, 4 weeks was significantly better with the use of TIRPs. In Fig. 5, the use of FallPry with TIRPs is presented with the various state abstraction methods (EWD, and the number of states (2, 3, or 4 bins), in which the TD4C-KL seem to outperform, especially with 2 or 3 bins. Figure 5, shows the mean results for the TIRPs based, including the various TIRPs' metrics of Binary (BIN), Horizontal Support (HS) and Mean Duration (MD), and the Features based, with the four types of classifiers that were used. There is a chart for each of the observation time periods. It can be seen that the 4 weeks has the best performance, especially with the XGB classifier and mean duration (which describes the average duration of the TIRP's instances). Thus, using Maitreya with 4 weeks observation time window, 2 bins, TD4C-KL discretization, MD representation and XGB model achieves the best performance (88.3% AUC, question B)

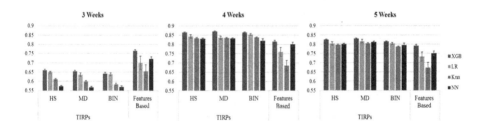

Fig. 5. FallPry preform better then the feature based approach when using both 4 and 5 weeks observation time windows. The XGB was significantly better than the KNN, NN and LR in all cases.

5   Discussion and Conclusions

Falls in elderly population is one of the leading causes for rapid deterioration and death. Being able to predict a fall, or assess the risk for a fall, is desirable in order to increase caution and ideally decrease the risk. Most of the literature so far that reported the intention to predict falls was in hospital environments, consisting on Electronic Health Records data, or in nursing homes, consisting on data collected infrequently and manually [5]. In this study, for the first time, as far as we know, falls prediction in care homes is performed based on such rich daily data of residents that was documented through a mobile app by the care homes' careres. Moreover, our datasets include residents' daily activities, nutrition and medical status which routinely documented from over 1,769 care homes and 68,200 residents. An experiment, reflecting real-life conditions, was performed that focused on evaluating the falls prediction a week ahead, given a sliding observation time window of 3, 4 or 5 weeks duration. Using 4 weeks observation time window performed significantly better than 3 or 5 weeks. Also, using the TD4CKL abstraction method with two states and the mean duration TIRP metric, with the XGB classifier achieved the highest performance. For future work we would like to reduce the number of variables, and understand better the predictors for the falls, as well as experiment with the frequency of the risk estimations.

References

1. Rao, S.S.: Prevention of falls in older patients. Am. Fam. Phys. **72**(1), 81–88 (2005)
2. Bloch, F., et al.: Estimation of the risk factors for falls in the elderly: can meta-analysis provide a valid answer? Geriatr. Gerontol. Int. **13**(2), 250–263 (2013)
3. Gillespie, L.D., et al.: Interventions for preventing falls in older people living in the community. Cochrane Database Syst. Rev. (9) (2012)
4. Kojima, G., et al.: Does the timed up and go test predict future falls among British community-dwelling older people? Prospective cohort study nested within a randomised controlled trial. BMC Geriatr. **15**(1), 38 (2015)
5. Marier, A., Olsho, L.E., Rhodes, W., Spector, W.D.: Improving prediction of fall risk among nursing home residents using electronic medical records. J. Am. Med. Inform. Assoc. **23**(2), 276–282 (2016)
6. Moskovitch, R., Shahar, Y.: Classification of multivariate time series via temporal abstraction and time intervals mining. Knowl. Inf. Syst. **45**(1), 35–74 (2014). https://doi.org/10.1007/s10115-014-0784-5
7. Close, J.C.: Prevention of falls in older people. Disabil. Rehabil. **27**(18–19), 1061–1071 (2005)
8. Kamel, H.K.: Q & A with the expert on: falls preventing falls in the nursing home. Ann. Long Term Care **14**(10), 42 (2006)
9. Lin, J., Keogh, E., Lonardi, S., Chiu, B.: A symbolic representation of time series, with implications for streaming algorithms. In: Proceedings of the 8th ACM SIGMOD Workshop on Research Issues in Data Mining and Knowledge Discovery, pp. 2–11, June 2003
10. Moskovitch, R., Shahar, Y.: Classification-driven temporal discretization of multivariate time series. Data Min. Knowl. Discov. **29**(4), 871–913 (2015)
11. Batal, I., Valizadegan, H., Cooper, G.F., Hauskrecht, M.: A pattern mining approach for classifying multivariate temporal data. In: 2011 IEEE International Conference on Bioinformatics and Biomedicine, pp. 358–365 (2011)
12. Moskovitch, R., Walsh, C., Wang, F., Hripcsak, G., Tatonetti, N.: Outcomes prediction via time intervals related patterns. In: 2015 IEEE International Conference on Data Mining, pp. 919–924 (2015)

13. Lu, Y., Comsa, I.S., Kuonen, P., Hirsbrunner, B.: Dynamic data aggregation protocol based on multiple objective tree in wireless sensor networks. In: 2015 IEEE Tenth International Conference on Intelligent Sensors, Sensor Networks and Information Processing (ISSNIP), pp. 1–7 (2015)
14. Min, K., Corso, J.J.: TASED-net: temporally-aggregating spatial encoder-decoder network for video saliency detection. In: Proceedings of the IEEE International Conference on Computer Vision 2019, pp. 2394–2403 (2019)
15. Keogh, E.J., Pazzani, M.J.: Scaling up dynamic time warping for datamining applications. In: Proceedings of the Sixth ACM SIGKDD International Conference on Knowledge Discovery and Data Mining, pp. 285–289, August 2000
16. Moskovitch, R., Polubriaginof, F., Weiss, A., Ryan, P., Tatonetti, N.: Procedure prediction from symbolic electronic health records via time intervals analytics. J. Biomed. Inform. **75**, 70–82 (2017)

Transitive Sequential Pattern Mining for Discrete Clinical Data

Hossein Estiri[1,2(✉)] ⓘ, Sebastien Vasey[3] ⓘ, and Shawn N. Murphy[1,2]

[1] Massachusetts General Hospital, Boston, MA 02144, USA
hestiri@mgh.harvard.edu
[2] Harvard Medical School, Boston, MA 02155, USA
[3] Department of Mathematics, Harvard University, Cambridge, MA 02138, USA

Abstract. Electronic health records (EHRs) contain important temporal information about disease progression and patients. However, mining temporal representations from discrete EHR data (e.g., diagnosis, medication, or procedure codes) for use in standard Machine Learning is challenging. We propose a transitive Sequential Pattern Mining approach (tSPM) to address the temporal irregularities involved in recording discrete records in EHRs. We perform experiments to compare the classification performance metrics for predicting "true" diagnosis between traditional sequential pattern mining (SPM) and the proposed tSPM algorithms across multiple diseases. We demonstrate that transitive approach is superior to the traditional SPM in mining temporal representations for diagnosis prediction.

Keywords: Transitive Sequential Pattern Mining · Electronic Health Records · Temporal representations

1 Introduction

In today's biomedical research, modern Machine Learning (ML) algorithms are being increasingly applied to data from electronic health records (EHRs). EHRs contain important temporal information about disease progression and patients. However, in the temporal nature of these data a multitude of convolutions are embodied that are created through what is known as the recording processes [11]. EHR observations are often acquired asynchronously across time (i.e., recorded at different time instants and irregularly in time) [2,5]. This property provides challenges for directly applying standard temporal analysis methods to clinical data recorded in EHRs. Specifically, traditional time-series analysis methods can not be directly applied to analyze the temporal dimensions of the discrete clinical data (e.g., diagnosis records) due to the irregular acquisition of these data. There are also concerns about the validity of diagnosis records in EHRs due to the payer-provider dynamics and healthcare processes [11].

Innovative methods that enable us to properly incorporate time and understand the complexities involved in the healthcare process can yield to interpretable findings from large scale clinical databases. Since the late 1990s, a body

© Springer Nature Switzerland AG 2020
M. Michalowski and R. Moskovitch (Eds.): AIME 2020, LNAI 12299, pp. 414–424, 2020.
https://doi.org/10.1007/978-3-030-59137-3_37

of work has accumulated on mining temporal representations. Temporal representation mining aims to transform temporal data into machine-readable representations, which can be used in standard Machine Learning (ML) techniques for temporal reasoning. Much of this work in medical domains explicitly has been thought of, formed, and developed around clinical measurements data that are commonly continuous in nature and have precise time stamps. From the data mining community, sequential pattern mining (SPM) approaches provide solutions for constructing temporal representations from discrete clinical data. However, SPM algorithms were primarily developed on transaction data, and thus, may require careful modifications for application in medicine.

In this paper, we propose modifications to the traditional SPM approach in a transitive sequential pattern mining (tSPM) algorithm. The goal in tSPM is to mine temporal data representations from discrete clinical data (e.g., diagnosis or medication codes) for application in downstream ML. We apply the tSPM and SPM to predicting "true" diagnosis records in the next visit for six diseases and demonstrate that the tSPM algorithm improves the prediction performance over the traditional SPM by up to 15%.

2   Related Work

2.1   Temporal Representation Mining

Temporal representation mining involves providing a machine-readable representation that formalizes the notion of time in the context of a set of events and temporal relationships [24]. In biomedical research, development and evolution of temporal representation approaches has been largely confined to the abstraction of temporal data from clinical measurements [23]. While numerical observations (e.g., laboratory test results) have explicit timestamps, precise time-stamped information is often unavailable in discrete clinical data such as records of diagnoses, medications, and procedures.

Sequential Pattern Mining (SPM) [1] is a viable approach for discrete data. The goal in SPM is to discover "relevant" sub-sequences from a large set of sequences (of events or items) with time constraints. The "relevance" is often determined by a user-specified occurrence frequency, known as the minimum support [14]. The frequent sequential pattern (FSP) problem is to find the frequent sequences among all sequences [6]. Apriori-based sequential pattern mining methods, such as sequential pattern discovery using equivalence classes (SPADE) [27] and sequential pattern mining (SPAM) [3], are popular in the healthcare domain. For example, Perer, Wang, and Hu (2015) used the SPAM algorithm for mining long sequences of events [21]. The apriori property is that if a sequence cannot pass the minimum support test (i.e., is not assumed frequent), all of its sub-sequences will be ignored. The goal in the apriori-based approach is to cut the number of features by finding the frequent temporal patterns among all patterns. For example, most frequent temporal patterns were utilized in Moskovitch & Shahar (2015); Batal et al. (2013, & 2016),

Moskovitch et al. (2015 & 2017), and Orphanou et al. (2018) [4,5,16–19]. For various reasons, temporal patterns that are mined based on frequency may not make clinical sense.

2.2 Discrete Clinical Event Prediction

Recurrent Neural Networks (RNNs) [22] and its variants such as Long Short-Term Memory (LSTM) and Gated Recurrent Unit (GRU) are being increasingly applied to model discrete clinical events [13]. Applied a LSTM model to synthetic EHR data (derived from MIMIC-III dataset [12]) to predict the next medication administration events, lab results events, procedure events, and physiological result events. Their study, however, does not model diagnosis records, potentially due to the limitations of the dataset. [8] proposed Med2Vec, a scalable two layer neural network for learning lower dimensional representations to predict future medical codes and Clinical Risk Groups (CRG) levels. Evaluation of this work, however, is limited to prediction of inpatient visits. [9] developed the REverse Time AttentIoN model (RETAIN) that mined sequences of primary care visits to predict if a patient will be diagnosed with heart failure. On the one hand, RETAIN has been tested only on 1 disease. On the other hand, all these models assume that the existence of a diagnosis code in the medical records reflects true diagnosis. These models, therefore, at best can mimic clinical data, which may not represent the truth about patients' health. In this study, we predict the possibility that the patient will be "truly" diagnosed for a disease based on her past medical records, in six different diseases.

3 Methodology

We aim to leverage the diagnosis and medication histories in patients' electronic medical records to predict the "true" diagnosis record at the next visit. This presents an important difference from the related work in this space that aim to predict the next diagnosis records. To do so, we mine two types of sequential patterns (i.e., representations) and train and test prediction models on computer- and expert-annotated labels.

3.1 Traditional Sequential Pattern Mining

We first constructed a baseline method that applies the traditional sequential pattern mining (SPM) algorithm, using EHR observations as features for disease prediction. Given a list $O_1, O_2, \ldots, O_n$ of diagnosis or medication observations, for each patient p, we have recorded the times $t_{i1}^p \leq t_{i2}^p \leq \ldots \leq t_{ik_i^p}^p$ at which the observation O_i was recorded.

In the SPM approach, the features are all possible *pairs* of distinct observations (O_i, O_j), $i \neq j$. We now explain how their frequencies are counted. For a given patient p and a given time t, let $t' > t$ be minimal such that for some i and some $\ell \leq k_i^p$, $t_{i\ell}^p = t'$. In words, t' is the first time strictly bigger than t

at which some observation is made for the patient (it could be several observations are made at the same time). Now, for a given patient p and a given index $i \in \{1, 2, \ldots, n\}$, let S_{pi} be the set of all pairs (j, ℓ'), with $j \in \{1, 2, \ldots, n\}$ and $\ell' \leq k_j^p$, such that observation j is made right after observation i at time $t_{j\ell'}^p$. Formally, $(j, \ell') \in S_{pi}$ if and only if $k_i^p, k_j^p \geq 1$ and there exists $\ell \leq k_i^p$ such that $(t_{i\ell})' = t_{j\ell'}$.

For each $i, j \leq n$, $i \neq j$, let $r_{ijp} = |S_{ip}|$. We think of the r_{ijp}'s as samples of a random variable X_{ij}. Our goal is then to predict the class label Y given $(X_{ij})_{i \neq j}$.

3.2 Transitive Sequential Pattern Mining

To account for irregularity of clinical records and recording processes, we propose a transitive sequential pattern mining (tSPM) algorithm. In the tSPM algorithm, the features are again all possible pairs of distinct observations (O_i, O_j), $i \neq j$. For a fixed patient p and $i \neq j \leq n$, we set r_{ijp} to be 1 if $k_i^p \geq 1$, $k_j^p \geq 1$ and $t_{i1}^p \leq t_{j1}^p$, and 0 otherwise. In other words, r_{ijp} is 1 if and only if both O_i and O_j were recorded for the patient and the *first* record of observation i was before, or at the same time as, the first record of observation j. Again, for each fixed $i \neq j$, we think of the r_{ijp}'s as samples of a random variable X_{ij} and our goal is to predict the class label Y given $(X_{ij})_{i \neq j}$.

The use of first record (rather than all records) is a major difference in the way sequential patterns are mined in tSPM to handle the repeated problem list entries.

It is important to emphasize that we call the sequential pairs in the tSPM approach transitive sequences as they embody distinctive modifications to the conventional sequential pattern mining (SPM). Imagine a sequential pattern where observation A happened right before B, and B happened right before C ($A \rightarrow B \rightarrow C$). SPM mines subsequences $A \rightarrow B$ and $B \rightarrow C$. To account for the potential biases in EHRs, the transitive sequencing algorithm mines subsequences $A^* \rightarrow B^*$, $B^* \rightarrow C^*$, but also $A^* \rightarrow C^*$ from the sequence $A^* \rightarrow B^* \rightarrow C^*$, where A^*, B^*, and C^* are the first records of A, B, C in the database.

3.3 Dimensionality Reduction

Since the numbers of pairs (O_i, O_j) with $i \neq j \leq n$ is exactly $\frac{n(n-1)}{2}$, the number of sequential features is quadratic in the number of observations. This leaves us with a highly dimensional vector of representations. The usual approach when using SPM is to keep only the first n most frequent features, where n is a hyperparameter. We call this approach FSPM.

We attempt to improve on this heuristic with a formal dimensionality reduction procedure that aims to minimize sparsity and maximize relevance (MSMR). To minimize sparsity, we removed any feature that has a prevalence smaller than one percent. On the remaining features we compute the empirical mutual information using the entropy of the empirical probability distribution [10, 15].

Mutual information provides a measurement of the mutual dependence between two random variables, which unlike most correlation measures can capture non-linear relationships [10,20]. We ranked the data representations based on their mutual information with the labeled outcome and kept the first 10'000 (in ties, we used prevalence to determine the ranking).

We further scrutinized the relevance using joint mutual information (JMI) [25]. The algorithm starts with a set S containing the top feature according to mutual information, then iteratively adds to S the feature X maximizing the *joint mutual information score*

$$J_{jmi}(X) = \sum_{X^* \in S} I(XX^*; Y)$$

Here, $I(Z; Y)$ denotes the mutual information between random variables Z and Y (a measure of the information shared by Z and Y—it can be expressed as the entropy of Z minus the entropy of Z given Y). The random variable XX^* is simply the random variable corresponding to the joint distribution of X and X^*. In the end, we select the first 40–500 features that were added to the set S.

The idea of using the joint mutual information score (as opposed to just the mutual information) is that it also takes into account the redundancy between the features: two features could each be highly relevant on their own, but also be strongly correlated. On the other hand computing joint mutual information is quadratic in the number of features, hence the initial dimensionality reduction using mutual information.

Joint mutual information belongs to a large zoo of feature selection scores J, see for example [7] for a theoretical analysis and some experimental results suggesting that JMI is a reasonable default choice. They compare the performance of 9 different feature selection scores across 22 data sets, and suggest JMI as the "best trade-off [...] of accuracy and stability". It is one of few scores that satisfies three theoretical criteria: reference to a conditional redundancy term, keeping the relative size of the redundancy term from swamping the relevancy term, and whether it is a low-dimensional approximation, hence usable with small sample sizes.

4   Experimental Setting

We ask three questions. The first question involves examining the efficiency of temporal representation mining algorithms; (1) can a classifier trained on transitive sequences (tSPM) improve true disease prediction over the classifier trained on traditional sequences (SPM)? As we discussed, a possible shortfall of traditional sequential pattern mining methods in clinical domains is the use of frequency-based criterion to mine temporal representations. The second question aims to examine the proposed MSMR dimensionality reduction algorithm; (2) does an entropy-based criteria improve the frequency-based criterion for feature selection? Finally, the third question examines whether together the proposed temporal representation mining and dimensionality reduction can improve true disease prediction; (3) Does tSPM+MSMR provide an overall better prediction that the Frequent traditional sequential pattern mining (FSPM)?

We used EHR data from the Mass General Brigham Biobank on six diseases: congestive heart failure (CHF), type 1 and 2 diabetes mellitus (T1DM and T2DM), rheumatoid arthritis (RA), chronic obstructive pulmonary disease (COPD), and ulcerative colitis (UC). In each disease cohort, the medical records end with an encounter that includes International Classification of Diseases, Ninth Revision, Clinical Modification (ICD-9-CM) code for the respective disease. We excluded data from the last encounter to predict. There is no time constraint for the next encounter that we aim to predict. For each of the diseases, we had a random subset of patients for whom we had curated gold-standard labels via in-depth expert review of clinical notes. The gold-standard labels identified if the patient truly had (or did not have) the disease. We used the gold-standard labels as the test sets. For the rest of the patients in each cohort, we had silver-standard labels that were curated by ML algorithms for training. The use of data for this study was approved by the Mass General Brigham Institutional Review Board (2017P000282). Table 1 shows the basic data statistics for each disease cohort.

Table 1. Basic statistics from the disease cohorts.

Disease	Cohort population	Unique records	Average depth	Average age	Test set size
CHF	6,857	25,480	16.24 yrs	68 yrs	49
COPD	5,107	29,880	9.14 yrs	60 yrs	61
RA	4,015	20,315	7.83 yrs	52 yrs	72
T1DM	3,107	23,001	8.59 yrs	57 yrs	58
T2DM	9,500	27,099	7.62 yrs	55 yrs	74
UC	1,560	14,682	6.68 yrs	46 yrs	45

From each representations, a nested set of the top 50, 100, 200, and 400 features obtained from the MSMR algorithm are used for prediction. We applied Logistic Regression with L1 regularization to the training sets to predict the true diagnosis of the target disease (Dx_i) at the next encounter, using bootstrap cross-validation. To evaluate the classifiers, we utilized the gold-standard labels on the test sets. We used the area under the receiver operating characteristic curve (AUC ROC) to compare the algorithms' prediction performances. We also compared the Youden's J statistic [26], that aims to capture both sensitivity and specificity in a single statistic. We computed the all possible Youden's J statistics on the receiver operating characteristic curve, and used the best J statistic a classifier had to offer.

5 Results

As expected, the number of tSPM sequences were about 10-fold the number of SPM sequences. Table 2 shows number of unique sequential patterns mined by

Table 2. Representations mined by each algorithm.

Disease	SPM representations	tSPM representations
CHF	3,555,577	30,969,742
COPD	3,520,689	37,137,257
RA	1,863,405	18,310,561
T1DM	2,209,541	22,294,391
T2DM	3,480,550	29,397,627
UC	863,097	9,062,784

the SPM and tSPM algorithms. The prediction performances from each algorithm are presented in Table 3. Results (Table 3) showed that on average, transitive sequential pattern mining (tSPM) improved equivalent SPM metrics by between 6% to 15% in AUC ROC and 7% to 10% in Youden's J statistics, regardless of the feature selection approach. Frequent transitive sequential representations mined in the FtSPM improved AUC ROC by 6 and Youden's J by 7% compared with the frequent sequential representations in the FSPM algorithm. Using the MSMR dimensionality reduction approach, the tSPM representations offered an average 15% improvement in AUC ROC and 10% in Youden's index over the SPM representations. Across the six diseases, there was some variability in the magnitude of improvements transitive sequencing provided over traditional sequencing. For instance, in COPD and RA, the difference between frequency-based tSPM (FtSPM) and SPM (FSPM) in AUC ROC was close to zero, while this performance difference was 25% in T2DM between the MSMR-based algorithms. Nevertheless, the overall improvement from using tSPM is supported by the results.

We also found that in the traditional sequencing (SPM), the improvement in prediction is almost non-existent. That is, the MSMR algorithm did not offer any benefit over using the most frequent traditional sequential representations. For transitive sequences, however, the MSMR algorithm provided an average of 9% improvement in the AUC ROC and 4% improvement in Youden's J over the frequency-based criterion. Across the six diseases, there was some level of variability in both magnitude and direction (positive/negative change) of the results that impacted the averages. For example, in COPD and RA, the MSMR algorithm for dimensionality reduction decreased the AUC ROC and Youden's J statistic obtained from the SPM algorithms. This resulted in the minimal difference in the effect of MSMR on representations mined from the traditional sequential pattern mining. Variability was less of an issue in the overall effectiveness of the MSMR algorithm on transitive sequences. In four of the six diseases, the transitive sequential representations selected by the MSMR algorithm significantly improved the classifiers' AUC ROC over using the most frequent transitive sequential representations.

Regarding the third question, we found that, on average, the proposed tSPM+MSMR approach improved the AUC ROC by 15% and Youden's J statis-

Table 3. AUC ROC and Youden's J statistics across diseases and by algorithms.

		AUC ROC	FSPM	FTSPM	SPM+MSMR	Youden's J statistic	FSPM	FTSPM	SPM+MSMR
CHF	FSPM	0.672				0.682			
	FtSPM	0.689	3%			0.671	-2%		
	SPM+MSMR	0.723	8%	5%		0.709	4%	6%	
	tSPM+MSMR	0.826	23%	20%	14%	0.792	16%	18%	12%
COPD	FSPM	0.751				0.726			
	FtSPM	0.740	-1%			0.767	6%		
	SPM+MSMR	0.644	-14%	-13%		0.682	-6%	-11%	
	tSPM+MSMR	0.788	5%	6%	22%	0.702	-3%	-8%	3%
RA	FSPM	0.692				0.671			
	FtSPM	0.694	0%			0.784	17%		
	SPM+MSMR	0.659	-5%	-5%		0.654	-3%	-17%	
	tSPM+MSMR	0.793	15%	14%	20%	0.759	13%	-3%	16%
T1DM	FSPM	0.888				0.846			
	FtSPM	0.926	4%			0.904	7%		
	SPM+MSMR	0.865	-3%	-7%		0.875	3%	-3%	
	tSPM+MSMR	0.904	2%	-2%	4%	0.904	7%	0%	3%
T2DM	FSPM	0.596				0.626			
	FtSPM	0.728	22%			0.705	13%		
	SPM+MSMR	0.647	9%	-11%		0.671	7%	-5%	
	tSPM+MSMR	0.808	36%	11%	25%	0.796	27%	13%	19%
UC	FSPM	0.782				0.783			
	FtSPM	0.832	6%			0.813	4%		
	SPM+MSMR	0.804	3%	-3%		0.800	2%	-2%	
	tSPM+MSMR	0.849	9%	2%	6%	0.863	10%	6%	8%
Mean	FtSPM		6%¹				7%¹		
	SPM+MSMR		0%‡	-6%			1%‡	-5%	
	tSPM+MSMR		15%*	9%‡	15%¹		12%*	4%‡	10%¹

¹ question 1: transitive sequencing (tSPM) vs. traditional sequencing (SPM)?
‡ question 2: MSMR vs. the frequency-based criterion
* question 3: transitive sequencing with MSMR vs. traditional frequency-based sequencing

tic by 12% over the FSPM. The only exception was in COPD where Youden's J decreased by 3% as compared with the FSPM (AUC ROC improved by 5%), in which case the frequent transitive sequential patterns provided the best overall performance. Given the benchmark of frequency-based sequential pattern mining (FSPM), both performance indices improved consistently by utilizing the MSMR driven transitive sequential patterns for predicting the diagnosis.

6   Conclusion

In this work, we argue that discrete clinical data in EHRs are fundamentally different from transaction data (for which the SPM algorithm and its often used frequency-based feature selection criteria were developed). We proposed a transitive sequential pattern mining (tSPM) algorithm as well as a dimensionality reduction algorithm (MSMR) to construct and apply temporal representations from discrete clinical data into standard Machine Learning. Our results showed that the tSPM representations improve prediction performance over the traditional sequential pattern mining (SPM). In other words, we showed that the assumption of transitivity improves prediction power in sequential pattern mining with EHR data. More research is needed to evaluate other use cases of transitive sequential pattern mining. We also showed that selecting features using the MSMR algorithm improved prediction using tSPM representations. Another novelty of this research is in its prediction task, where we aim to predict a verified disease diagnosis record. This is different and possibly harder than the prior clinical prediction tasks, such as [9], that aim to predict the next diagnosis record. We believe that it is naive to assume that the diagnosis records in EHRs are always correct and reflect clinical practice. The tSPM and MSMR algorithms together provide methodological pathway to making sense of data in electronic health records.

Acknowledgements. The work in this paper was supported by NIH grant R01-HG009174. The content of the paper is solely the responsibility of the authors and does not necessarily represent the official views of NIH.

References

1. Agrawal, R., Srikant, R., et al.: Mining sequential patterns. In: ICDE, vol. 95, pp. 3–14 (1995)
2. Albers, D.J., Hripcsak, G.: Estimation of time-delayed mutual information and bias for irregularly and sparsely sampled time-series. Chaos, Solitons Fractals **45**(6), 853–860 (2012)
3. Ayres, J., Flannick, J., Gehrke, J., Yiu, T.: Sequential pattern mining using a bitmap representation. In: Proceedings of the Eighth ACM SIGKDD International Conference on Knowledge Discovery and Data Mining, KDD 2002, pp. 429–435. ACM, New York (2002)
4. Batal, I., Cooper, G.F., Fradkin, D., Harrison, J., Moerchen, F., Hauskrecht, M.: An efficient pattern mining approach for event detection in multivariate temporal data. Knowl. Inf. Syst. **46**(1), 115–150 (2015). https://doi.org/10.1007/s10115-015-0819-6
5. Batal, I., Valizadegan, H., Cooper, G.F., Hauskrecht, M.: A temporal pattern mining approach for classifying electronic health record data. ACM Trans. Intell. Syst. Technol. **4**(4) (2013)

6. Berlingerio, M., Bonchi, F., Giannotti, F., Turini, F.: Mining clinical data with a temporal dimension: a case study. In: 2007 IEEE International Conference on Bioinformatics and Biomedicine (BIBM 2007), pp. 429–436, November 2007
7. Brown, G., Pocock, A., Zhao, M.J., Luján, M.: Conditional likelihood maximisation: a unifying framework for information theoretic feature selection. J. Mach. Learn. Res. **13**, 27–66 (2012)
8. Choi, E., et al.: Multi-layer representation learning for medical concepts. In: Proceedings of the 22nd ACM SIGKDD International Conference on Knowledge Discovery and Data Mining, KDD 2016, pp. 1495–1504. Association for Computing Machinery, New York, August 2016
9. Choi, E., Bahadori, M.T., Sun, J., Kulas, J., Schuetz, A., Stewart, W.: RETAIN: an interpretable predictive model for healthcare using reverse time attention mechanism. In: Lee, D.D., Sugiyama, M., Luxburg, U.V., Guyon, I., Garnett, R. (eds.) Advances in Neural Information Processing Systems, vol. 29, pp. 3504–3512. Curran Associates, Inc. (2016)
10. Cover, T.M., Thomas, J.A.: Elements of Information Theory. Wiley, Hoboken (2012)
11. Hripcsak, G., Albers, D.J., Perotte, A.: Exploiting time in electronic health record correlations. J. Am. Med. Inform. Assoc. **18**(Suppl 1), i109–15 (2011)
12. Johnson, A., Pollard, T., Shen, L., et al.: MIMIC-III, a freely accessible critical care database. Sci. Data **3**, 160035 (2016). https://doi.org/10.1038/sdata.2016.35
13. Lee, J.M., Hauskrecht, M.: Recent context-aware LSTM for clinical event time-series prediction. In: Riaño, D., Wilk, S., ten Teije, A. (eds.) AIME 2019. LNCS (LNAI), vol. 11526, pp. 13–23. Springer, Cham (2019). https://doi.org/10.1007/978-3-030-21642-9_3
14. Mabroukeh, N.R., Ezeife, C.I.: A taxonomy of sequential pattern mining algorithms. ACM Comput. Surv. **43**(1), 41 p. (2010). Article 3. https://doi.org/10.1145/1824795.1824798
15. Meyer, P.E.: Information-theoretic variable selection and network inference from microarray data. Ph.D. thesis, Université Libre de Bruxelles (2008)
16. Moskovitch, R., Choi, H., Hripcsak, G., Tatonetti, N.: Prognosis of clinical outcomes with temporal patterns and experiences with one class feature selection. IEEE/ACM Trans. Comput. Biol. Bioinform. **14**(3), 555–563 (2017)
17. Moskovitch, R., Polubriaginof, F., Weiss, A., Ryan, P., Tatonetti, N.: Procedure prediction from symbolic electronic health records via time intervals analytics. J. Biomed. Inform. **75**, 70–82 (2017)
18. Moskovitch, R., Shahar, Y.: Classification-driven temporal discretization of multivariate time series. Data Min. Knowl. Disc. **29**(4), 871–913 (2014). https://doi.org/10.1007/s10618-014-0380-z
19. Orphanou, K., Dagliati, A., Sacchi, L., Stassopoulou, A., Keravnou, E., Bellazzi, R.: Incorporating repeating temporal association rules in Naïve Bayes classifiers for coronary heart disease diagnosis. J. Biomed. Inform. **81**, 74–82 (2018)
20. Paninski, L.: Estimation of entropy and mutual information. Neural Comput. **15**(6), 1191–1253 (2003)
21. Perer, A., Wang, F., Hu, J.: Mining and exploring care pathways from electronic medical records with visual analytics. J. Biomed. Inform. **56**, 369–378 (2015)
22. Rumelhart, D.E., Hinton, G.E., Williams, R.J.: Others: learning representations by back-propagating errors. Cogn. Model. **5**(3), 1 (1988)
23. Stacey, M., McGregor, C.: Temporal abstraction in intelligent clinical data analysis: a survey. Artif. Intell. Med. **39**(1), 1–24 (2007)

24. Sun, W., Rumshisky, A., Uzuner, O.: Temporal reasoning over clinical text: the state of the art. J. Am. Med. Inform. Assoc. **20**(5), 814–819 (2013)
25. Yang, H., Moody, J.: Data visualization and feature selections: new algorithms for non-Gaussian data. In: Advances in Neural Information Processing Systems, vol. 12 (1999)
26. Youden, W.J.: Index for rating diagnostic tests. Cancer **3**(1), 32–35 (1950)
27. Zaki, M.J.: Parallel sequence mining on shared-memory machines. J. Parallel Distrib. Comput. **61**(3), 401–426 (2001)

Clinical Practice Guidelines

A Verified, Executable Formalism for Resilient and Pervasive Guideline-Based Decision Support for Patients

Nick L. S. Fung[1]([✉]), Marten J. van Sinderen[1], Valerie M. Jones[1],
and Hermie J. Hermens[1,2]

[1] University of Twente, 7500 AE Enschede, The Netherlands
{l.s.n.fung,m.j.vansinderen,v.m.jones}@utwente.nl
[2] Roessingh Research and Development, 7500 AH Enschede, The Netherlands
h.hermens@rrd.nl

Abstract. We present an executable formalism for clinical practice guidelines, with the aim of providing pervasive and evidence-based decision support to patients. Unlike traditional formalisms that capture the control flow between tasks, we focus on data flow, with tasks modeled as processes that execute in parallel. By parallelizing and distributing guideline knowledge, each device that constitutes the patient's pervasive healthcare system can provide decision support independently, avoiding single points of failure. This distribution also enables dynamic system re-configurations, increasing its resilience against evolving requirements and changing communications environments.

Our model recognizes four types of processes: Monitoring, Analysis, Decision and Effectuation. These processes were specified using (axiomatic) set theory and implemented as a set of libraries on top of Rosette, which supports execution of the formalism and verification of it using constraint solvers. The formalism was also tested by formalizing a complete clinical guideline for diabetes management, which yielded a Rosette program that was then tested on simulated patient data. The major point of clinical relevance is enhancing the quality and safety of decision support delivered to patients.

Keywords: Computerized clinical practice guidelines · Pervasive healthcare · Knowledge representation · Data flow modeling · Formal specification · Verification and validation · Diabetes management

1 Introduction

Pervasive healthcare systems, which aim to support patients anytime, anywhere, have potential to address the healthcare challenges arising from increased prevalence of chronic diseases, an aging population and shortage of healthcare resources [1]. For example, instead of relying on infrequent checkups, diabetic patients may use such systems to help constantly monitor and control their blood

© Springer Nature Switzerland AG 2020
M. Michalowski and R. Moskovitch (Eds.): AIME 2020, LNAI 12299, pp. 427–439, 2020.
https://doi.org/10.1007/978-3-030-59137-3_38

glucose levels. Such systems should, however, ensure high quality care, which in hospital settings is increasingly supported by the use of clinical practice guidelines (CPGs), especially computerized CPGs that can be executed automatically by knowledge-based systems (KBSs).

We aim to bring computerized CPGs to free-living settings to provide pervasive guideline-based decision support to patients. Clinical KBSs are typically deployed on fixed infrastructures, e.g. hospital servers, but for pervasive healthcare systems, components may be distributed across multiple personal devices and may require dynamic reconfiguration in response to changing requirements and unreliable communications environments. Here we present a new verified formalism for representing clinical guidelines which models clinical tasks as processes executing in parallel. In this way, guideline knowledge and reasoning can be flexibly distributed across system components, allowing them to operate independently and thereby avoid a single point of failure.

Section 2 presents background and related work while Sect. 3 presents our formalism. We present a reference implementation of the formalism in Sect. 4 and demonstrate an application of the formalism to a complete clinical guideline in Sect. 5. Section 6 discusses the findings and clinical relevance. Conclusions are found in Sect. 7.

2   Background and Related Work

CPGs are typically formalized as task-network models [5], i.e. hierarchical plans that comprise constructs for decisions and actions as well as embedded sub-plans. To enable automated execution, decisions are generally modeled as decision trees or tables [9], with data processing specified using expression languages (e.g. GELLO) [5].

Regardless of the specific formalism, task-network models encapsulate the control flow (i.e. the logical ordering) between different tasks over time [5]. As a result, they assume a centralized system architecture in which a supervisory component controls the execution of guidelines (i.e. the application of guidelines to patient data). To reduce reliance on such components, Shalom et al. in 2015 proposed a projection mechanism whereby self-contained portions of a clinical guideline are identified for execution in parallel with the overall plan [10]. This mechanism was implemented in the MobiGuide patient guidance system and was demonstrated in the atrial fibrillation and gestational diabetes domains [6]. The MobiGuide system contains two decision support systems: a front-end system running on the patient's smartphone to execute guideline fragments locally; and a back-end system running on hospital servers to execute the overall guideline and "project" the appropriate portions to recipient devices [10].

While this projection mechanism can be extended to an arbitrary number of "local" devices, it still requires a supervisory component to project guideline fragments. To remove the reliance on a centralized controller, formalisms based on production rules may be adopted instead, the main exemplars of which include the Arden Syntax [8] and the OpenEHR Guideline Definition Language [11]. In general, these rules encapsulate the data flow of clinical guidelines, thus they do not exhibit control flow dependencies and can therefore be executed in parallel. However, they also do not intrinsically support the distinction between different types of tasks featured in clinical guidelines, such as diagnosing a patient's condition and making a therapeutic decision.

Therefore, while our formalism also focuses on data flow rather than control flow, we base our formalism on previous work [3] which introduced an informal conceptual data flow model of disease management. Our model comprises four types of data flow processes interacting with the environment (e.g. the patient):

- Monitoring (M), the process of making observations about the patient.
- Analysis (A), the process of making assessments about the state of the patient.
- Decision (D), the process of deciding on the appropriate therapeutic plan.
- Effectuation (E), the process of executing the decided plan, which may involve performing an action or controlling the execution of a process (possibly itself).

Our MADE model (Fig. 1) was demonstrated to be a useful conceptual tool for analyzing clinical guidelines and designing decision support systems in the context of pervasive healthcare [3]. To support interoperability in pervasive healthcare systems, we also derived from it a reference information model (the MADE RIM) for representing clinical data [2]. Building on this work, we present in this paper a formalism for representing guideline knowledge as interconnected MADE processes.

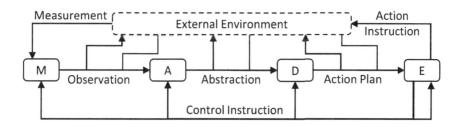

Fig. 1. The MADE model for disease management.

3   The MADE Guideline Formalism

3.1   Overarching Model of Clinical Guidelines

Our MADE formalism is specified using (axiomatic) set theory and comprises 26 set definitions, 13 function signatures and 34 logical invariants. This paper focuses on the set definitions, which specify the data types that constitute our formalism. In particular, we abstract away the technical aspects of pervasive healthcare systems and model a clinical guideline as a set of processes that are fully connected with each other, with data flowing instantaneously between them.

In turn, each process is modeled to comprise four components: a unique ID, a data state to store previously input data, a control state which determines when the process is activated and a specification of the instructions for the process when it runs. This leads to the following two set definitions:

$$Guideline = \mathcal{P}(Process) \tag{1}$$

$$Process = Id \times DataState \times ControlState \times InstSpec \tag{2}$$

Processes are modeled to execute at each time instant, and during execution each process updates its states and outputs a set containing 0 or 1 data item depending on its states, its input data, the current date-time as well as its specified instructions. In accordance to Fig. 1, six types of data are distinguished, which together constitute the MADE RIM as specified in [2]. Here we present the specification of $InstSpec$, of which four types are distinguished, one for each type of MADE process:

$$Data = Measurement \cup Observation \cup Abstraction \cup ActionPlan$$
$$\cup \; ActionInstruction \cup ControlInstruction \tag{3}$$

$$InstSpec = MSpec \cup ASpec \cup DSpec \cup ESpec \tag{4}$$

For ease of understanding, the behavior of each type of MADE process is presented in this paper using natural language English, although it is in fact specified using function signatures and logical invariants. For example, the execution of a process as described above can be formally specified as the following function:

$$execute : Process \times \mathcal{P}(Data) \times DateTime \rightarrow Process \times \mathcal{P}(Data), \text{ such that} \tag{5}$$
$$\forall p \in Process, \; d_{in} \in \mathcal{P}(Data), \; t \in DateTime.$$
$$(\neg isProcessActivated(\pi_3(p), t) \Rightarrow d_{out} = \{\}) \wedge |d_{out}| \leq 1 \; \wedge$$
$$\pi_1(p) = \pi_1(p_{out}) \wedge \pi_4(p) = \pi_4(p_{out}) \tag{6}$$

Here $\pi_i(x)$ refers to the ith element of x, (p_{out}, d_{out}) is the result of $execute(p, d_{in}, t)$ and $isProcessActivated$ is a function that determines whether the process is activated or not given its control state and current date-time. Thus, Eq. 6, which is a logical invariant, specifies that a process may only output data when dictated by its control state and that at most 1 data item can be generated. If the process is not activated, it can still update its data state and control state, but under no circumstances can it be transformed into another process by changing its identifier or its instructions.

3.2 Model of Monitoring Processes

Two types of Monitoring processes are distinguished to reflect the two types of observations specified in [2], viz. observed properties and observed events. In our formalism, observed properties are generated by performing digital signal processing, e.g. noise filtering, on input measurements. Thus for Monitoring processes which output observed properties, the specification comprises:

- A time window indicating the duration beyond which data is considered irrelevant.
- A mathematical function that operates on the measurements that have been filtered using the time window, returning the value of the output property.
- An output type identifying the specific type of observed property to generate.

$$MSpec = PropertySpec \cup EventSpec, \text{ where} \qquad (7)$$
$$PropertySpec = TimeWindow \times ValueFunction \times OutputType \qquad (8)$$
$$TimeWindow = Duration \qquad (9)$$
$$ValueFunction = \mathcal{P}(Measurement) \rightarrow PropertyValue \qquad (10)$$
$$OutputType = Id \qquad (11)$$

Whenever a Monitoring process for observed properties is activated, all input data (including those stored in its data state) are filtered using the time window to remove data that are either not measurements or not relevant at the current date-time. The process then feeds the remaining measurements into its value function and outputs an observed property with the specified output type and computed value.

Unlike observed properties, observed events can exhibit a start and an end, and they can only be assigned a boolean value to indicate whether they occurred or not [2]. Thus Monitoring processes for observed events are specified to comprise:

- A time window and predicate specifying the conditions indicating the event's start.
- A time window and predicate specifying the conditions indicating the event's end.
- An output type identifying the specific type of observed event to output.

$$EventSpec = EventTrigger \times EventTrigger \times OutputType, \text{ where} \qquad (12)$$
$$EventTrigger = TimeWindow \times TriggerPredicate \qquad (13)$$
$$TriggerPredicate = \mathcal{P}(Measurement) \rightarrow Boolean \qquad (14)$$

As with the time window in *PropertySpec*, the time windows in *EventSpec* are used to filter out irrelevant measurements; the remaining measurements are input into the corresponding predicates to determine if the start or end of the event is detected. If the start is detected, the Monitoring process will then search through the historical data to determine when the event ended; during this time period, the event did not happen and therefore its value would be false. Similarly, if the end is detected, the process will search for the date-time starting from which the event occurred.

3.3   Model of Analysis Processes

To generate abstractions from observations, Analysis processes comprise:

- An output type identifying the specific type of abstraction to output.
- A set of abstraction triplets, each containing:
  - A time window for filtering out expired measurements.
  - A predicate on the filtered observations that specifies the condition under which an abstraction should be generated.
  - An abstraction function that accepts a set of observations as input and returns the appropriate value for the output abstraction (if one is generated).

$$ASpec = OutputType \times \mathcal{P}(TimeWindow \times AbstractionPredicate$$
$$\times AbstractionFunction), \text{ where} \tag{15}$$
$$AbstractionPredicate = \mathcal{P}(Observation) \rightarrow Boolean \tag{16}$$
$$AbstractionFunction = \mathcal{P}(Observation) \rightarrow AbstractionValue \tag{17}$$

Like observed events, abstractions are valid over a date-time range [2]. However, unlike observed events, abstractions are not generated by detecting start and end conditions. Whenever an Analysis process is activated, it iterates through its abstraction triplets, and for each triplet it applies the abstraction predicate and abstraction function to the data that has been filtered through the corresponding window. If a predicate is satisfied, the iteration terminates and the process outputs an abstraction with the specified type and corresponding value—otherwise, the process does not generate any abstraction. The valid date-time range of the output abstraction is computed by finding the longest period (starting from the current date-time) during which the predicate remains satisfied and the computed value remains unchanged.

3.4   Model of Decision Processes

Decision processes comprise a plan template from which a new action plan can be instantiated as well as a set of decision criteria governing the conditions under which that action plan should be enacted. In our formalism, decision tables are adopted for decision-making, such that the decision criteria are specified as predicates over the input abstractions. Furthermore, reflecting the specification of action plans, plan templates are modeled to comprise a set of instruction templates, of which three types are distinguished, one for each type of scheduled instruction in an action plan [2]:

- Homogeneous action instructions for actions that exhibit a rate and duration.
- Culminating action instructions for actions that exhibit an end goal.
- Control instructions for controlling the execution of MADE processes by setting their schedules and/or status (whether they should be running or paused).

$$DSpec = PlanTemplate \times \mathcal{P}(DecisionCriterion), \text{ where} \qquad (18)$$
$$PlanTemplate = PlanType \times \mathcal{P}(ControlTemplate$$
$$\cup \; HomogeneousActionTemplate \cup \; CulminatingActionTemplate) \qquad (19)$$
$$DecisionCriterion = \mathcal{P}(Abstraction) \rightarrow Boolean \qquad (20)$$

In contrast to Monitoring and Analysis processes, Decision processes do not require a time window; the only relevant abstractions are those that are valid at the current date-time. These abstractions are checked against the decision criteria; if any criterion is triggered, an action plan will be instantiated from the plan template. The schedule of each instruction in the plan is computed from the current date-time and the relative schedule of the corresponding instruction template.

3.5   Model of Effectuation Processes

Since action plans already contain the complete details of all instructions to be executed and when, Effectuation processes simply specify which instructions they are responsible for effectuating. More specifically, Effectuation processes comprise:

- A specification of the target scheduled instructions to be effectuated by the process. Targets are identified by type of action plan, type of target instruction as well as a predicate on the scheduled instruction.
- An output type indicating the specific type of instruction to output.

$$ESpec = \mathcal{P}(TargetScheduledInstruction) \times OutputType, \text{ where} \qquad (21)$$
$$TargetScheduledInstruction = PlanType \times InstructionType$$
$$\times InstructionPredicate \qquad (22)$$
$$InstructionPredicate = ScheduledInstruction \rightarrow Boolean \qquad (23)$$

When activated, an Effectuation process filters out any action plans that are not valid at the current date-time and extracts, from the remaining action plans, all relevant scheduled instructions based on the specified targets. From these relevant scheduled instructions, the process then determines if any should be effectuated at the current date-time (given their schedule). If yes, then an instruction will be generated with the specified output type, which may possibly be more specific than the scheduled instruction. For example, an action plan may require an hour's (unspecified) exercise daily while an Effectuation process may instantiate this as an instruction to walk on a treadmill.

4   Reference Implementation

To ensure that the formalism is executable and to clarify any ambiguities, a reference implementation of the formalism was developed as a set of libraries on top of the language Rosette. Clinical guidelines can be formalized using this reference implementation into Rosette programs (Fig. 2), which would comprise interconnected MADE processes. Subsequently, by executing those programs, clinical guidelines can be applied onto (simulated) patient data to demonstrate the semantics of the formalism.

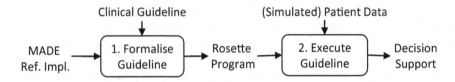

Fig. 2. The overall procedure for using the reference implementation.

The source code is available at https://github.com/nlsfung/MADE-Language. To summarize, each data type in the formalism is implemented as a structure (which is analogous to classes in objected-oriented programs), while behaviors are implemented as procedures. Furthermore, elements of data types are generally implemented as fields in the corresponding structures, with the exception of *InstSpec* which is implemented using interfaces (to ensure that it remains constant over time). Clinical guidelines can then be formalized by extending the structures and implementing the interfaces; during execution, these structures would be instantiated into concrete data items.

For example, a process to analyze ketonuria may be formalized as the following structure. It inherits the data state and control state fields from `analysis-process` and implements the `gen:analysis` interface to return the appropriate output type (`ketonuria`) and abstraction triplets (denoted by x).

```
(struct analyze-ketonuria analysis-process ()
#:methods gen:analysis [
(define (analysis-process-output-type self) ketonuria)
(define (analysis-process-output-specification self) x)])
```

Apart from an interpreter for executing programs, Rosette also provides access to off-the-shelf constraint solvers to analyze them [12]. This allowed the reference implementation to be verified against 34 logical invariants derived for the formalism (e.g. Eq. 6). Each invariant was translated into an assertion in Rosette, which was then checked (using a constraint solver) whether a concrete counter-example can be found that violates it. If yes, then the assertion is not valid. Otherwise, as is the case with the 34 invariants, it provides evidence (but not a definitive proof) that the assertion is valid.

5   Case Study: Gestational Diabetes Guideline

The MADE formalism was tested by formalizing a complete clinical guideline for gestational diabetes (GD) [7] that has previously been adopted to evaluate the MobiGuide system [4]. The result was a Rosette program (also available on GitHub) that comprises 0 Monitoring, 4 Analysis, 22 Decision and 29 Effectuation processes. As an example, we focus on the guideline fragment shown in Fig. 3, which relates to the decision to increase the carbohydrates intake of a GD patient. To highlight the identified MADE processes, we annotated the extract by underlining followed by an inserted [M] for Monitoring, [A] for Analysis and [D] for Decision processes.

```
''... The patient measures ... ketones in the urine every day
at fasting conditions [M]. ... The results of ketonuria could
be: a) positive (++); b) positive (+); c) negative (+/-);
d) negative (-); e) negative (--). ... In case of ketonuria
detection (the number of ketonuria measurements with result
''positive'' is equal or higher than 3 in a period of time of
one week) [A]:- If the patient was COMPLIANT with the prescribed
diet [A], the nurse decides to increase the carbohydrates intake
either at dinner or at bedtime ... by 1 unit [D] ... ''
```

Fig. 3. Extract from the GD guideline [7] annotated by [3].

Since urinary ketone levels must be monitored manually by the GD patient, only two processes were formalized for this example. The first is the analysis of ketonuria, which requires a window of 7 days and outputs a ketonuria abstraction with value positive if more than two urinary ketone values are + or ++. The second is the decision to increase carbohydrates intake, which was confirmed by clinicians to be automatable. It accepts as input ketonuria and diet compliance abstractions and outputs an action plan to increase carbohydrates intake at dinner if ketonuria is positive and diet is compliant.

These two processes are specified as follows. In practice, both processes were formalized into Rosette code as with the rest of the GD guideline, all of which was then tested using simulated data (see the Appendix for examples). However, for conciseness, their specifications are presented here using mathematical notation, with constants (denoted using small caps) replacing instances of low-level structures (such as durations).

$$AnalyseKetonuria \subset Analysis, \text{ such that} \qquad (24)$$

$$\forall p \in AnalyseKetonuria. \; \pi_1(p) = \text{Analyse Ketonuria} \; \wedge$$

$$\pi_1(\pi_4(p)) = \text{Ketonuria} \wedge \pi_2(\pi_4(p)) = \{(\text{One Week},$$

$$d_{in} \mapsto |\{d \mid d \in d_{in} \wedge d \in UrinaryKetoneLevel \wedge \pi_4(d) \in \{+, ++\}\}| \geq 3,$$

$$d_{in} \mapsto \text{Positive})\}$$

$DecideIncreaseCarbohydrates \subset Decision$, such that $\qquad$ (25)

$\qquad \forall p \in DecideIncreaseCarbohydrates.$

$\pi_1(p) = \text{DECIDE INCREASE CARBOHYDRATES} \wedge$

$\pi_1(\pi_4(p)) = (\text{DIETARY PLAN}, \{\text{INCREASE DINNER CARB. INTAKE}\}) \wedge$

$\pi_2(\pi_4(p)) = \{d_{in} \mapsto (\exists d \in d_{in}.\ d \in Ketonuria \wedge \pi_4(d) = \text{POSITIVE}) \wedge$

$\qquad (\exists d \in d_{in}.\ d \in DietCompliance \wedge \pi_4(d) = \text{COMPLIANT})\}$

6 Discussion

6.1 Expressiveness of the Formalism

Experience gained from the case study showed that the MADE formalism has sufficient expressiveness to represent automatable portions of clinical guidelines. In the future, we plan to evaluate this further by, for example, comparing against existing guideline formalisms. However, it can already be observed that our formalism does not support partial specifications for tasks that must be manually performed. For example, since urinary ketone levels are measured manually, the guideline does not explicate how these measurements should be processed (see Fig. 3). For this reason, this and all other measurement tasks could not be formalized into Monitoring processes, which we believe would also apply to other clinical guidelines.

Furthermore, our formalism does not support the personalization of clinical guidelines according to individual patient preferences, which is an important usability feature for providing decision support to patients [6]. In our case study for example, dinnertime must be made explicit (e.g. 7 pm) to formalize the decision process to increase carbohydrates intake; individual adjustments to dinnertime can only be made by directly accessing and changing the formalized guideline. However, we believe the MADE formalism provides a solid foundation on which such extensions can be added, and we will continue to evaluate and improve the formalism using different clinical guidelines.

6.2 Clinical Relevance

The GD guideline was developed by a team of expert clinicians, knowledge engineers and researchers [4], and its clinical relevance has been established in patient trials of the MobiGuide system [6]. However, the MobiGuide system comprises a fixed number (viz. 2) of KBSs, while our aim is to support an arbitrary distribution of knowledge. Therefore, in collaboration with clinicians, patients and other stakeholders, we plan to fully evaluate our formalism and its clinical value by implementing and testing "n-ary" guideline-based pervasive healthcare systems for GD and other clinical applications. Such a system would adopt the MADE reference information model [2] to share data between devices (including EHR repositories) and would implement an optimization algorithm to distribute guideline knowledge so as to maximize system resilience.

While our formalism allows parallelization of clinical guidelines at the knowledge level, an equally valid alternative may be to parallelize them at the source code level, such as by adopting research results from the well-established area of high-performance computing. However, we believe our formalism can offer the advantages of increased transparency and amenability to analysis. In particular, we are currently investigating how to apply formal verification not only on the guideline formalism itself, but also on computerized CPGs expressed in the formalism, such as to ensure that mutually exclusive MADE processes would never be activated at the same time.

7   Conclusions

Guideline-based pervasive healthcare systems can extend evidence-based healthcare beyond the traditional healthcare setting. It is all the more crucial then to have demonstrable system resilience, quality of clinical information and correct operational logic in a highly distributed environment. To this end, the MADE formalism was developed to represent clinical guidelines in the context of pervasive healthcare and is specified using axiomatic set theory to avoid ambiguity and to allow formal analysis. In particular, the reference implementation was formally verified against its specification using Rosette.

Due to its mathematical foundation, the MADE formalism also opens up possibilities for formally verifying clinical guidelines, which we currently investigate. Furthermore, while the formalism has been tested by formalizing a clinical guideline, its clinical relevance must be further evaluated, such as by conducting field studies using a fully implemented system. The overall objective is to improve clinical correctness of guidelines and safety of their implementations as computerized CPGs.

Appendix

The complete formalized guideline for GD was tested by applying it onto simulated patient data. For example, Fig. 4 shows some urinary ketone levels that may be used to test the analysis of ketonuria for GD patients (*AnalyseKetonuria*), which is specified in Eq. 24. Here, the urinary ketone levels are positive from time points 3 to 6, and the Analysis process is activated at time point 7. Each time point is separated by one day, thus at time point 7, there are three or more urinary ketone levels in the past 7 days. Therefore, as expected, a ketonuria abstraction is generated with value positive. This abstraction is valid until time point 11; beyond this point, there are less than 3 positive urinary ketone levels in a 7-day window given the available data.

As another example, Fig. 5 shows a schematic of the data that may be used to test the decision to increase carbohydrates intake (*DecideIncreaseCarbohydrates*), which is specified in Eq. 25. Here, ketonuria is positive from time points 1 to 5 while diet is compliant at time point 2 and from time points 5 to 7. Furthermore, the decision process is activated at even time

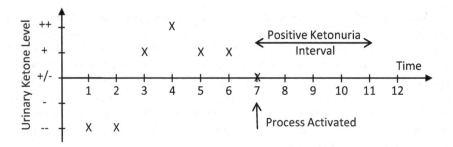

Fig. 4. Example data for testing the analysis of positive ketonuria.

points only, thus on execution, the process outputs an action plan at time point 2 only and not at time point 5, which is as expected (Inv. 6).

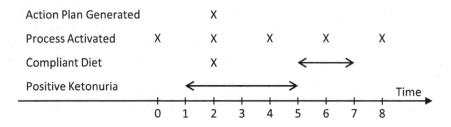

Fig. 5. Example data for testing the decision to increase carbohydrates intake.

References

1. Bardram, J.E.: Pervasive healthcare as a scientific discipline. Method Inform. Med. **47**, 178–185 (2008)
2. Fung, N.L.S., Jones, V.M., Hermens, H.J.: The MADE reference information model for interoperable pervasive telemedicine systems. Method Inform. Med. **56**(2), 180–186 (2017)
3. Fung, N.L.S., Jones, V.M., Widya, I., Broens, T.H.F., Larburu, N., et al.: The conceptual MADE framework for pervasive and knowledge-based decision support in telemedicine. Int. J. Knowl. Syst. Sci. **7**(1), 25–39 (2016)
4. García-Sáez, G., Rigla, M., Martínez-Sarriegui, I., Shalom, E., Peleg, M., et al.: Patient-oriented computerized clinical guidelines for mobile decision support in gestational diabetes. J. Diabetes Sci. Technol. **8**, 238–246 (2014)
5. Peleg, M., Tu, S., Bury, J., Ciccarese, P., Fox, J., et al.: Comparing computer-interpretable guideline models: a case-study approach. J. Am. Med. Inform. Assoc. **10**, 52–68 (2003)
6. Peleg, M., Shahar, Y., Quaglini, S., Fux, A., García-Sáez, G., et al.: MobiGuide: a personalized and patient-centric decision-support system and its evaluation in the atrial fibrillation and gestational diabetes domains. User Model User-adapt Interact. **27**(2), 159–213 (2017)

7. Rigla, M., Tirado, R., Caixàs, A., Pons, B., Costa, J.: Gestational diabetes guideline CSPT. Technical report, MobiGuide Project (FP7-287811) (2013). Version 1.0, 12/02/2013. Internal document
8. Samwald, M., Fehre, K., de Bruin, J., Adlassnig, K.P.: The Arden Syntax standard for clinical decision support: experiences and directions. J. Biomed. Inform. **45**(4), 711–718 (2012)
9. Seyfang, A., Miksch, S., Marcos, M.: Combining diagnosis and treatment using ASBRU. Int. J. Med. Inform. **68**, 49–57 (2002)
10. Shalom, E., et al.: Implementation of a distributed guideline-based decision support model within a patient-guidance framework. In: Riaño, D., Lenz, R., Miksch, S., Peleg, M., Reichert, M., ten Teije, A. (eds.) KR4HC 2015. LNCS (LNAI), vol. 9485, pp. 111–125. Springer, Cham (2015). https://doi.org/10.1007/978-3-319-26585-8_8
11. The openEHR Foundation: Guideline Definition Language v2 (GDL2), CDS Release 2.0.0 edn. (2019). https://specifications.openehr.org/releases/CDS/latest/GDL2.html. Accessed 13 July 2020
12. Torlak, E., Bodik, R.: Growing solver-aided languages with Rosette. In: Proceedings Onward! 2013, pp. 135–152. Association for Computing Machinery, New York (2013)

A CIG Integration Framework to Provide Decision Support for Comorbid Conditions Using Transaction-Based Semantics and Temporal Planning

William Van Woensel[1]([⊠]), Samina Abidi[1,2], Borna Jafarpour[1], and Syed Sibte Raza Abidi[1]

[1] NICHE Research Group, Faculty of Computer Science, Dalhousie University, Halifax, Canada
{william.van.woensel,samina.abidi,ssrabidi}@dal.ca
[2] Department of Community Health and Epidemiology, Dalhousie University, Halifax, Canada

Abstract. Managing comorbid conditions, i.e., patients with multiple medical conditions, is quite challenging for Clinical Decision Support Systems (CDSS) based on computerized Clinical Practice Guidelines (CPG). In case of comorbidity, CDSS will need to recommend treatments from multiple different CPG, which may adversely interact (e.g., drug-disease interactions), or introduce inefficiencies. A-priori, static integration of computerized comorbid CPG is insufficient for clinical practice. In this paper, we present a solution for dynamic integration of CPG in response to evolving health profiles. Using Description and Transaction Logics, we define a set of CIG integration semantics for encoding integration decisions that cope with comorbidity issues at execution-time. These dynamic, transaction-based semantics are well-suited to roll back prior decisions when no longer safe or efficient; or, inversely, apply new decisions when relevant. Moreover, comorbid CIG integration should consider temporal properties of CIG tasks—at execution-time, these properties will be influenced by a range of temporal constraints. Given all temporal constraints, optimal task schedules will be calculated that will determine the feasibility of CIG integration decisions.

Keywords: Clinical guidelines · Decision support systems · Comorbidity

1 Introduction

Clinical Practice Guidelines (CPG) are evidence-based recommendations for guiding diagnosis, prognosis, and treatment of a specific illness [1]. By computerizing CPG as Computer Interpretable Guidelines (CIG), using formalisms such as PROforma [2] or SDA* [3], they can be utilized by Clinical Decision Support (CDS) systems to deliver care recommendations based on the latest patient medical data. To manage often-occurring comorbidities, domain experts manually review disease-specific CPG for comorbidity situations, such as adverse health interactions, and then reconfigure the

© Springer Nature Switzerland AG 2020
M. Michalowski and R. Moskovitch (Eds.): AIME 2020, LNAI 12299, pp. 440–450, 2020.
https://doi.org/10.1007/978-3-030-59137-3_39

workflows to yield a single, comorbidity-compliant clinical workflow [4, 5]. Many works in the literature have introduced formal methods to (semi) automatically integrate CIG for comorbid care (e.g., [5, 6]), some of which focusing on temporal aspects [7, 8]. These systems focus on generating a single, static, comorbidity-compliant CIG, which will then be followed at execution-time. But, we observe that continuing clinical safety and efficiency of CIG integration decisions, which were made at design-time, are often affected by execution-time clinical events and evolving health profiles:

- HTN guidelines recommend Thiazide Diuretics due to their effectiveness at controlling BP. But, these are known to worsen glycemic control in case of HTN/Diabetes comorbidity. During treatment, patients need to be closely monitored for Hypokalemia, possibly reducing treatment dosage. In case of late-stage renal impairment, Loop Diuretics rather than a Thiazide Diuretics are recommended [9].
- Guidelines for Venous Thromboembolism (VTE) prescribes treatment with Warfarin, whereas antibiotics (e.g., Erythromycin) are recommended for Respiratory Tract Infection (RTI). But, these antibiotics potentiate the anticoagulant effect of Warfarin, increasing the risk of bleeding. In case of VTE/RTI comorbidity, treatment may commence if bleeding risk is monitored [10] and Warfarin dose is reduced when needed.
- COPD guidelines recommend X-rays or CT-scans for differential diagnosis; whereas CT Pulmonary Angiography (PA) tests are recommended for chronic PE. In case of COPD/PE comorbidity [11], CT-PA scan results can be re-used for COPD diagnosis, rendering further X-Rays or CT-scans redundant. But, this may require delaying the CT-PA chest scan to ensure its results can be safely re-used for COPD diagnosis.

We make the following observations. Firstly, when health conditions evolve at execution-time (e.g., hypokalemia; bleeding risk), new CIG integration decisions are often needed (e.g., Loop diuretics; reducing dosage), whereas prior decisions should be rolled back (e.g., Thiazide diuretics). Secondly, at execution-time, temporal task timings are influenced by a range of temporal constraints—originating from CIG workflows (e.g., sequential relations) and inherent properties (e.g., duration), but also real-time delays (e.g., delayed scans) and integration decisions (e.g., re-use of scan results).

We propose a dynamic approach for the execution-time integration of multiple comorbid CIG. In particular, we define *dynamic, transaction-based CIG integration semantics*, which allow (*a*) rolling back any integration decision when conditions no longer hold; and, inversely (*b*) applying new decisions when conditions become valid. To capture the impact of temporal aspects, we calculate *optimal task schedules* whenever new medical data comes in, based on all relevant temporal constraints. In turn, these schedules will determine the feasibility of CIG integration decisions— possibly leading to new integration decisions, or prior ones being rolled back. At design-time, a clinician will instantiate so-called *CIG integration policies*, implemented by our transaction-based integration semantics, to cope with specific comorbidity issues. We introduce a framework based on Description Logic [12] and Transaction Logic [13], which specifies (*1*) a DL-based CIG Integration Ontology (CIG-IntO) with core concepts, and (*2*) transaction-based semantics for CIG integration policies (available online [14]).

2   CIG Integration Approach

Our CIG Integration Framework provides tools for a clinician to, at design-time (*a*) identify the set of clinical tasks, called *CIG integration points*, which impact clinical safety or efficiency in comorbid situations; (*b*) instantiate *CIG integration policies* for these points for resolving the comorbidity issue. At execution-time, these policies will be used to ensure clinical safety and efficiency by resolving the comorbid issue. We supply Medical Linked Open Data (MLOD) [14] to aid the clinician in these tasks.

Figure 1 shows the core *CIG-IntO* concepts and properties:

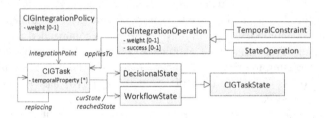

Fig. 1. Core classes and properties in CPG-IntO.

To resolve the comorbid situation, CIG integration policies will be informed by, and operate on (**a**) a *multi-level state machine*, which manages the execution-time lifecycles of their CIG integration points (*WorkflowState* and *DecisionalState* classes); and (**b**) the *temporal properties* of their CIG integration points (*temporalProperty* properties). Hence, two execution-time CIG operations are available, namely task state operations (*StateOperation*) and temporal constraints (*TemporalConstraint*). Each operation has an associated truth value (*success* property) and importance (*weight* property).

Below, we elaborate on the clinical task lifecycles and temporal properties.

2.1   Clinical Task Lifecycles

Figure 2 shows the multi-level state machine that model the lifecycles of clinical tasks. Below, we elaborate on the constituent state machines.

Workflow State Machine

In the workflow state machine (Fig. 2, top), during regular execution, a CIG task travels from *inactive* to *active* when it is next in line for execution; to the *started* state when selected for execution; and finally proceeds to the *completed* state when completed. Two sources of events govern the task lifecycle i.e., the clinician and external services (e.g., Lab Information Systems, Electronic Medical Records). Jafarpour et al. [15] presented a set of *CIG execution semantics* for moving tasks to the *active* state.

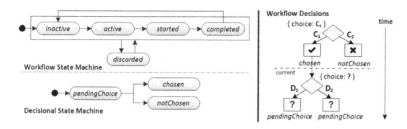

Fig. 2. Multi-level state machine.

Using the *discard* operation, a CIG integration policy can discard a clinical task, which involves moving it to the *discarded* state[1]. The *replace* operation replaces a task in the CIG workflow, which involves moving the replaced task to the *discarded* state, and attaching a reference to its "replacing" task (Fig. 1). A state operation can be successful or not (*success* property) depending on a conflict resolution strategy; for brevity, we do not elaborate on this aspect (see [16] for details).

Decisional State Machine

At execution-time, the decisional state of a workflow will proceed in unforeseeable ways, depending on medical test results, clinician/patient choices, or other data referred by decision nodes; this information only becomes available as the workflow progresses. We capture this uncertainty using a decisional state machine (Fig. 2, bottom).

As illustrated in Fig. 2 (right), a clinical task will have a *pendingChoice* state when no choice has yet been made at the *nearest preceding* decision node—i.e., it is still unknown whether the task will be executed. Once a choice is made, all subsequent tasks in the non-chosen branch(es) travel to the *notChosen* state, i.e., these will not be executed; while tasks in the chosen branch travel to the *chosen* state up until the next decision node, i.e., these tasks are in line for execution.

This decisional lifecycle captures the information needed for effectively coordinating comorbid CIG integration at execution-time. Some integration decisions will need to be made even when uncertainty exists on their integration points—we give examples of this in Sect. 3.2. Once it is known that one of the clinical tasks will not be executed (i.e., *notChosen* state), such integration decisions will need to be reverted.

2.2 Temporal Task Properties

Each CIG task is associated with a set of temporal properties, including (*a*) inherent temporal properties, such as duration, and its max. allowable delay, which are given by domain experts; and (*b*) current task activation and completion times estimated during execution (these are best-effort estimates, based on durations and min. delays of preceding clinical tasks; and will be updated as CIG workflows progress). In general, this temporal information indicates when treatments should commence and conclude; when medical tests should take place, and hence be ordered, from clinical institutions; etc.

---

[1] As a *discard* operation can be reverted, a task can travel both to and from the *discarded* state.

For our purposes, CIG integration policies are informed by, and can act on, these temporal properties to resolve their comorbid situation (Table 1).

Table 1. Temporal properties of CIG tasks.

Inherent temporal properties of clinical tasks (*periods of time*)	
duration	Time required to complete the clinical task
maxAbsDelay	Max. period a task may be delayed relative to its original activation time
maxRelDelay	Max. period a task may be delayed after its prior task was completed
minRelDelay	Min. period a task should be delayed after its prior task was completed
resultValidPeriod	Period during which a clinical task's results (if any) are considered valid
Calculated temporal properties of clinical tasks (*single points in time*)	
activeTime	Estimated activation time based on workflows, integration policies and delay
startTime	Marks the time a clinician flagged an activated task for execution
completeTime	Estimated completion time based on workflows, integration policies and delays

(Note that we could not find uses for *minAbsDelay*, i.e., min. period a task should be delayed relative to its original activation time). Using declarative *temporal constraints*, a CIG integration policy can restrict the temporal properties of CIG tasks. Also, a set of temporal constraints will ensure that inherent workflow and task restrictions are met (e.g., durations, min. or max. delays). A temporal constraint will have a fuzzy or crisp truth value, which depends on their feasibility given other temporal constraints. Below, we show 2 out of 9 currently supported temporal constraints:

(1) $time_2 = time_1 + d$
This constraint dictates that task time ($time_2$) must lie exactly at distance d after $time_1$:

$$\{time_2 = time_1 + d : 1 \mid else : 0\}$$

E.g., we will use this constraint to encode a task's completion time: $completeTime(T) = startTime(T) + duration(T)$

(2) $time_2$ **within** d **of** $time_1$
This constraint dictates that $time_2$ should lie within distance d of $time_1$:

$$\begin{cases} time_2 < time_1 = 0 \mid time_2 \leq time_1 + d = 1 \\ else = 1 - (time_2 - (time_1 + d))/max_{time} \end{cases}$$

If $time_2$ lies within d of $time_1$, 1 is returned; if $time_2$ lies before $time_1$, 0 is returned; else, the value is inversely related to the distance between $time_2$ and ($time_1 + d$) (normalized by max_{time}, i.e., max. workflow duration and durations of proceeding tasks).

E.g., we apply this temporal constraint to enforce the *maxRelDelay* property: $startTime(T_n)$ **within** $maxRelDelay(T)$ **of** $completeTime(T_{n-1})$.

3   CIG Integration Policies

In this section, we start by describing several CIG integration policies to integrate comorbid CIG at execution-time; these are able to resolve the presented comorbid situations (Sect. 1) by leveraging the set of CIG integration operations (Sect. 2).

Then, we proceed with formalizing their semantics (Sect. 3.1 and 3.2), and describe the planning of an optimal comorbid care plan (Sect. 3.3).

ReplaceTasksPolicy
This policy replaces a clinical task with safer, or more efficient, alternative. In line with clinical practice, a replacement may be conditional on up-to-date health profiles. E.g., the replacement policy for HTN/Diabetes comorbidity is written as follows (using Terse RDF Triple Language (Turtle); default namespace is the CIG-IntO ontology):

```
Thiazide_Diuretics_Diabetes a :ReplaceTasksPolicy ;
  :taskToReplace :prescribe_Thiazide_diuretics .
  :replacement [
   :condition [ :observation :hypokalemia ] ; :task :low_dose_Thiazide_diuretics ] ;
  :replacement [
   :condition [:illness :end_stage_renal_impairment]; :task :prescribe_Loop_diuretics] .
```

$$(1)$$

The policy specifies the prescription of Thiazide diuretics as integration point, and lists two replacements: when Hypokalemia is observed, lower-dosage diuretics should be prescribed; in case of end-stage renal impairment, loop diuretics are instead needed.

The following represents a replacement policy for the VTE/RTI comorbidity:

```
:Warfarin_Erythromycin a :EvtCondReplacePolicy ;
  :taskToReplace :prescribe_Warfarin ; :eventTask :prescribe_Erythromycin ;
  :replacement [ :time :during ;
     :task :adjust_Warfarin_during_Erythromycin ; :exitCond [ :INR_value :normal ] ] ;
  :replacement [ :time :after ;
     :task :adjust_Warfarin_after_Erythromycin ; :exitCond [ :INR_value :normal ] ] .
```

$$(2)$$

This policy specifies the prescription of Warfarin as integration point, and indicates an *eventTask* that replacements will be relative to, in this case, the prescription of Erythromycin: (*a*) during the event, the Warfarin dose should be adjusted based on the patient's INR value, until the value returns to normal; (*b*) after the Erythromycin treatment, a similar activity should take place until the INR value becomes normal again. We note that, in these examples, CIG tasks are given meaningful names for brevity (e.g., "prescribe X"; "low dose Y"). In the actual system, CIG tasks are annotated with the action (e.g., treatment), drug (e.g., warfarin), concrete dosage, and so on.

RedunTasksPolicy

This policy discards tasks in case their results are subsumed for a given purpose (e.g., observations), and hence made redundant, by another comorbid task. E.g., the redundant task policy for the COPD/PE comorbidity is written as follows:

$$\text{:CTPA\_CT a :RedunTasksPolicy ; :weight :resourceWeight ;} \quad (3)$$
$$\text{:essenTask :CTPA\_scan ; :redunTask :CT\_scan , :X\_rays .}$$

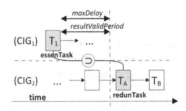

Fig. 3. Example workflows with *RedunTasksPolicy* (⊃ symbol indicates redundancy relation).

The policy specifies two integration points: CT-PA scan is the "subsuming" task; X-rays or CT-scans are "redundant" task(s) to be discarded. A weight category is assigned, with a pre-defined weight that reflects importance (i.e., safety > resource > preference).

However, in clinical practice, the ability to re-use the test results of essential tasks will be limited, as they only remain valid for a limited time (*resultValidPeriod* temporal property); a CT-PA scan will not remain valid for long in case the illness is evolving. In Fig. 3, since T_A falls outside the result validity period of T_1, essential task T_1 could be delayed until task T_A lies within the validity period (inversely, redundant task T_A could be sped up, if possible). At that point, task T_A can be discarded since the results of T_1 can still be safely reused. But, CIG tasks cannot be delayed indefinitely as to not compromise patient safety (*maxDelay* temporal property).

3.1   Formalization of CIG Integration Policies

A CIG integration policy should be applied whenever all its conditions are met at execution-time; and, whenever any condition no longer holds, its applied operations should be reverted. Further, integration operations will require changes in the Knowledge Base (KB): i.e., assigning new states to tasks or changing temporal properties. Hence, to formalize CIG integration policies, we require a logic with (*a*) explicit semantics for KB changes; and (*b*) support for rolling back applied changes. Transaction Logic (𝒯𝓇) [13], as a logic of change with atomic transactions, is a good candidate. We created a dynamic extension of 𝒯𝓇 [14] that will, at runtime (*1*) rollback a transaction once a condition no longer holds, and (2) (re-)apply a transaction when all its conditions hold.

Transactions are special types of rules that include KB updates (called transitions). The logic introduces an operator called serial conjunction (symbol $\otimes$) that allows specifying pre- and post-conditions, thus giving $\mathcal{T}r$ an all-or-nothing flavor. For instance, the following transaction implements conditional replacements (see (1)):

$$cond(E, V) := replacement(E, CR) \otimes condition(CR, C) \otimes Satisfied(C) \otimes task(CR, R)$$
$$\otimes\ ins : replacement(E, R) \tag{4}$$

In case a policy has a replacement with a satisfied condition, it will be inserted into the KB. The following transaction supports the *tmp:withinPeriod* temporal constraint:

$$tmp{:}withinPeriod(t_2, d, t_1) := tmp{:}apWithinPeriod(t_2, d, t_1) \otimes tmp{:}isWithinPeriod(t_2, d, t_1) \tag{5}$$

The transaction applies the temporal constraint and then checks whether it was satisfied (i.e., as a post-condition); since its successfulness depends on a planning component (see Sect. 3.3), this will only be known after it was applied.

3.2   Representing CIG Integration Semantics Using Transaction Logic

A $\mathcal{T}r$ program consists of a KB state, a transaction base P, i.e., set of $\mathcal{T}r$ rules, and a transition base $\mathcal{B}$, i.e., set of transitions. Together with the two transactions above, the below transactions are the part of P needed for representing CIG integration policies:

$$state{:}transit(t, s) := curState(t, c) \otimes WorkflowState(c) \otimes del{:}state(t, c) \otimes ins{:}state(t, s) \tag{6}$$

$$state{:}replace(t_1, t_2) := state{:}transit(t_1, \text{'discarded'}) \otimes ins{:}replacing(t_2, t_1) \tag{7}$$

The *transit* transaction will delete the workflow state of task t and insert its new state. In addition, the *replace* transaction will insert a *replacing* relation between the two tasks. For brevity, we do not define the transaction base $\mathcal{B}$ here; these are prefixed with their general purpose (e.g., "ins", "del", "tmp") in our transactions.

We are now ready to define the integration semantics of our integration policies.

$$replaceTask(P, U) := ReplaceTasksPolicy(P) \otimes intPt(P, T) \otimes replacement(P, R)$$
$$\otimes\ state : replace(T, R) \tag{8}$$

Given a *ReplaceTasksPolicy* instance, its integration point (*T*) will be replaced by the given replacement *R*. The "*cond*" transaction (4) supports dynamic conditional replacements, e.g., as needed by HTN/Diabetes comorbidity. E.g., in case another conditional replacement is asserted, the prior replacement assertion will be *rolled back* (4); which, in turn, will result in *rolling back* the prior *replace* operation (7).

$$evtReplaceTask(P, T, U) := EvtReplacePolicy(P) \otimes eventTask(P, E) \otimes curState(E,'started')$$
$$\otimes intPt(P, T) \otimes curState(T, 'chosen') \otimes replacement(P, R) \otimes time(R,'during') \otimes task(R, U) \otimes$$
$$entryCond(R, C) \otimes Satisfied(C_1) \otimes exitCond(R, C_2) \otimes NotSatisfied(C_2) \otimes state:replace(T, U)$$

$$(9)$$

For an event-based replacement policy (see (2)), while (a) event task E is in the started state, and task-to-replace T is in a "chosen" branch; and (b) a "during" replacement has a satisfied entry-condition C_1 and non-satisfied exit-condition C_2; T will be replaced with task U. (An "after" replacement is represented analogously.) E. g., the replace operation will be *rolled back* once the entry-condition no longer holds, or the exit-condition is valid; or, in the inverse case, the replace operation will be *re-applied*.

$$redunTask(P, U) := RedunTasksPolicy(P) \otimes essen(P, E) \otimes curState(E,'chosen') \otimes$$
$$redun(P, R) \otimes curState(R, 'pendingOrChosen') \otimes activeTime(R, A) \otimes completeTime(E, C) \otimes$$
$$resultValidPeriod(E, P) \otimes tmp:withinPeriod(A, P, C) \otimes state:discard(R)$$

$$(10)$$

For a redundant task policy, when essential task E is part of a "chosen" branch, and redundant task R is part of a pending or chosen branch, the transaction will try to ensure that R's activation time lies within the result validity period of E's completion time, using the *tmp:withinPeriod* transaction (see Fig. 3). (In practice, this will likely result in delaying E, or speeding up R if possible.) In that case, task R will be discarded.

These dynamic transactions will ensure that KB updates and temporal constraints are properly applied or reverted. E.g., if E is not part of a chosen branch, or the result validity period can no longer be adhered, R should *not* be discarded, since R is no longer redundant in that case. Further, it suffices that R is part of a pending branch; if we wait until R is in the chosen state, it may be too late to delay E. To cope with conflicts between CIG integration policies themselves (e.g., when the cause of a discard operation is itself discarded), we introduced an intra-policy conflict resolution scheme [16].

3.3   Planning Component for Supporting CIG Temporal Constraints

The set of CIG temporal constraints, issued for resolving comorbid situations and encoding inherent CIG task/workflow properties, will be utilized to find an *optimal CIG task schedule*. This schedule will specify timings for CIG tasks that maximize the collective truth values of the weighted temporal constraints. Whenever new medical data becomes available, a new CIG task schedule will be calculated at execution-time.

In turn, a generated CIG task schedule will be used to associate truth values with issued temporal constraints—i.e., reflecting the degree to which it adheres to the temporal constraint—and will hence determine the feasibility of temporal constraints. E.g., this schedule will determine the truth value of the *tmp:withinPeriod* constraint (5); in case this truth value falls below a configured threshold, its post-condition will be considered false and the remainder of the transaction will be rolled back. We generate a task schedule by applying the Hill Climbing algorithm to find a solution state that best suits an objective function, which, in our case, is defined by the set of weighted temporal constraints. The importance, or weight, of temporal constraints will determine how likely a CIG task schedule will satisfy the constraint. This is where clinical pragmatics come into play: exceeding the max. delay for some tasks can still be deemed acceptable; or, some result-validity periods may be found more flexible than others. A clinician can change associated weights of temporal constraints while using the system.

Table 2. Evaluation of CIG integration framework performance.

Comorbid CIG integration scenario	Performance (ms)
RedunTasksPolicy (COPD-PE comorbidity)	8 ms
operation: discard redundant CP scan; delay essential CT-PE scan	7 ms; 8 ms
update: validity period exceeded; task is part of non-chosen branch	
ReplacePolicy (HTN-Diabetes comorbidity)	5 ms; 5 ms
operation: reduce thiazide dosage; or replace with loop diuretics	
EventCondReplacePolicy (VTE-RTI comorbidity)	5 ms
operation: during Erythromycin treatment, adjust Warfarin dosage	
update: during, INR returns to normal; after, INR is not normal	4 ms; 4 ms

4   Evaluation of the CIG Integration Framework

To measure the performance of CIG integration, we (*a*) modeled the comorbid CIG; (*b*) instantiated CIG integration policies; (*c*) executed the CIG integration framework for a set of comorbidity scenarios; (*d*) we emulated external execution-time updates (e.g., changes in decisional state, real-time delays). We executed each experiment 10 times on a PC equipped with 8 GB of RAM and an Intel® Core™ i7-3520 CPU.

In prior work, we evaluated the validity, utility and clarity of our approach [16]. Table 2 shows the detailed performance results.

Loading the comorbid CIG and instantiated *CIG-IntO* took an avg. ca. 15 ms. We consider these performance times to be acceptable for a consumer-grade PC.

5   Conclusions and Future Work

The importance of CIG, and, more recently, coping with multiple comorbid CIG, has been reflected by their popularity in health informatics.

We presented an innovative, execution-time approach to comorbid CIG integration. To meet the challenges of typical execution-time comorbidity scenarios, we (*a*) define dynamic, transaction-based CIG integration semantics, which easily allow rolling back prior integration policies when some conditions no longer hold, and, inversely (re-) applying integration policies when all conditions hold; and (*b*) calculate optimal CIG task schedules at execution-time, based on all relevant temporal constraints and currently available data, which determine the feasibility of CIG integration policies.

We target more comprehensive methods to identify integration points between comorbid CIG, in addition to our MLOD resources [14], such as done by Zamborlini et al. [17]. Although our system currently supports patient preferences as well, this was not elaborated nor evaluated in this paper. Future work also involves identifying other comorbidity considerations, i.e., not yet covered by our integration policies.

References

1. Brush, J.E., Radford, M.J., Krumholz, H.M.: Integrating clinical practice guidelines into the routine of everyday practice. Crit. Pathways Cardiol. **4**(3), 161–167 (2005)
2. Sutton, D.R., Fox, J.: The syntax and semantics of the PROforma guideline modeling language. J. Am. Med. Inf. Assoc. **10**, 433–443 (2003)
3. Riano, D.: The SDA model: a set theory approach. In: Twentieth IEEE International Symposium on Computer-Based Medical Systems (CBMS 2007), pp. 563–568. IEEE (2007)
4. National Institute for Health and Care Excellence (NICE): Chronic kidney disease in adults (2014)
5. Abidi, S.R.: A knowledge management framework to develop, model, align and operationalize clinical pathways to provide decision support for comorbid diseases (2010)
6. Riaño, D., Collado, A.: Model-based combination of treatments for the management of chronic comorbid patients. In: Peek, N., Marín Morales, R., Peleg, M. (eds.) AIME 2013. LNCS (LNAI), vol. 7885, pp. 11–16. Springer, Heidelberg (2013). https://doi.org/10.1007/978-3-642-38326-7_2
7. Wilk, S., Michalowski, M., Michalowski, W., Rosu, D., Carrier, M., Kezadri-Hamiaz, M.: Comprehensive mitigation framework for concurrent application of multiple clinical practice guidelines. J. Biomed. Inform. **66**, 52–71 (2017)
8. Anselma, L., Piovesan, L., Terenziani, P.: Temporal detection and analysis of guideline interactions. Artif. Intell. Med. **76**, 40–62 (2017)
9. KDIGO Chronic Kidney Disease Guidelines. https://kdigo.org/wp-content/uploads/2017/02/KDIGO_2012_CKD_GL.pdf
10. Witt, D.M., Clark, N.P., Kaatz, S., Schnurr, T., Ansell, J.E.: Guidance for the practical management of warfarin therapy in the treatment of venous thromboembolism. J. Thromb. Thrombolysis **41**(1), 187–205 (2016). https://doi.org/10.1007/s11239-015-1319-y
11. Hawkins, N.M., Virani, S., Ceconi, C.: Heart failure and chronic obstructive pulmonary disease: the challenges facing physicians and health services. Eur. Heart J. **34**, 2795–2807 (2013)
12. Baader, F., Calvanese, D., McGuinness, D.L., Nardi, D., Patel-Schneider, P.F.: The Description Logic Handbook: Theory, Implementation, and Applications (2003)
13. Bonner, A.J., Kifer, M.: An overview of transaction logic. Theor. Comput. Sci. **133**, 205–265 (1994)
14. Van Woensel, W.: Comorbid CIG integration repository. https://github.com/william-vw/comorbid-cig.git
15. Jafarpour, B., Abidi, S.S.R., Abidi, S.R.: Exploiting semantic web technologies to develop OWL-based clinical practice guideline execution engines. IEEE J. Biomed. Health Inform. **20**, 388–398 (2016)
16. Jafarpour, B., Abidi, S.R., Van Woensel, W., Abidi, S.S.R.: Execution-time integration of clinical practice guidelines to provide decision support for comorbid conditions. Artif. Intell. Med. **94**, 117–137 (2019)
17. Zamborlini, V., Hoekstra, R., Silveira, M.D., Pruski, C., ten Teije, A., van Harmelen, F.: Inferring recommendation interactions in clinical guidelines. Seman. Web **7**, 421–446 (2016)

Information Retrieval

Searching for Pneumothorax in Half a Million Chest X-Ray Images

Antonio Sze-To[1(✉)] and Hamid Tizhoosh[1,2]

[1] KIMIA Lab, University of Waterloo, Waterloo, ON N2L 3G1, Canada
{hy2szeto,tizhoosh}@uwaterloo.ca
[2] Vector Institute, Toronto, ON M5G 1M1, Canada

Abstract. Pneumothorax, a collapsed or dropped lung, is a fatal condition typically detected on a chest X-ray by an experienced radiologist. Due to shortage of such experts, automated detection systems based on deep neural networks have been developed. Nevertheless, applying such systems in practice remains a challenge. These systems, mostly compute a single probability as output, may not be enough for diagnosis. On the contrary, content-based medical image retrieval (CBIR) systems, such as image search, can assist clinicians for diagnostic purposes by enabling them to compare the case they are examining with previous (already diagnosed) cases. However, there is a lack of study on such attempt. In this study, we explored the use of image search to classify pneumothorax among chest X-ray images. All chest X-ray images were first tagged with deep pretrained features, which were obtained from existing deep learning models. Given a query chest X-ray image, the majority voting of the top K retrieved images was then used as a classifier, in which similar cases in the archive of past cases are provided besides the probability output. In our experiments, 551,383 chest X-ray images were obtained from three large recently released public datasets. Using 10-fold cross-validation, it is shown that image search on deep pretrained features achieved promising results compared to those obtained by traditional classifiers trained on the same features. To the best of knowledge, it is the first study to demonstrate that deep pretrained features can be used for CBIR of pneumothorax in half a million chest X-ray images.

Keywords: Deep learning · Chest X-ray images · Content-Based Image Retrieval (CBIR) · Image search · Pneumothorax

1 Introduction

Pneumothorax is a life-threatening emergency condition that can lead to death of the patient [5]. It is an urgent situation [16] where air enters the pleural space, i.e. the space between the lungs and the chest wall [5]. An illustration of pneumothorax, and a sample chest X-ray image are provided in Fig. 1.

Supported by Vector Institute Pathfinder Project.

M. Michalowski and R. Moskovitch (Eds.): AIME 2020, LNAI 12299, pp. 453–462, 2020.
https://doi.org/10.1007/978-3-030-59137-3_40

Pneumothorax is typically detected on chest X-ray images by qualified radiologists [16]. As it is time-consuming and expensive to train qualified radiologists [8], the supply of qualified radiologist is rather limited. Since an incorrect or delayed diagnosis can cause harm to patients [8], it is vital to develop computer-aided approaches to assist radiologists.

Due to its recent success, an increasing number of studies have adopted deep learning or deep neural networks (DNNs) to detect pneumothorax or other thoracic diseases in chest X-ray images [6,11,15]. Since the availability of ChestXray8 (or its later version ChestXray14) [15], one of the largest publicly available chest X-ray datasets, it is possible to train DNNs as classifiers to output a probability for certain thoracic diseases. Its performance has been reported to achieve or exceed the level of qualified radiologists on certain diseases such as pneumonia [11]. However, a single probability output may not be enough for convincing diagnosis.

As an alternative, image search not only provides a probabilistic output but also similar cases from the past cases. Retrieving similar images given a query image for medical applications is an application of Content-based Image Retrieval (CBIR) [14] for medical images. It is also known as Content-Based Medical Image Retrieval (CBMIR) [1]. CBMIR can help doctors in retrieving similar images and case histories for understanding the specific patient's disease or injury status and can also help to exploit the information in corresponding medical reports [1]. It may also help radiologists in preparing the report for particular diagnosis [1]. While deep learning methods have been applied to image retrieval tasks in recent studies [14], there is less attention on exploring deep learning methods for CBMIR tasks [10].

In this study, we explored the use of image search, based on features obtained from DNNs i.e. deep features, to detect pneumothorax among more than 550,000 chest X-ray images obtained from three large recently released labelled datasets, namely ChestX-ray14 [15], CheXpert [6] and MIMIC-CXR [2,7]. In our experiments, all chest X-ray images were first tagged with DenseNet121 deep features [4]. Given a query chest X-ray image, the majority voting of the top K retrieved X-ray images was then used as a classifier.

2   Methodology

The proposed method of using image search as a classifier comprises of three phases (Fig. 2): 1) Tagging images with features (all images in the database are tagged with deep pretrained features), 2) Receiving a query image (tagging with features and calculating its distance with all other features in the database to find the most similar images), and 3) Classification (majority voting among the labels of retrieved images).

Phase 1: Tagging Images with Deep Pretrained Features – In this phase, all chest X-ray images in the database are tagged with deep pretrained features. To represent a chest X-ray image as a feature vector with a fixed dimension, the output of the fully-connected layer before the classification layer of a deep

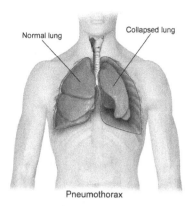
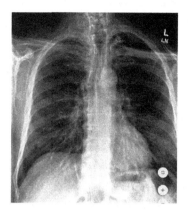

Fig. 1. Left: A graphical illustration of pneumothorax [12]. Right: A X-ray image of Pneumothorax in left lung visible thorough slight contrast different as a result of the lung collapse, obtained from CheXpert [6].

convolutional neuron network (DCNN) is used. These values are denoted as deep features [18], or technically deep pretrained features if the DCNN is pretrained with other datasets. In other words, the DCNN is considered as a feature extractor to convert a chest X-ray image into an n-dimensional feature vector. In this study, following [11], DenseNet121 [4], a DCNN with 121 layers pretrained on ImageNet [13] dataset, is adopted among existing models for converting a chest X-ray image into a feature vector with 1024 dimensions.

Phase 2: Image Search – In this phase, the query chest X-ray image is first tagged with deep pretrained features. Then, the distance between the deep pretrained features of the query chest X-ray image and those of the chest X-ray images in the database are computed. The chest X-ray images having the shortest distance with those of the query chest X-ray image are subsequently retrieved. In this study, Euclidean distance, which is the most widely used distance metric in k-NN [3], is used for computing the distance between the deep features of two given chest X-ray images.

Phase 3: Classification – In this phase, the majority voting of the labels of retrieved chest X-ray images is used as a classification decision. For example, given a query chest X-ray image, the top K most similar chest X-ray images are retrieved. If $\lceil K/2 \rceil$ chest X-ray images are labelled with pneumothorax, the query image is classified as pneumothorax with a class likelihood of $\lceil K/2 \rceil / K$.

3 Experiments and Results

In this section, we describe the experiments that investigate the use of image search to classify pneumothorax among half a million chest X-ray images, with comparison to traditional classifiers such as Random Forest (RF) [9]. We first

Phase 1: Tagging Images with Deep Pretrained Features

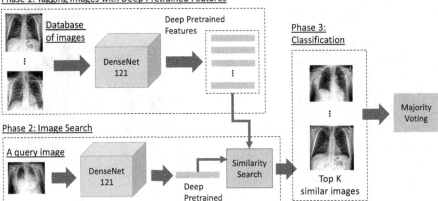

Fig. 2. An overview of using image search as a classifier to detect pneumothorax in chest X-ray images. The method is composed of three phases. Phase 1: tagging images with deep pretrained features. All images in the database are tagged with deep pretrained features. Following [11], DenseNet121 [4], pretrained on ImageNet [13] dataset, is adopted among existing models for converting a chest X-ray image into a feature vector with 1024 dimensions. Phase 2: image search. The query image is first tagged with deep pretrained features. Then, the distance between the query features and all other features in the database are computed to find the most similar images. Phase 3: classification. The majority voting of the retrieved images is used as a classifier.

describe the datasets collected and preprocessing procedure, then the experiments, followed by the analysis.

3.1 Data Collection

Three large public datasets of chest X-ray images were collected. The first is MIMIC-CXR [2,7], a large public dataset of 371,920 chest X-rays associated with 227,943 imaging studies. Only 248,236 frontal chest X-ray images in the training set were used in this study. The second dataset is CheXpert [6], a public dataset for chest radiograph interpretation consisting of 224,316 chest radiographs of 65,240 patients. Only 191,027 frontal chest X-ray images in the training set were used in this study. The third dataset is ChestX-ray14 [15], a public dataset of 112,120 frontal-view X-ray images of 30,805 unique patients. All chest X-ray images in this dataset were used in this study. In total, 551,383 frontal chest X-ray images were used in this study. The labels refer to the entire image; the collapsed lungs are not highlighted in any way.

3.2 Dataset Preparation and Preprocessing

Dataset 1 is a dataset composing of 34,605 pneumothorax chest X-ray images and 160,003 normal chest X-ray images. The pneumothorax images were

Table 1. A summary of chest X-ray images in the Dataset 1 through combination of three public datasets.

	MIMIC-CXR [2,7]	CheXpert [6]	ChestX-ray14 [15]	Total
+ve: Pneumothorax	11,610	17,693	5,302	34,605
−ve: Normal	82,668	16,974	60,361	160,003
Total	94,278	34,667	65,663	194,608

Table 2. A summary of chest X-ray images in the Dataset 2 through combination of three public datasets.

	MIMIC-CXR [2,7]	CheXpert [6]	ChestX-ray14 [15]	Total
+ve: Pneumothorax	11,610	17,693	5,302	34,605
−ve: Non-pneumothorax	236,626	173,334	106,818	516,778
Total	248,236	191,027	112,120	551,383

obtained from the collected frontal chest x-ray images with the label "Pneu-mothorax" = 1. They were considered as positive (+ve) class. The normal images were obtained from the collected frontal chest x-ray images with label the "No Finiding" = 1. These chest X-ray images were considered as negative (−ve) class. A summary is provided in Table 1. **Dataset 2** is a dataset composing of 34,605 pneumothorax chest x-ray images and 516,778 non-pneumothorax chest x-ray images. The pneumothorax image were obtained from the collected frontal chest X-ray images with the label "Pneumothorax" = 1. They were considered as posi-tive (+ve) class. The non-pneumothorax images were obtained from the collected frontal chest X-ray images without the label "Pneumothorax" = 1, meaning that they contain cases such as normal, pneumonia, edema, cardiomegaly, pneumonia and more. They were considered as negative (−ve) class. A summary is provided in Table 2. It should be noted that ChestX-ray14 [15] dataset was raised with a concern that its chest X-ray images with chest tubes were frequently labelled with Pneumothorax [17,19]. In our experiments, through combining ChestX-ray14 with CheXpert [6], and MIMIC-CXR [2,7] datasets, this concern was mit-igated to address the bias.

3.3   Implementation and Parameter Setting

In this study, DenseNet121 [4] was implemented via the deep learning library Keras (http://keras.io/) v2.2.4 with Tensorflow backend. Its model weights were obtained through the default setting of Keras. All images were resize to 224×224 before inputting to the network. Also, Random Forest [9] was implemented using the latest version (v0.20.3) of the machine learning library scikit-learn with the parameter 'class_weight' set to 'balanced'. All other parameters were set default unless further specified. All experiments were run on a computer with 64.0 GB

DDR4 RAM, an Intel Core i9-7900X @3.30 GHz CPU (10 Cores) and one GTX 1080 graphics card. These settings were used in all experiments unless further specified.

3.4   Experiment 1

In this experiment, we studied the performance of image search to classify pneumothorax among pneumothorax and normal chest X-ray images. All chest X-ray images were first tagged with deep pretrained features. A standard 10-fold cross-validation was then adopted. All chest X-ray images (with deep pretrained features tagged) were divided into 10 sections. In each fold, one section of chest X-ray images was used as validation set, while the remaining chest X-ray images were used as training set. The process was repeated 10 times, such that in each fold a different section of chest X-ray images was used as the validation set.

For image search, given a chest X-ray image (from the validation set), it was conducted to search in the training set and used the majority voting of the top K retrieved chest X-ray images to classify. One experiment was conducted with $K = 11$ and another was with $K = 51$. For Random forest (RF), with number of trees (t) setting as 11 and 51 respectively, it was trained on the deep pretrained features of the training set and evaluated on those of the validation set in each fold.

For performance evaluation, following [11], the area under Receiver operating characteristics (ROC) curve was computed for each fold. A comparison of average area under ROC curve on the results on Dataset 1 under 10-fold cross-validation is summarized in Table 3. For completeness and transparency, the result obtained in each fold is demonstrated. For statistical analysis, two-sample t-test (two-tailed, unequal variance) was conducted on the ROC obtained by RF $(t = 11)$ and image search $(K = 11)$. The p-value was $\mathbf{1.68E{-}13 < 0.05}$, indicating that image search $(K = 11)$ obtained a higher ROC, with statistically significance, than that obtained by RF $(t = 11)$. Similarly, the same test was conducted on the ROC obtained by RF $(t = 51)$ and image search $(K = 51)$. The p-value was $\mathbf{5.88E{-}3 < 0.05}$, indicating that image search $(K = 51)$ obtained a higher ROC, with statistically significance, than that obtained by RF $(t = 51)$. A ROC curve was provided on fold-1 was provided in Fig. 3.

3.5   Experiment 2

In this experiment, we study the performance of image search to classify if a chest X-ray image has pneumothorax, without any prior knowledge. Experimental procedure that was similar to the previous experiment was conducted. A comparison of area under ROC curve on the results on Dataset 2 is summarized in Table 4.

For statistical analysis, two-sample t-test (two-tailed, unequal variance) was conducted on the ROC obtained by RF $(t = 11)$ and image search $(K = 11)$. The p-value was $\mathbf{3.78E{-}16 < 0.05}$, indicating that image search $(K = 11)$ obtained a higher ROC, with statistically significance, than that obtained by RF $(t = 11)$.

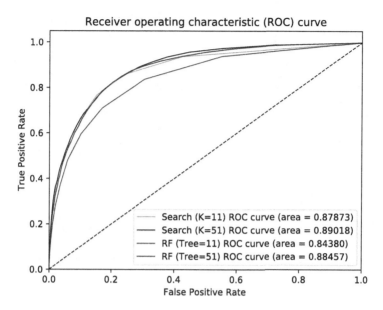

Fig. 3. A comparison of the prediction performance in terms of the area under the receiver operating characteristic (ROC) curve on Dataset 1 for Fold-1.

Table 3. A comparison of area under ROC curve on Dataset 1 under 10-fold cross-validation between Image Search and Random Forest (RF) trained with deep pretrained features. The p-value of two-sample t-test (two-tailed, unequal variance) between RF ($t = 11$) and Image Search ($K = 11$) is **1.68E−13** < 0.05, between RF ($t = 51$) and Image Search ($K = 11$) is **5.88E−3** < 0.05.

Fold	RF ($t = 11$)	Image search ($K = 11$)	RF ($t = 51$)	Image search ($K = 51$)
1	0.84380	**0.87873**	0.88457	**0.89018**
2	0.84906	**0.87918**	0.88690	**0.89245**
3	0.84611	**0.88000**	0.88744	**0.89214**
4	0.84652	**0.88118**	0.88788	**0.89222**
5	0.84658	**0.88156**	0.88760	**0.89399**
6	0.84911	**0.87950**	0.88709	**0.89192**
7	0.84627	**0.87220**	0.88464	**0.88807**
8	0.84210	**0.87121**	0.87630	**0.88449**
9	0.85349	**0.87956**	0.89008	**0.89351**
10	0.84815	**0.87461**	0.88572	**0.88719**
Mean	0.84703	**0.87777**	0.88582	**0.89062**

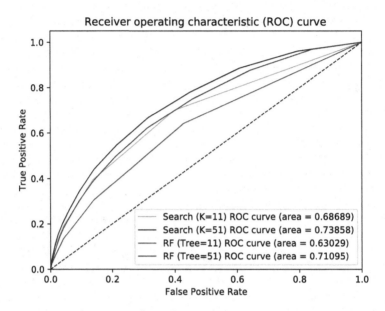

Fig. 4. A comparison of the prediction performance in terms of the area under the receiver operating characteristic (ROC) curve on Dataset 2 for Fold-1.

Table 4. A comparison of area under ROC curve on Dataset 2 under 10-fold cross-validation between image search and random forest (RF) trained with deep pretrained features. The p-value of two-sample t-test (two-tailed, unequal variance) between RF $(t = 11)$ and Image Search $(K = 11)$ is **3.78E−16 < 0.05**, between RF $(t = 51)$ and Image Search $(K = 51)$ is **5.85E−12 < 0.05**.

Fold	RF $(t = 11)$	Image search $(K = 11)$	RF $(t = 51)$	Image search $(K = 51)$
1	0.63029	**0.68689**	0.71095	**0.73858**
2	0.63106	**0.69318**	0.71645	**0.74032**
3	0.62390	**0.69205**	0.70187	**0.73909**
4	0.63325	**0.69434**	0.71316	**0.74593**
5	0.62453	**0.68891**	0.70969	**0.74218**
6	0.62249	**0.68670**	0.70402	**0.73942**
7	0.63237	**0.69885**	0.71536	**0.75248**
8	0.63431	**0.68261**	0.71104	**0.73979**
9	0.62619	**0.69531**	0.70674	**0.74389**
10	0.63761	**0.68842**	0.71350	**0.74463**
Mean	0.62960	**0.69073**	0.71028	**0.74263**

Similarly, the same test was conducted on the ROC obtained by RF $(t = 51)$ and image search $(K = 51)$. The p-value was $\mathbf{5.85E{-}12} < \mathbf{0.05}$, indicating that image search $(K = 51)$ obtained a higher ROC, with statistically significance, than that obtained by RF $(t = 51)$. The ROC curve on Fold-1 is provided in Fig. 4.

4   Conclusions

In this study, we explored the use of image search, based on deep pretrained features, in classifying pneumothorax among more than half a million chest X-rays. The experiments showed that content-based medical image retrieval system, such as image search, is a potentially viable cost-effective solution. To the best of knowledge, it is the first study to demonstrate that deep pretrained features can be used for CBIR of pneumothorax in half a million chest X-ray images. The setting of image search can be deployed both as a semi-automated and automated solution in the practice of diagnostic radiology. Compared with traditional classification, image search results might be clinically more practical as they are supported by the reports and history of evidently diagnosed cases, representing a virtual "second opinion" for diagnostic purposes, a factor that may provide more confidence for a reliable diagnosis.

References

1. Das, P., Neelima, A.: An overview of approaches for content-based medical image retrieval. Int. J. Multimed. Inf. Retrieval **6**(4), 271–280 (2017). https://doi.org/10.1007/s13735-017-0135-x
2. Goldberger, A.L., et al.: PhysioBank, PhysioToolkit, and PhysioNet: components of a new research resource for complex physiologic signals. Circulation **101**(23), e215–e220 (2000)
3. Hu, L.-Y., Huang, M.-W., Ke, S.-W., Tsai, C.-F.: The distance function effect on k-nearest neighbor classification for medical datasets. SpringerPlus **5**(1) (2016). Article number: 1304. https://doi.org/10.1186/s40064-016-2941-7
4. Huang, G., Liu, Z., Van Der Maaten, L., Weinberger, K.Q.: Densely connected convolutional networks. In: Proceedings of the IEEE Conference on Computer Vision and Pattern Recognition, pp. 4700–4708 (2017)
5. Imran, J.B., Eastman, A.L.: Pneumothorax. JAMA **318**(10), 974–974 (2017)
6. Irvin, J., et al.: CheXpert: a large chest radiograph dataset with uncertainty labels and expert comparison. arXiv preprint arXiv:1901.07031 (2019)
7. Johnson, A.E., et al.: MIMIC-CXR: a large publicly available database of labeled chest radiographs. arXiv preprint arXiv:1901.07042 (2019)
8. Ker, J., Wang, L., Rao, J., Lim, T.: Deep learning applications in medical image analysis. IEEE Access **6**, 9375–9389 (2017)
9. Liaw, A., Wiener, M., et al.: Classification and regression by randomForest. R news **2**(3), 18–22 (2002)
10. Qayyum, A., Anwar, S.M., Awais, M., Majid, M.: Medical image retrieval using deep convolutional neural network. Neurocomputing **266**, 8–20 (2017)

11. Rajpurkar, P., et al.: CheXNet: radiologist-level pneumonia detection on chest x-rays with deep learning. arXiv preprint arXiv:1711.05225 (2017)
12. Richfield, D.: Medical gallery of Blausen medical 2014. Wiki J. Med. **1**(2), 9–11 (2014)
13. Russakovsky, O., et al.: ImageNet large scale visual recognition challenge. Int. J. Comput. Vis. **115**(3), 211–252 (2015). https://doi.org/10.1007/s11263-015-0816-y
14. Tzelepi, M., Tefas, A.: Deep convolutional learning for content based image retrieval. Neurocomputing **275**, 2467–2478 (2018)
15. Wang, X., Peng, Y., Lu, L., Lu, Z., Bagheri, M., Summers, R.M.: ChestX-Ray8: hospital-scale chest x-ray database and benchmarks on weakly-supervised classification and localization of common thorax diseases. In: Proceedings of the IEEE Conference on Computer Vision and Pattern Recognition, pp. 2097–2106 (2017)
16. Zarogoulidis, P., et al.: Pneumothorax: from definition to diagnosis and treatment. J. Thorac. Dis. **6**(Suppl 4), S372 (2014)
17. Zech, J.R., Badgeley, M.A., Liu, M., Costa, A.B., Titano, J.J., Oermann, E.K.: Variable generalization performance of a deep learning model to detect pneumonia in chest radiographs: a cross-sectional study. PLoS Med. **15**(11), 1–17 (2018)
18. Zhang, R., Isola, P., Efros, A.A., Shechtman, E., Wang, O.: The unreasonable effectiveness of deep features as a perceptual metric. In: Proceedings of the IEEE Conference on Computer Vision and Pattern Recognition, pp. 586–595 (2018)
19. Zhang, Y., Wu, H., Liu, H., Tong, L., Wang, M.D.: Mitigating the effect of dataset bias on training deep models for chest x-rays. arXiv preprint arXiv:1910.06745 (2019)

A New Local Radon Descriptor for Content-Based Image Search

Morteza Babaie$^{(\boxtimes)}$ iD, Hany Kashani, Meghana D. Kumar,
and H. R. Tizhoosh iD

Kimia Lab, University of Waterloo, Waterloo, ON, Canada
{mbabaie,kashani,meghana,tizhoosh}@uwaterloo.ca

Abstract. Content-based image retrieval (CBIR) is an essential part of computer vision research, especially in medical expert systems. Having a discriminative image descriptor with the least number of parameters for tuning is desirable in CBIR systems. In this paper, we introduce a new simple descriptor based on the histogram of local Radon projections. We also propose a very fast convolution-based local Radon estimator to overcome the slow process of Radon projections. We performed our experiments using pathology images (KimiaPath24) and lung CT patches and test our proposed solution for medical image processing. We achieved superior results compared with other histogram-based descriptors such as LBP and HoG as well as some pre-trained CNNs.

Keywords: Image retrieval · Local radon · Medical imaging

1 Introduction

Over the past decades, there has been a dramatic increase in capturing and storing data in the form of digital images. In medical imaging, for example, the volume of stored data is expected to exceed more than $2,314$ Exabytes (10^9 GB) by 2020 which is an exponential growth from 153 Exabytes in 2013 [16]. Demands to extract information from these massive archives is growing day to day. Content-based image retrieval (CBIR) system undoubtedly is considered a necessary way to extract this information. CBIR has long been a subject of great interest in a wide range of fields, from searching pictures of celebrities on the Internet to helping radiologists and pathologists to make a more accurate diagnosis by providing them access to the images of similar cases. In general, CBIR is referred to searching a dataset to retrieve similar images to a query image [24]. In the medical community, content-based medical image retrieval (CBMIR) refers to the same tasks in the medical image domain. However, in CBMIR the semantic gap between algorithms and the experts in capturing the similarity is more crucial [2]. In addition, the size and quantity of medical images are quite overwhelming. A desirable searching method is expected to return similar images for any query image in a reasonable time. To this end, we used a practical and well-known technique in the medical domain, called Radon transform, to describe the

© Springer Nature Switzerland AG 2020
M. Michalowski and R. Moskovitch (Eds.): AIME 2020, LNAI 12299, pp. 463–472, 2020.
https://doi.org/10.1007/978-3-030-59137-3_41

visual features in medical images. We utilized Local Radon Projections (LRP) in a very fast and accurate way. Local Radon Descriptor (LRD) has been applied to medical images with large window size and a relatively high number of parameters to tune [4]. However, implementations of Radon transform on the large patches are computationally slow because it interpolates the pixel values from multiple pixels. In this work, we estimated the Radon transform in small patches by convolving a series of designed filters in the whole image and then reading small patches to achieve a fast method to apply in CBMIR. Also, the presented version of LRP is quite simple and no parameter to tune.

The rest of this paper is organized as follows: The next section explores related works in CBMIR and descriptors fields as well as the literature related to the Radon Transform and filter-based image processing. The methodology section describes details of the LRP and its fast implementation. And finally, the results of our experiment are presented at the end of the paper followed by conclusions.

2   Related Works

Artificial Intelligence (AI) can play an important role in the field of medical image analysis. In recent years, there has been an increasing interest in applying AI in the medical domain to help clinicians with the diagnosis and treatment of the diseases [6]. Detection, classification, segmentation, and retrieval are the mainstream fields that are currently subject to research in various domains of medical imaging [10]. All the above-mentioned approaches are mainly designed to work on a specific organ and modality (e.g., analysis methods on brain images produced by MRI [8] or lung segmentation in X-ray images). However, these methods generally need carefully labeled medical image data which is not a feasible task, due to the expense related to the specialist's involvement and the huge size of datasets [13]. CBIR, unlike all other above-mentioned methods, has the ability to work efficiently by a small amount of labeled data and even raw data in large datasets [27]. However, labeled data can help to overcome the semantic gap (perhaps the greatest challenge in CBIR) between algorithms and expert regarding perceiving the image similarity [12].

Historically, early research in the field of image search focused on text-based systems that return similar images within the same anatomical region, with the same orientation and using the same modality, based on textual annotations of each image. Obviously, annotating all parts of the image and complex visual contents is not feasible. As a result, the retrieval performance is limited by the quality and extent of the annotations [20]. Over the past two decades, major advances in AI, machine vision, and computer hardware have made it possible to apply CBIR in the medical domain with the possibility of search among a huge number of images. Nevertheless, CBMIR is different based on the sensitive semantic gap, the large amount of data, and the lack of labeled datasets [31].

One of the main steps to overcome the challenges can be addressed by appropriate feature extraction. Features can be divided into two main categories: low-level and high-level [1]. In the low-level feature extraction category, patch-based methods can be divided into the whole image descriptors (Local Mesh Patterns and Local Binary Patterns, LBP) [21] and keypoint-based descriptors such as SIFT, SURF and ORB [15]. However, in the medical domain, feature detection can easily fail to provide keypoints with acceptable quantity, quality, and distribution [25]. To overcome this problem, dense sampling may be applied. For example dense-SIFT and dense-SURF (applying path descriptors in the whole image) [19] have been implemented in the medical field frequently.

High-level features, on the other hand, can provide more accurate results due to utilizing the labeled data and reducing the semantic gap. In many works, learning methods have been applied to the low-level features to map or modify them in CBMIR [23]. Also using the information in the dense layers of deep network architectures has recently become quite popular in CBMIR. The idea is that if a network can achieve good results in classification, then it should be able to extract all necessary features like corners, edges and any other necessary element in every single layer to provide useful information for the last layer [11]. Surprisingly, dense layers of pre-trained networks such as VGG and AlexNet which have been trained using very large datasets like ImageNet, can also provide discriminating features and perform well in CBMIR. They have the ability to extract useful information from any input image [17]. As a matter of fact, the quality of features extracted from a pre-trained network might increase by retraining the network, using new labeled data in the applied domain [14].

3   Methodology

This section describes Radon transform as well as the fast and effective method we propose to calculate local Radon projections based on convolution kernels. We also explain the steps and methods to assemble the histogram based on these local Radon projections.

3.1   Radon Transform

In general, Radon transform is described by integrating the projection values of a scene, an image or body parts from various directions [22]. The inverse Radon transform along with filtered back-projection has been used to reconstruct the image of internal body parts from projections which were captured at different directions [9]. Applying the inverse Radon on values received from an X-ray machine or other modalities is a well-established field in medical imaging. Given the spatial intensities $f(x, y)$, the Radon transform can be formulated as

$$\mathbf{R}(\rho, \theta) = \int\limits_{-\infty}^{+\infty} \int\limits_{-\infty}^{+\infty} f(x, y) \delta(\rho - x cos\theta - y sin\theta) dx dy, \tag{1}$$

where $\delta(\cdot)$ is the Dirac delta function. In Fig. 1, the Radon transform is illustrated for a small window.

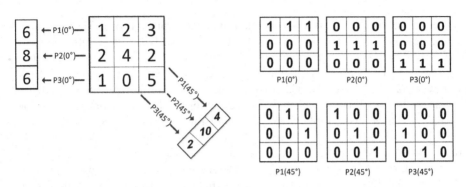

Fig. 1. Left: Two estimated Radon projections of a 3 × 3 sample window at 0° and 45°. Right (Up/Down): Filters to produce 0°/45° Radon for each 3 × 3 local windows.

3.2   Convolutional Local Radon Projections

Recently, there has been renewed interest in using Radon transform as an image descriptor [5,7,28]. Most methods apply Radon transform globally on the entire image or on a relatively large patches [4]. However, calculating the Radon transform in quite small local windows would be of interest. We used the local Radon in 3 × 3 windows, which resulted in a fast and innovative convolution-based method that calculated Radon in small neighbors. For each direction, we needed a summation of pixels along that direction. For instance, if we wanted to calculate the zero-degree Radon for each pixel of a given image in 3 × 3 windows, we added all 3 rows in 3 × 3 windows (specified by dashed arrows in Fig. 1).

As depicted in Fig. 1, each kernel could easily add one row. Convolving these 3 kernels, resulting is 3 images with same size gives us 3 digits for each pixel which is equivalent to zero-degree Radon. We designed other kernels for 45°, 90° and 135° as well. Because there is just one pixel in the first and last element of the Radon projections in 45° and 135°, we ignored them to get the same length vector for all 4 directions (the remaining elements are marked by arrows in Fig. 1 for 45°).

3.3   LRP – Local Radon Patterns

Figure 2 depicts our proposed method to create the LRP descriptor. Each given image was convolved by three sets of 3 × 3 kernels to obtain the local Radon projections in each direction. Resulting in 12 kernels were applied for 4 equidistant directions. Each kernel outcome was an image with the same size as the input image. We concatenated these projections together as explained in Fig. 2.

In the next step, we binarized these 12 numbers to count their equivalent integers in a histogram. We tried two different methods to create a binary vector: Median thresholding and the Min-Max method [30]. The median method is binarizing 12 numbers based on thresholding the vector by its median value.

Let's suppose $X = [x_1, x_2, \ldots, x_n]$, then the median method can be formulated as

$$B(i) = \begin{cases} 1, & \text{if } x(i) \geq \text{median}(X) \\ 0, & \text{otherwise} \end{cases}$$

On the other hand, the Min-Max method tries to capture the signal's-concatenated Radon projections- shape, by assigning one if the next adjunct element is greater than the current element and zero if otherwise:

$$B(i) = \begin{cases} 1, & \text{if } x(i) \geq x(i+1) \\ 0, & \text{otherwise} \end{cases}$$

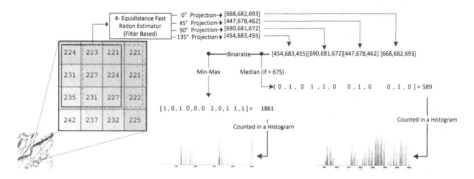

Fig. 2. Visualization of the proposed method: local Radon projections in all directions are calculated by corresponding kernels.

4 Experiments and Results

In order to evaluate our descriptor performance, we selected two publicly available medical datasets. These datasets have been used in related publications.

4.1 KimiaPath24 Dataset

The KimiaPath24 dataset consists patches of 24 whole slides images (WSIs) in pathology domain representing diverse body parts with different texture and stains. The images were captured by TissueScope LE 1.0[1] bright field using a 0.75 NA lens. For each image, one can determine the resolution by checking the description tag in the header of the file. For instance, if the resolution is 0.5 μm, then the magnification is 20x. The dataset offers 27,055 training patches and

---

[1] http://www.hurondigitalpathology.com/tissuescope-le-3/.

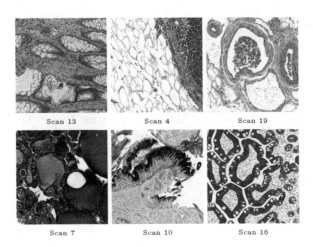

Scan 13   Scan 4   Scan 19

Scan 7   Scan 10   Scan 16

Fig. 3. Sample images from KimiaPath24 dataset.

1,325 (manually selected) test patches of size 1000×1000 ($0.5\,\mu\text{m} \times 0.5\,\mu\text{m}$) [3]. The test patches have been removed (whitened) in WSIs such that they cannot be mistakenly used for training. The color (*staining*) is neglected; all patches are saved as grayscale images. The dataset has a total of $n_{\text{tot}} = 1,325$ patches P_s^j that belong to 24 sets $\Gamma_s = \{P_s^i | s \in S, i = 1, 2 \ldots, n_{\Gamma_s}\}$ with $s = 0, 1, 2, \ldots, 23$. Looking at the set of retrieved images R for any experiment, the **patch-to-scan accuracy** η_p can be given as

$$\eta_\text{p} = \frac{1}{n_{tot}} \sum_{s \in S} |R \cap \Gamma_s|. \tag{2}$$

As well, we calculate the **whole-scan accuracy** η_W as

$$\eta_\text{W} = \frac{1}{24} \sum_{s \in S} \frac{|R \cap \Gamma_s|}{n_{\Gamma_s}}. \tag{3}$$

Hence, the total accuracy η_{total} (patch-to-scan and whole-scan accuracy)can be defined as $\eta_{\text{total}} = \eta_\text{p} \times \eta_\text{W}$. The dataset and the code for accuracy calculations can be downloaded from the web[2]. Figure 3 shows sample patches from KimiaPath24 dataset. We resized images to 250×250 for all methods (for deep network slightly smaller).

4.2 CT Emphysema Dataset

Computed tomography emphysema database was introduced by Sorensen et al. [26] to classify lung CT images. A part of this database includes 168 square patches that have been manually annotated in a subset of the slices with an

---

[2] http://kimia.uwaterloo.ca/.

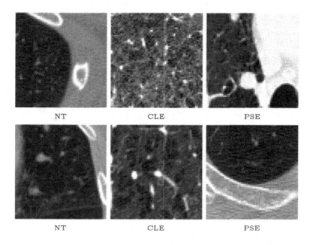

Fig. 4. Sample images from CT emphysema data-set.

in-plane resolution of $0.78 \times 0.78\,\mathrm{mm}^2$, slice thickness of $1.25\,\mathrm{mm}$, tube voltage equal $140\,\mathrm{kV}$, and a tube current of 200 mAs. The 512×512 pixel slices depict the upper, middle, and lower part of the lung of each patient. The 168 patches, of size 61×61 pixels, are from three different classes, NT (normal tissue, 59 observations), CLE (centrilobular emphysema, 50 observations), and PSE (paraseptal emphysema, 59 observations). The NT patches were annotated in *never smokers*, and the CLE and PSE ROIs were annotated in *healthy smokers* and *smokers with COPD* (chronic obstructive pulmonary disease) in areas of the leading pattern. Figure 4 shows examples for NT, CLE and PSE classes from the CT Emphysema dataset. Given the set of correctly classified images $\mathcal{C}$, the accuracy A_{CT} can be calculated as

$$A_{\mathrm{CT}} = \frac{|\mathcal{C}|}{168}. \tag{4}$$

4.3   Results

Table 1 represents our experimental results. We performed image search and selected the label of the best match image (top-1 accuracy). For the CT Emphysema dataset the leave-one-out strategy was chosen. While for the KimaPath24 test images are searched on the training set. The LRP method does not have any parameter to tune except the method of binarization. On the other hand, for LBP and HoG presented results, we did exhaustive search to tune their parameters. As Table 1 suggests, both binarization methods provided superior results in comparison to other methods. Moreover, LRP benefits from fast kernel-based calculation time. LRP takes $0.01\,\mathrm{s}$ for a 1000×1000 while the optimized ELP (encoded local projections) takes $90\,\mathrm{s}$ for the same image size mainly due to the Radon computation time.

Table 1. The best results for all datasets and descriptors when city block (L_1), Euclidean (L_2), Chi-squared (χ^2) and cosine (cos) distances are used for direct similarity measurements. For LBP and HOG, the best results were achieved for each dataset via exhaustive parameter search.

CT emphysema dataset		
Method	A_{CT}	$|\mathbf{h}|$
LRP (Our method)$_{MinMax}$	82.14%	2048
LRP (Our method)$_{median}$	81.32%	4096
ELP [29]$_{median}$	80.95%	256
LBP$^{uri}_{(12,3),\chi^2}$	80.36%	18
VGG-Deep$_{L_2}$[18]	69.64%	4096
HOG$_{L_2}$	65.47%	1215
Kimia Path24 dataset		
Method	$\{\eta_p, \eta_W\}$	$|\mathbf{h}|$
LRP (Our method)$_{median}$	{75.38%, 72.02%}	4096
LRP (Our method)$_{MinMax}$	{74.12%, 73.43%}	2048
ELP [29]$_{median}$	{71.16%, 68.05%}	1024
VGG-Deep$_{cos}$[18]	{70.11%, 68.13%}	4096
LBP$^u_{(24,2),L_1}$	{65.55%, 62.56%}	555
HOG$_{L_1}$	{17.58%, 16.76%}	648

5 Conclusions

In this work, we introduced a fast and simple histogram-based descriptor based on the local Radon transform and applied it in two medical datasets with different modalities. In the im developed local Radon descriptor) in a much shorter time.

References

1. A new method of content based medical image retrieval and its applications to CT imaging sign retrieval. J. Biomed. Inform. **66**, 148–158 (2017)
2. Alzu'bi, A., Amira, A., Ramzan, N.: Semantic content-based image retrieval: a comprehensive study. J. Vis. Commun. Image Represent. **32**, 20–54 (2015)
3. Babaie, M., et al.: Classification and retrieval of digital pathology scans: a new dataset. In: The IEEE Conference on Computer Vision and Pattern Recognition (CVPR) Workshops, July 2017
4. Babaie, M., Tizhoosh, H.R., Khatami, A., Shiri, M.: Local radon descriptors for image search. In: 2017 Seventh International Conference on Image Processing Theory, Tools and Applications (IPTA), pp. 1–5. IEEE (2017)

5. Babaie, M., Tizhoosh, H.R., Zhu, S., Shiri, M.E.: Retrieving similar x-ray images from big image data using radon barcodes with single projections. In: Proceedings of the 6th International Conference on Pattern Recognition Applications and Methods, ICPRAM 2017, Porto, Portugal, 24–26 February 2017, pp. 557–566 (2017). https://doi.org/10.5220/0006202105570566

6. Bankman, I.: Handbook of Medical Image Processing and Analysis. Elsevier, Amsterdam (2008)

7. Bathina, Y.B., Medathati, M., Sivaswamy, J.: Robust matching of multi-modal retinal images using radon transform based local descriptor. In: Proceedings of the 1st ACM International Health Informatics Symposium, pp. 765–770. ACM (2010)

8. Bauer, S., Wiest, R., Nolte, L.P., Reyes, M.: A survey of MRI-based medical image analysis for brain tumor studies. Phys. Med. Biol. **58**(13), R97 (2013)

9. Beylkin, G.: Imaging of discontinuities in the inverse scattering problem by inversion of a causal generalized radon transform. J. Math. Phys. **26**(1), 99–108 (1985)

10. de Bruijne, M.: Machine learning approaches in medical image analysis: from detection to diagnosis. Med. Image Anal. **33**, 94–97 (2016)

11. Donahue, J., et al.: DeCAF: a deep convolutional activation feature for generic visual recognition. In: International Conference on Machine Learning, pp. 647–655 (2014)

12. Ghosh, P., Antani, S., Long, L.R., Thoma, G.R.: Review of medical image retrieval systems and future directions. In: 2011 24th International Symposium on Computer-Based Medical Systems (CBMS), pp. 1–6. IEEE (2011)

13. Greenspan, H., van Ginneken, B., Summers, R.M.: Guest editorial deep learning in medical imaging: overview and future promise of an exciting new technique. IEEE Trans. Med. Imaging **35**(5), 1153–1159 (2016)

14. Kalra, S.: Content-based image retrieval of gigapixel histopathology scans: a comparative study of convolution neural network, local binary pattern, and bag of visual words. Master's thesis, University of Waterloo (2018)

15. Kashif, M., Deserno, T.M., Haak, D., Jonas, S.: Feature description with SIFT, SURF, BRIEF, BRISK, or FREAK? A general question answered for bone age assessment. Comput. Biol. Med. **68**, 67–75 (2016)

16. Kejariwal, A., Kulkarni, S., Ramasamy, K.: Real time analytics: algorithms and systems. Proc. VLDB Endow. **8**(12), 2040–2041 (2015)

17. Kieffer, B., Babaie, M., Kalra, S., Tizhoosh, H.: Convolutional neural networks for histopathology image classification: training vs. using pre-trained networks. arXiv preprint arXiv:1710.05726 (2017)

18. Kumar, M.D., Babaie, M., Tizhoosh, H.R.: Deep barcodes for fast retrieval of histopathology scans. In: 2018 International Joint Conference on Neural Networks (IJCNN), pp. 1–8. IEEE (2018)

19. Li, S., Kang, X., Fang, L., Hu, J., Yin, H.: Pixel-level image fusion: a survey of the state of the art. Inf. Fusion **33**, 100–112 (2017)

20. Long, F., Zhang, H., Feng, D.D.: Multimedia information retrieval and management. Technological Fundamentals and Applications (2003)

21. Murala, S., Wu, Q.J.: Local mesh patterns versus local binary patterns: biomedical image indexing and retrieval. IEEE J. Biomed. Health Inform. **18**(3), 929–938 (2014)

22. Radon, J.: On the determination of functions from their integral values along certain manifolds. IEEE Trans. Med. Imaging **5**(4), 170–176 (1986)

23. Rahman, M.M., Antani, S.K., Thoma, G.R.: A medical image retrieval framework in correlation enhanced visual concept feature space. In: 22nd IEEE International Symposium on Computer-Based Medical Systems, CBMS 2009, pp. 1–4. IEEE (2009)

24. Rui, Y., Huang, T.S., Chang, S.F.: Image retrieval: past, present, and future. J. Vis. Commun. Image Represent. (1997). Citeseer

25. Sargent, D., Chen, C.I., Tsai, C.M., Wang, Y.F., Koppel, D.: Feature detector and descriptor for medical images. In: Medical Imaging 2009: Image Processing, vol. 7259, p. 72592Z. International Society for Optics and Photonics (2009)

26. Sorensen, L., Shaker, S.B., De Bruijne, M.: Quantitative analysis of pulmonary emphysema using local binary patterns. IEEE Trans. Med. Imaging **29**(2), 559–569 (2010)

27. Tizhoosh, H., Babaie, M.: Representing medical images with encoded local projections. IEEE Trans. Biomed. Eng. 1 (2018). https://doi.org/10.1109/TBME.2018.2791567

28. Tizhoosh, H.R.: Barcode annotations for medical image retrieval: a preliminary investigation. In: 2015 IEEE International Conference on Image Processing (ICIP), pp. 818–822. IEEE (2015)

29. Tizhoosh, H.R., Babaie, M.: Representing medical images with encoded local projections. IEEE Trans. Biomed. Eng. **65**(10), 2267–2277 (2018)

30. Tizhoosh, H.R., Zhu, S., Lo, H., Chaudhari, V., Mehdi, T.: MinMax radon barcodes for medical image retrieval. In: Bebis, G., et al. (eds.) ISVC 2016. LNCS, vol. 10072, pp. 617–627. Springer, Cham (2016). https://doi.org/10.1007/978-3-319-50835-1_55

31. Wei, C.H., Li, C.T., Wilson, R.: A content-based approach to medical image database retrieval. In: Database Modeling for Industrial Data Management: Emerging Technologies and Applications, pp. 258–292. IGI Global (2006)

Bioinformatics

Analysis of Viability of TCGA and GTEx Gene Expression for Gleason Grade Identification

Matthew Casey[1]([✉])(iD) and Nianjun Zhou[2]

[1] Ardsley High School, Ardsley, NY 10522, USA
`mattcasey02@gmail.com`
[2] IBM, 1101 Kitchawan Rd, Yorktown Heights, NY 10598, USA
`jzhou@us.ibm.com`

Abstract. Gleason grade is a critical indicator for determining patient treatment for prostate cancer. In this paper, we analyze the viability of RNA sequencing gene expression data for Gleason grade identification. We combine datasets from the TCGA (sampled from cancer patients) and GTEx (sampled from healthy patients) databases. Using mutual information techniques, we reduce the dimensionality from 19046 genes to only the 20 most predictive genes. Then, we apply an unsupervised approach to analyze the separability of the grades of cancer. We use the t-SNE algorithm to map features into two dimensions and apply a Gaussian Mixture Model (GMM) for clustering. The result shows a clear visual separability between cancer and healthy samples. However, the grades of cancer themselves are not visually separable. Also, we apply the Mann-Whitney U test to compare the statistical similarity of the different Gleason grades and find that most grades are similar to each other. We further apply a random forest model to estimate the Gleason grade. The results show that the model accurately predicts whether a sample comes from healthy or cancer tissue. However, the model is weak in classifying the Gleason grade. The best performing model has a weighted macro-averaged F1 score of 0.66, improving on a baseline score of 0.22 obtained by random guessing. Our results indicate that the difference in gene expression among Gleason grades is relatively small compared to the difference between healthy and cancer samples. Thus, gene expression alone cannot be used for Gleason grade identification.

Keywords: RNA-sequencing · Gene expression · Gleason grade · Gaussian mixture model · Grid search · Mutual information · T-SNE · Dimensionality reduction · Hypothesis testing

1 Introduction and Related Work

Prostate cancer is one of the leading causes of cancer-related death among men, with over 33,000 deaths projected for 2020 [1]. Early detection of and accurate diagnosis of the disease are ongoing areas of research that aim to aid doctors in selecting a proper treatment plan for patients. Patients suspected of having prostate cancer may be recommended for a biopsy examination. In a standard 12-core biopsy procedure, cells are removed from the prostate in twelve different places [2]. The cells are then analyzed by a pathologist to identify the severity of cancer. Primary and secondary Gleason scores are

© Springer Nature Switzerland AG 2020
M. Michalowski and R. Moskovitch (Eds.): AIME 2020, LNAI 12299, pp. 475–485, 2020.
https://doi.org/10.1007/978-3-030-59137-3_42

assigned based on observed cell abnormalities. The primary Gleason score represents the most common cell morphology. Cells with a score of 1 have the most well-differentiated tumor pattern, and thus are relatively similar to healthy cells. Cells with a score of 5 have very poor or no differentiation, and thus the most mutation. The secondary Gleason score represents the second most common cell morphology. These two scores are then added together to determine the overall Gleason grade of the patient [3].

The grade is used to outline treatment plans for patients, and thus proper identification is essential. Higher Gleason grade has been found to be correlated with metastasis rate, mortality rate, and biochemical recurrence rate [4–7]. Thus, a patient with a Gleason grade of 1 may be put on active surveillance, while a patient with a Gleason grade of 3 may be recommended for radiation therapy [8].

Current diagnostic techniques are not very accurate at identifying Gleason scores and grades. Doctors underestimate Gleason score in 47.8% of patients [9] and only accurately diagnose Gleason grade in 57% of patients [10]. It is critical to improve grade identification accuracy so that appropriate treatment is recommended and potential harms (including mental, physical, and financial strain placed on patients) are minimized [11].

Artificial intelligence-based diagnosis is a potential promising addition to classical diagnosis by a pathologist. Though there has been prior work using artificial intelligence for Gleason grade identification, it has only employed digital slide images [12]. These studies have been reasonably successful in cancer prediction, and many have been able to achieve equal or higher accuracy than doctors themselves. However, no studies have analyzed the gene expressions of cells to identify the Gleason grade of a patient. RNA sequencing data provides an improved approach to gene expression and makes this type of study possible.

RNA sequencing gene expression values for certain specific genes have been shown to be positively correlated with Gleason grade progression, and thus show promise to be used for artificial intelligence-based identification [13]. Recently, there have been a few databases created with patient RNA sequencing data for both healthy patients and patients diagnosed with cancer. Namely, the TCGA (The Cancer Genome Atlas) database from the National Institute of Health (NIH), which provides RNA sequencing data for thirty-three different types of cancer [14]. This dataset provides a wealth of research possibilities but is weakened since it contains limited healthy data. The GTEx (Genotype-Tissue Expression) dataset addresses this problem by providing RNA sequencing data from healthy patients in many of the same tissues analyzed by the TCGA [15]. In our previous work, we used the TCGA dataset in conjunction with artificial intelligence to create a model to diagnose patients with prostate cancer [16].

In this paper, we explore whether RNA sequencing data can be used for Gleason grade identification. The application value for clinicians is high, as Gleason grade is directly related to treatment determinations. To determine the proper treatment path, the doctor needs to know the severity, and thus the Gleason grade of cancer [8].

The paper is structured as follows. Section 2 discusses the acquisition and processing of data. Section 3 discusses unsupervised clustering, statistical analysis, and results. Section 4 discusses machine learning modeling and its results. Section 5 summarizes the work and discusses possible directions for future work. Section 6 is an appendix containing the genes selected and parameters used in the grid search.

2 Data Acquisition and Preprocessing

This section details the preparation of the data for analysis and modeling. We explain how the gene expression datasets are acquired and organized. All source code can be found at https://github.com/mattcasey02/Gleason-Grade-Analysis.

2.1 Data Acquisition and Preprocessing

We use the TCGA-Assembler 2 tool to download the clinical dataset containing information about primary and secondary Gleason score and the type of tissue (cancer or healthy) [17]. We also download the TCGA and GTEx prostate gene expression datasets. These datasets are not natively compatible, so we use the compatible versions published by Wang et al. in 2018 [18].

For the convenience of future analysis, we categorize samples into TCGA Cancer, TCGA Healthy, and GTEx sets. The TCGA Cancer set contains TCGA samples taken from cancer tissue. The TCGA Healthy set contains TCGA samples taken from healthy tissue of cancer patients. The GTEx set contains the GTEx samples, taken from healthy tissue of healthy patients. In our dataset, there are 426 samples in the TCGA Cancer set, 48 in the TCGA Healthy set, and 106 in the GTEx set. There is an imbalance of about 3:1 of cancer to healthy samples.

We use the Gleason scores found in the clinical data to assign a Gleason grade to each sample in the TCGA Cancer set. For the convenience of research, we label all samples in the healthy sets as grade 0. Table 1 details the summary of our dataset. We further combine the grades to create a second, concerted dataset. Multi-class classification can become very difficult when a large number of groups are involved, so we believe that we can improve the multi-class model accuracy with fewer groups. Grades 1 and 2 are combined to represent "low risk," grade 3 represents "moderate risk," and grades 4 and 5 are combined to represent "high risk." We now have two processed datasets, one with labels using grade value 0–5 and another using ID 0–3 as labels. These are referred to as the 6-group and 4-group datasets, respectively.

Table 1. Gleason score to grade conversion

Gleason score components	Grade	Number of samples
Healthy	0	154
Sum $\leq$ 6	1	38
Primary: 3; Secondary: 4	2	122
Primary: 4; Secondary: 3	3	79
Sum = 8	4	60
Sum = 9 or 10	5	127

2.2   Dimensionality Reduction with Mutual Information

High dimensionality hinders the performance of classifiers as they are not able to utilize the information contained in the data adequately, and as such, have a difficult time making accurate classifications [19]. The datasets start with 19046 genes. To reduce their dimensionality, we use the mutual information metric. The mutual information metric quantifies the amount of information gained about one variable by observing the other [20]. A high mutual information value for a particular gene would indicate that knowing the expression value of that gene greatly helps in determining what grade or group the sample belongs to.

We determine the mutual information between each gene and the target, and then select the top genes with the highest mutual information values to be used for classification. Reducing the number of genes allows the classifiers to more readily access the critical information contained in the dataset. The information, in turn, improves the accuracy of classification. Having too few genes, however, limits the classifiers' accuracy, as they have less information available to them. We try various numbers of genes (5,20,30,100) and find that machine learning models trained on 20 genes have the highest average F1 scores. Using the top 20 genes provides the best balance between having too few genes and too many genes. All further analysis presented in this paper will thus use the top 20 genes. The 20 genes selected can be found in Sect. 6.1.

3   Unsupervised Clustering and Statistical Analysis

This section describes the unsupervised clustering and statistical tests used to analyze the separability between the gene expression values for the grades of prostate cancer. We analyze the similarity of the various grades of cancer to better understand how gene expression changes as the grade increases.

3.1   Unsupervised Clustering with Gaussian Mixture Model

The first step in analyzing the separability of the data is the use of a clustering algorithm, the Gaussian Mixture Model (GMM). This algorithm tries to separate a set of data based on the number of clusters selected by us. It does this by assuming that each group within the data follows a Gaussian distribution and tries to find the distributions that best cluster the data [21]. We choose the GMM algorithm instead of other clustering algorithms, such as K-Means, because it provides a 'soft' assignment to a cluster. It allows us to see the probability density estimation, allowing us to better understand the predicted likelihood that a sample belongs to each group. The result of the clustering gives more meaningful insight into how two grades may be overlapped.

We start with the 20 genes selected by mutual information and reduce them to two dimensions using the t-Distributed Stochastic Neighbor Embedding (t-SNE) algorithm [22]. This dimensionality reduction process is done for both the 6-group dataset and the 4-group dataset. The GMM is then applied to each of the two datasets.

The output of the GMM models is shown in Fig. 1 for the two datasets. In Fig. 1, we color each sample based on the grade or group IDs. The color-coding helps us to examine the separability of the samples in terms of their grade and group. The contour lines drawn from the GMM's density estimation are shown and are used to see where the GMM believes the clusters are centered.

The visualizations show that, for the most part, cancer and healthy data are visually separable into different clusters. However, there is no visual separability between the various cancer grades. Even after grouping the data, the same results are obtained.

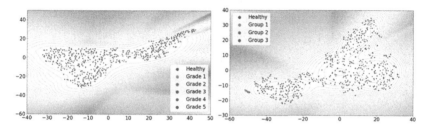

Fig. 1. Gaussian mixture model visualization for 6-Group (left) and 4-Group (right) datasets

3.2 Statistical Similarity of Samples of Different Grade

We also examine the statistical similarity of the gene expressions for the different Gleason grades. We reduce our 20 gene-features into one dimension using the t-SNE technique, as discussed before. This data compression allows us to apply traditional statistical techniques. Before these techniques can be applied, however, we must analyze the normality of the data. According to current literature, to most accurately test for normality, both visual techniques and normality tests should be used [23].

We first use the Quantile-Quantile (Q-Q) plot, which graphs the quantiles of the given dataset against the quantiles of a theoretical Gaussian normal distribution. We further use the Shapiro-Wilk test for normality. The Shapiro-Wilk test is the recommended test to use when assessing normality [23]. From the results of the Q-Q plot and the Shapiro-Wilk test, we find that most of the grades/groups are not normally distributed. Thus, a non-parametric test is used for further analysis.

3.3 Mann-Whitney U Test

The Mann-Whitney U test is used in place of an unpaired t-test because it does not require normally distributed data. The null hypothesis of this test is that the two groups are drawn from the same population.

Table 2 shows p-values for each comparison between grades. All grades are found to be different from GTEx ($p < .001$). All the grades are found to be similar to each other except grades 2 and 3 ($p = .026$), and grades 3 and 5 ($p = .009$).

Table 3 shows p-values for each comparison between the combined grades. The combined grade 1 + 2 is similar to grade 3 (p = .059) and is similar to the combined grade 4 + 5 (p = .617).

The results of the comparisons with TCGA healthy data show TCGA healthy data is different from all the grades, but is similar to GTEx (p = .130), as would be expected.

Table 2. Mann-Whitney U p-values for grade comparisons

	Grade 1	Grade 2	Grade 3	Grade 4	Grade 5
GTEx	<.001	<.001	<.001	<.001	<.001
Grade 1		.129	.767	.258	.106
Grade 2			.026	.778	.982
Grade 3				.075	.009
Grade 4					.815

Table 3. Mann-Whitney U p-values for combined grade comparisons

	Grade 1 + Grade 2	Grade 3	Grade 4 + Grade 5
GTEx	<.001	<.001	<.001
Grade 1 + Grade 2		.059	.617
Grade 3			.009

4   Predictive Modeling

In this section, we discuss the creation of predictive models for the 6-group and 4-group datasets. The purpose of this section is to analyze the relationship between gene expression and Gleason grade using a supervised machine learning approach. To find a better predictive model, we leverage the grid-search technique to tune the parameters of the predictive model [24].

4.1   Model Creation and Selection

We apply a stratified 80%/20% train/validation split for each dataset. We select four different types of models: Support Vector Machine (SVM) with linear kernel, SVM with the radial kernel, k-Nearest Neighbors (KNN), and Random Forest (RF).

While using the grid search for model tuning, we use 5-fold stratified cross-validation. We use the stratified version so that each fold has the same class imbalance. We test 95 different parameter combinations (the full list can be found in Sect. 6.2). Since it is equally important for all grades to be correctly identified, a model would ideally have similar performance across all grades. We thus choose to use the macro-averaged F1 score for comparing the models. Macro-averaged F1 score gives equal weight, and therefore importance, to all classes. This means that the chosen model has minimal or no preference for classifying any specific class.

4.2 Model Evaluation

The parameters returned by the grid search are used to train a machine learning model. The performance of the model on the validation set is evaluated with several metrics: precision, recall, and F1 score for each individual class as well as overall micro, macro, and weighted macro-precision, recall, and F1 score. As a baseline of comparison, we construct a dummy model to represent random guessing in a stratified approach, meaning its predictions follow the class distribution of the data.

4.3 Results of the Machine Learning Models

The hyperparameters found through the grid search to be the best for each dataset are found in Table 4. These models are referred to as the 6-group and 4-group datasets' tuned random forest models, respectively. The performance of these models on the validation set is shown in Tables 5 and 6. Our model's performance is to the left of the slash, and the random-guessing dummy model's performance is to the right of the slash.

Our 6-group tuned random forest model performs better than the dummy model in the micro, macro, and weighted macro metrics and at classifying each individual grade, except for grade 4. Overall, this model is very accurate at classifying healthy samples, with an F1 score of 0.93, but is much worse at classifying the various cancer grades.

Our 4-group tuned random forest model performs better than the dummy model in the micro, macro, and weighted macro metrics and at classifying each individual class. Our model is nearly as good as the 6-group model in classifying healthy samples, and is better overall at classifying the cancer samples, as was expected.

Table 4. Best hyperparameters for the 6-Group and 4-Group datasets

	Model	Number of estimators	Max depth	Min. samples to split an internal node	Min. samples to be leaf node
6-Group	Random Forest	20	None	10	1
4-Group	Random Forest	10	None	10	2

Table 5. Performance of 6-Group dataset's tuned random forest model and dummy model

	Precision	Recall	F1 Score	Number of samples
Healthy	0.86/0.32	1.00/0.35	0.93/0.34	31
Grade 1	0.50/0.00	0.12/0.00	0.20/0.00	8
Grade 2	0.39/0.15	0.62/0.12	0.48/0.14	24
Grade 3	0.17/0.09	0.06/0.06	0.09/0.07	16
Grade 4	0.33/0.23	0.08/0.25	0.13/0.24	12
Grade 5	0.48/0.26	0.60/0.18	0.54/0.27	25
Micro-Avg	0.55/0.22	0.55/0.22	0.55/0.22	116
Macro-Avg	0.46/0.18	0.42/0.18	0.39/0.18	116
Weighted Macro-Avg	0.51/0.21	0.55/0.22	0.50/0.21	116

Table 6. Performance of 4-Group dataset's tuned random forest model and dummy model

	Precision	Recall	F1 Score	Number of samples
Healthy	0.91/0.30	1.00/0.26	0.95/0.28	31
Grade 1 + Grade 2	0.56/0.30	0.62/0.34	0.59/0.32	32
Grade 3	0.50/0.22	0.12/0.12	0.20/0.16	16
Grade 4 + Grade 5	0.60/0.40	0.68/0.46	0.63/0.42	37
Micro-Avg	0.67/0.33	0.67/0.33	0.67/0.33	116
Macro-Avg	0.64/0.30	0.61/0.30	0.59/0.29	116
Weighted Macro-Avg	0.66/0.32	0.67/0.33	0.65/0.32	116

5   Conclusion and Future Work

This work explores the relationship between gene expression and Gleason grade. We employ statistical analysis to compare the samples from different grades of cancer. Furthermore, we explore the viability of using RNA sequencing gene expression data in a machine learning model to predict the Gleason grade of a sample. With the advent of increased availability of genetic data, these types of studies are important to find the boundaries of its application and identify additional features required to improve model performance.

The statistical analysis shows that in terms of gene expression, healthy and cancer samples are visually separable and different from each other ($p < .001$). The individual grades/combined grades, however, are not visually separable from each other. In addition, all the grades are all found to be similar to each other except grades 2 and 3 ($p = .026$), and grades 3 and 5 ($p = .009$). Further, the combined grade 1 + 2 is similar to grade 3 ($p = .059$) and is similar to the combined grade 4 + 5 ($p = .617$).

The machine learning models are accurately able to classify healthy vs. cancer, with an F1 score of 0.93. Although the models do improve on stratified random guessing in all grades and combined grades except grade 4, they are weak in identifying the Gleason grade very accurately. These results suggest that the genetic differences between healthy and cancerous tissues are considerable. However, the genetic differences between the Gleason grades are minor.

The results of the statistical analysis and classification modeling are consistent. Both results show that it is easy to use gene features to separate healthy samples from cancer samples, but it is not easy to use gene features to identify cancer grade. Thus, gene expression data alone is not viable to improve Gleason grade identification.

Our results also provide valuable insight into the relationship between gene expression and Gleason grade progression. Previously, the belief was that as grade progressed, there would be an increased expression in certain specific genes [13]. We find that the gene expressions of genes selected to differentiate between the different grades do not change as grade increases, or change too subtly to be identified by the models. This result suggests that higher grade is not linked to increased gene expression, but another variable.

Our analysis is limited by the genes we selected. It is possible that one or a few genes are linked to increased grade, and they were not selected by the mutual information approach. Future work should consider genes with established connections to grade progression, and analyze their viability for use in a machine learning model. Prior studies have identified genes that increase in expression as grade increases [13]. In addition, models should use genes known to have a biological role in prostate cancer progression.

The sample size of our dataset also limits our analysis. The statistical analysis yielded some strange results such as grade 3 being different from grade 2 (p = .026), grade 5 (p = .009), and grade 4 + 5 (p = .009). This is curious since grade 3 is the intermediate grade, and thus if all other grades are similar, then grade 3 should be similar to them as well. It is possible that the low sample size of grade 3 provided erroneous results in the statistical analysis. Future work should explore this finding and see if the same results are received with a larger sample size dataset.

Finally, future work should explore other data that can be used in conjunction with gene expression data for identification of Gleason grade. One aspect that we do not consider is patient clinical data, including race, age, and family history. Clinical data has previously been successfully applied in machine learning for prostate cancer diagnosis [25]. Many opportunities need to be explored before gene expression data can be ignored for use in Gleason grade identification.

6 Appendix

6.1 Selected Genes

Table 7 details the twenty genes that are selected for use in this study. These are the twenty genes with the highest mutual information values.

Table 7. Top 20 genes selected for use in this study

RCBTB2-1102	AOX1-316	NDRG2-57447	EZH2-2146
SERPINA5-5104	RP11-122A3.2-0	SEC16A-9919	APOBEC3C-27350
GSTP1-2950	HECTD4-283450	FOXA1-3169	TPX2-22974
RAB9B-51209	HPN-3249	MYBL2-4605	CTD-3193O13.9-0
ABHD6-57406	TRPM4-54795	AKR1B1-231	PPP1R14B-26472

6.2 Grid Search Parameters

Table 8 details the various hyperparameters tested and the values chosen for testing. Values were chosen based on examples found in the Scikit-learn documentation.

Table 8. Grid search parameters

Model	Parameters			
	C			
Linear SVM	1			
	10			
	100			
	1000			
	C			Gamma
Radial SVM	1			.001
	10			.0001
	100			
	1000			
KNN	Number of Neighbors			
	Each Number from 1 to 29			
	Number of Estimators	Max Depth	Min. Samples to Split an Internal Node	Min. Samples to be a Leaf Node
Random Forest	10	5	2	1
	20	None	5	2
	40		10	4

References

1. Siegel, R., Miller, K., Jemal, A.: Cancer statistics, 2020. CA Cancer J. Clin. **70**(1), 7–30 (2020)
2. Presti, J.: Prostate biopsy: how many cores are enough? Urol. Oncol. Semin. Original Invest. **21**, 135–140 (2003)
3. Kryvenko, O., Epstein, J.: Changes in prostate cancer grading: including a new patient-centric grading system. Prostate **76**(5), 427–433 (2016)

4. Leapman, M., et al.: Application of a prognostic Gleason grade grouping system to assess distant prostate cancer outcomes. Eur. Urol. **71**(5), 750–759 (2017)
5. He, J., Albertsen, P., Moore, D., Rotter, D., Demissie, K., Lu-Yao, G.: Validation of a contemporary five-tiered gleason grade grouping using population-based data. Eur. Urol. **71** (5), 760–763 (2017)
6. Loeb, S., Folkvaljon, Y., Robinson, D., Lissbrant, I., Egevad, L., Stattin, P.: Evaluation of the 2015 gleason grade groups in a nationwide population-based cohort. Eur. Urol. **69**(6), 1135–1141 (2016)
7. Beckmann, K., et al.: Oncological outcomes in an Australian cohort according to the new prostate cancer grading groupings. BMC Cancer **17**(1), 537 (2017). https://doi.org/10.1186/s12885-017-3533-9
8. National Cancer Institute: Prostate Cancer Treatment (PDQ®): Health Professional Version. National Cancer Institute (US) (2002)
9. Serefoglu, E., Altinova, S., Ugras, N., Akincioglu, E., Asil, E., Balbay, M.: How reliable is 12-core prostate biopsy procedure in the detection of prostate cancer? J. Can. Urol. Assoc. **7** (5-6), E293 (2013)
10. Danneman, D., et al.: Accuracy of prostate biopsies for predicting Gleason score in radical prostatectomy specimens: nationwide trends 2000–2012. BJU Int. **119**(1), 50–56 (2017)
11. Albertsen, P.: The unintended burden of increased prostate cancer detection associated with prostate cancer screening and diagnosis. Urology **75**(2), 399–405 (2010)
12. Ström, P., et al.: Pathologist-level grading of prostate biopsies with artificial intelligence (2019)
13. Bibikova, M., et al.: Expression signatures that correlated with Gleason score and relapse in prostate cancer. Genomics **89**(6), 666–672 (2007)
14. Network, T., et al.: The cancer genome atlas pan-cancer analysis project. Nat. Genet. **45**(10), 1113 (2013)
15. Lonsdale, J., et al.: The genotype-tissue expression (GTEx) project. Nat. Genet. **45**(6), 580–585 (2013)
16. Casey, M., Chen, B., Zhou, J., Zhou, N.: A machine learning approach to prostate cancer risk classification through use of RNA sequencing data. In: Chen, K., Seshadri, S., Zhang, L. J. (eds.) BIGDATA 2019. LNCS, vol. 11514, pp. 65–79. Springer, Cham (2019). https://doi.org/10.1007/978-3-030-23551-2_5
17. Wei, L., Jin, Z., Yang, S., Xu, Y., Zhu, Y., Ji, Y.: TCGA-assembler 2: software pipeline for retrieval and processing of TCGA/CPTAC data. Bioinformatics **34**(9), 1615–1617 (2018)
18. Wang, Q., et al.: Data descriptor: unifying cancer and normal RNA sequencing data from different sources. Sci. Data **5**, 180061 (2018)
19. Altman, N., Krzywinski, M.: The curse(s) of dimensionality. Nat. Methods **15**(6), 399–400 (2018)
20. Zeng, G.: A unified definition of mutual information with applications in machine learning. Math. Probl. Eng. **2015**, article ID 201874, 12 p. (2015). https://doi.org/10.1155/2015/201874
21. Biernacki, C., Celeux, G., Govaert, G.: Assessing a mixture model for clustering with the integrated completed likelihood. IEEE Trans. Pattern Anal. Mach. Intell. **22**(7), 719–725 (2000)
22. Van Der Maaten, L., Hinton, G.: Visualizing data using t-SNE (2008)
23. Ghasemi, A., Zahediasl, S.: Normality tests for statistical analysis: a guide for non-statisticians. Int. J. Endocrinol. Metab. **10**(2), 486–489 (2012)
24. Feurer, M., Hutter, F.: Hyperparameter Optimization (2019)
25. Takeuchi, T., Hattori-Kato, M., Okuno, Y., Iwai, S., Mikami, K.: Prediction of prostate cancer by deep learning with multilayer artificial neural network. Can. Urol. Assoc. J. **13**(5), E145–E150 (2019)

Assessing the Impact of Distance Functions on K-Nearest Neighbours Imputation of Biomedical Datasets

Miriam S. Santos[1,3(✉)], Pedro H. Abreu[1], Szymon Wilk[2], and João Santos[3]

[1] CISUC, Department of Informatics Engineering, University of Coimbra,
Coimbra, Portugal
{miriams,pha}@dei.uc.pt
[2] Institute of Computing Science, Poznan University of Technology,
Poznan, Poland
szymon.wilk@cs.put.poznan.pl
[3] IPO-Porto Research Centre, Porto, Portugal
joao.santos@ipoporto.min-saude.pt

Abstract. In healthcare domains, dealing with missing data is crucial since absent observations compromise the reliability of decision support models. K-nearest neighbours imputation has proven beneficial since it takes advantage of the similarity between patients to replace missing values. Nevertheless, its performance largely depends on the distance function used to evaluate such similarity. In the literature, k-nearest neighbours imputation frequently neglects the nature of data or performs feature transformation, whereas in this work, we study the impact of different heterogeneous distance functions on k-nearest neighbour imputation for biomedical datasets. Our results show that distance functions considerably impact the performance of classifiers learned from the imputed data, especially when data is complex.

Keywords: Missing data · Heterogeneous data · Data imputation · Distance functions · K-nearest neighbours · Biomedical data

1 Introduction

A common data quality problem in healthcare domains is the presence of Missing Data, which consists of absent observations in patients' medical records [9]. Dealing with missing data is of outstanding importance, since absent observations may jeopardise algorithms' predictions, compromising the reliability of patient-oriented models for decision making. K-nearest neighbours (KNN) imputation is a popular imputation technique in healthcare domains, since it takes advantage of the similarity between patients to produce accurate estimates for imputation. Furthermore, it is a nonparametric method which does not require any assumptions about the data [16], has proven to preserve the data distribution [15] and allows for a great interpretability and explainability, crucial in healthcare domains [3].

© Springer Nature Switzerland AG 2020
M. Michalowski and R. Moskovitch (Eds.): AIME 2020, LNAI 12299, pp. 486–496, 2020.
https://doi.org/10.1007/978-3-030-59137-3_43

Nevertheless, KNN performance largely depends on the distance function used to evaluate such similarity. For heterogeneous data, typical solutions include feature transformation, although leading to loss of information (e.g., discretisation of continuous features) or increased dimensionality (e.g., one-hot encoding) and the use of heterogeneous distance functions that handle different scales [18]. However, besides their heterogeneous nature and susceptibility to missing data, biomedical data is also prone to other difficulty factors, such as data imbalance, the presence of subconcepts in data (small disjuncts), class overlap, and noisy data [2,5], which make them especially complex domains where choosing suitable distance functions becomes a more strenuous and critical task.

In related work, KNN imputation frequently neglects the nature of data or performs feature transformation [16]. Considering KNN classification, either the studies consider only complete datasets (or derisory amounts of missing values) or the nature of data is ignored [7]. This work studies the impact of different heterogeneous distance functions on KNN imputation, evaluating their effect on the performance of classifiers constructed from biomedical datasets with different characteristics (we focus on tree-based classifiers constructed with Classification and Regression Trees - CART). The purpose of this research is two-fold: *1) Determining if distance functions impact KNN imputation of biomedical datasets and whether the type of features affected by missing data influences the classification performance* and *2) Determining whether observed classification performance is related to the characteristics of biomedical datasets*. Regarding *1)*, it is important to state that we evaluate the impact of distance functions on imputation indirectly by focusing on the performance of resulting tree-based classifiers. In other works, we focus on how accurate are the resulting classifiers, rather than how well the imputation reconstructs the data. Regarding *2)* we explore if there are scenarios (data characteristics) where the choice of distance function considerably influences the obtained results. To that end, a benchmark of biomedical datasets with different characteristics was collected and missing data was generated following 4 different variants and percentages (5 to 30%). Then, data imputation was performed using 7 different distance functions and imputation results were evaluated through the analysis of a classifier learned from the imputed data.

To the authors knowledge, no study has yet investigated the impact of different distance functions on the imputation of biomedical data with different characteristics and its effect on classification performance, which constitutes the main contribution of this work. Furthermore, we explored recent distance measures never before studied for imputation purposes, such as SIMDIST and MDE (Sect. 2), extended MDE to handle categorical data, studied redefinitions of popular distance functions (HEOM and HVDM), often overlooked in related work, and proposed yet another redefinition of HVDM, which constitute additional contributions.

2   Heterogeneous Distance Functions for Missing Data

All distance functions measure the distance between two patterns $\mathbf{x}_A$ and $\mathbf{x}_B$ through a sum of their individual distances for each j-th feature, $d_j(x_{Aj}, x_{Bj})$, as $D(\mathbf{x}_A, \mathbf{x}_B) = \sqrt{\sum_{j=1}^{p} d_j(x_{Aj}, x_{Bj})^2}$; yet they differ on the computation of individual d_j distances and treatment of missing values. p represents the total number of features and x_{Aj}, x_{Bj} are two values of feature j. The mathematical formulation of all distance functions may be found in the Appendix. In what follows, we describe each studied distance function referring to the formulas presented in the Appendix.

HEOM and HVDM: The definition of $d_j(x_{Aj}, x_{Bj})$ for Heterogeneous Euclidean-Overlap Metric (HEOM) and Heterogeneous Value Difference Metric (HVDM) depends on the type of feature j (Eqs. 1 and 2) [18]. For categorical features, HEOM defines d_j as an overlap metric, d_O (Eq. 3) whereas HVDM uses d_{vdm} (Eq. 4). For continuous features, HEOM uses the normalised Euclidean distance d_N (Eq. 5), whereas HVDM considers d_{diff} (Eq. 6). However, d_O, d_{diff}, d_N and d_{vdm} are only computed if both x_{Aj} and x_{Bj} are observed; otherwise, $d_j(x_{Aj}, x_{Bj}) = 1$.

HEOM-R, HVDM-R and HVDM-S: HEOM-R and HVDM-R [8] consider missing values as "special values": if both x_{Aj} and x_{Bj} are missing, then $d_j(x_{Aj}, x_{Bj}) = 0$ (Eq. 7). In addition, we propose another redefinition of HVDM: missing values are considered an "special" category and d_{vdm} is applied when only x_{Aj} or x_{Bj} are missing and j is categorical, referred to as HVDM-S (Eq. 9).

SIMDIST: SIMDIST defines a similarity measure S (Eq. 8), where s_{ABj} is an intermediate similarity between patterns according to j, s_j represents the mean similarity among all patterns according to j and z is a normalisation function: $z(a) = \frac{a}{a+1}$ [4]. For categorical features, s_{ABj} is defined by Eq. 10, whereas for continuous features, s_{ABj} is determined by Eq. 11. $S_{ABj} = \frac{1}{2}$ when x_{Aj} or x_{Bj} are missing which is the equivalent of replacing the missing similarity by the mean similarities of all patterns according to j (Eq. 8). The individual similarities S_{ABj} are then transformed to distances $D_{ABj} = 1 - S_{ABj}$ and aggregated to produce $D(\mathbf{x}_A, \mathbf{x}_B)$.

MDE: When both values are observed, Mean Euclidean Distance (MD_E) [1] is defined as the Euclidean distance (Eq. 12). When either x_{Aj} or x_{Bj} are missing, MD_E is approximated as the mean distance of each value of j to the observed value (Eq. 14). When both values are missing, MD_E is approximated as the mean distance between all values of j (Eq. 16). To allow a proper weighting of continuous features with different ranges, a min-max normalisation, $z_i = \frac{x_i - min(x_j)}{max(x_j) - min(x_j)}$ is applied before the Euclidean distance is computed. When first proposed, MD_E considered only continuous features. Therefore, starting from the overlap distance, d_O (Eq. 3) we extended MD_E for categorical features, MD_O: when both values are known, MD_O is the same as d_O; when one value is missing, MD_O is computed as the mean distance between all elements in x_j and the observed value (Eq. 15); when both values are missing MD_O is determined as the mean distance between all elements in x_j (Eq. 17). After the individual distances are computed, their aggregation is performed as for the remaining distances.

3 Experimental Setup

We started by collecting 31 complete and binary-classification datasets from open-source repositories (UCI Machine Learning, KEEL, KAGGLE, OPENML), comprising different biomedical contexts, sample sizes, number of features, type of features (continuous and categorical), imbalance ratios (IR) and data characteristics (given by complexity measures, as detailed later in Sect. 4). Each dataset was divided into 5 folds following a stratified crossvalidation (SCV) approach (as some datasets have a lower number of minority examples, using 10 folds would result in test sets with a very small amount of minority examples or the need to repeat minority examples across folds). Then, missing data was introduced in the training set, following 4 different variants, herein referred to as Weighted-Plain (PLAIN), Weighted-All (WA), Weighted-Continuous (WA-CONT) and Weighted-Categorical (WA-CAT). The same missing rate was inserted in both classes according to the IR of each dataset (hence the "weighted" designation), to guarantee that missing data is affecting both classes proportionally to their distribution. However, the features affected by missing data differ for each type. PLAIN generation does not control for the number or type of features where missing values are placed. In this case, missing data is generated over the entire dataset with no restrictions, simulating a scenario more likely to be found in real-world domains. Nevertheless, WA approach generates the same percentage of missing values for each feature (all features are equally affected by missing data), whereas WA-CONT and WA-CAT approaches generate the same amount of missing data for all continuous and categorical features, respectively. The goal of comparing different generations variants of missing data is to determine if the type of features (continuous or categorical) affected by missing data influenced the choice of a proper distance function for imputation. Also, missing data was generated at 4 different rates (5, 10, 20 and 30%) under a Missing Completely At Random (MCAR) mechanism. MCAR mechanism was considered for a rigorous control of the missing generation. As missing data is synthetically generated in a multivariate scheme (several features, if not all, are affected by missing data), for each variant and rate, and the considered datasets comprise heterogeneous features, other mechanisms could be compromised due to existing limitations of generation approaches for categorical data [13]. After the injection of missing data, the datasets with missing values were either directly classified with Classification and Regression Trees (CART) model (BASELINE approach) or first imputed with KNN ($k = 1, 3, 5, 7$) and then classified with CART. CART model is also a non-parametric classifier and relatively fast to construct and to provide classification results. Furthermore, it handles missing data directly

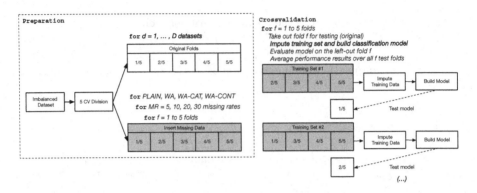

Fig. 1. Stratified crossvalidation and missing data generation: missing data is injected after the splitting of the data into training and test sets, for each fold. The same splits are used for all methods (both for training and testing stages).

through the use of surrogate splits (without discarding any patterns from the dataset or assuming a particular missing mechanism) thus allowing a comparison between learning a model with missing data or complete data (via imputation) [17]. Regarding the chosen distance functions, we started by considering distance functions that handled the three components of the problem (continuous, categorical and missing data) or required minimal adjustments, as described in Sect. 2. On that note, MDE, although lacking the treatment of categorical data, was extended and included given its similarity with HEOM and SIMDIST, yet using the data distribution to handle missing data (as is performed for continuous data). Similarly, we propose HVDM-S as a modification of HVDM, only altering the treatment of missing data and maintaining the remaining aspects regarding continuous and categorical data [12]. Finally, classification performance is evaluated using Accuracy, Sensitivity, Specificity, Precision, F-measure, G-mean and AUC (although for simplicity we present the most relevant metrics for imbalanced domains: Sensitivity, F-measure and G-mean) [11,14]. For each dataset, 10 repetitions of the crossvalidation procedure were performed, resulting in a 10×5 SCV approach (Fig. 1). Overall, 31 *datasets* $\times$ 10 *SCV repetitions* $\times$ 4 *missing variants* $\times$ 4 *missing rates* $\times$ 7 *distance functions* $\times$ 4 *k values* (imputed datasets) + 31 *datasets* $\times$ 10 *SCV Versions* $\times$ 4 *missing variants* $\times$ 4 *missing rates* (for BASELINE approach) sums up to an equivalent of 143,840 datasets evaluated.

4   Results and Discussion

Tables 1 and 2 report on the average sensitivity ranks obtained for CART, considering training sets with missing values (BASELINE) and training sets imputed with each of the 7 considered distances (KNN with $k = 1$ and $k = 3$, respectively). Furthermore, results are grouped by missing data variant (PLAIN, WA, WA-CAT and WA-CONT) and missing rate (5% to 30%). Overall, for both $k = 1$

Table 1. CART average sensitivity ranks per missing rate (MR), and variant ($k = 1$). The best values in each row are marked in bold and underlined. **B**: BASELINE.

	MR	B	HEOM	HEOM-R	HVDM	HVDM-R	HVDM-S	MDE	SIMDIST
PLAIN Datasets	5%	5.31	4.50	4.58	3.95	5.18	4.24	**<u>3.79</u>**	4.45
	10%	5.52	4.35	5.31	5.03	4.63	**<u>3.48</u>**	3.73	3.95
	20%	4.32	4.66	5.06	4.77	5.37	**<u>3.32</u>**	4.10	4.39
	30%	4.55	4.47	4.94	4.44	5.35	**<u>3.10</u>**	3.56	5.60
WA Datasets	5%	**<u>3.55</u>**	4.98	4.89	4.63	4.61	4.65	**<u>3.55</u>**	5.15
	10%	5.24	4.81	4.87	4.42	5.06	**<u>3.16</u>**	4.18	4.26
	20%	5.10	5.05	4.87	4.34	4.21	**<u>3.63</u>**	3.77	5.03
	30%	5.23	4.27	5.08	3.97	5.21	**<u>3.60</u>**	3.71	4.94
WA-CAT Datasets	5%	5.57	4.93	4.12	3.91	**<u>3.86</u>**	4.10	5.03	4.47
	10%	5.02	4.59	5.00	4.64	5.21	**<u>3.12</u>**	3.64	4.79
	20%	4.91	4.38	5.14	4.00	4.78	**<u>3.60</u>**	4.83	4.36
	30%	4.97	4.71	5.02	4.45	4.81	**<u>3.52</u>**	4.29	4.24
WA-CONT Datasets	5%	5.31	4.26	4.63	4.08	4.39	4.39	5.03	**<u>3.92</u>**
	10%	4.92	4.79	4.73	4.56	4.37	4.37	4.24	**<u>4.02</u>**
	20%	5.31	4.18	4.71	5.10	**<u>4.08</u>**	**<u>4.08</u>**	4.19	4.35
	30%	**<u>3.92</u>**	4.56	4.71	4.97	4.63	4.63	4.19	4.39

and $k = 3$, HVDM-S is globally the top performing approach, independently of the generation variant. For $k = 1$, where KNN imputation has a more local behaviour, HVDM-S is consistently the best approach for most missing rates ($>5\%$) in all variants, only surpassed by SIMDIST when missing data is generated exclusively on continuous features. This suggests that although HVDM-S handles efficiently both continuous, categorical and missing values, the strategy used by SIMDIST to handle continuous values might be superior. For $k = 3$, HVDM-S surpasses the remaining approaches for higher missing rates ($>10\%$), with MDE showing competitive results for WA datasets in lower rates (5 and 10%). Furthermore, for $k = 3$, HVDM-S presents a lower average rank for higher missing rates than for $k = 1$; whereas as the value of k increases further, $k = \{5, 7\}$, KNN imputation shows a more global behaviour, and differences among the approaches become less clear, although HVDM-S remains in the top performing approaches, especially when missing data is generated across the entire dataset (PLAIN and WA) approaches. This impact of k (differences between approaches becoming more smoothed) is expected given that with a larger k-neighborhood, the local properties of KNN which grant it its greatest advantage (taking advantage of the similarity between patterns) become negligible. As the analysis of ranks does not provide information of the classification results directly, we analyse also several important performance metrics for complex, imbalanced data, such as Sensitivity, F-measure and G-mean, as shown in Table 3. As follows, HVDM-S is the top performing approach across all metrics and missing rates, and its superiority becomes more evident for higher missing rates (20% and 30%).

However, despite HVDM-S presents the highest performance results, it becomes clear from the analysis of Table 3 that the classification performance is overall poor, even if data is imputed. As we focus solely on the analysis of the

Table 2. CART average sensitivity ranks per missing rate (MR), and variant ($k = 3$). The best values in each row are marked in bold and underlined. **B**: BASELINE.

	MR	B	HEOM	HEOM-R	HVDM	HVDM-R	HVDM-S	MDE	SIMDIST
PLAIN Datasets	5%	5.76	4.56	4.52	4.26	4.61	3.76	**3.74**	4.79
	10%	6.40	4.66	3.94	5.18	3.87	**3.44**	4.31	4.21
	20%	5.40	5.26	4.65	4.29	4.52	**3.48**	3.89	4.52
	30%	6.39	5.02	4.02	4.24	5.00	**2.95**	3.56	4.82
WA Datasets	5%	4.44	4.90	4.47	5.06	5.31	3.82	**3.60**	4.40
	10%	5.60	4.06	4.76	4.50	4.44	4.47	**3.94**	4.24
	20%	5.50	4.61	5.11	4.89	4.56	**3.34**	3.76	4.23
	30%	6.21	5.16	3.84	4.15	4.21	**3.53**	4.32	4.58
WA-CAT Datasets	5%	5.34	4.28	4.83	3.59	**3.52**	4.31	5.07	5.07
	10%	4.79	4.53	4.93	4.60	4.86	4.00	4.64	**3.64**
	20%	5.72	3.93	4.76	4.29	4.67	**3.76**	5.00	3.86
	30%	4.86	3.98	4.66	5.50	5.12	**3.34**	4.45	4.09
WA-CONT Datasets	5%	5.29	4.85	**3.89**	4.48	4.35	4.35	4.23	4.55
	10%	5.87	4.35	4.32	4.42	4.37	4.37	**3.66**	4.63
	20%	5.27	5.00	4.35	4.27	**3.98**	**3.98**	4.19	4.94
	30%	4.98	4.66	4.37	4.82	**4.16**	**4.16**	4.29	4.55

Table 3. CART performance results (mean $\pm$ standard deviation) on *PLAIN Datasets* without imputation (BASELINE) and with KNN ($k = 3$) imputation using specific distances for distinct missing rates (MR). The best values for each performance metric are marked in bold and underlined.

Distance	MR	Sens	F-measure	G-mean	MR	Sens	F-measure	G-mean
BASELINE		0.468 ± 0.331	0.472 ± 0.326	0.536 ± 0.300		0.460 ± 0.334	0.463 ± 0.331	0.524 ± 0.306
HEOM		0.482 ± 0.324	0.483 ± 0.317	0.553 ± 0.283		0.479 ± 0.332	0.475 ± 0.319	0.541 ± 0.292
HEOM-R		0.482 ± 0.324	0.483 ± 0.317	0.554 ± 0.283		0.483 ± 0.329	0.480 ± 0.316	0.548 ± 0.288
HVDM	5%	0.480 ± 0.328	0.480 ± 0.320	0.549 ± 0.288	10%	0.475 ± 0.331	0.472 ± 0.320	0.539 ± 0.295
HVDM-R		0.480 ± 0.327	0.479 ± 0.320	0.549 ± 0.286		0.482 ± 0.325	0.478 ± 0.314	0.547 ± 0.285
HVDM-S		**0.485** ± 0.328	**0.485** ± 0.320	**0.556** ± 0.286		**0.487** ± 0.329	**0.481** ± 0.316	**0.550** ± 0.288
MDE		**0.485** ± 0.329	0.484 ± 0.319	0.552 ± 0.288		0.480 ± 0.332	0.474 ± 0.319	0.544 ± 0.290
SIMDIST		0.481 ± 0.328	0.482 ± 0.320	0.551 ± 0.286		0.483 ± 0.332	0.478 ± 0.320	0.544 ± 0.294
BASELINE		0.461 ± 0.337	0.459 ± 0.332	0.516 ± 0.311		0.436 ± 0.334	0.437 ± 0.334	0.489 ± 0.317
HEOM		0.463 ± 0.326	0.454 ± 0.315	0.519 ± 0.293		0.450 ± 0.320	0.437 ± 0.309	0.505 ± 0.288
HEOM-R		0.469 ± 0.320	0.463 ± 0.311	0.529 ± 0.289		0.461 ± 0.314	0.445 ± 0.302	0.514 ± 0.279
HVDM	20%	0.470 ± 0.327	0.460 ± 0.314	0.526 ± 0.292	30%	0.462 ± 0.321	0.444 ± 0.307	0.512 ± 0.285
HVDM-R		0.466 ± 0.329	0.455 ± 0.315	0.522 ± 0.292		0.456 ± 0.321	0.441 ± 0.309	0.507 ± 0.289
HVDM-S		**0.479** ± 0.323	**0.468** ± 0.310	**0.539** ± 0.284		**0.476** ± 0.321	**0.456** ± 0.303	**0.532** ± 0.276
MDE		0.476 ± 0.328	0.465 ± 0.314	0.534 ± 0.291		0.470 ± 0.324	0.452 ± 0.307	0.523 ± 0.284
SIMDIST		0.468 ± 0.331	0.458 ± 0.319	0.523 ± 0.296		0.456 ± 0.325	0.440 ± 0.311	0.509 ± 0.289

effect of data imputation, the datasets did not suffer any pre-processing, such as data oversampling, outlier removal or cleaning approaches. As biomedical datasets are often complex by nature, presenting a considerable imbalance ratio and associated problems such as small disjuncts, overlap and outliers, among others, we moved to a more detailed analysis of the characteristics of the collected datasets, with the objective to determine if some datasets were in fact complex and whether that complexity could be related to differences in performance for selected distance functions. To that end, several data complexity measures where computed for each dataset. These measures regard key proper-

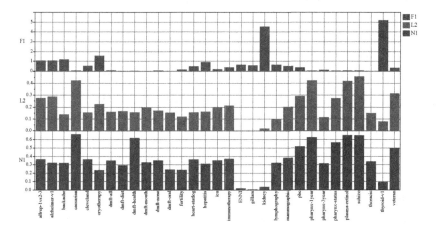

Fig. 2. Data complexity measures of the considered datasets: F1, L2 and N1.

ties of datasets such as geometry/topology (L3, N4), class overlap (F1, F2, F3) and class separability (L1, L2, N1, N2 and N3) and have proved to accurately provide important meta-information on the learning abilities of classifiers, especially in imbalanced domains [14]. We found the most informative measures to be related to class overlap (F1) and class separability (L2 and N1), as presented in Fig. 2. F1 captures the highest discriminative power of all features in data and lower values indicate more complex problems. In turn, L2 and N1 focus on the characteristics of the decision boundary between classes, where L2 measures the error rate of a support vector machine with linear kernel and N1 measures the fraction of data points connected to the opposite class by an edge in a minimum spanning tree. On contrary to F1, higher values of L2 and N1 indicate more complex problems. Accordingly, the top most complex datasets are *caesarian*, *dmft-health*, *pharynx-1year*, *pharynx-status*, *plasma-retinol*, *schizo*, and *veteran* (Fig. 2), which were further analysed. Figure 3 compares the mean performance ($k = 1$ and 3, for a more local behaviour of distances) of each dataset for PLAIN variant and a missing rate of 30%, where differences are more relevant (PLAIN variant is also the most likely to encounter in real-world domains where missing data is scattered throughout the entire dataset). For simplicity, and to determine clinical relevance, we focus on a direct comparison of HVDM-S with the BASELINE and the most common used approaches in the literature for healthcare data, i.e., HEOM and HVDM [6,10,11]. However, results for the remaining distances follow a similar trend (for MDE, some datasets obtain similarly performances to HVDM-S, as expected from Table 3). An analysis of Fig. 3 reveals that HVDM-S provides a substantial improvement in sensitivity results for more complex datasets (especially in comparison to HEOM). This suggests that choosing a proper distance function for imputation is important to produce quality training sets and that choice is even more important when data is complex, as determining the most similar patterns becomes crucial to obtain better classification results.

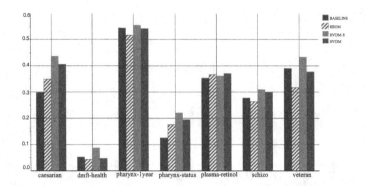

Fig. 3. CART sensitivity results for most complex datasets, considering a PLAIN variant and 30% of missing data (considering $k = 1$ and 3 for a more local analysis).

5   Conclusions, Limitations and Future Work

From the experimental results, the following conclusions may be derived:

- Distance functions impact KNN imputation, where HVDM-S has proved to be a feasible and robust approach for the imputation of heterogeneous biomedical data, independently of the type of features affected by missing data (generation variant);
- HVDM-S shows a particular good behaviour when compared to more common distance function (HEOM and HVDM) for more complex datasets, indicating that choosing a proper distance function becomes crucial when data is complex;
- Missing Data should be considered as yet another data difficulty factor for imbalanced domains, as it influences the computation of distances and assignment of nearest neighbours, becoming specially critical when other factors are present in data;

Findings of the present study should be interpreted considering that the included datasets are somewhat standard in terms of sample size (number of examples and features) and that the missing data variants are based on MCAR. Future work will involve including larger datasets, such as MIMIC-III and further missing data mechanisms, as MAR and MNAR. Finally, the evaluation of imputation performance, i.e., studying the impact of distance functions on the reconstruction of data, is also a subject for future research.

Acknowledgements. This work was supported in part by the project NORTE-01-0145-FEDER-000027 (Norte Portugal Regional Operational Programme – Norte 2020) and in part by the FCT Research Grant SFRH/BD/138749/2018.

Appendix

Table 4 presents the mathematical formulation for all distance functions described in Sect. 2.

Table 4. Mathematical formulation of heterogeneous distance functions that handle missing data.

HEOM

$$d_j(x_{Aj}, x_{Bj}) = \begin{cases} 1, & \text{if } j \text{ is missing in } x_{Aj} \text{ or } x_{Bj}, \\ d_O(x_{Aj}, x_{Bj}), & \text{if } j \text{ is a categorical feature,} \\ d_N(x_{Aj}, x_{Bj}), & \text{if } j \text{ is a continuous feature} \end{cases} \quad (1)$$

$$d_O(x_{Aj}, x_{Bj}) = \begin{cases} 0, & \text{if } x_{Aj} = x_{Bj} \\ 1, & \text{otherwise} \end{cases} \quad (3)$$

$$d_N(x_{Aj}, x_{Bj}) = \frac{|x_{Aj} - x_{Bj}|}{max(x_j) - min(x_j)} \quad (5)$$

HVDM

$$d_j(x_{Aj}, x_{Bj}) = \begin{cases} 1, & \text{if } j \text{ is missing in } x_{Aj} \text{ or } x_{Bj}, \\ d_{vdm}(x_{Aj}, x_{Bj}), & \text{if } j \text{ is a categorical feature,} \\ d_{diff}(x_{Aj}, x_{Bj}), & \text{if } j \text{ is a continuous feature} \end{cases} \quad (2)$$

$$d_{vdm}(x_{Aj}, x_{Bj}) = \sqrt{\sum_{c=1}^{C} \left| \frac{N_{x_{Aj},c}}{N_{x_{Aj}}} - \frac{N_{x_{Bj},c}}{N_{x_{Bj}}} \right|^2} \quad (4)$$

$$d_{diff}(x_{Aj}, x_{Bj}) = \frac{|x_{Aj} - x_{Bj}|}{4\sigma_{x_j}} \quad (6)$$

HEOM-R/HVDM-R/HVDM-S

$$d_j(x_{Aj}, x_{Bj}) = \begin{cases} 1, & \text{if } j \text{ is missing only on } x_{Aj} \text{ or } x_{Bj}, \\ 0, & \text{if } j \text{ is missing in both } x_{Aj} \text{ and } x_{Bj} \end{cases} \quad (7)$$

$$d_j(x_{Aj}, x_{Bj}) = \begin{cases} 0, & \text{if } x_{Aj} \text{ and } x_{Bj} \text{ are both missing,} \\ 1, & \text{if } x_{Aj} \text{ or } x_{Bj} \text{ are missing and } j \text{ is continuous,} \\ d_{vdm}(x_{Aj}, x_{Bj}), & \text{if } x_{Aj} \text{ or } x_{Bj} \text{ are missing and } j \text{ is categorical} \end{cases} \quad (9)$$

SIMDIST

$$s_{ABj} = \begin{cases} \frac{1}{2}, & \text{if either } x_{Aj} \text{ or } x_{Bj} \text{ are missing,} \\ z\left(\frac{s_{ABj}}{s_j}\right), & \text{if both } x_{Aj} \text{ and } x_{Bj} \text{ are observed} \end{cases} \quad (8)$$

$$s_{ABj} = \begin{cases} 0, & \text{if } x_{Aj} \neq x_{Bj}, \\ 1 - P_{ij}, & \text{if } x_{Aj} = x_{Bj} \end{cases} \quad (10)$$

$$s_{ABj} = 1 - \frac{|x_{Aj} - x_{Bj}|}{max(x_j) - min(x_j)} \quad (11)$$

MDE

$$MD_E(x_{Aj}, x_{Bj}) = (x_{Aj} - x_{Bj})^2 \quad (12)$$

$$MD_O(x_{Aj}, x_{Bj}) = \begin{cases} 0, & \text{if } x_{Aj} = x_{Bj} \\ 1, & \text{otherwise} \end{cases} \quad (13)$$

$$MD_E(x_{Aj}, x_{Bj}) = E\left((x - x_{3j})^2\right) = \int p(x)(x - x_{Bj})^2 dx = (x_{Bj} - \mu_x)^2 + \sigma_x^2 \quad (14)$$

$$MD_O(x_{Aj}, x_{Bj}) = \sum_x p(x)\, d_O(x, x_{Bj}) = \sum_{x \neq x_{Bj}} p(x) = 1 - p(x_{Bj}) \quad (15)$$

$$MD_E(x_{Aj}, x_{Bj}) = \int \int p(x)p(y)(x - y)^2 dx dy = \left(E(x) - E(y)\right)^2 + \sigma_x^2 + \sigma_y^2 = 2\sigma_x^2 \quad (16)$$

$$MD_O(x_{Aj}, x_{Bj}) = \sum_x \sum_y p(x)p(y)\, d_O(x, y) = \sum_x \sum_{x \neq y} p(x)p(y) = 1 - \sum_x p^2(x) \quad (17)$$

Notes: x_j represents all values of the j-th feature. In the formulae of MD_E and MD_O, we consider $x = x_j$.
Thus, μ_x and σ_x are equivalent to μ_{x_j} and σ_{x_j} (mean and standard deviation of all elements in j).
Similarly, we consider the auxiliary variable $y = x_j$. $E(x)$ corresponds to the expected value of x and $p(x)$ is the probability distribution of x.
For SIMDIST, P_{ij} is the fraction of patterns that takes value x_{ij} for j.
In practice, P_{ij} is the fraction of examples that assume value x_{Aj} or x_{Bj} for j, since for this computation they are equal.

References

1. AbdAllah, L., Shimshoni, I.: K-means over incomplete datasets using mean Euclidean distance. MLDM 2016. LNCS (LNAI), vol. 9729, pp. 113–127. Springer, Cham (2016). https://doi.org/10.1007/978-3-319-41920-6_9
2. Abreu, P.H., Santos, M.S., Abreu, M.H., Andrade, B., Silva, D.C.: Predicting breast cancer recurrence using machine learning techniques: a systematic review. ACM Comput. Surv. (CSUR) **49**(3), 1–40 (2016)
3. Amorim, J.P., Domingues, I., Abreu, P.H., Santos, J.: Interpreting deep learning models for ordinal problems. In: ESANN (2018)
4. Belanche Muñoz, L.A., Hernández González, J.: Similarity networks for heterogeneous data. In: ESANN 2012, pp. 215–220 (2012)
5. Das, S., Datta, S., Chaudhuri, B.B.: Handling data irregularities in classification: foundations, trends, and future challenges. Pattern Recogn. **81**, 674–693 (2018)
6. García-Laencina, P., Abreu, P.H., Abreu, M.H., Afonoso, N.: Missing data imputation on the 5-year survival prediction of breast cancer patients with unknown discrete values. Comput. Biol. Med. **59**, 125–133 (2015)
7. Hu, L.-Y., Huang, M.-W., Ke, S.-W., Tsai, C.-F.: The distance function effect on k-nearest neighbor classification for medical datasets. SpringerPlus **5**(1), 1–9 (2016). https://doi.org/10.1186/s40064-016-2941-7
8. Juhola, M., Laurikkala, J.: On metricity of two heterogeneous measures in the presence of missing values. Artif. Intell. Rev. **28**(2), 163–178 (2007)
9. Pereira, R.C., Santos, M.S., Rodrigues, P.P., Abreu, P.H.: MNAR imputation with distributed healthcare data. In: Moura Oliveira, P., Novais, P., Reis, L.P. (eds.) EPIA 2019. LNCS (LNAI), vol. 11805, pp. 184–195. Springer, Cham (2019). https://doi.org/10.1007/978-3-030-30244-3_16
10. Sáez, J.A., Krawczyk, B., Woźniak, M.: Handling class label noise in medical pattern classification systems. J. Med. Inform. Technol. **24** (2015)
11. Santos, M.S., Abreu, P.H., García-Laencina, P., Simão, A., Carvalho, A.: A new cluster-based oversampling method for improving survival prediction of hepatocellular carcinoma patients. J. Biomed. Inform. **58**, 49–59 (2015)
12. Santos, M.S., Abreu, P.H., Wilk, S., Santos, J.: How distance metrics influence missing data imputation with k-nearest neighbours. Pattern Recogn. Lett. **136**, 111–119 (2020)
13. Santos, M.S., Pereira, R.C., Costa, A., Soares, J., Santos, J., Abreu, P.H.: Generating synthetic missing data: a review by missing mechanism. IEEE Access **1**(1), 1–18 (2019)
14. Santos, M.S., Soares, J.P., Abreu, P.H., Araújo, H., Santos, J.: Cross-validation for imbalanced datasets: avoiding overoptimistic and overfitting approaches [research frontier]. IEEE Comput. Intell. Mag. **13**(4), 59–76 (2018)
15. Santos, M.S., Soares, J.P., Henriques Abreu, P., Araújo, H., Santos, J.: Influence of data distribution in missing data imputation. In: ten Teije, A., Popow, C., Holmes, J.H., Sacchi, L. (eds.) AIME 2017. LNCS (LNAI), vol. 10259, pp. 285–294. Springer, Cham (2017). https://doi.org/10.1007/978-3-319-59758-4_33
16. Tutz, G., Ramzan, S.: Improved methods for the imputation of missing data by nearest neighbor methods. Comput. Stat. Data Anal. **90**, 84–99 (2015)
17. Twala, B., Cartwright, M.: Ensemble missing data techniques for software effort prediction. Intell. Data Anal. **14**(3), 299–331 (2010)
18. Wilson, R., Martinez, T.: Improved heterogeneous distance functions. J. Artif. Intell. Res. **6**, 1–34 (1997)

Correction to: HYPE: Predicting Blood Pressure from Photoplethysmograms in a Hypertensive Population

Ariane Morassi Sasso ⓘ, Suparno Datta ⓘ, Michael Jeitler ⓘ,
Nico Steckhan ⓘ, Christian S. Kessler ⓘ, Andreas Michalsen,
Bert Arnrich, and Erwin Böttinger

Correction to:·
Chapter "HYPE: Predicting Blood Pressure
from Photoplethysmograms in a Hypertensive Population"
in: M. Michalowski and R. Moskovitch (Eds.):
Artificial Intelligence in Medicine, **LNAI 12299,**
https://doi.org/10.1007/978-3-030-59137-3_29

The original version of this chapter was revised. The conclusion section was corrected and reference was added.

The updated version of this chapter can be found at
https://doi.org/10.1007/978-3-030-59137-3_29

Author Index

Printed in the United States
by Baker & Taylor Publisher Services